Climate Change

Multidecadal and Beyond

World Scientific Series on Asia-Pacific Weather and Climate

This series will continue the collaboration with the World Meteorological Organization Working Group on Tropical Meteorology Research, a component of the World Weather Research Programme, and other international weather and climate programs in publication projects.

The Editorial Board welcomes book or program/project proposals relevant to weather and climate or related fields, including interdisciplinary topics. Interested scientists please contact Professor Chih-Pei Chang (cpchang@nps.edu)

Published

Vol. 1: East Asia and Western Pacific Meteorology and Climate: Selected Papers of the Fourth Conference
edited by C.-P. Chang, Guoxiong Wu, Ben Jou and C. Y. Lam

Vol. 2: East Asian Monsoon
edited by C.-P. Chang

Vol. 3: Observation, Theory, and Modeling of Atmospheric Variability Selected Papers of Nanjing Institute of Meteorology Alumni in Commemoration of Professor Jijia Zhang
edited by Xun Zhu, Xiaofan Li, Ming Cai, Shuntai Zhou, Yuejian Zhu, Fei-Fei Jin, Xiaolei Zou and Minghua Zhang

Vol. 4: Global Perspectives on Tropical Cyclones: From Science to Mitigation
edited by Johnny C. L. Chan & Jeffrey D. Kepert

Vol. 5: The Global Monsoon System: Research and Forecast (2nd Edition)
edited by C.-P. Chang, Yihui Ding, R. H. Johnson, G. N.-C. Lau, Bing Wang and Tetsuzo Yasunari

Vol. 6: Climate Change: Multidecadal and Beyond
edited by Chih-Pei Chang, Michael Ghil, Mojib Latif and John M Wallace

Forthcoming

Vol. 7: Indo-Pacific Climate Variability and Predictability
edited by Swadhin Behera and Toshio Yamagata

Vol. 8: Aridity Trend in Northern China
edited by Congbin Fu and Huiting Mao

WORLD SCIENTIFIC SERIES ON ASIA-PACIFIC WEATHER AND CLIMATE

Vol. 6

Climate Change

Multidecadal and Beyond

edited by

Chih-Pei Chang

National Taiwan University, Taipei & Naval Postgraduate School, Monterey

Michael Ghil

Ecole Normale Supérieure, Paris & UCLA

Mojib Latif

Helmholtz Centre for Ocean Research, Kiel

John M Wallace

University of Washington, Seattle

World Scientific

NEW JERSEY • LONDON • SINGAPORE • BEIJING • SHANGHAI • HONG KONG • TAIPEI • CHENNAI • TOKYO

Published by

World Scientific Publishing Co. Pte. Ltd.
5 Toh Tuck Link, Singapore 596224
USA office: 27 Warren Street, Suite 401-402, Hackensack, NJ 07601
UK office: 57 Shelton Street, Covent Garden, London WC2H 9HE

British Library Cataloguing-in-Publication Data
A catalogue record for this book is available from the British Library.

World Scientific Series on Asia-Pacific Weather and Climate — Vol. 6
CLIMATE CHANGE: MULTIDECADAL AND BEYOND

ISBN 978-981-4579-92-6

Printed in Singapore by Mainland Press Pte Ltd.

Preface

Climate change, and particularly global warming, is one of the top priorities in scientific research because of its potential impacts on society and the economy. Finding out what changes have happened or are happening, what the causes of these changes are, how and when the changes may happen in different regions are major challenges of scientists.

Most climate change meetings and expert reports have focused on projections of future climate changes based on computer models. All climate models are approximations of the physical processes governing the earth climate. As in all fields of empirical science, these model results need to be validated through a combination of physical understanding and observational and mathematical analyses. Mathematical modeling turns the climate science into an experimental science, and critical data analysis and verification are essential for the development of the modeling.

During the past decade, rapid advances have been made in climate science. However, two important issues remain in understanding the climate change: (1) How to separate the variability due to anthropogenic forcing and natural variability; and (2) How to delineate the multi-decadal and longer time scales when high-quality observational records are only available during the recent 100 years.

In September 2012, the National Taiwan University (NTU) Center for Advanced Study in Theoretical Science invited some of the world's leading experts in the areas of observational analysis, mathematical theory, and modeling to discuss these two issues in the NTU International Science Conference on Climate Change. One outcome of the meeting is the 23 chapters in this book each of which were finalized after extensive peer reviews and revisions. An important message come from this research is that internal climate variability on decadal and longer time scales is so large that not properly accounting for it may lead to false estimates of the climate's sensitivity to anthropogenic forcing.

We are grateful to the editors Professors Chih-Pei Chang, Michael Ghil, Mojib Latif, and Mike Wallace and all the authors who made this book possible. Special thanks are due to Professors Chung-Hsiung Sui and Shouhong Wang for their contributions to the organization of the NTU conference, and to the President of National Taiwan University and the National Science Council/Ministry of Science and Technology for their support. We also wish to thank Ms. Wei-Jhen Tsai for coordinating the original conference program, and to Ms. Kathryn Hua Hsu whose diligence and patience were invaluable in the editing of this book.

Chang-Shou Lin
Department of Mathematics
Director, Center for Advanced Study
in Theoretical Science
National Taiwan University
Taipei, Taiwan

Hung-Chi Kuo
Department of Atmospheric Sciences
Director, Research Center for Climate
Change and Sustainable Development

January 2015

Contents

Chapter 1

Attribution of Climate Change in the Presence of Internal Variability

John M. Wallace*, Clara Deser†, Brian V. Smoliak*, and Adam S. Phillips†

*Department of Atmospheric Sciences, University of Washington, Seattle, WA 98195, USA

†Climate and Global Dynamics Division, National Center for Atmospheric Research, Boulder, CO 80305, USA

wallace@atmos.washington.edu

Spontaneous, internally-generated variability of the climate system is pervasive. On the multidecadal time scale it dominates the variability of surface air temperature averaged over extratropical land areas as large as the contiguous United States, and it modulates the rate of rise of global mean temperature in response to the buildup of greenhouse gases. Unforced variability is one of the factors that imposes limitations on the degree of confidence that can be attached to assessments and predictions of human-induced climate change. This chapter summarizes results of some recent studies based on the analysis of large ensembles of numerical integrations run with a suite of different atmospheric initial conditions but with the same prescribed external forcing scenario. The future trajectory of the real climate system is, in some sense, like the trajectory of an individual member of such an ensemble. The diversity of the trends among the different ensemble members is a part of the irreducible uncertainty inherent in projections of future climate change. It is shown how statistical methods can be used to diagnose the causes of this diversity, most of which is in response to member-to-member diversity in the atmospheric circulation trends, as reflected in the associated patterns of the sea-level pressure trends. Interactions between the atmosphere, oceans, and land also contribute to the variability of surface air temperature trends on the multidecadal time scale, as discussed in several chapters of this book. It is argued that in the face of such large uncertainties in the attribution of climate change in the extratropics, more attention should be focused on climate change in the tropics, where the greenhouse warming signal stands out more clearly, and on the broader suite of environmental issues that impact food security and the viability of ecosystems.

1. Introduction

Many questions concerning the nature and causes of climate variability on the multidecadal time scale are still unresolved. For example, there is no consensus within the scientific community as to whether time-varying forcing associated with aerosols or whether a strengthening of the Atlantic Meridional Overturning Circulation was mainly responsible for the mid-20th century hiatus and the recent slowdown in the rate of global warming. Nor is it clear why the Arctic has experienced rapid warming during the past decade while surface air temperatures over the Northern Hemisphere as a whole warmed less than in the two prior decades, or why wintertime temperatures over the Northern Hemisphere continents poleward of 40°N warmed three times as rapidly as global-mean (land plus ocean) annual-mean surface air temperature during the late 20th century (*e.g.*, see Trenberth *et al.*, 2007). These large spatial and

Climate Change: Multidecadal and Beyond
edited by Chih-Pei Chang et al.

temporal differences in the rate of warming stem from the fact that the climate system is varying on the multidecadal time scale in response to its own internal variability as well as to a variety of natural and anthropogenic forcings. It is often difficult to distinguish between internally generated low frequency climate variability and human-induced climate change. In the words of the Technical Summary of the IPCC's Fourth Assessment Report, "Difficulties remain in attributing temperature changes at smaller than continental scales and over time scales less than 50 years." (Solomon *et al.*, 2007). These ambiguities can be expected to persist until the signature of human-induced climate change becomes large enough to stand out clearly above the natural "background variability", as is projected to occur in the second half of this century (Deser *et al.*, 2012a).

The causes of surface air temperature (SAT) trends over the continents can be formally separated into the four categories listed in Fig. 1, which are arranged in the form of a 2×2 matrix, the columns separating thermodynamically-versus dynamically-induced SAT variability and the rows separating forced versus free variability. In this terminology, *thermodynamically-induced* refers to SAT changes induced by time-varying radiative fluxes or by time-varying fluxes of sensible and latent heat at the Earth's surface, exclusive of any concomitant changes in the atmospheric circulation and *dynamically-induced* denotes SAT changes attributable to changes in the atmospheric circulation, irrespective of their cause. The term *forced* refers to responses to externally imposed changes in the Earth's energy balance, including both anthropogenic influences and natural forcings such as volcanic eruptions, solar variability and, on long time scales, orbital changes. Forced, thermodynamically-induced trends in global-mean temperature (the upper left box in Fig. 1) plays a central role in projections of human-induced climate change. However, for attribution of regional and perhaps even global climate trends on the multidecadal time scale, the other three categories need to be considered as well.

	thermodynamically induced	dynamically induced
Forced anthropogenic and natural		
Free		

Fig. 1. A scheme for categorizing the factors that contribute to trends in regional surface air temperature. See text for explanation.

If the global atmospheric circulation changes systematically in response to human-induced global warming or due to natural causes it will result in further SAT changes that can be said to be dynamically-induced (the upper right box in Fig. 1). Examples of dynamically-induced climate change include a circulation-induced poleward amplification of the temperature trend at the Earth's surface (Alekseev *et al.*, 2005), a widening of the tropical Hadley cells (Lu *et al.*, 2007), a poleward shifting of the extratropical storm tracks (Yin, 2005), and a systematic weakening of the tropical circulations (Vecchi and Soden, 2007). It has been proposed that the "robust response to global warming" includes a suite of circulation-related changes that can be inferred from basic conservation laws (Held and Soden, 2006). Other examples of externally forced dynamical responses include changes in the wintertime circulation over high northern latitudes induced by large volcanic eruptions (Robock and Mao, 1994; Shindell *et al.*, 2001) and changes in the monsoon circulations in response to the changing meridional profile of insolation induced by orbital changes (Kutzbach, 1981).

Spontaneously occurring changes in the amplitude and polarity of preferred atmospheric circulation patterns such as the Northern and

Southern Hemisphere annular modes (Wallace and Thompson, 2002) or the patterns observed in association with ENSO (Nitta and Yamada, 1989; Trenberth and Hurrell, 1994; Zhang *et al.*, 1997), as represented by the lower right box in Fig. 1, should induce regional SAT trends on the multidecadal time scale. The amplitude of the dynamically-induced SAT changes tend to be much larger over land than over sea, because of its lower heat capacity. Hence, if the dynamically-induced SAT trends project strongly upon the land-sea distribution, they may contribute to the hemispherically or globally averaged temperature trend; *e.g.*, if they were to change in a manner so as to cool the oceans and warm the land, that would constitute a positive contribution. Dynamically-induced warming in response to atmospheric circulation changes has been invoked to account for the rapidity of the wintertime warming over Eurasia and North America poleward of 40°N during the late 20th Century (Hurrell, 1996; Wallace *et al.*, 1995, 1996; Bracco *et al.*, 2004).

In most of the existing literature on the impacts of the unforced (or internal) variability of the climate system it is assumed that the associated regional or global temperature trends over land are mediated by changes in the atmospheric circulation. However, it is also possible that a component of the SAT trends could be thermodynamically-induced (the lower left box in Fig. 1). For example, it has been suggested that the ENSO cycle affects SAT not only through planetary-scale teleconnections (the lower right box), but also by modulating the surface energy fluxes over the equatorial Pacific cold tongue region (Yulaeva *et al.*, 1994; Chiang and Sobel, 2002). It has recently been proposed the sea surface temperature variations over high latitudes of the Atlantic Ocean and the Arctic Ocean that occur in association with unforced variations in the intensity of the meridional overturning circulation on the multidecadal time scale might be capable of causing even larger SAT variations over Eurasia on the multidecadal time scale (Semenov *et al.*, 2010).

In the world of models, the distinctions between the four boxes are clear, at least in principle. Anthropogenically-forced climate change can be distinguished from internally generated, free climate variability based on a suite of simulations, performed with a single model in which each ensemble member is started from a different set of initial conditions and run with the same prescribed, time-varying external forcings. In principle, the trends in the ensemble-mean of the simulations can be identified with the externally forced climate change "signal" and the departures of the trends in the individual realizations from the ensemble mean trends are attributable to the internal variability of the simulated climate system. This methodology is useful when the number of individual realizations is large enough to ensure a high level of statistical significance. However, thus far, ensembles sizes in the CMIP simulations have been so small that it has been necessary to use multi-model ensemble means to obtain statistically significant results, in which case internal variability and model-to-model differences both contribute to the departures of the trends in the individual realizations from the ensemble-mean trend, rendering attribution problematical. In principle, thermodynamically versus dynamically forced variability can likewise be distinguished through experimentation with climate models, but in practice the design of experiments that permit one kind of forcing but not the other is not straightforward.

Just how applicable the results derived from the model world are to the real world depends upon how well the models run with and without various prescribed external forcings are able to simulate the internally generated low frequency variability of the climate system. With only one observed climate trajectory that can be used as a basis for validating the models, this question can only be addressed in a probabilistic way. Validating climate models requires a robust characterization of the low frequency variability in the historical climate record, including temporal means, variance and covariance statistics, and spectra. To obtain robust estimates of these quantities (i.e., estimates with a sufficient number of statistical degrees of freedom) it is necessary to restrict the analysis to frequencies at least an order of magnitude higher than one cycle over the length of the ~100-year-long historical record, *e.g.*, by applying a high pass filter. Attribution of variability with frequencies lower than this cutoff frequency is inherently ambiguous. Multidecadal variability falls within this "twilight zone" in which attribution can be performed only in a probabilistic way.

Another factor that limits our ability to diagnose the decadal-scale variability in the climate record is the fact that inherently stochastic variability on the interannual time scale associated, for example, with the ENSO cycle or with large excursions of the Northern and Southern Hemisphere annular modes is capable of inducing substantial sampling variability on the multidecadal time scale. For example, it has been questioned whether the so-called Pacific Decadal Oscillation (PDO: Mantua *et al.*, 1997), which is alleged to be a multidecadal phenomenon is merely a manifestation of such stochastic, sampling variability (Newman *et al.*, 2003).

Regardless of the mechanisms that give rise to it, multidecadal climate variability mediates the rate of rise of global-mean temperature. Performing a "dynamical adjustment" to remove, or at least reduce the contribution of these circulation changes that contribute to or detract from the rate of rise in global-mean temperature simplifies the space-time structure of the surface air temperature record and renders it more spatially and seasonally coherent (Wallace *et al.*, 1995; Thompson *et al.*, 2009).

The existence of internally generated sampling variability is well known and is discussed extensively in reviews of Barnett *et al.* (2005) and Hegerl *et al.* (2007) and the references therein. Various approaches have been used to estimate the uncertainty in past and projected SAT trends that is attributable to such "climate noise," but the emphasis in these studies has generally been on establishing the statistical significance of historical or projected climate trends rather than on the characteristics of the noise itself. In many studies the noise is represented as formless error bars flanking the observed, reconstructed, or projected "signal." In reality, climate noise exhibits distinctive space-time structure that can mimic that of human-induced climate change. Later in this chapter we will recall two relatively recent instances in which apparent secular trends deemed as having been beyond the range of natural variability have subsequently reversed, suggesting that they were, in fact, internally generated.

In this chapter we will summarize the state of our knowledge of internally generated interdecadal variability of the climate system. In the next section we will show how the interplay between free and forced climate variability complicates the attribution of global warming and regional climate impacts. In Section 3 we show examples of the dynamical contribution to SAT trends and we demonstrate how performing a dynamical adjustment can simplify the representation of climate change in the historical record. In Section 4 we offer a few brief comments relating to the internally generated thermodynamically-induced variability. In the final section we summarize and discuss the implications of these results more generally.

2. On the Role of Internally-Generated Climate Variability in Climate Change

In this section we will discuss the role of internally-generated climate variability in climate change on space scales ranging from regional to global. We will begin in Section 2.1 by considering the diversity and spatial patterns of surface air temperature (SAT), sea level pressure (SLP) and precipitation trends in a 40 member ensemble of simulations run with Version 3 of the Community Climate System Model (CCSM3) forced with the A1B greenhouse gas scenario initialized in 2000 and integrated forward in time through 2060. Details of the model simulations may be found in Deser *et al.* (2012a). Then in Section 2.2 we will show and discuss observations and simulations of global warming during the 20th Century. In interpreting these results it should be noted that existing climate models exhibit substantial diversity with respect to their mean climates and the level of internally generated climate variability. Hence, the results presented here, which are derived from only two of these models, should be regarded as illustrative, but not necessarily representative of the behavior of the real climate system.

2.1. *Insights derived from projections of future trends*

As background for the discussion in this subsection we show in Fig. 2a,c the ensemble mean of the 56 year (2005–2060) December-February (DJF) SLP and SAT trends in the 40 member CCSM3 ensemble in raw (hPa per 56 years; °C per 56 years) and standardized form (Fig. 2b,d respectively). The SLP trends in the ensemble mean map range up to several hPa (*e.g.*, off the west coast of Canada) but with the exception of a few patches in the tropics, they are generally much less than one standard deviation and therefore may be interpreted as sampling fluctuations. In contrast, the ensemble mean SAT trend (Fig. 2d) exhibits a robust global warming signal, with an average warming of 2–3 standard deviations over the Northern Hemisphere

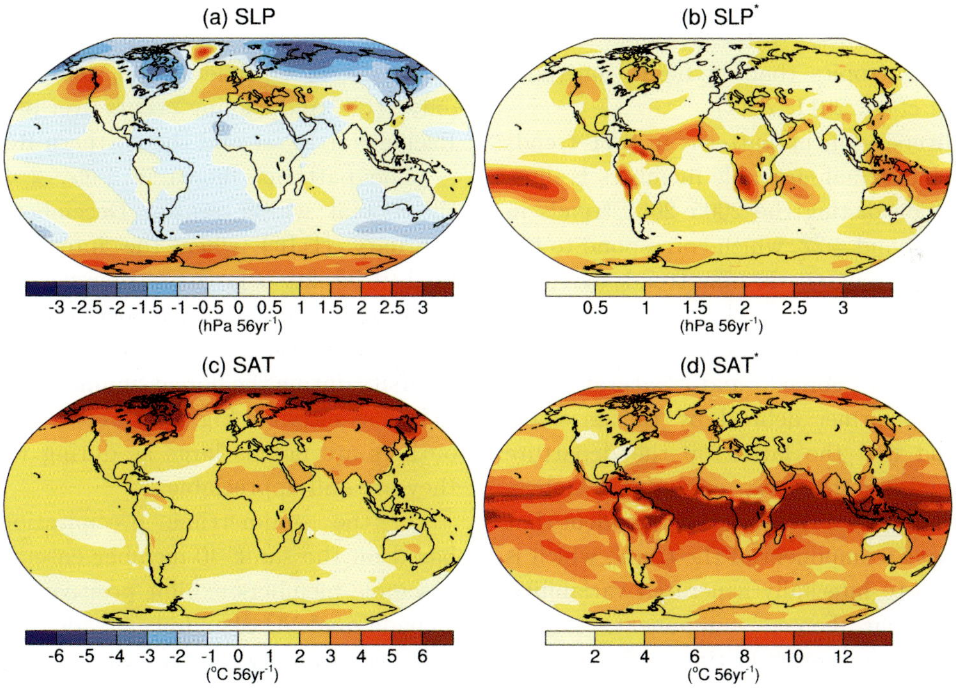

Fig. 2. DJF sea level pressure (SLP) and surface air temperature (SAT) trends for 2005–2060 in a 40-member ensemble of simulations with the CCSM3 climate model. Left panels (a) Ensemble-mean SLP trend; (b) Ensemble-mean SAT trend. Right panels (c) Standardized ensemble-mean SLP trend (SLP*) computed by dividing the ensemble mean trend at each grid point by the standard deviation of the 40-member ensemble trends at the same grid point and (d) Standardized ensemble-mean SAT trend (SAT*). From Deser *et al.* (2012a).

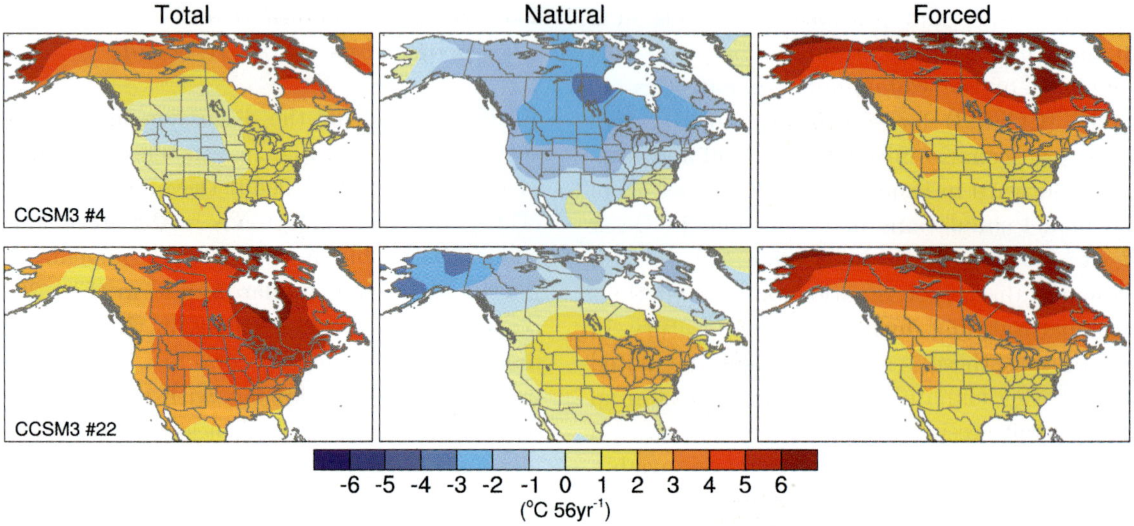

Fig. 3. Partitioning of the 56-year temperature trends in Ensemble Members #4 and #22 into a component forced by the buildup of greenhouse gases and a residual trend attributable to internally generated variability. Adapted from Deser *et al.* (2013).

continents. In agreement with observations, the "signal to noise ratio" of the warming trend is larger in the tropics than at higher latitudes (Mahlstein *et al.*, 2011, 2012).

The diversity of the patterns in SAT trends in the 40 individual ensemble members is illustrated by contrasting the 2005–2060 trends in Members #4 and #22, which correspond to the weakest and strongest warming trends averaged over the contiguous United States. In Fig. 3 the trends for these two ensemble members are partitioned into "Forced" and "Natural" components, the former defined by the ensemble mean trend and the latter by the departure of the trends in these two ensemble members from the ensemble mean trend. It is evident that the "Natural" component of the trend, which is attributable to internally generated variability, is as large as the "Forced" component.

Figure 4 shows selected time series of SAT in Ensemble Members #4 and #22. The global-mean SAT trend (land areas only) is quite comparable in the two ensemble members but the regional and local trends over the contiguous United States (CONUS) have a profoundly different character. In Member #22 average warming over the CONUS over the 56 year period is $\sim$3°C, whereas in Member #4 it is only $\sim$1°C and cooling occurs over parts of the Pacific Northwest and the northern Rockies.

The patterns shown in Figs. 2–4 are for the boreal winter season December through February (DJF) when the internally generated variability is greatest. However, the spatial patterns in individual members of the CCSM3 40-member ensemble also exhibit substantial diversity during the boreal summer June-August (JJA), as shown in Fig. 5. The contrasts between outliers are not as large as during winter but they are still appreciable.

In the summertime precipitation trends based on the same 40-member ensemble, most ensemble members exhibit positive trends over Canada and Alaska, but a wide member-to-member diversity of the rainfall trends over the continental United States. The two examples shown in Fig. 6 portray sharply contrasting future rainfall trends over the Great Plains and Midwest relative to the 2005–2008 rainfall climatology in that ensemble member.

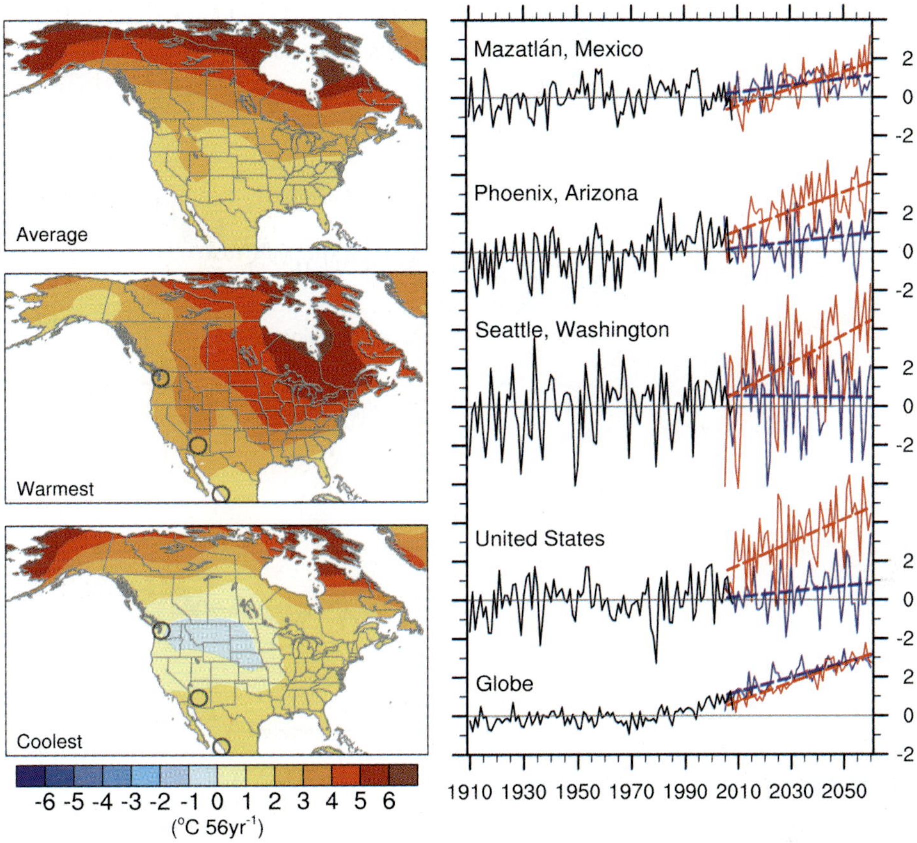

Fig. 4. (Left panels) DJF temperature trends during 2005–2060 (°C per 56 years). Top panel shows the 40-member ensemble mean; middle and bottom panels show the corresponding trends in Members #4 and #22. Right panels show DJF time series for selected regions and specific locations as indicated by the labeling and marked by the open circles in the left panels). The black segments of the curves show observed records from 1910–2008 (minus the long-term mean); the red and blue segments show model projections for 2005–2060 based on the ensemble members with the largest and smallest projected trends, respectively, for each region or location. Dashed red and blue lines show the best-fit linear trends in these same ensemble members. For visual clarity, time mean SAT in the model projections are adjusted to match to observations averaged over their common period of record 2005–2008. From Deser *et al.* (2012b).

Large ensembles are required to detect statistically significant precipitation trends in simulations with climate models and there is no guarantee that nature will conform to the ensemble-mean projection. It follows that unless the CCSM is seriously underestimating the role of soil moisture feedbacks, regional precipitation trends forced by the buildup of greenhouse gases are not likely to become detectable above the internally generated background variability in the observations until well beyond 2060.

2.2. *Observed and simulated 20th century trends*

Historical reconstructions of the climate of the 20th century provide an even more graphic illustration of the inherent difficulties in comparing an individual realization (in this case, the historical record) with an ensemble of numerical simulations. Figure 7 shows 1970–2005 wintertime surface air temperature trends for individual ensemble members of suites of historical simulations conducted with the CCSM4 and ECHAM5

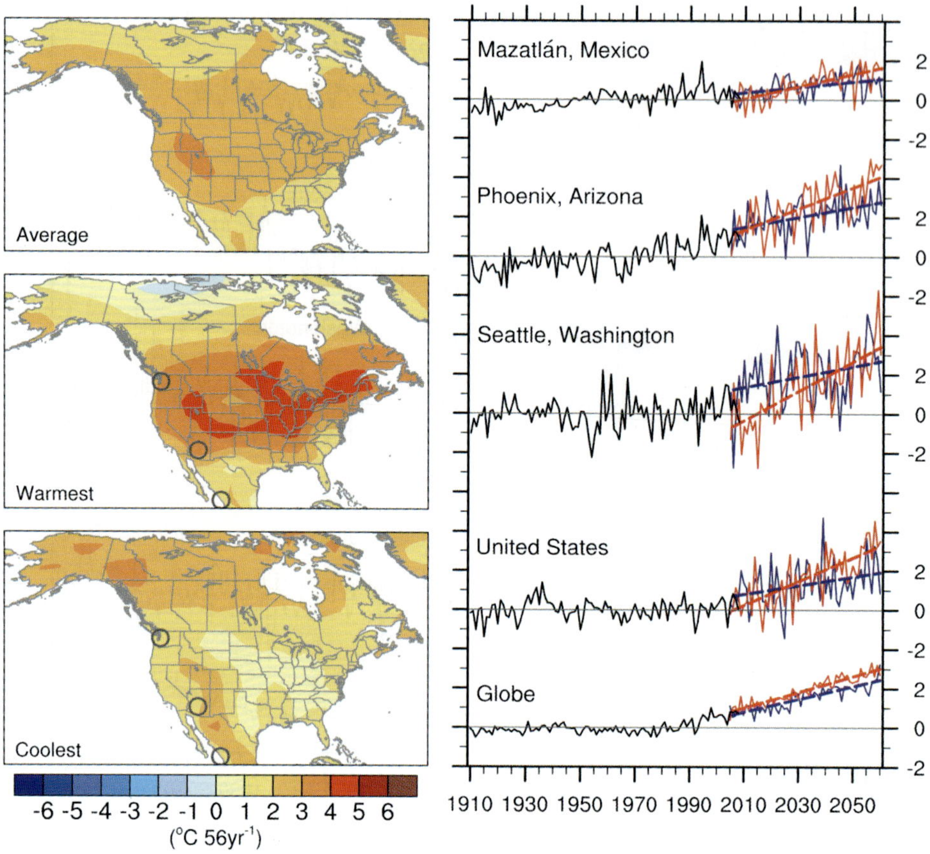

Fig. 5. As in Fig. 4 but for Ensemble Members #3 and #31, the most contrasting runs in JJA. From Deser *et al.*, (2012b).

models. These initial-condition historical ensembles follow a similar experimental design as the 21st Century ensembles discussed above (Deser *et al.*, 2013). CCSM4, introduced in 2010, is more advanced than CCSM3 in several important respects (Gent *et al.*, 2011). The patterns for the historical reconstructions are even more diverse than the ones shown in Figs. 4–6 because the interval over which the trend is computed is 36, rather than 56 years, and is thus subject to smaller anthropogenic forcing and larger sampling variability.

In the context of these simulations, the observational record is just one of the many possible outcomes of the time dependent forcing that the climate system has been subject to over the past century. Let us now consider the observed trends in greater detail. Figure 8 shows time series of observed and simulated trends in (a) global mean surface temperature including both land and ocean, (b) global-mean surface air temperature over land and (c) cold season surface air temperature poleward of 40°N. The simulations in this case are based on the reference interval 1965–2000 (5 years earlier than the one used in the simulations described in Fig. 7, but 30 out of the 36 years of the two records are overlapping. Both intervals 1965–2000 and 1970–2005 were marked by pronounced warming that was more rapid over land than over the oceans, as shown in Fig. 8a,b and the high latitude Northern Hemisphere continents warmed

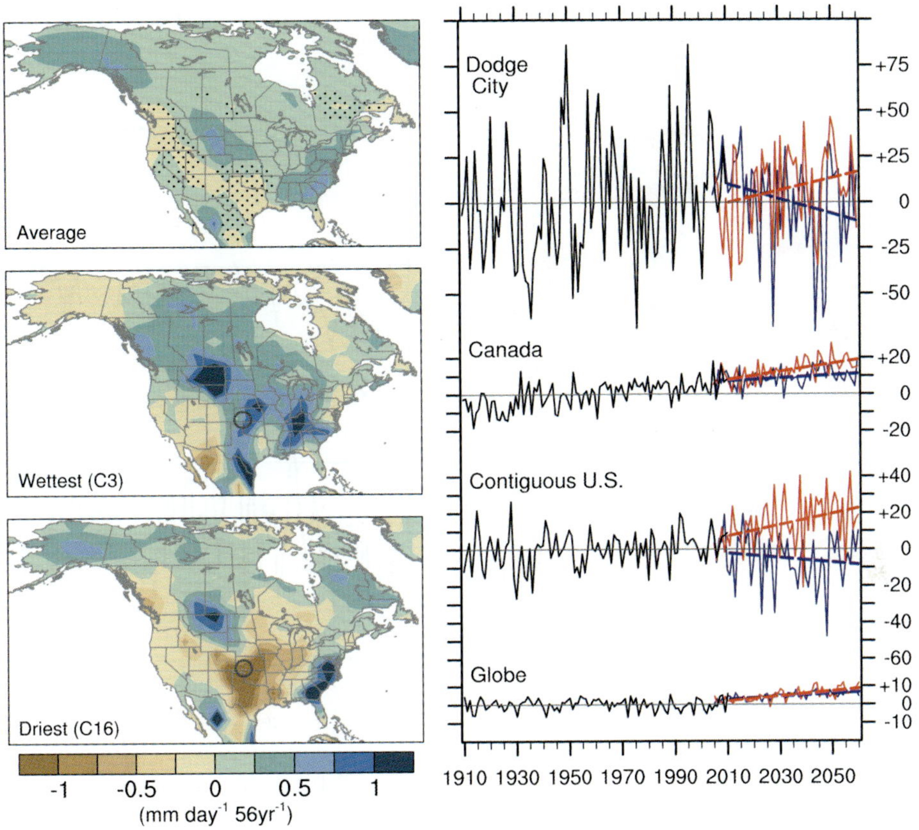

Fig. 6. As in Fig. 4 but for summer (JJA) precipitation.

about three times as much as annual mean GST during the boreal cold season (Fig. 8c).

The GST time series based on the multi-member ensemble mean of the CMIP simulations closely tracks the observations during the reference period 1965–2000 but it fails to capture the flattening of the curve in the past decade. As of the end of 2012 the projections were ∼0.4°C higher than the observed GST.

Figure 9 shows the spatial distribution of the observed 1965–2000 SAT trends during the boreal cold and warm seasons (a,c) together with multi-model ensemble means of numerical simulations that were used as a basis for the Fourth Assessment Report of the IPCC (b,d; Randall *et al.*, 2007; Meehl *et al.*, 2007): the simulations include nine different models with ensemble sizes ranging from 1 to 5. Consistent with the modeling results presented in the previous subsection, the multi-model ensemble-mean trends are much more spatially homogeneous than the observed trends, particularly over high northern latitudes during the boreal cold season. The observed warming trends over Siberia and Canada during this interval, which ranged up to ∼3°C per 36 years, were an order of magnitude larger than the mean rate of global (GST) warming during the 20th century (∼0.08°C per decade). They are also larger and extend farther southward than the corresponding multi-model ensemble mean trends shown in the right hand column of Fig. 9. Hence, it seems quite likely that the enhancement of the warming was either thermodynamically- or dynamically-induced by

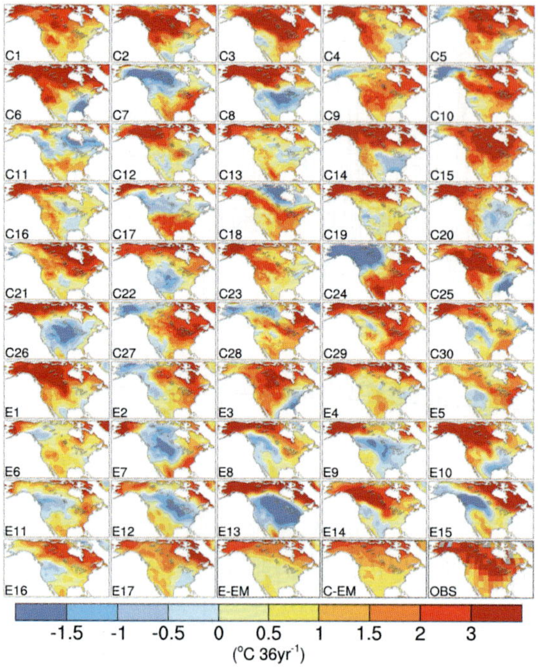

Fig. 7. DJF surface air temperature trends for the reference interval 1970–2005 in a 30-member ensemble of the CCSM4 (C1–C30) and 17-member ensemble performed with the ECHAM model (E1–E17). Trends are expressed in °C per 36 years. The ensemble-mean trend and the observed trend for this period are shown at the end of the sequence of maps.

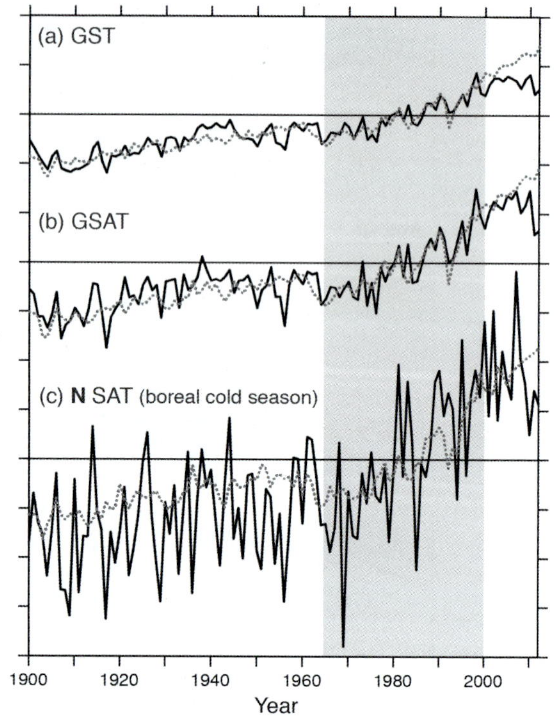

Fig. 8. Observed (*solid*) and multi-model ensemble mean (MMEM) (*dotted*) temperature anomaly time series with respect to the 1965–2000 reference period (indicated by light gray shading) for (a) annual-mean, global-mean surface temperature (GST), (b) annual-mean, global mean land temperature (GSAT), and (c) boreal cold season-mean (Nov.–Apr.) land temperature poleward of 40°N. Tick marks on the abscissa denote intervals of 0.5°C. Observations from NOAA merged ocean land surface temperature dataset. Model output based on AR4 (CMIP3) historical simulations in which ozone depletion and volcanic aerosols are included as part of the forcing (1900–1999) and SRES A1B simulations (2000–2012). Adapted from Wallace *et al.* (2012).

the free (internally generated) variability of the climate system; i.e., the categories listed in the bottom row of Fig. 1.

Over the 92-year-long reference interval 1920–2011 the regional SAT trends shown in Fig. 10, expressed in °C per 92 years, are smaller than those for the shorter reference interval shown in the previous figure, but they still range up to nearly 3°C per century, about three times the GST trend. The heterogeneity of the trends during the boreal winter shown in Fig. 10a, and the fact that they are so much larger than the corresponding warm season trends shown in Fig. 10b suggest that even on this extended time scale, the internal variability of the climate system makes an important contribution to the observed trends.

3. Assessing the Dynamically-Induced Variability

As noted in the Introduction, dynamically-induced variability of SAT and rainfall (i.e., variability attributable to the time-varying atmospheric circulation) is not identical to the internally generated variability for two reasons: (1) anthropogenic forcing can induce changes in the atmospheric general circulation and (2) internal variability of the climate system would

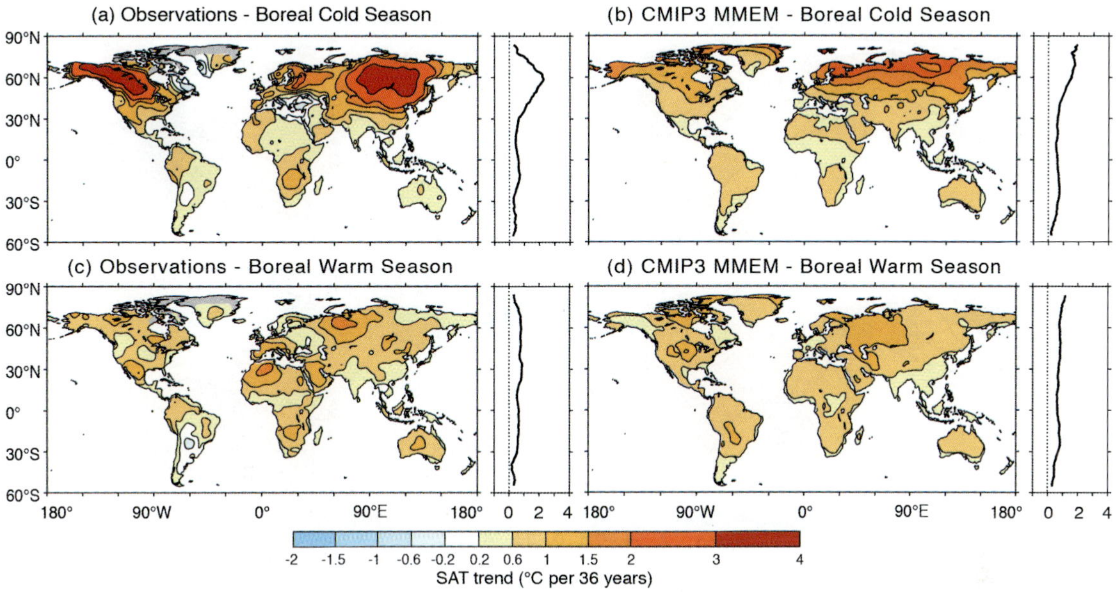

Fig. 9. Maps and zonally averaged meridional profiles showing rates of warming (i.e., the linear trend) over the historical reference interval 1965–2000 expressed in units of °C per 36 years: (a,b) boreal cold season; (c,d) boreal warm season. Left column (a,c) based on observations and right column (b,d) the multi-model ensemble mean (MMEM) of the AR4 simulations. From Wallace *et al.* (2012).

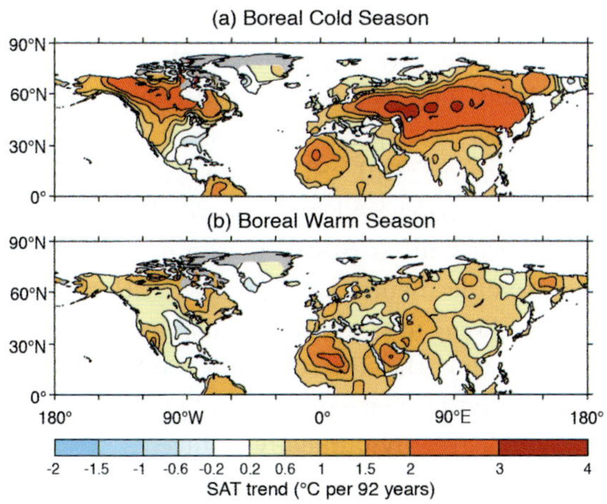

Fig. 10. Observed rates of warming in SAT over the historical reference interval 1920–2011 (i.e., the linear trend) expressed in units of °C per 92 years: (a) boreal cold season; (b) boreal warm season. Based on the NCDC MLOST dataset.

be capable of inducing changes in SAT, even in the absence of circulation changes. These distinctions are reflected in the categories of causal mechanisms in Fig. 1. But despite these caveats, it appears that much of the internally generated variability in the SAT field is mediated by changes in the atmospheric circulation. To illustrate this point, we reexamine Ensemble Members #4 and #22 from the CCSM3 40-member ensemble, whose SAT fields were depicted in Figs. 4, 5, and 6, but here in Fig. 11a,b we show them together superimposed on the corresponding trends in the sea level pressure (SLP) field. The SLP trend in Member #4 is in the sense as to favor offshore flow along the west coast from British Columbia to California, reducing the moderating influence of the influx of marine air masses upon winter temperatures over most of North America. In contrast, the trend in Member #22 (Fig. 11b) is in the sense as to increase the onshore flow over western Canada. That the strength of the onshore flow along the Pacific seaboard exerts a strong influence on wintertime SAT over the continental US is borne out in the Member #22 minus Member #4 SLP difference map shown in Fig, 11c, and the regression map

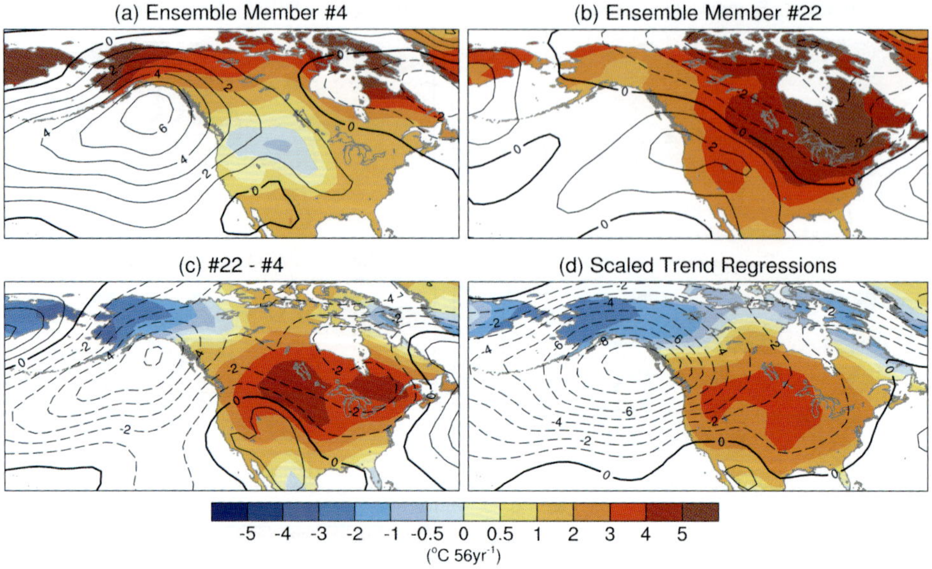

Fig. 11. (a,b) 2005–2060 DJF SAT trends in Ensemble Members 4 and 22 of the CCSM3 40-member ensemble, (c) The difference between (a) and (b). (d) Trends in the individual ensemble members (exclusive of #4 and #22) regressed on the 38 corresponding raw SAT trends averaged over the continental US. SAT trends are indicated by colored shading and SLP trends by contours. Contour interval 1 hPa per 56 years. The zero contour is bold and dashed contours indicate SLP falls.

based on all the other ensemble members shown in Fig. 11d. It is notable that the SLP trends in these "outlier" members of the 40-member ensemble are much larger than the ensemble-mean SLP trend shown in Fig. 2a. The same must be true of randomly selected members because at most grid points the ensemble-mean 56-year trend is less than 0.5 standard deviation of the trends in the individual ensemble members, as shown in Fig. 2b. In agreement with this conclusion, based on an analysis of simulations performed with CMIP3 models, Oshima *et al.* (2012) reported that the contribution of internal climate variability the diversity of century-long SLP trends over the North Pacific in the ensemble members is as large as the ensemble mean trend itself and that internal variability accounts for virtually all of the diversity of the 50-year SLP trends in the individual ensemble members during the first half of the 20th century.

Is it merely chance coincidence that the SLP trends in Ensemble Members #4 and #22 are dynamically consistent with the contrasting SAT trends over the continental United States? To address this question we show in Fig. 11d the pattern of DJF SLP trends regressed on the standardized SAT trends averaged over the continental United States in the other individual ensemble members. That the SLP patterns in Figs. 11c,d are broadly similar confirms the inference that the differences in the SAT trends in the individual ensemble members are at least to some degree dynamically induced.

Analogous results for ensemble members with contrasting summertime 2005–2060 SAT trends are shown in Fig. 12. These runs exhibit contrasting patterns of SLP trends over the North Pacific and over the Canadian Arctic, but distinctions between the warm and cold patterns are not as clearly discernible over the contiguous US. That the JJA SLP trends in the other 38 ensemble members regressed upon the respective SAT trends averaged over the continental United States yields (Fig. 12d) yields a pattern

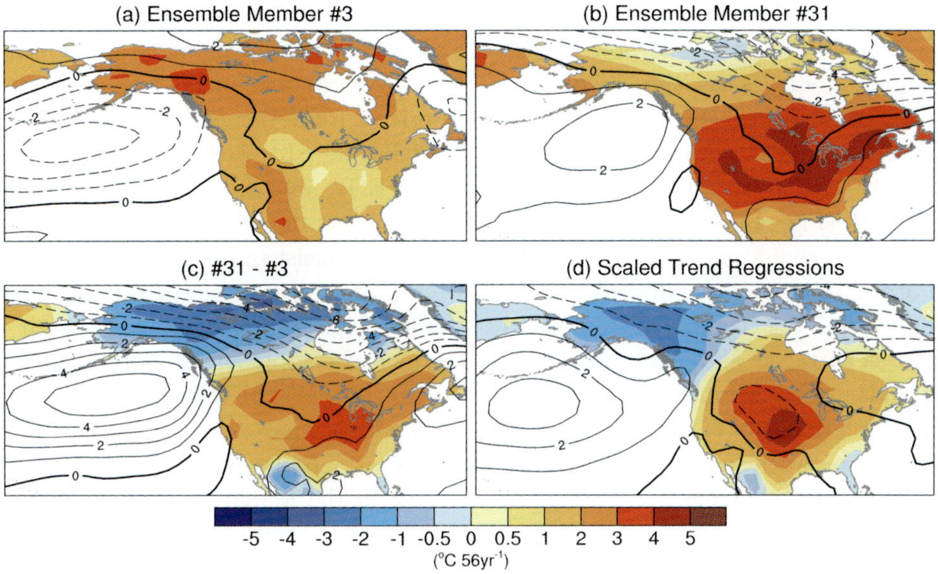

Fig. 12. As in Fig. 11 but for SAT and SLP in JJA, contrasting the two ensemble members that exhibit the weakest (#3) and strongest (#31) SAT trends averaged over the contiguous US.

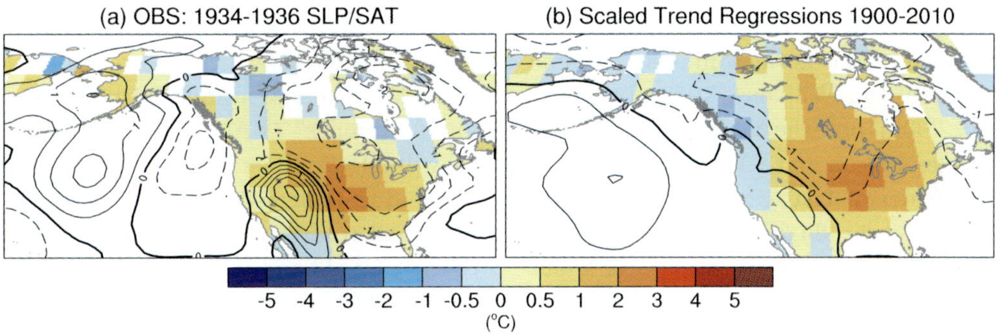

Fig. 13. (a) Pattern of SLP anomalies (contoured) and SAT anomalies (shaded) during the US Great Plains Dust Bowl summers, the average of JJA 1934, 1935, and 1936. The reference period for the anomalies is 1900–2010. (b) SLP and SAT regressed on the detrended time series of JJA SAT averaged over the continental US. The regression coefficients are based on the period of record 1900–2010, exclusive of JJA 1934, 1935, and 1936. Contour interval 0.5 hPa. Based on the 20th Century Reanalyses (Compo *et al.*, 2011).

similar to the difference map (Fig. 12c) is suggestive of a dynamical influence upon the SAT trends in the individual ensemble members in summer as well as in winter.

Patterns of SLP and SAT anomalies observed during anomalously hot summers over the contiguous US the so-called "Dust Bowl) and, more generally during hot summers over CONUS are shown in Fig. 13. The anomalous upstream anticyclone over the North Pacific and the anomalously low SLP along the Arctic coast are not as prominent in the historical data as they are in the simulated SLP patterns shown in the previous figure. Apart from the anomalies over the regions of elevated terrain over the southwestern US, which is mainly a SAT signature, the SLP anomalies are quite small. It is difficult to understand how the low level circulation anomalies associated with such weak features could have been responsible for the large SAT anomalies either in the simulations (Fig. 12) or in extreme years in the historical record (Fig. 13). Feedbacks involving soil moisture could greatly increase the amplitude of the internal variability of SAT, at least in the model. The land surface tends to become drier as it warms and, conversely, summer SAT tends to be warmer over dry land surfaces. Both relationships contribute to a negative correlation between temperature and precipitation. In the CCSM3 ensemble members with strong summertime warming trends over the continental US tend to be marked by negative precipitation trends and vice versa: the correlation coefficient between JJA SAT trends and precipitation trends averaged over the continental US among the 40 individual ensemble members is −0.82. Dry months also tend to be warmer than normal in the observations (Madden and Williams, 1978; Trenberth and Shea, 2005) but the negative correlation between precipitation and SAT is not as strong. A drying tendency favors a tendency toward reduced latent heat fluxes and enhanced sensible heat fluxes at the Earth's surface, which favors warming. Such land surface feedbacks can serve to amplify and prolong dynamically-induced heat waves (*e.g.*, see Black *et al.*, 2004 or Dole *et al.*, 2012), rendering internally generated dynamically- and thermodynamically-induced diversity in the temperature trends difficult to separate, even in large ensembles of simulations.

3.1. *Further insights derived from projections of future SAT trends*

The 40-member ensemble of CCSM3 simulations in Deser *et al.* (2012a) has been subjected to a more formal analysis of the dynamical contribution to the diversity of the wintertime SAT trends. The left hand columns of Fig. 14 show the leading Northern and Southern Hemisphere empirical orthogonal functions (EOFs) derived from the extratropical DJF and JJA 2005–2060 SLP trends. In the climate literature EOF analysis is usually performed in the space/time domain and the "modes" obtained from it consist of an ordered set of spatial patterns (EOFs) whose time-varying amplitudes and polarities are described by their respective principal component (PC) time series. In Deser *et al.* (2012a) the analysis is performed in the domain of space/ensemble number: the EOFs are spatial patterns but in their analysis the PCs refer to the amplitude and polarity of various 2005–2060 trend EOFs as expressed in the individual ensemble members. Despite the differences in which that data are analyzed, the DJF SLP patterns and the Northern Hemisphere JJA pattern shown in Fig. 14 closely resemble the leading EOFs of the month-to-month variability in historical time series (Thompson and Wallace, 2000; not shown), which are commonly referred to as the "Northern and Southern Hemisphere annular modes" (NAM and SAM, respectively; we will refer to them by those names). They correspond to preferred modes of variability in the historical record and in extended control runs of the CCSM3 and other climate models (Miller *et al.*, 2006).

The CCSM is a coupled model in which ENSO and longer term climate variations resulting from large-scale atmosphere-ocean interactions come into play. In order to distinguish between the internally generated variability that is attributable to changes in the atmospheric circulation alone and variability that is a consequence of atmosphere-ocean interactions, the SAT trends in the 40-member ensemble of CCSM3 runs are compared with trends in non-overlapping 56 year segments of a 10,000 year control integration of the Community Atmospheric Model (CAM3), the same atmospheric model that is used in the coupled CCSM3 integrations. In the control integration

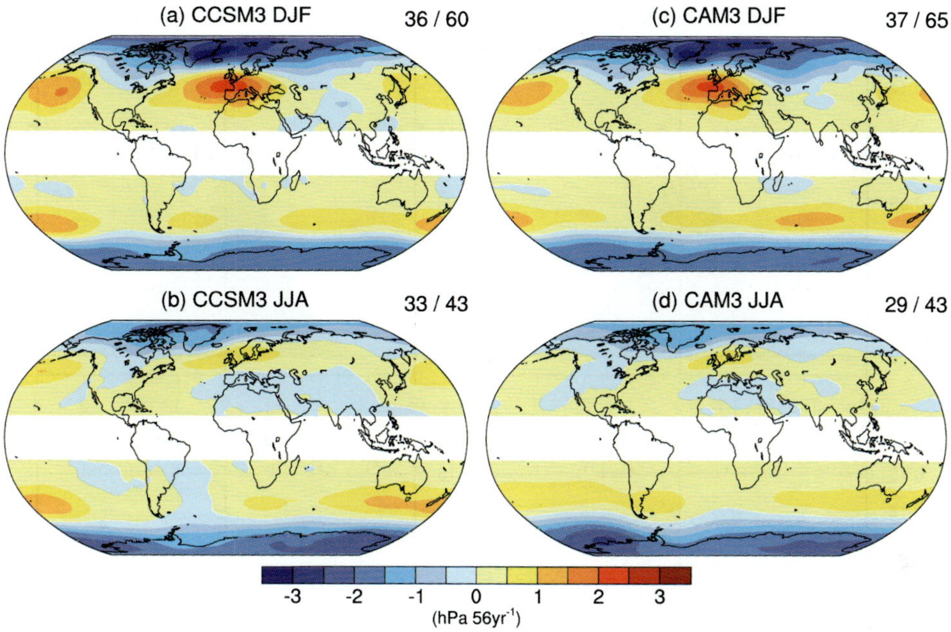

Fig. 14. The leading EOFs of extratropical SLP trends from the 40-member CCSM3 ensemble (a) DJF and (c) JJA; and in the CAM3 ensemble (b) DJF and (d) JJA. Trends are computed over the period 2005–2060 for CCSM3 and for 56-year non-overlapping segments for CAM3. EOF analysis is performed for each hemisphere separately but plotted on a single map. The percent variance explained by each EOF is given in the upper right corner of each panel, with the first number denoting the NH and the second number the SH (for example, for CCSM3 in DJF, NH EOF1 accounts for 36% of the NH variance and SH EOF1 accounts for 60% of the SH variance). From Deser *et al.* (2012a).

greenhouse gas and aerosol concentrations are held constant and SST is prescribed in accordance with the seasonally varying climatology but the treatment of land surface feedbacks is the same as in CCSM3. Results are shown in the right hand panels of Fig. 14. If the atmosphere-ocean coupled variability were making an important contribution to the diversity of the trends among the CCSM3 ensemble members, one would expect the amplitude of the annular modes and other preferred patterns of variability in the SLP field to be larger in the CCSM3 ensemble members than in the 56-year segments of the CAM simulations. That the EOFs obtained from the CCSM3 and CAM3 runs are similar in shape, comparable in amplitude, and explain comparable fractions of the total SLP variance suggests that the diversity of the SLP trends in the CCSM3 ensemble members is mainly due to the internal variability of the atmospheric circulation in the absence of any coupling to the ocean.

The ensemble mean of the trends in the 56-year segments of the CAM3 control run is indistinguishable from zero. However, the projections of the patterns of SLP trends in the 40 CCSM3 simulations onto EOF1 of the control run shown in Fig. 15 exhibit a statistically significant positive bias indicative of falling SLP in both polar cap regions and rising SLP over much of middle latitudes. SLP changes in this sense are commonly referred to as a trend toward the "high index polarity" of the NAM and the SAM. Nearly all the climate models used in the IPCC projections exhibit a trend in this sense (Meehl *et al.*, 2007). This shift toward the high index polarity of the annular modes is

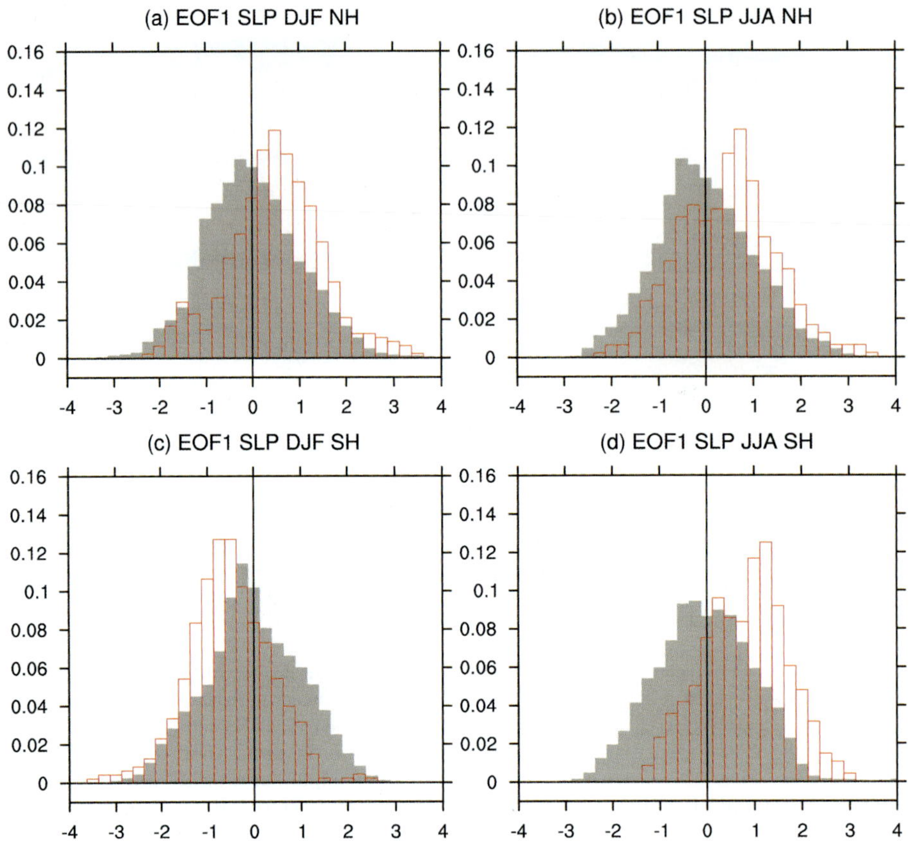

Fig. 15. Histograms of the SLP 2005–2060 trend projections onto EOF1 from the CAM3 control integration broken down by hemisphere and season as indicated. The red open bars show results from the 40-member CCSM3 and the gray filled bars from the 178-member Community Atmospheric Model (CAM3), which is used as a control. The x axis is in units of standard deviations of the CAM3 control integration, and the y axis is frequency (number of ensemble members divided by the total number of ensemble members). From Deser *et al.* (2012a).

evidently a robust feature of the ensemble-mean response to the buildup of greenhouse gases in the CCSM3 (i.e., the upper right box in Fig. 1), but it is expressed with different strengths in different ensemble members and in some members the trend is in the opposite sense. The diversity of the SLP trends in the individual members of the ensemble, as reflected in the width of the frequency distributions in Fig. 15, is a measure of a kind of structural uncertainty inherent in the projected SLP trends for 2005–2060.

The low level circulation pattern implied by EOF1 of the SLP trend induces a spatially varying SAT trend of the form shown in Fig. 16, which is obtained by averaging the SAT trends over all ensemble members, weighting each in accordance with the projection of its pattern of SLP trend upon EOF1 (i.e., by its value of PC1). The patterns of induced SAT trend in the two hemispheres are dynamically consistent with EOF1 of the SLP trend and they resemble the SAT patterns obtained by regressing the SAT field upon time varying (SLP) indices of the annular modes (Thompson and Wallace, 2000). The regression patterns in Fig. 16, in effect, define the anomaly in the trend of SAT in each member of the ensemble (i.e., the departure

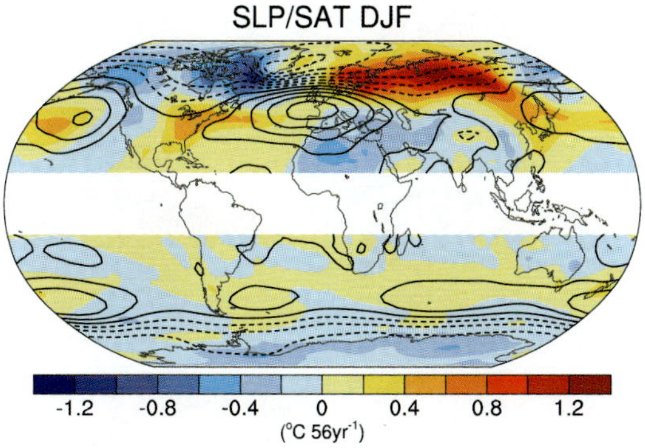

Fig. 16. Surface temperature regressed upon the scores of EOF1 of extratropical SLP trends from the 40-member CCSM3 ensemble in DJF, shown in Fig. 15. Contours show the SLP trend EOF (contour interval 0.6 hPa per 56 years; negative values are dashed). Trends are computed over the period 2005–2060. EOF and regression analyses are performed for each hemisphere separately but plotted on a single map. From Deser *et al.* (2012a).

from the ensemble-mean SAT trend) that can be attributed to the anomalous trend in the NAM/SAM in each ensemble member (again, anomalous with reference to the ensemble mean trend). Subtracting out that portion of the SAT trend at each grid point in each ensemble member that is attributable to the anomalous trend in the NAM/SAM in that ensemble member constitutes a *dynamical adjustment* that serves to reduce the diversity of the trends among the ensemble members. A *dynamically adjusted* global or hemispheric SAT trend field can be created by subtracting the dynamical contribution from the raw SAT trend field for each member. In principle, a complete dynamical adjustment of the SAT trends in the individual members of the ensemble can be obtained by the applying dynamical adjustments not only for the NAM/SAM, but for all the significant EOFs of the SLP trends. In practice, applying this procedure to an ensemble with a finite number of members will inevitably tend to overestimate the dynamical contribution. The challenge is to limit the degree of "overfitting."

3.2. *Estimating the dynamical adjustment for individual ensemble members*

In order to avoid serious overfitting of the diversity (or variance) of the trends in the individual ensemble members it is necessary to apply analysis techniques that are parsimonious i.e., that involve regression equations containing very few SLP predictors. Deciding how best to perform the analysis is an area of active research. Here we will briefly describe some of the linear analysis tools that might be used in performing such an analysis.

Consider the set of SAT trends $T(n)$, $(n = 1, N)$ at a specified grid point, one belonging to each of the N ensemble members. A simple, conservative approach is to apply least squares regression (LSR), using the SLP trend field in a prescribed spatial domain to define a single "predictor" $T^*(n)$ of $T(n)$. The first step is to regress the standardized SLP trend field upon $T(n)$ to obtain a single predictor pattern $P(x)$. Here the standardization is performed in "ensemble-space"; i.e., the standard deviation of SLP trend at each grid point is based on the sample consisting of the N ensemble members. The second step is to project the standardized SLP trend field in each ensemble member upon $P(x)$ to obtain a "score" $S(n)$ that provides a relative measure of the polarity and the strength with which the predictor pattern $P(x)$ is expressed in that ensemble member. The final step is to scale the scores $S(n)$ by fitting them to $T(n)$ using the method of least squares. The rescaled scores are the predictors $T^*(n)$, which constitute the dynamical adjustment. Spatial averaging and the steps in LSR are commutative, so this procedure can equally well be applied directly to an ensemble of spatially averaged SAT trends such as GST(n).

For each specified grid point or area average time series, LSR yields a single correlation coefficient or "predictor" $T^*(n)$ of the SAT trend in the nth ensemble member and is thus as parsimonious as any method can be. The LSR approach can be extended to multiple predictors by the method of partial least squares regression (PLSR). In PLSR, the first predictor $T_1{}^*(n)$ is obtained by LSR, as described above. A residual data set is then created by regressing $T_1{}^*(n)$ out of both the SLP trend field in each of the individual ensemble members (i.e., the field from which the "predictor patterns" are chosen) and the ensemble of grid point or area-average SAT trends $T(n)$ (the "predictand"). Then LSR is applied to these residuals to obtain a second predictor $T_2{}^*(n)$ that is orthogonal to $T_1{}^*(n)$ by construction and a second set of residuals. The procedure can be repeated as many times as desired. PLSR is widely used in other fields such as econometrics, chemometrics, neuroscience, and computer science, and is beginning to be more widely used in geophysics (Smoliak *et al.*, 2010). It has been used by Wallace *et al.* (2012) to estimate the dynamical contribution to the wintertime SAT trends over the Northern Hemisphere continents, as discussed in the next subsection.

The above procedure can be computationally intensive when it is applied *pointwise* (i.e., to every grid point in the SAT field) because of the high degree of redundancy inherent in the calculations, especially when the grid is fine. Under some circumstances it may be preferable to expand the SAT trend field in terms of EOFs and to perform LSR or PLSR on a subset of the resulting PC time series. The same considerations that apply to pointwise or area-wise LSR/PLSR are also applicable when these forms of analysis are performed in PC space.

Figure 17 shows raw and dynamically adjusted wintertime SAT trends for Members #4 and #22 of the 40-member ensemble conducted with CCSM3, the ensemble members with the smallest and largest SAT trends over CONUS. In this example the dynamical adjustment was computed by performing pointwise PLSR on the SAT field over North America. The patterns $P(x)$ in the SLP field are for

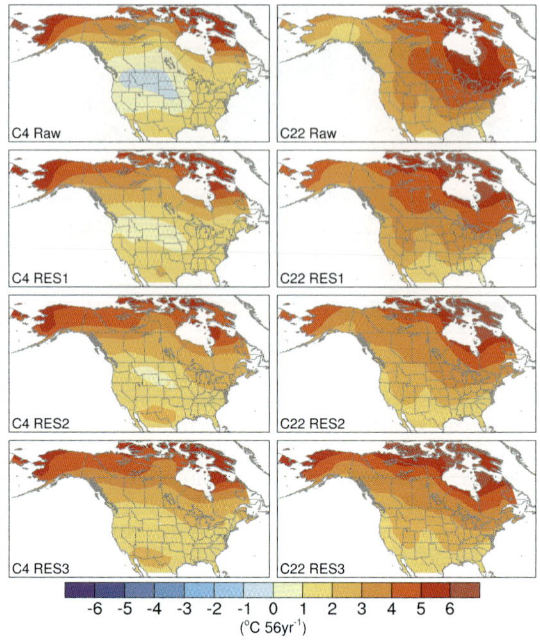

Fig. 17. Examples of raw and dynamically adjusted 2005–2060 SAT trends for ensemble members 4 (*left*) and 22 (*right*) of the 40-member CCSM3 ensemble, the members that exhibited the smallest and largest SAT trends averaged over the contiguous US. The top row of panels show the raw patterns, the second shows the adjusted patterns with one pass of PLSR, the third row with two passes, and the bottom row after three passes.

the domain 10° to 90°N and from the Date Line to the Greenwich Meridian. The dynamical adjustment has the strongest influence on outlier ensemble members such as these. With each pass the contrast between the two ensemble members becomes less pronounced.

Figure 18 shows dynamically adjusted trends for the individual members of the CCSM4 and ECHAM ensembles, whose raw trends were shown in Fig. 7. Application of the dynamical adjustment substantially reduces the diversity of the SAT trend patterns in the ensemble members, bringing them more into line with the ensemble-mean trends. Averaged over the North American domain, applying this dynamical adjustment with two passes accounts for 75% of the variance of the SST trends in the 30-member CCSM4 ensemble. That it leaves

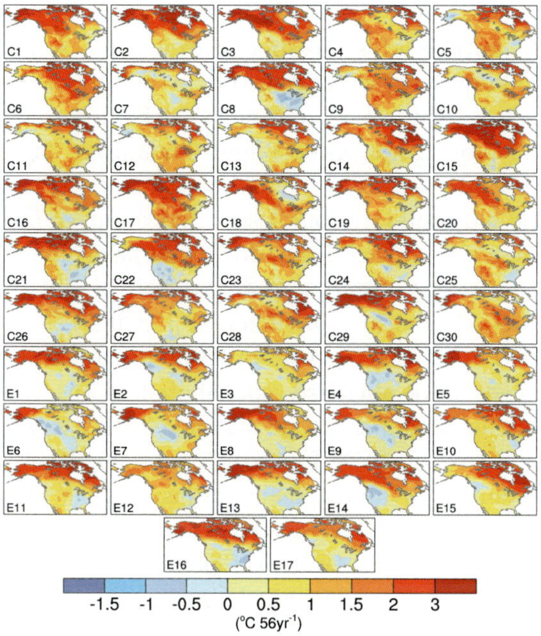

Fig. 18. As in Fig. 7 but dynamically adjusted with two passes of PLSR using the SLP field in the domain 10° to 90°N and from the Date Line to the Greenwich Meridian as the predictor field.

25% of the variance unexplained suggests that our dynamical adjustment scheme, which relies exclusively on the SLP field, might not be capturing all the dynamically-induced variance.

When applied to regions the size of the CONUS, the dynamical adjustment based on SLP appears to be even more effective in reducing the diversity of the SAT trends in summer than in winter, accounting for 81% of the variance of the SST trends in the 30-member CCSM4 ensemble with only two passes.

3.3. *Dynamically adjusting SAT trends in the historical record*

To adapt the methods described in the previous subsection to estimating the dynamical contribution to the observed SAT trends it is necessary to use temporal variability within the historical record as a surrogate for the diversity of the individual ensemble members in the simulations. In this case the goal is to explain the multidecadal variability of the trends observed within a single time series using SLP patterns derived from the analysis of that same time series. The strong late 20th Century warming trend during the boreal cold season over the continental interiors documented in Section 2.2 is arguably the most dramatic feature in the historical record. Was it dynamically induced? If so, is the adjustment large enough to affect estimates of the rise in annual mean GST during this period?

As background for addressing this question, Fig. 19 shows the global pattern of November-April SLP trends from 1965 to 2000. Compared to other sampling intervals of comparable length, the trends during this interval are particularly strong and spatially coherent, with pressure falls in both polar regions and pressure rises in midlatitudes, indicative of a shift toward the high index polarity of the NAM and the SAM. The trend in the PNA pattern was associated with a spontaneous, abrupt shift toward a more El Niño-like state of the sea surface temperature distribution and related atmospheric circulation patterns equatorial Pacific in 1976–77 (Nitta and Yamada, 1989; Trenberth and Hurrell, 1994; Zhang *et al.*, 1997).

Shindell *et al.* (2001) investigated the cause of the strong NAM-related trends during this interval and they concluded, on the basis of

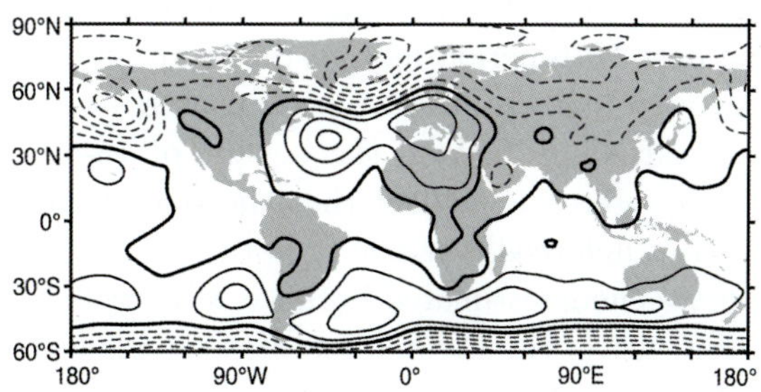

Fig. 19. Observed 1965–2000 SLP trends for the boreal cold season (Nov.–Apr.), based on the 20th Century Reanalyses. The contour interval is 1 hPa per 36 years. Negative values are dashed. Zero contour bold. From Wallace *et al.* (2012).

numerical simulations with an array of external forcings, that the observed trend in the NAM during this interval was beyond the range of internal variability and was likely forced by a change in the meridional heating gradient in the stratosphere induced by the buildup of greenhouse gases. This explanation would account for the prominence of the NAM signature in the 1965–2000 SLP trend but does not explain the reversal in the trend in the NAM that began in the mid- 1990s and became more clearly apparent after the turn of the millennium (see NCAR's Climate Data Guide, http://climatedataguide.ucar.edu, for a NAM index from 1899 to present). Subsequent numerical experiments by Bracco *et al.* (2004) and Deser and Phillips (2009) based on ensembles of integrations with two atmospheric GCMs forced with the observed evolution of SSTs, greenhouse gases, ozone and aerosols suggest that the prominent NAM signature in Fig. 19 is likely a manifestation of atmospheric internal variability that happened to be particularly strong during this 36-year interval.

In a similar manner, Trenberth and Hoar (1996, 1997) investigated the cause of the late 20th Century trend toward the positive polarity of the PNA pattern; i.e., the prevalence of an El Niño-like state of the tropical Pacific after the mid-1970s. On the basis of a statistical analysis of the historical record of the Southern Oscillation Index (SOI) they concluded in their second paper that the behavior of the SOI during the post 1976 period had been "highly unusual and very unlikely to have been caused by natural variability". With the benefit of 15-years of hindsight it is evident that the role of natural (i.e., unforced) variability of the climate system was underestimated in this case as well: the prevalence of the El Niño-like state of the equatorial Pacific has not persisted into the 21st Century (Hu *et al.*, 2013).

In contrast to the trends in the NAM and the PNA pattern, the late 20th Century trend in the SAM is widely regarded as having been at least partially forced by a decrease in concentrations of stratospheric ozone due to the buildup of CFCs during this period — the so-called "Antarctic ozone hole" (Thompson and Solomon, 2002; Gillett and Thompson, 2003; Polvani *et al.*, 2011). It has been suggested that the more subtle ozone losses that occurred over the Arctic during this period might have induced an analogous trend in forcing the NAM, but the case is less clear (Hartmann *et al.*, 2000).

The NAM- and PNA/ENSO-related SLP trends have obviously contributed to the wintertime warming of Eurasia and North America poleward of 40°N from 1965 to 2000. Results obtained by dynamically adjusting the time series of SAT averaged over all the Northern Hemisphere land grid points poleward of 40°N, using two PLSR Northern Hemisphere SLP predictors, are shown in Fig. 20. The

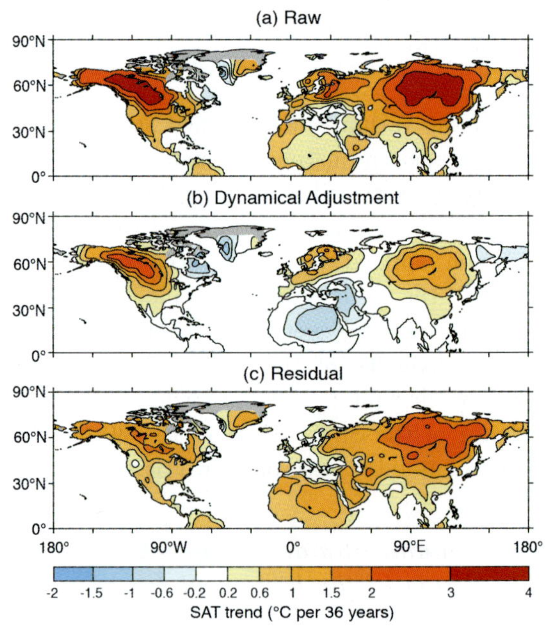

Fig. 20. 1965–2000 SAT trends for the boreal cold season (Nov.–Apr.): (a) the raw trend pattern, (b) the dynamical contribution, estimated by partial least squares regression (PLSR) with two predictors for each grid point, and (c) the dynamically adjusted trend pattern, calculated by subtracting the dynamical contribution in (b) from the raw trend pattern in (a).

dynamical contribution to the SAT trend resembles the SAT signature of the NAM. The residual SAT trend is weaker than the raw trend and more closely resembles the corresponding warm season trend shown in Fig. 9c. Yet it is clear that the dynamical adjustment shown in Fig. 20 does not account for all of the enhanced warming over the interiors of the continents during the boreal cold season relative to the model simulations shown in Fig. 9b. Either the methodology described in the previous subsection is failing to capture all the dynamically-induced variability or there is a contribution from the thermodynamically-induced internal variability discussed in the next subsection and in Chapter 7.

To place the dynamical adjustment for 1965–2000 in a global context, Table 1 shows a summary of how it affects the trends in global mean SAT over land (GSAT) and over the entire globe (GST). Applying the dynamical adjustment reduces the warming over the high latitude Northern Hemisphere land masses during the boreal cold season from 1.72°C to 1.02°C over this 36-year interval, eliminating most of the excess warming relative to the warm season and relative to the land lying south of 40°N. The magnitude of the adjustment (0.70°C) is diluted by averaging over all land (to 0.28°C) and over the year (to 0.14°C) and by combining land and ocean data (to 0.04°C in GST). Hence, changing extratropical circulation patterns during the boreal winter accounted for less than 10% of the global warming signal during this interval. Compared to the *accelerated warming*; i.e., the rate of rise of annual mean GST during this interval minus the mean rate of rise of GST during the past century of 0.08°C per decade or 0.29°C per 36 years, it amounts to about 15%. Hence, though appreciable and of first order importance for the attribution of high latitude Northern Hemisphere wintertime temperature trends over land, the dynamical contribution explains only a small fraction of the accelerated global warming during the late 20th Century and even after it is removed, land (GSAT) warmed twice as rapidly as ocean (GSST) during this period.

Table 1. Observed 1965–2000 SAT trends over land for boreal cold and warm seasons November–April and May–October expressed in °C per 36 years. The italicized numbers refer to dynamically adjusted trends as explained in Section 4. GSAT refers to global-mean SAT, GSST to global-mean SST and GST refers to global mean surface temperature including both land and sea. Data based on the National Climatic Data Center historical merged land-ocean land surface temperature analysis (MLOST), with its land component, GHCNv3, and ocean component, ERSSTv3b. Adapted from Wallace *et al.* (2012).

	Cold	Warm	Annual mean
N: 40°N–90°N	1.72 (*1.02*)	0.79	1.26 (*0.91*)
S: 60°S–40°N	0.70	0.69	0.70
GSAT (Land)	1.03 (*0.80*)	0.72	0.88 (*0.76*)
GSST (Ocean)	0.35	0.37	0.36
GST	0.57 (*0.49*)	0.48	0.52 (*0.49*)

The importance of the dynamical adjustment tends to decline with the length of the interval over which the trends are computed, but even for the longer interval 1920–2011, it remains quite important during the boreal cold season, as seen in Fig. 21. Applying the dynamical adjustment to the cold season trends and averaging them with the warm season trends is seen to yield a relatively simple spatial pattern suggestive of reduced warming over China and the southeastern United States possibly in response to the buildup of aerosols.

Applying the dynamical adjustment to the cold season trends substantially improves the coherence between cold and warm season SAT time series both for the global mean and for regional means, as illustrated in Fig. 22. In particular, it accounts for most of the excess warming during the boreal cold season relative to the warm season over the course of the 20th Century. The positive correlation between the warm and cold season time series on a year-to-year basis is indicative of a season-to-season memory that transcends the very short thermal adjustment time of the land surface. Features such as the cooling following the 1991 eruption

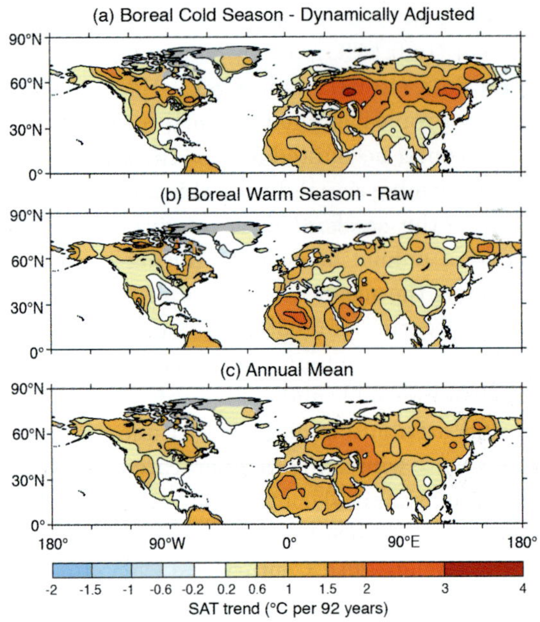

Fig. 21. SAT trend patterns for the reference interval 1920–2011: (a) dynamically adjusted trend for the boreal cold season estimated by PLSR as in Fig. 20, (b) raw trend for the warm season, and (c) annual mean trend estimated by averaging the dynamically adjusted cold season trend and the raw warm season trend.

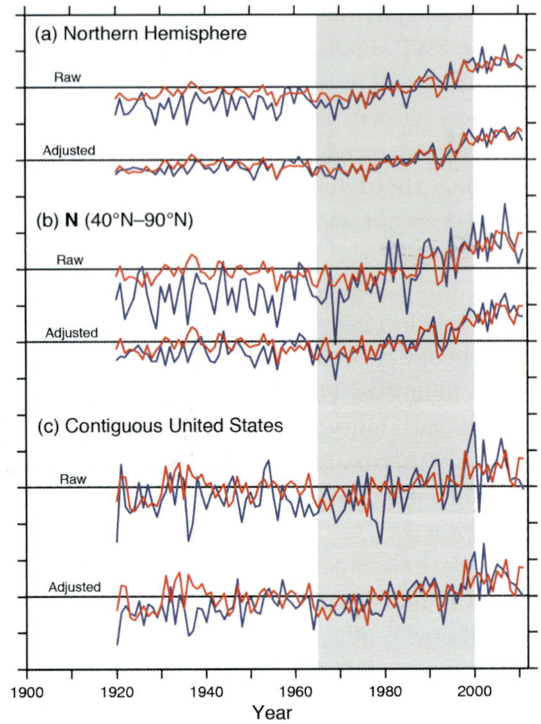

Fig. 22. Time series of (red) warm season and (blue) cold season SAT spatially averaged as indicated. The top pair of curves in each set shows raw time series and the bottom pair shows the dynamically adjusted cold season series and raw warm season series. Adapted from Wallace *et al.* (2012).

of Mt. Pinatubo are more clearly evident in the adjusted data.

4. Does Thermodynamically-Induced Variability Play a Role?

Of the four categories of processes listed in Fig. 1 that contribute to climate trends, thermodynamically-induced internal variability is the most difficult to quantify because it requires an understanding of variations in the oceanic global overturning circulation and the land biosphere and hydrosphere, and in the cryosphere, all of which are on the frontiers of climate science. It is not clear how well the models used in IPCC assessments, including the ones referred to in this chapter, simulate the relevant processes on the multidecadal time scale. For example, it is quite possible that the diversity of the trends in Fig. 7, large though it is, could be underestimated for failure of the models to realistically represent the variability of the ocean circulation for lack of adequate treatment of the physics or proper initialization. In a similar manner, changes in vegetation and ground hydrology that are ignored or not accurately represented in the models could conceivably render the summertime trends more diverse than the examples included in this chapter.

Multidecadal variations in the strength of the meridional overturning circulation in the North Atlantic (MOC) affect the poleward heat transport by the Gulf Stream and its extension to subpolar latitudes and they regulate the rate of exchange of heat between the oceanic mixed layer and the layers below. The temperature of the oceanic mixed layer responds rapidly to changes in the surface energy balance, through which the global warming signal is transmitted together with regional signals associated with atmospheric circulation anomalies. When the MOC is anomalously strong, enhanced poleward heat transports warm the surface waters of the subpolar North Atlantic and parts of the Arctic, which in turn, transfer heat to the atmospheric boundary layer, where it may be advected downstream by the winds. The MOC strength also modulates the rate at which heat that accumulates in the ocean mixed later is transferred into the deeper ocean (Meehl *et al.*, 2013). Hence, variations in the strength of the MOC could act as a time-varying thermodynamic forcing of SAT that is likely to be strongest over the higher latitudes during the boreal cold season, when the sea to air fluxes of sensible and latent heat are strongest.

In the absence of reliable direct measurements of variations in the strength the MOC, much of what is known about its role in multidecadal climate variability is inferred from numerical simulations of the oceanic response to prescribed surface forcing as in Alvarez-Garcia *et al.* (2008). The induced MOC-related variations in the oceanic circulation can feed back on the surface fluxes and thereby force the atmosphere on the interdecadal time scale. Numerical experiments by Semenov *et al.* (2010) with a coupled model with interactive MOC-related variability suggest that the modulations in the fluxes may have been strong enough to force significant SAT anomalies over the high latitude continents, irrespective of any related circulation changes. If so, the resulting thermodynamic contribution to the high latitude warming would appear in the residual SAT trend field in Fig. 20c.

A prominent feature of the multidecadal variability in the strength of the MOC is the so-called Atlantic Multidecadal Oscillation (AMO) or Atlantic Multidecadal Variability (AMV), which exhibits a period on the order of 70 years, as discussed in Chapter 9. There and in Chapters 8 and 12 it is argued that the AMO is a manifestation of internal variability of the coupled atmosphere-ocean climate system that would exist even in the absence of external forcing. It has further been argued that AMO-related variations in the strength of the MOC during the 20th century are responsible for much of the multidecadal variability in the rate of rise of GST; i.e., the relatively rapid rise in the 1920s and 30s, the mid-century hiatus, the resumption of the rapid warming toward the end of the century and perhaps even the recent second hiatus (DelSole *et al.*, 2011; Wu *et al.*, 2011; Gulev *et al.*, 2013). For further discussion, the reader is referred to Chapter 12 by K. K. Tung and Chapter 7 by Latif *et al.*

Another way in which the exchange of water between the mixed layer and the deep ocean on a decadal time scales could exert a thermodynamic influence global surface temperature is through the wind-driven zonal overturning circulation in the equatorial Pacific, which pumps cold water from below the thermocline into the warm ocean mixed layer. The stronger the equatorial trade winds, the stronger the zonal overturning circulation and the sea surface temperature (SST) contrast between the cooler eastern Pacific and the so-called Indo-Pacific warm pool" in the west. Conversely, the stronger the east-west SST contrast, the stronger the trade winds. The existence of this strong positive feedback gives rise to pronounced variability of SST and wind on time scales ranging from seasons to decades. It plays an essential role in the ENSO cycle, but it would exist even in a hypothetical atmosphere coupled to a passive ocean mixed layer (Kitoh *et al.*, 1999; Dommenget, 2010; Clement *et al.*, 2011). Please see Chapter 11 for further discussion of the modal structures that emerge in a global aqua-planet model in which the ocean is represented by mixed layer model. Based on experiments with numerical models, Meehl *et al.* (2011) and Kosaka and Xie (2013) have shown evidence that an intensification of this zonal overturning circulation and the resulting cooling of the tropical eastern Pacific contributed to the pronounced slowdown in the rate of global warming that occurred around the beginning of the current millennium.

Despite the array of statistical and modeling evidence that has been presented in support of the role of the ocean in thermodynamic forcing, many climate scientists remain convinced that the irregularities in the rate of warming in the historical record are attributable to variations in the time-varying anthropogenic forcing by aerosols. For example, on the basis of numerical experiments with a state-of-the-art climate system model, Booth *et al.* (2012) conclude that aerosol emissions and periods of volcanic activity explain 76 per cent of the simulated multidecadal variance in detrended 1860–2005 North Atlantic sea surface temperatures and that after 1950, their simulated variability is within observational estimates. Numerical experiments reported in the attribution chapter of the IPCC's Fourth Assessment Report (Hegerl *et al.*, 2007) and highlighted in the Technical Summary also portray the late 20th Century warming in GST as being entirely in response to anthropogenic forcing. On the other hand, on the basis of ob-servations and climate simulations conducted as part of the IPCC's Fifth Assessment Report, Kavvada *et al.* (2012) concluded that unless low-frequency, internally generated variability is incorporated into climate simulations, the decadal predictability and the accuracy of climate projections under different climate change scenarios will remain compromised. For further discussion of these issues see Zhang *et al.* (2013) and Chapters 8, 9, and 12 of this book.

Another arena in which thermodynamically forced internal variability of the climate system might come into play is through the feedbacks associated with the fluxes of latent and sensible heat by soil moisture and vegetation during the warm season (or in the tropics, during the growing season). Although land surface processes act on rather short time scales, the existence of positive feedbacks can serve to amplify the variability on all time scales. Another aspect of this problem on a regional scale is the existence of human-induced external forcings not directly related to global warming; *e.g.*, cultivation, irrigation, deforestation and afforestation, and inadvertent desertification.

5. Summary and Discussion

In this chapter we have categorized SAT trends as being externally forced versus internally generated (i.e., forced versus free) and as being dynamically versus thermodynamically induced, as elaborated in Fig. 1. The distinction between forced and free is illuminated by analysis of numerical simulations in which a single model is run with a single set of time dependent external forcings but starting from a suite of different atmospheric initial conditions, yielding an ensemble of time-dependent climate scenarios. The dynamically induced component of the SAT trend is isolated by linear regression using the SLP field as a "predictor" and the individual ensemble members as samples. When ensembles of climate scenarios are not available, the regression can be performed in the time domain. In principle, this methodology can be applied to variables other than SAT and variables other than SLP can be used as "predictors" of the dynamically-induced trends.

The relative importance of the forced and free components of the trends depends upon the interval τ over which the trends are computed. If the forced component can be viewed as increasing linearly with τ and the dynamically-induced variability can be modeled as white noise whose contribution to the variance of the trend is inversely proportional to τ, then the prominence of the forced component (i.e., the ratio of the trend to the standard deviation of the variability about the trend) should vary as $\tau^{3/2}$. The "signal" in the SAT trends is also latitude dependent (i.e., much more prominent in the tropics than in the extratropics) and dependent upon the spatial averaging. Here we have focused mainly on SAT trends over the continental US within 36 and 56 year intervals. The individual ensemble members in large ensembles of simulations are shown to exhibit a remarkable amount of diversity, most of which is dynamically induced. That this dynamical contribution to the variability is as large in an atmospheric model run with SST prescribed in accordance with its seasonally varying climatology as in the coupled runs suggests that the dynamically induced component of the trends is mainly attributable to atmosphere's own internal variability rather than to coupled atmosphere-ocean interactions.

Whether the real climate system exhibits as much dynamically-induced variability as the climate models examined in this study remains to be determined. There are indications that the temporal variance of the atmospheric circulation in the CAM3 is somewhat overestimated (Deser *et al.*, 2012a) and this may also be true of CCSM4. If this is the case, one might expect the dynamically-induced variability to be overestimated as well. On the other hand, it is possible that the coupled atmosphere-ocean system exhibits multidecadal variability that is not fully captured by the models (Deser *et al.*, 2012c; Danabasoglu *et al.*, 2012), in which case, the uncertainty inherent in projections of 56-year SAT trends might be underestimated.

Dynamically-induced atmospheric variability accounts for ~0.7 of the 1.7°C the warming of the Northern Hemisphere continents from 1965 to 2000, but this has had only a small effect on the globally averaged warming. A potentially more important contributor to multidecadal variations in the rate of rise of GST is the variability in the strength of the MOC, which modulates the fluxes of sensible and latent heat at the air-sea interface over the subpolar North Atlantic and parts of the Arctic. Just how much of the spatial and temporal variability in the rate of warming over the continents is MOC-related how much of it is induced by spatial and temporal inhomogeneities in the forcing by greenhouse gases and aerosols has important implications for estimates of climate sensitivity. If the enhanced late 20th century warming was, indeed, a reflection of the internal variability of the climate system, the climate models purported to replicate it in ensemble-mean simulations may be getting the right answer for the wrong reason; i.e., they may be overly sensitive to the radiative forcing.

We have shown that sampling variability constitutes a large part of the uncertainty inherent in projections of regional climate change over the next 50 years, even in averages over areas as large as the continental US. The uncertainty in estimates of the forced response can be narrowed by performing ensembles of simulations and/or by applying dynamical adjustments to the individual ensemble members. However, even if the forced variability were known exactly, the inherently unpredictable, internally generated sampling variability in the future trajectory of the climate system would still remain because it is only one member of an ensemble of possible time-dependent scenarios that could result from a single prescribed external forcing. During the summer season the internally generated SLP variability is smaller than during winter but the thermodynamic consequences of changing atmospheric circulation patterns could be amplified by hydrologic and terrestrial biosphere feedbacks. Deterministic, multidecadal, MOC-related variability is an additional source of uncertainty that could modulate or even temporarily reverse the sign of short term GST trends, as discussed in Chapters 8 and 12.

The large sampling variability inherent in local or regional interdecadal and multidecadal temperature trends has implications for the attribution of record high temperature events and heat waves, whose frequency of occurrence is closely tied to the mean temperature (Wergen and Krug, 2010; Lau and Nath, 2012; Coumou *et al.*, 2013). To the extent that inherently stochastic trends in atmospheric circulation patterns modulate local seasonal mean temperatures, they also raise or lower the frequency of occurrence of extreme events, resulting in clusters of extreme events in particular geographical "hot spots" while extreme event statistics other regions may appear to be unaffected by climate change.

Opponents of environmental protection exploit the uncertainties inherent in projections of future climate change to cast doubt on the immediacy, seriousness, and policy relevance of human-induced environmental degradation and to portray the scientific community as "crying wolf." The sampling issues discussed in this chapter afford the so-called "climate skeptics" some degree of aid and comfort, but not nearly as much as they derive from misstatements about attribution of extreme events and exaggerated projections of regional, near term climate change. Until the signal of human-induced climate change emerges more clearly above background variability, a more compelling case for environmental protection, including reductions the emissions of greenhouse gases, can be made by focusing on the impacts of human-induced warming in the tropics, where the "signal to noise ratio" is much stronger (Mahlstein *et al.*, 2011, 2012) and on the combined impacts of climate change, looming shortages of fresh water (Pearce, 2007; Brown, 2011) and loss of topsoil (Montgomery, 2007) on food security.

References

Alekseev, V. A., P. L. Langen, and J. R. Bates, 2005: Polar amplification of surface warming on an aquaplanet in "ghost forcing" experiments without sea ice feedbacks. *Clim. Dyn.*, **24**, 655–666.

Álvarez-García, F., M. Latif, and A. Biastoch, 2008: On multidecadal and quasi-decadal North Atlantic variability. *J. Climate*, **21**, 3433–3452.

Barnett, T., F. Zwiers, G. Hegerl, M. Allen, T. Crowley, N. Gillett, K. Hasselmann, P. Jones, B. Santer, R. Schnur, P. Stott, K. Taylor and S. Tett, 2005: Detecting and attributing external influences on the climate system: a review of recent advances. *J. Climate*, **18**, 1291–1314.

Black, E., M. Blackburn, G. Harrison, B. J. Hoskins and J. Methven, 2004: Factors contributing to the summer 2003 European heatwave. *Weather*, **59**, 217–223.

Booth, B. B. B., N. J. Dunstone, P. R. Halloran, T. Andrews, N. Bellouin, 2012: Aerosols implicated as a prime driver in twentieth century North Atlantic climate variability. *Nature*, **484**, 228–232.

Bracco, A., F. Kucharski, R. Kallumal and F. Molteni, 2004: Internal variability, external forcing and climate trends in multi-decadal AGCM ensembles. *Clim. Dyn.*, **23**, 659–678.

Brown, L. R., 2011: *World on the Edge: How to Prevent Economic and Environmental Collapese.* W. W. Norton Co., 240 pp.

Chiang, J. C. H., and A. Sobel, 2002: Tropical tropospheric temperature variations caused by ENSO and their influence on the remote tropical climate. *J. Climate*, **15**, 2616–2631.

Clement, A., P. DiNezio, and C. Deser, 2011: Rethinking the Ocean's role in the Southern Oscillation. *J. Climate*, **24**, 4056–4072.

Compo, G. P., and Coauthors, 2011: The Twentieth Century Reanalysis project. *Quart J. Roy. Meteorol. Soc.*, **137**, 1–28.

Coumou, D., A. Robinson, and S. Rahmstorf, 2013: Global increase in record-breaking monthly-mean temperatures. *Climatic Change*, doi: 10.1007/s10584-012-0668-1.

Danabasoglu, G., S. G. Yeager, Y.-Oh Kwon, J. J. Tribbia, A. S. Phillips, and J. W. Hurrell, 2012: Variability of the Atlantic meridional overturning circulation in CCSM4. *J. Climate*, **25**, 5153–5172.

DelSole, T., M. K. Tippett, and J. Shukla, 2011: A significant component of unforced multidecadal variability in the recent acceleration of global warming. *J. Climate*, **24**, 909–926.

Deser, C., R. Knutti, S. Solomon, and A. S. Phillips, 2012b: Communication of the role of natural variability in future North American climate. *Nature: Climate Change*, **2**, 775–779.

———, and A. S. Phillips, 2009: Atmospheric circulation trends, 1950–2000: The relative roles of sea surface temperature forcing and direct atmospheric radiative forcing. *J. Climate*, **22**, 396–413.

———, ———, M. Alexander, and B. V. Smoliak, 2013: Projecting North American climate over the next 50 years: Uncertainty due to internal variability. *J. Climate*, submitted.

———, ———, V. Bourdette, and H. Teng, 2012a: Uncertainty in climate change projections: The role of internal variability. *Clim. Dyn.*, **38**, 527–546.

———, ———, R. A. Tomas, Y. Okumura, M. A. Alexander, A. Capotondi, J. D. Scott, Y.-O. Kwon, and M. Ohba, 2012c: ENSO and Pacific decadal variability in Community Climate System Model version 4. *J. Climate*, **25**, 2622–2651.

Dole, R., M. P. Hoerling, J. Perlwitz, J. Eischeid, P. Pegion. T. Zhang, X. -W. Quan, T. Xu, and D. Murray, 2012: Was there a basis for anticipating the 2011 Russian heat wave? *Geophys. Res. Lett.*, **38**, L06702.

Dommenget, D., 2010: The slab ocean El Niño. *Geophys. Res. Lett.* **37**, L20701.

Gent, P. R., *et al.*, 2011: The Community Climate System Model version 4. *J. Climate*, **24**, 4973–4991.

Gillett, N. P., and D. W. J. Thompson, 2003: Simulation of recent Southern Hemisphere climate change. *Science*, **302**, 273–275.

Gulev, S. K., M. Latif, N. Keenlyside, W. Park, and K. P. Koltermann, 2013: North Atlantic Ocean control on surface heat flux on multidecadal timescales. *Nature*, **499**, 464–467.

Hartmann, D. L., J. M. Wallace, V. Limpasuvan, D. W. J. Thompson and J. R. Holton, 2000: Can ozone depletion and global warming interact to produce rapid climate change? *Proc. Nat. Acad. Sci.*, **97**, 1412–1417.

Hegerl, G. C., *et al.*, 2007: Understanding and attributing climate change. *Climate Change 2007: The Physical Science Basis. Contribution of Working Group I to the Fourth Assessment Report of the IPCC*, Solomon, S., *et al.*, Eds., Cambridge University Press, Cambridge, 663–745.

Held, I. M., and B. J. Soden, 2006: Robust responses of the hydrological cycle to global warming. *J. Climate*, **19**, 5686–5699.

Hu, Z.-Z., A. Kumar, H.-L. Ren, H. Wang, M. L'Heureux, F.-F. Jin, 2013: Weakened interannual variability in the tropical Pacific Ocean since 2000. *J. Climate*, **26**, 2601–2613.

Hurrell, J. W., 1996: Influence of variations in extratropical wintertime teleconnections on Northern Hemisphere temperature. *Geophys. Res. Lett.*, **23**, 665–668.

Kitoh, A., T. Motoi, and H. Koide, 1999: SST variability and its mechanism in a coupled atmosphere-mixed layer ocean model. *J. Climate*, **12**, 1221–1239.

Kosaka, Y., and S.-P. Xie, 2013: Recent global-warming hiatus tied to equatorial Pacific surface cooling. *Nature*, doi: 10.1038/nature12534.

Kutzbach, J. E., 1981: Monsoon climate of the early Holocene: Climate experiment with the Earth's orbital parameters for 9000 years ago. *Science*, **214**, 59–61.

Lau, N.-C., M. J. Nath, 2012: A model study of heat waves over North America: Meteorological aspects and projections for the twenty-first century. *J. Climate*, **25**, 4761–4784.

Lu, J., G. A. Vecchi, and T. Reichler, 2007: Expansion of the Hadley cell under global warming. *Geophys Res. Lett.*, **34**, L06805.

Madden, R. A., and J. Williams, 1978: The correlation between temperature and precipitation in the United States and Europe. *Mon. Wea. Rev.*, **106**, 142–147.

Mahlstein, I., G. Hegerl, and S. Solomon, 2012: Emerging local warming signals in observational data. *Geophys. Res. Lett.*, **39**, L21711.

———, R. Knutti, S. Solomon, and R. W. Portmann, 2011: Early onset of significant local warming in low latitude countries. *Environ. Res. Lett.*, **6**, 034009.

Mantua, N. J., S. R. Hare, Y. Zhang, J. M. Wallace, and R. C. Francis, 1997: An interdecadal oscillation with impacts on salmon production. *Bull. Amer. Meteor. Soc.*, **78**, 1069–1079.

Meehl, G. A., J. M. Arblaster, J. T. Fasullo, A. Hu, and K. E. Trenberth, 2011: Model-based evidence of deep-ocean heat uptake during surface-temperature hiatus periods. *Nature Climate Change*, **1**, 360–364.

———, A. Hu, J. Arblaster, J. Fasullo and K. E. Trenberth, 2013: Externally forced and internally generated decadal climate variability associated with the interdecadal Pacific oscillation. *J. Climate, in press.*

———, *et al.*, 2007: Global climate projections. *Climate Change 2007: The Physical Science Basis. Contribution of Working Group I to the Fourth Assessment Report of the IPCC*, Solomon, S., *et al.*, Eds., Cambridge University Press, 747–845.

Miller, R. L., G. A. Schmidt, and D. T. Shindell, 2006: Forced annular variations in the 20th century Intergovernmental Panel on Climate Change Fourth Assessment Report models. *J. Geophys. Res.*, **111**, D18101.

Montgomery, D. R., 2007: *Dirt: The Erosion of Civilizations.* University of California Press, 295 pp.

Newman, M., G. P. Compo and M. A. Alexander, 2003: ENSO-forced variability of the Pacific decadal oscillation. *J. Climate*, **16**, 3853–3857.

Nitta, T. and S. Yamada, 1989: Recent warming of tropical sea surface temperature and its relationship to the Northern Hemisphere circulation. *J. Meteor. Soc. Japan*, **67**, 187–193.

Oshima, K., Y. Tanimoto, and S.-P. Xie, 2012: Regional patterns of wintertime SLP change over the North Pacific and their uncertainty in CMIP3 multi-model projections. *J. Met. Soc. Japan*, **90A**, 385–396.

Pearce, F., 2007: *When the Rivers Run Dry: Water, the Defining Crisis of the 21st Century.* Beacon Press, 336 pp.

Polvani, L., D. Waugh, G. J.-P. Correa, and S.-W. Son, 2011: Stratospheric ozone depletion: The main driver of 20th century atmospheric circulation changes in the Southern Hemisphere. *J. Climate*, **24**, 795–812.

Randall, D.A., *et al.*, 2007: Climate models and their evaluation. *Climate Change 2007: The Physical Science Basis. Contribution of Working Group I to the Fourth Assessment Report of the IPCC*, Solomon, S., *et al.*, Eds., Cambridge University Press, 589–662.

Robock, A., and J. Mao, 1994: The volcanic signal in surface temperature observations. *J. Climate*, **8**, 1086–1103.

Semenov, V. A., M. Latif, D. Dommenget, N. S. Keenlyside, A. Strehz, T. Martin and W. Park, 2010: The impact of North Atlantic-Arctic multidecadal variability on Northern Hemisphere surface air temperature. *J. Climate*, **23**, 5668–5677.

Shindell, D. T., G. A. Schmidt, R. L. Miller, D. Rind, 2001: Northern Hemisphere winter climate response to greenhouse gas, ozone, solar, and volcanic forcing. *J. Geophys. Res.*, **106**, 7193–7210.

Smoliak, B. V., J. M. Wallace, M. T. Stoelinga and T. P. Mitchell, 2010: Application of partial least squares regression to the diagnosis of year-to-year variations in Pacific Northwest snowpack and Atlantic hurricanes. *Geophys. Res,. Lett.*, **37**, L03801.

Solomon, S., *et al.*, 2007: Technical summary. *Climate Change 2007: The Physical Science Basis. Contribution of Working Group I to the Fourth Assessment Report of the IPCC*, Solomon, S., *et al.*, Eds. Cambridge University Press, 63pp.

Thompson, D. W. J., and S. Solomon, 2002: Interpretation of recent Southern Hemisphere climate change. *Science*, **296**, 895–899.

———, and J. M. Wallace, 2000: Annular modes in the extratropical circulation. Part 1: month-to-month variability. *J. Climate*, **13**, 1000–1016.

———, ———, P. D. Jones and J. J. Kennedy, 2009: Identifying signatures of natural climate variability in time series of global mean surface temperature: methodology and insights. *J. Climate*, **22**, 6120–6141.

Trenberth, K. E., and T. J. Hoar, 1996: The 1990–1995 El Niño-Southern Oscillation event: Longest on record. *Geophys. Res. Lett.*, **23**, 57–60.

———, and ———, 1997: El Niño and climate change. *Geophys. Res. Lett.*, **24**, 3057–3060.

———, and J. W. Hurrell, 1994: Recent obsrerved interdecadal climate changes in the Northern Hemisphere. *Bull. Amer. Meteorol. Soc.*, **71**, 988–993.

———, and D. J. Shea, 2005: Relationships between precipitation and surface temperature. *Geophys. Res. Lett.*, **32**, L14703.

———, *et al.*, 2007: Observations: Surface and atmospheric climate change. *Climate Change 2007: The Physical Science Basis. Contribution of Working Group I to the Fourth Assessment Report of the IPCC*, Solomon, S., *et al.*, Eds., Cambridge University Press, 235–335.

Vecchi, G. A., and B. J. Soden, 2007: Global warming and the weskening of the tropical circulation. *J. Climate*, **20**, 4316–4340.

Wallace, J. M., Q. Fu, B. V. Smoliak, P. Lin, and C. M. Johanson, 2012: Simulated versus observed patterns of warming over the extratropical Northern Hemisphere continents during the cold season. *Proc. Natl. Acad. Sci.*, **109**, 14337–14332.

———, and D. W. J. Thompson, 2002: Annular modes and climate prediction. *Physics Today*, **55**, 28–33.

———, Y. Zhang, and L. Bajuk, 1996: Interpretation of interdecadal trends in Northern Hemisphere surface air temperature. *J. Climate*, **9**, 249–259.

———, ———, and J. A. Renwick, 1995: Dynamic contribution to hemispheric mean temperature trends, *Science*, **270**, 780–783.

Wergen, G. and J. Krug, 2010: Record breaking temperatures reveal a warming climate. *Europhysics Lett.*, **92**, 30008.

Wu, Z., N. E. Huang, J. M. Wallace, B. V. Smoliak, and X. Chen, 2011: On the time-varying trend in global-mean surface temperature. *Clim. Dyn.*, **37**, 759–773.

Yin, J. H., 2005: A Consistent poleward shift of the storm tracks in simulations of 21st century climate. *Geophys. Res. Lett.*, **32**, L18701.

Yulaeva, E., and J. M. Wallace, 1994: The signature of ENSO in global temperature and precipitation fields derived from the microwave sounding unit. *J. Climate*, **7**, 1719–1736.

Zhang, R., T. L. Delworth, R. Sutton, D. Hodson, K. Dixon. I. M. Held, Y. Kushnir, J. Marshall, Y. Ming, R. Msadek, J. Robson, T. Rosati, M. Ting, G. Vecchi, 2013: Have aerosols caused the observed Atlantic multi-decadal variability? *J. Atmos. Sci.*, **70**, 1135–1144.

Zhang, Y., J. M. Wallace and D. S. Battisti, 1997: ENSO-like interdecadal variability: 1900–1993. *J. Climate*, **10**, 1004–1020.

Chapter 2

A Mathematical Theory of Climate Sensitivity or, How to Deal With Both Anthropogenic Forcing and Natural Variability?

Michael Ghil

Ecole Normale Supérieure, 75005 Paris, France, and
University of California, Los Angeles, CA 90095, USA
*ghil@lmd.ens.fr**

Recent estimates of climate evolution over the coming century still differ by several degrees. This uncertainty motivates the work presented here. There are two basic approaches to apprehend the complexity of climate change: deterministically nonlinear and stochastically linear, i.e., the Lorenz and the Hasselmann approach. The grand unification of these two approaches relies on the theory of random dynamical systems. We apply this theory to study the random attractors of nonlinear, stochastically perturbed climate models. Doing so allows one to examine the interaction of internal climate variability with the forcing, whether natural or anthropogenic, and to take into account the climate system's non-equilibrium behavior in determining climate sensitivity.

This non-equilibrium behavior is due to a combination of nonlinear and random effects. We give here a unified treatment of such effects from the point of view of the theory of dynamical systems and of their bifurcations. Energy balance models are used to illustrate multiple equilibria, while multi-decadal oscillations in the thermohaline circulation illustrate the transition from steady states to periodic behavior. Random effects are introduced in the setting of random dynamical systems, which permit a unified treatment of both nonlinearity and stochasticity. The combined treatment of nonlinear and random effects is applied to a stochastically perturbed version of the classical Lorenz convection model.

Climate sensitivity is then defined mathematically as the derivative of an appropriate functional or other function of the system's state with respect to the bifurcation parameter. This definition is illustrated by using numerical results for a model of the El Niño–Southern Oscillation. The concept of a hierarchy of models is the thread that runs across this chapter, and the robustness of elementary bifurcations across such a hierarchy is emphasized.

1. Introduction

The global climate system is composed of a number of subsystems — atmosphere, biosphere, cryosphere, hydrosphere and lithosphere — each of which has distinct characteristic times, from days and weeks to centuries and millennia. Each subsystem, moreover, has its own internal variability, all other things being constant, over a fairly broad range of time scales. These ranges overlap between one subsystem and another. The interactions between the subsystems thus give rise to climate variability on all time scales. We outline here the rudiments of the way in which *dynamical systems theory* provides an understanding of this vast range of variability. Such an understanding proceeds through the study of successively more complex

*Additional E-address: ghil@atmos.ucla.edu.

Climate Change: Multidecadal and Beyond
edited by Chih-Pei Chang et al.

patterns of behavior. These spatio-temporal patterns are studied within narrower ranges of time scales, such as intraseasonal, interannual, interdecadal and multi-millennial. The main results of dynamical systems theory that have demonstrated their importance for the study of climate variability involve *bifurcation theory* and the *ergodic theory* of dynamical systems. More recently, the theory of random dynamical systems has made substantial contributions as well.

Before proceeding with the outline of this chapter, we point out that the successive reports of the Intergovernmental Panel on Climate Change [IPCC: Houghton *et al.* (1991); Solomon *et al.* (2007)] have mainly considered the contributions of uncertainty in the representation of various physical processes to the uncertainty in the climate's equilibrium sensitivity, as defined in Sec. 5.1 below. The well-known result of the IPCC reviews of numerous studies in this direction is that cloud feedbacks are the most important contributor to said uncertainty.

General circulation models (GCMs), on which the IPCC reports rely more and more, do not have a sufficient degree of internal variability — largely but not exclusively due to their still relatively low numerical resolution [*e.g.*, Dijkstra and Weijer (2005)]. Thus it is cloud feedbacks, and their paramount effect on equilibrium sensitivity, that seem to dominate the IPCC estimates of future climate change as well.

The main purpose of this chapter is to illustrate the way in which natural climate variability could affect these estimates. We do not attempt, at this time, to provide a quantitative evaluation of the extent to which this variability's effect would modify IPCC conclusions, whether they relate to cloud feedbacks or to other types of feedbacks. Among the latter, the oceanic circulation also plays an important role.

In the next section, we describe the climate systems dominant balance between incoming solar radiation, dominated by short waves, and outgoing terrestrial radiation, dominated by long waves. This balance is consistent with the existence of multiple equilibria of surface temperatures.

Such *multiple equilibria* are also present for other balances of climatic actions and reactions. Thus, on the intraseasonal time scale, the thermal driving of the mid-latitude westerly winds is countered by surface friction and mountain drag. Multiple equilibria typically arise from saddle-node bifurcations of the governing equations. Transitions from one equilibrium to another may result from small and random pushes — a typical case of minute causes having large effects in the long term.

In Sec. 3, we sketch the oceans' overturning circulation between cold regions, where water is heavier and sinks, and warm regions, where it is lighter and rises. The effect of temperature on the water masses density and, hence, motion is in competition with the effect of salinity: density increases, through evaporation and brine formation, compete further with decreases in salinity and, hence, density through precipitation and river run-off. These competing effects can also give rise to two distinct equilibria.

In the present-day oceans, a *thermohaline* circulation prevails, in which the temperature effects dominate. In the remote past, about 50 Myr ago, a *halothermal* circulation may have obtained, with salinity effects dominating. In a simplified mathematical setting, these two equilibria arise by a pitchfork bifurcation that breaks the problems mirror symmetry. On shorter time scales, of decades-to-millennia, oscillations of intensity and spatial pattern in the thermohaline circulation seem to be the dominant mode of variability. We show how interdecadal oscillations in the oceans' circulation arise by Hopf bifurcation.

In Sec. 4, we address the way that faster processes, modeled as random effects, can interact with the slower, nonlinear ones. The combined treatment of the nonlinear and stochastic processes can reveal amazingly fine structure in the climate systems behavior, but also — and rather surprisingly — add robustness and predictability to the results.

In Sec. 5, we discuss the way that climate sensitivity can be defined in the stochastic vs. the deterministic context. Concluding remarks follow in Sec. 6.

2. Energy-Balance Models and the Modeling Hierarchy

The concepts and methods of the theory of deterministic dynamical systems (Andronov and Pontryagin, 1937; Arnol'd, 1983; Guckenheimer and Holmes, 1983) have been applied first to simple models of atmospheric and oceanic flows, starting about fifty years ago (Lorenz, 1963; Stommel, 1961). More powerful computers now allow their application to fairly realistic and detailed models of the atmosphere, ocean, and the coupled atmosphere–ocean system. We start therefore by presenting such a hierarchy of models.

This presentation is interwoven with that of the successive bifurcations that lead from simple to more complex solution behavior for each climate model. Useful tools for comparing model behavior across the hierarchy and with observations are provided by ergodic theory (Eckmann and Ruelle, 1985; Ghil *et al.*, 2008a). Among these, advanced methods for the analysis and prediction of uni- and multivariate time series play an important role.

2.1. *Radiation balance and energy-balance models (EBMs)*

The concept of a modeling hierarchy in climate dynamics was introduced by Schneider and Dickinson (1974); it is discussed in greater detail by Ghil and Robertson (2000) and by Dijkstra and Ghil (2005). At present, the best-developed hierarchy is for atmospheric models. These models were originally developed for weather simulation and prediction on the time scale of hours to days. Currently they serve in a stand-alone mode or coupled to oceanic and other models to address climate variability on all time scales.

The first rung of the modeling hierarchy for the atmosphere is formed by zero-dimensional (0-D) models, where the number of dimensions, from zero to three, refers to the number of independent space variables used to describe the model domain, i.e., to physical-space dimensions. Such 0-D models essentially attempt to follow the evolution of global surface-air temperature as a result of changes in global radiative balance:

$$c\frac{\mathrm{d}\bar{T}}{\mathrm{d}t} = R_\mathrm{i} - R_\mathrm{o}, \tag{1a}$$

$$R_\mathrm{i} = \mu Q_0\{1 - \alpha(\bar{T})\}, \tag{1b}$$

$$R_\mathrm{o} = \sigma m(\bar{T})(\bar{T})^4. \tag{1c}$$

Here R_i and R_o are incoming solar radiation and outgoing terrestrial radiation. The heat capacity c is that of the global atmosphere, plus that of the global ocean or some fraction thereof, depending on the time scale of interest: one might only include in c the ocean mixed layer when interested in subannual time scales but the entire ocean when studying paleoclimate.

The rate of change of $\bar{T}$ with time t is given by $\mathrm{d}\bar{T}/\mathrm{d}t$, while Q_0 is the solar radiation received at the top of the atmosphere, σ is the Stefan-Boltzmann constant, and m is an insolation parameter, equal to unity for present-day conditions. To have a closed, self-consistent model, the planetary reflectivity or albedo α and grayness factor m have to be expressed as functions of $\bar{T}$; $m = 1$ for a perfectly black body and $0 < m < 1$ for a grey body like planet Earth.

There are two kinds of one-dimensional (1-D) atmospheric models, for which the single spatial variable is latitude or height, respectively. The former are so-called energy-balance models (EBMs), which consider the generalization of the model (2.1) for the evolution of surface-air temperature $T = T(x, t)$, say,

$$c(x)\frac{\partial\bar{T}}{\partial t} = R_\mathrm{i} - R_\mathrm{o} + D. \tag{2}$$

Here the terms on the right-hand side can be functions of the meridional coordinate x (latitude, co-latitude, or sine of latitude), as well as of time t and temperature T. The horizontal heat-flux term D expresses heat exchange between latitude belts; it typically contains first and second partial derivatives of T with respect to x. Hence the rate of change of local temperature T with respect to time also becomes a partial derivative, $\partial \bar{T}/\partial t$. Such models were introduced independently by Budyko (1969) and by Sellers (1969).

The first striking results of theoretical climate dynamics were obtained in showing that Eq. (2) could have two stable steady-state solutions, depending on the value of the insolation parameter μ, cf. Eq. (1b). This multiplicity of stable steady states, or physically possible climates of our planet, can be explained, in its simplest form, in the 0-D model given by Eq. (1).

The physical explanation resides in the fact that — for a fairly broad range of μ-values around $\mu = 1.0$ — the curves for $R_{\rm i}$ and $R_{\rm o}$ as a function of $\bar{T}$ intersect in 3 points. One of these points corresponds to the present climate (highest $\bar{T}$-value), and another one to an ice-covered planet (lowest $\bar{T}$-value); both of these are stable, while the third one (intermediate $\bar{T}$-value) is unstable. To obtain this result, it suffices to make two assumptions: (i) that $\alpha = \alpha(\bar{T})$ is a piecewise-linear function of $\bar{T}$, with high albedo at low temperature, due to the presence of snow and ice, and low albedo at high $\bar{T}$, due to their absence; and (ii) that $m = m(\bar{T})$ is a smooth, increasing function of $\bar{T}$ that attempts to capture in its simplest from the greenhouse effect of trace gases and water vapor.

The bifurcation diagram of a 1-D EBM, like the one of Eq. (2), is shown in Fig. 1. It displays the model's mean temperature $\bar{T}$ as a function of the fractional change μ in the insolation $Q = Q(x)$ at the top of the atmosphere. The S-shaped curve in the figure arises from two back-to-back *saddle-node bifurcations*.

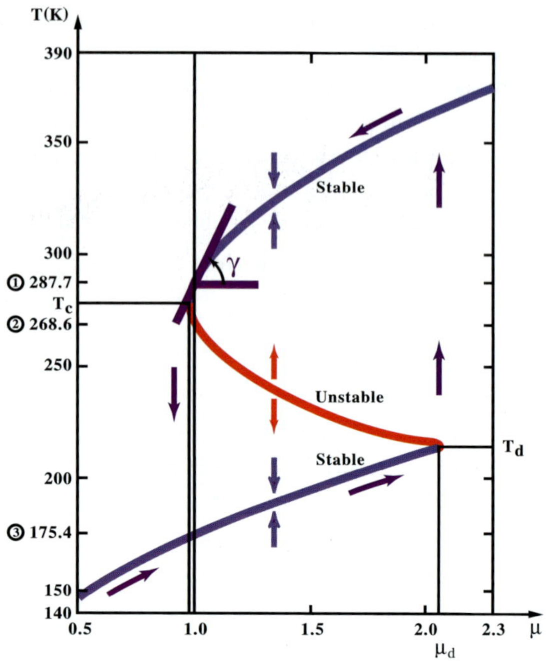

Fig. 1. Bifurcation diagram for the solutions of an energy-balance model (EBM), showing the global-mean temperature $\bar{T}$ vs. the fractional change μ of insolation at the top of the atmosphere. The arrows pointing up and down at about $\mu = 1.4$ indicate the stability of the branches: towards a given branch if it is stable and away if it is unstable. The other arrows show the hysteresis cycle that global temperatures would have to undergo for transition from the upper stable branch to the lower one and back. The angle γ gives the measure of the present climates sensitivity to changes in insolation. [After Ghil and Childress (1987) with permission from Springer-Verlag.]

The normal form of the first one is

$$\dot{X} = \mu - X^2. \tag{3}$$

Here X stands for a suitably normalized form of $\bar{T}$ and $\dot{X}$ is the rate of change of X, while μ is a parameter that measures the stress on the system, in particular a normalized form of the insolation parameter.

The uppermost branch corresponds to the steady-state solution $X = +\mu^{1/2}$ of Eq. (3) and it is stable. This branch matches rather well Earth's present-day climate for $\mu = 1.0$; more precisely the steady-state solution

$T = T(x; \mu)$ of the full 1-D EBM (not shown) matches closely the annual mean temperature profile from instrumental data over the last century.

The intermediate branch starts out at the left as the second solution, $X = -\mu^{1/2}$, of Eq. (3) and it is unstable. It blends smoothly into the upper branch of a coordinate-shifted and mirror-reflected version of Eq. (3), say

$$\dot{X} = \mu - X^2. \quad (4)$$

This branch, $X = X_0 + (\mu_0 - \mu)^{1/2}$, is also unstable. Finally, the lowermost branch in Fig. 1 is the second steady-state solution of Eq. (4), $X = X_0(\mu_0 - \mu)^{1/2}$, and it is stable, like the uppermost branch. The lowermost branch corresponds to an ice-covered planet at the same distance from the Sun as Earth.

The fact that the upper-left bifurcation point (μ_c, T_c) in Fig. 1 is so close to present-day insolation values created great concern in the climate dynamics community in the mid-1970s, when these results were obtained. Indeed, much more detailed computations (see below) confirmed that a reduction of about 2–5% of insolation values would suffice to precipitate Earth into a deep freeze. The great distance of the lower-right bifurcation point (μ_d, T_d) from present-day insolation values, on the other hand, suggests that one would have to nearly double atmospheric opacity, say, for the Earth's climate to jump back to more comfortable temperatures.

The results here follow Ghil (1976). Held and Suarez (1974) and North (1975) obtained similar results, and a detailed comparison between EBMs appears in Chapter 10 of Ghil and Childress (1987). The robustness of a saddle-node bifurcation has been documented by Wetherald and Manabe (1975, their Fig. 5 and associated text) in a slightly simplified GCM, coupled to a swamp ocean. Likewise, North *et al.* (1983) have found the manifestation of a hysteresis cycle among such bifurcations — their "small–ice-cap instability" — in a two-dimensional (2-D) EBM resolving the continents and seasons. Please see the next subsection for the rung occupied by each of these two models in the climate modeling hierarchy.

2.2. *Other atmospheric processes and models*

The 1-D atmospheric models in which the details of radiative equilibrium are investigated with respect to a height coordinate z (geometric height, pressure, etc.) are often called *radiative-convective models* (Ramanathan and Coakley, 1978). This name emphasizes the key role that convection plays in vertical heat transfer. While these models preceded historically EBMs as rungs on the modeling hierarchy, it was only recently shown that they, too, could exhibit multiple equilibria (Li *et al.*, 1997). The word (stable) equilibrium, here and in the rest of this article, refers simply to a (stable) steady state of the model, rather than a to true thermodynamic equilibrium.

2-D atmospheric models are also of two kinds, according to the third space coordinate that is not explicitly included. Models that resolve explicitly two horizontal coordinates, on the sphere or on a plane tangent to it, tend to emphasize the study of the dynamics of large-scale atmospheric motions. They often have a single layer or two. Those that resolve explicitly a meridional coordinate and height are essentially combinations of EBMs and radiative-convective models and emphasize therewith the thermodynamic state of the system, rather than its dynamics.

Yet another class of horizontal 2-D models is the extension of EBMs to resolve zonal, as well as meridional surface features, in particular land-sea contrasts (North *et al.*, 1983). We shall see in Sec. 3.2 how such a 2-D EBM is used, when coupled to an oceanic model.

Schneider and Dickinson (1974) and Ghil and Robertson (2000) discuss additional types of 1-D and 2-D atmospheric models and give references to these and to the types discussed above, along with some of their main applications. Finally, to encompass and resolve the main atmospheric phenomena with respect to all three spatial coordinates, GCMs occupy the pinnacle of the modeling hierarchy.

The dependence of mean zonal temperature on the insolation parameter μ (the normalized "solar constant") — as obtained for 1-D EBMs and shown in Fig. 1 here — was confirmed, to the extent possible, by using a simplified GCM, coupled to a swamp ocean model. More precisely, forward integrations with a GCM cannot confirm the presence of the intermediate, unstable branch. Nor was it possible in the mid-70s, when this numerical experiment was done, to reach the deep-freeze stable branch, because of the GCMs computational limitations. But the parabolic shape of the upper, present-daylike branch near the upper-left bifurcation point in our figure, cf. Eq. (3), was well supported by the GCM simulations.

Ghil and Robertson (2000) also describe the separate hierarchies that have grown over the last quarter-century in modeling the ocean and the coupled ocean-atmosphere system. More recently, an overarching hierarchy of earth-system models that encompass all the subsystems of interest, atmosphere, biosphere, cryosphere, hydrosphere and lithosphere has been developing. Eventually, the partial results about each subsystem's variability, outlined in this section and the next one, will have to be verified from one rung to the next of the Earth-system modeling hierarchy.

In the meantime, we have described in the present section here the robustness of bifurcation results for the multiple equilibria associated with the Earth system's energy balance, from 0-D EBMs all the way to full, 3-D GCMs. In the next section, we discuss the robustness of bifurcation results on both multiple equilibria and oscillations, across a modeling hierarchy for the ocean's circulation.

3. Oscillations in the Oceans' Thermohaline Circulation

3.1. *Theory and simple models*

Historically, the thermohaline circulation (THC) was first among the climate systems major processes to be studied using a very simple mathematical model. Stommel (1961) formulated a two-box model and showed that it possessed multiple equilibria.

A sketch of the Atlantic Oceans THC and its interactions with the atmosphere and cryosphere on long time scales is shown in Fig. 2. These interactions can lead to climate oscillations with multi-millennial periods, such as the Heinrich events, and are summarized in the figure's caption. An equally schematic view of the global THC is provided by the widely known "conveyor belt" diagram. The latter diagram captures greater horizontal, 2-D detail but it does not commonly include the THC's interactions with water in both its gaseous and solid phases, which our Fig. 2 here does include.

Basically, the THC is due to denser water sinking, lighter water rising, and water-mass continuity closing the circuit through near-horizontal flow between the areas of rising and sinking. The effects of temperature and salinity on the ocean waters density, $\rho = \rho(T, S)$, oppose each other: the density ρ *decreases* as T increases and it *increases* as S increases. It is these two effects that give the *thermohaline* circulation its name, from the Greek words for T and S. In high latitudes, ρ increases as the water loses heat to the air above and, if sea ice is formed, as the water underneath is enriched in brine. In low latitudes, ρ increases due to evaporation but decreases due to sensible heat flux into the ocean.

For the present climate, the temperature effect is stronger than the salinity effect, and ocean water is observed to sink in certain areas of the high-latitude North Atlantic and

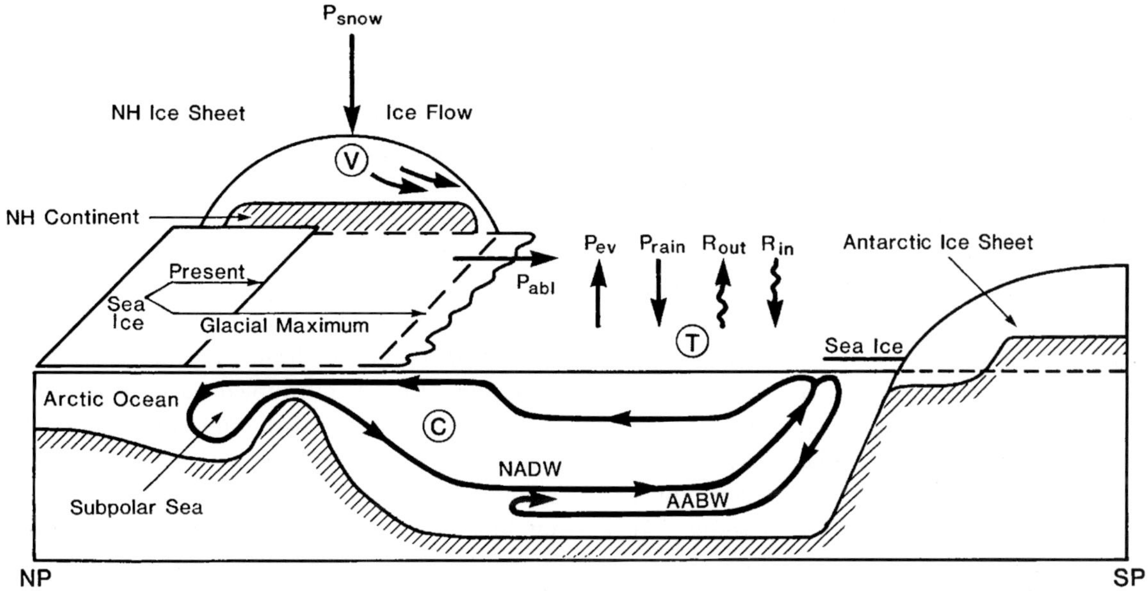

Fig. 2. Diagram of an Atlantic meridional cross section from North Pole (NP) to South Pole (SP), showing mechanisms likely to affect the thermohaline circulation (THC) on various time-scales. Changes in the radiation balance $R_{\text{in}}-R_{\text{out}}$ are due, at least in part, to changes in extent of Northern Hemisphere (NH) snow and ice cover V, and to how these changes affect the global temperature T; the extent of Southern Hemisphere ice is assumed constant, to a first approximation. The change in hydrologic cycle expressed in the terms $P_{\text{rain}}-P_{\text{evap}}$ for the ocean and $P_{\text{snow}}-P_{\text{abl}}$ for the snow and ice is due to changes in ocean temperature. Deep-water formation in the North Atlantic Subpolar Sea (North Atlantic Deep Water: NADW) is affected by changes in ice volume and extent, and regulates the intensity C of the THC; changes in Antarctic Bottom Water (AABW) formation are neglected in this approximation. The THC intensity C in turn affects the systems temperature, and is also affected by it. [After Ghil *et al.* (1987) with permission from Springer-Verlag.]

Southern Ocean — with very few and limited areas of deep-water formation elsewhere — and to rise everywhere else. Thus, in a *thermohaline* regime, T is more important than S and hence comes before it.

During some remote geological times, deep water may have formed in the global ocean near the equator; such an overturning circulation of opposite sign to that prevailing today has been dubbed *halothermal*, S before T. The quantification of the relative effects of T and S on the oceanic water masses buoyancy in high and low latitudes is far from complete, especially for paleocirculations; the association of the latter with salinity effects that exceed the thermal ones (Kennett and Stott, 1991) is thus rather tentative.

To study the reversal of the abyssal circulation, due to the opposite effects of T and S, Stommel (1961) considered a two-box model, with two pipes connecting the two boxes. He showed that the system of two nonlinear, coupled ordinary differential equations that govern the temperature and salinity differences between the two well-mixed boxes has two stable steady-state solutions; these two steady states are distinguished by the direction of flow in the upper and lower pipe.

Stommel's paper was primarily concerned with distinct local convection regimes, and hence vertical stratifications, in the North Atlantic and the Mediterranean or the Red Sea, say. Today, we mainly think of one box as representing the low latitudes and the other one the high latitudes in the global THC (Marotzke, 2000; Dijkstra and Ghil, 2005).

The next step in the hierarchical modeling of the THC is that of 2-D meridional plane models, in which the temperature and salinity fields are governed by coupled nonlinear partial differential equations with two independent space variables, latitude and depth, say. Given boundary conditions for such a model that are symmetric about the Equator, as the equations themselves are, one expects a symmetric solution, in which water either sinks near the poles and rises everywhere else (thermohaline) or sinks near the Equator and rises everywhere else (halothermal). These two symmetric solutions would correspond to the two equilibria of Stommel's box model of 1961.

In fact, *symmetry breaking* can occur, leading gradually from a symmetric two-cell circulation to an antisymmetric one-cell circulation. In between, all degrees of dominance of one cell over the other are possible. A situation lying somewhere between the two seems to resemble most closely the meridional overturning diagram of the Atlantic Ocean in Fig. 2.

This symmetry breaking can be described by a *pitchfork bifurcation*:

$$\dot{X} = \mu - X^3. \tag{5}$$

Here X stands for the amount of asymmetry in the solution, so that $X = 0$ is the symmetric branch, and μ is a parameter that measures the stress on the system, in particular a normalized form of the buoyancy flux at the surface. For $\mu < 0$ the symmetric branch is stable, while for $\mu > 0$ the two branches $X = \pm\mu^{1/2}$ inherit its stability.

In the 2-D THC problem, the left cell dominates on one branch, while the right cell dominates on the other: for a given value of μ, the two stable steady-state solutions — on the $\{X = +\mu^{1/2}\}$ branch and on the $\{X = -\mu^{1/2}\}$ branch — are mirror images of each other. The idealized THC in Fig. 2, with the North Atlantic Deep Water extending to the Southern Oceans polar front, corresponds to one of these two branches. In theory, therefore, a mirror-image circulation, with the Antarctic Bottom Water extending to the North Atlantics polar front, is equally possible.

3.2. *Bifurcation diagrams for GCMs*

Bryan (1986) was the first to document transition from a two-cell to a one-cell circulation in a simplified ocean GCM with idealized, symmetric forcing. Results of coupled ocean-atmosphere GCMs, however, have led to questions about the realism of more than one stable THC equilibrium. The situation with respect to the THC's pitchfork bifurcation (5) is thus subtler than it was with respect to Fig. 1 for radiative equilibrium.

In the previous section, atmospheric GCMs — both 2-D (North *et al.*, 1983) and 3-D (Wetherald and Manabe, 1975) — confirmed essentially the EBM results. The results obtained in climbing the rungs of the modeling hierarchy for the THC appear, however, to be still in need of further clarification.

To facilitate this clarification, Dijkstra (2007) defined an integral indicator Σ associated with the multiple-equilibria regime of the Atlantic Ocean's THC,

$$\Sigma = M_{\mathrm{ov}}(\theta_{\mathrm{N}}) - M_{\mathrm{ov}}(\theta_{\mathrm{S}}). \tag{6}$$

In the above definition, the sign of $M_{\mathrm{ov}}(\theta)$, positive or negative, states whether the Atlantic exports salt or fresh water across a particular parallel θ; here θ_{N} passes through the southern tip of Greenland, and θ_{s} passes through the southern tip of Africa. The present notation differs slightly from the original one, for clarity of presentation. When $\Sigma > 0$, the Atlantic's THC is in the multiple-equilibria regime, i.e., a thermohaline and a halothermal circulation can coexist.

Recent observational studies (Bryden *et al.*, 2011) and ocean reanalyses (Hawkins *et al.*, 2011) confirm that the Atlantic Ocean's THC is actually in this regime at the present time. On the other hand, the ocean GCMs used in the third Coupled-Model Intercomparison Program (CMIP3) — on which the conclusions of the IPCC's Fourth Assessment Report (Solomon *et al.*, 2007) were based — tended to have a systematic bias in their evaporation-minus-precipitation fluxes (Drijfthout *et al.*, 2011), which led to the wrong sign of Σ in these models.

Thus a failure of high-end models in the hierarchy to confirm results obtained at the hierarchy's lower end does not necessarily imply it's the simpler models that are wrong. To the contrary, such a failure might well indicate that the high-end models, no matter how detailed, can still be quite imperfect.

Internal variability of the THC with smaller and more regular excursions than the huge and totally irregular jumps associated with bistability was studied intensively in the late 1980s and the 1990s. These studies placed themselves on various rungs of the modeling hierarchy, from box models through 2-D models and all the way to ocean GCMs. A summary of the different kinds of oscillatory variability found in the latter appears in Table 1. Such oscillatory behavior seems to match more closely the instrumentally recorded THC variability, as well as the paleoclimatic records for the recent geological past, than bistability.

The (multi)millennial oscillations interact with variability in the surface features and processes shown in Fig. 2. Chen and Ghil (1996), in particular, studied some of the interactions between atmospheric processes and the THC. They used a so-called hybrid coupled model, namely a (horizontally) 2-D EBM, coupled to a rectangular-box version of the North Atlantic rendered by a low-resolution ocean GCM. This hybrid models regime diagram is shown in Fig. 3(a). A steady state is stable for high values of the coupling parameter or of the EBMs diffusion parameter d. Interdecadal oscillations with a period of 40–50 years are self-sustained and stable for low values of these parameters.

The self-sustained THC oscillations in question are characterized by a pair of vortices of opposite sign that grow and decay in quadrature with each other in the oceans upper layers. Their centers follow each other anti-clockwise through the northwestern quadrant of the models rectangular domain. Both the period and the spatio-temporal characteristics of the oscillation are thus rather similar to those seen in a fully coupled GCM with realistic geometry. The transition from a stable equilibrium to a stable limit cycle, via Hopf bifurcation, in this hybrid coupled model, is shown in Fig. 3(b).

4. Randomness and Nonlinearity

4.1. *What to expect*

The geometric (Arnol'd, 1983; Guckenheimer and Holmes, 1983) and the ergodic (Eckmann and Ruelle, 1985) theory of *deterministic* dynamical systems represent significant achievements of the 20th century. The foundations of the *stochastic* calculus in its second half (Doob, 1953) also led to the birth of a rigorous

Table 1. Oscillations in the oceans' thermohaline circulation.

Time scale	Phenomena	Mechanism[a]
Decadal	• Local migration of surface anomalies in the northwest corner of the ocean basin • Gyre advection in mid-latitudes	• Localized surface density anomalies due to surface coupling • Gyre advection
Centennial	Loop-type, meridional circulation	Conveyor-belt advection of density anomalies
Millennial	Relaxation oscillation, with "flushes" and superimposed decadal fluctuations	Bottom water warming, due to strong braking effect of salinity forcing

[a]Complete references to these mechanisms are given in Ghil (1994); they include both simplified and full ocean GCMs.

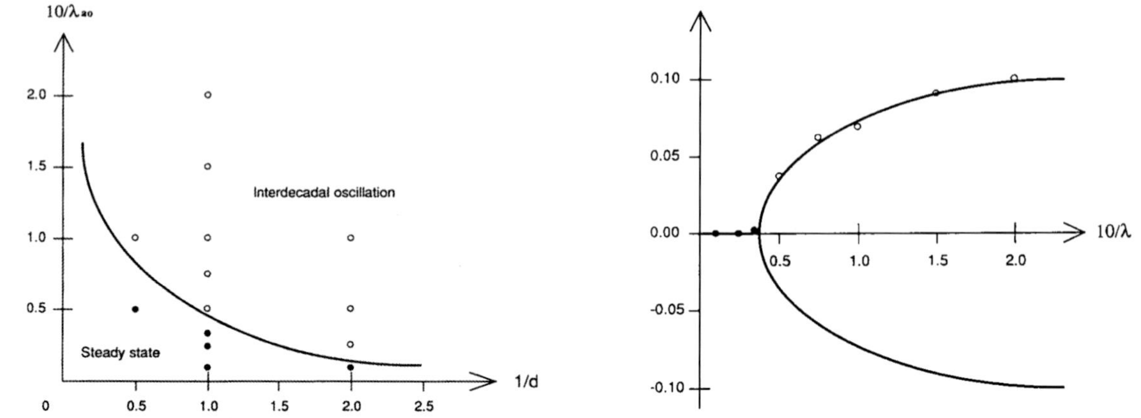

(a) Regime diagram (b) Bifurcation diagram

Fig. 3. Dependence of THC solutions on two parameters in a hybrid coupled model; the two parameters are the atmosphere-ocean coupling coefficient λ_{ao} and the atmospheric thermal diffusion coefficient d. (a) Schematic regime diagram. The full circles stand for the models stable steady states, the open circles for stable limit cycles, and the solid curve is the estimated neutral stability curve between the former and the latter. (b) Hopf bifurcation curve at fixed $d = 1.0$ and varying λ_{ao}; this curve was obtained by fitting a parabola to the models numerical-simulation results, shown as full and open circles. [From Chen and Ghil (1996) with permission from the American Meteorological Society.]

theory of time-dependent random phenomena. Historically, theoretical developments in climate dynamics have been largely motivated by these two complementary approaches, based on the work of E. N. Lorenz and that of K. Hasselmann [*e.g.*, Lorenz (1963); Hasselmann (1976)], respectively.

It now seems clear that these two approaches complement, rather than exclude each other. Incomplete knowledge of small-, subgrid-scale processes, as well as computational limitations will always require one to account for these processes in a stochastic way. As a result of sensitive dependence on initial data and on parameters, numerical weather forecasts, as well as climate projections are both expressed these days in probabilistic terms. In addition to the intrinsic challenge of addressing the nonlinearity along with the stochasticity of climatic processes, it is thus more convenient — and becoming more and more necessary — to rely on a model's (or set of models') *probability density function* (PDF) rather than on its individual, point-wise simulations or predictions.

We summarize here results on the surprisingly complex statistical structure that characterizes stochastic nonlinear systems. This complex structure does provide meaningful physical information that is not described by the PDF alone; it lives on a *random attractor*, which extends the concepts of a *strange attractor* and of the *invariant measure* that is supported by it (Ruelle, 2006), from the deterministic to the stochastic framework.

The expository review here follows the literature of random dynamical systems, as covered for instance by Arnold (1998). Our approach emphasizes the pathwise behavior of the system for individual realizations of the noise over the more usual approach of restricting oneself to the properties of the system's PDF when averaged over a hypothetical ensemble of realizations. This approach might be less familiar even for mathematically sophisticated, but non-specialist readers. For a more complete presentation of this pathwise approach, intended for climate dynamicists as well as for other non-specialists, we refer to Ghil *et al.* (2008a) and Chekroun *et al.* (2011b).

4.2. *What one finds*

On the road to including random effects, one needs to realize first that the climate system — as well as any of its subsystems, and on any time scale — is not closed: it exchanges energy, mass and momentum with its surroundings, whether other subsystems or the interplanetary space and the solid earth. The typical applications of dynamical systems theory to climate variability so far have only taken into account exchanges that are constant in time, thus keeping the model — whether governed by ordinary, partial or other differential equations — *autonomous*; i.e., the models had coefficients and forcings that were constant in time.

Succinctly, one can write such a system as

$$\dot{\mathbf{X}} = \mathbf{f}(\mathbf{X}; \boldsymbol{\mu}), \tag{7}$$

where $\mathbf{X}$ now may stand for any state vector or climate field, while $\mathbf{f}$ is a smooth function of $\mathbf{X}$ and of the vector of parameters $\boldsymbol{\mu}$, but does not depend explicitly on time. This characteristic of being autonomous greatly facilitated the analysis of model solutions' properties. For instance, two distinct trajectories, $\mathbf{X}_1(t)$ and $\mathbf{X}_2(t)$, of a well-behaved, smooth autonomous system cannot pass through the same point in phase space, which helps describe the systems phase portrait. So does the fact that we only need to consider the behavior of solutions $\mathbf{X}(t)$ as we let time t tend to $+\infty$: the resulting sets of points are — possibly multiple — equilibria, periodic solutions, and chaotic sets. In the language of dynamical systems theory, these are called, respectively: *fixed points, limit cycles*, and strange attractors.

We know only too well, however, that the seasonal cycle plays a key role in climate variability on many time scales, while orbital forcing is crucial on the Quaternary time scales of many millennia, and now anthropogenic forcing is of utmost importance on interdecadal time scales. How can one take into account such time-dependent forcings, and analyze the non-autonomous systems, written succinctly as

$$\dot{\mathbf{X}} = \mathbf{f}(\mathbf{X}, t; \boldsymbol{\mu}), \tag{8}$$

to which they give rise? In Eq. (8), the dependence of $\mathbf{f}$ on t may be periodic, $\mathbf{f}(\mathbf{X}, t+P) = \mathbf{f}(\mathbf{X}, t)$, as in various El Niño–Southern Oscillation (ENSO) models, where the period $P = 12$ months, or monotone, $\mathbf{f}(\mathbf{X}, t+\tau) \geq \mathbf{f}(\mathbf{X}, t)$ for $\tau \geq 0$, as in studying scenarios of anthropogenic climate forcing.

To illustrate the fundamental character of the distinction between an autonomous system like (7) and a non-autonomous one like (8), consider the simple scalar version of these two equations:

$$\dot{X} = -\beta X, \tag{9}$$

and

$$\dot{X} = -\beta X + \gamma t, \tag{10}$$

respectively. We assume that both systems are dissipative, i.e., $\beta > 0$, and that the forcing is monotone increasing, $\gamma \geq 0$, as would be the case for anthropogenic forcing in the industrial era. Lorenz (1963) pointed out the key role of dissipativity in giving rise to strange, but attracting solution behavior, while Ghil and Childress (1987) emphasized its importance and pervasive character in climate dynamics. Clearly the only attractor for the solutions of Eq. (9), given any initial point $X(0) = X_0$, is the fixed point $X = 0$, attained as $t \to +\infty$.

For the non-autonomous case of Eq. (10), though, this *forward-in-time approach* yields blow-up as $t \to +\infty$, for any initial point. To make sense of what happens in the case of time-dependent forcing, one introduces instead the *pullback approach*, in which solutions are allowed to still depend on the time t at which we observe them, but also on a time s from which the solution is started, $X(s) = X_0$; presumably $s \ll t$.

With this little change of approach, one can easily verify that

$$|X(s,t;X_0)-\mathcal{A}(t)| \to 0 \quad \text{as} \quad s \to -\infty, \tag{11}$$

for all t and X_0, where

$$\mathcal{A}(t) = \frac{\gamma(t-1/\beta)}{\beta}. \tag{12}$$

We thus obtain, in this pullback sense, the intuitively obvious result that the solutions, if started far enough in the past, all approach the attractor set $\mathcal{A}(\omega)$, which has a linear growth in time and thus follows the linear forcing.

Let us return now to the more general, nonlinear case of Eq. (8) and add not only deterministic time dependence $\mathbf{f}(\mathbf{X}, t)$, but also random forcing (Ledrappier and Young, 1988; Arnold, 1998),

$$\mathrm{d}\mathbf{X} = \mathbf{f}(\mathbf{X}, t)\mathrm{d}t + \mathbf{g}(\mathbf{X})\mathrm{d}\eta, \tag{13}$$

where $\eta = \eta(t;\omega)$ represents a Wiener process — and $\mathrm{d}\eta(t)$ is commonly referred to as "white noise — while ω labels the particular realization of this random process. The case $\mathbf{g}(\mathbf{X}) =$ **const.** is the case of additive noise, while in the case of $\partial\mathbf{g}(\mathbf{X})/\partial\mathbf{X} \neq 0$ we speak of multiplicative noise. The distinction between $\mathrm{d}t$ and $\mathrm{d}\omega$ in Eq. (13) is necessary since, roughly speaking and following Einstein's treatment of Brownian motion (Einstein, 1905), it is the variance of a Wiener process that is proportional to time and thus $\mathrm{d}\eta \sim (\mathrm{d}t)^{1/2}$.

In the case of random forcing, we can illustrate the concepts introduced by the simple example of Eqs. (11, 12) above by the random attractor $\mathcal{A}(\omega)$ (yellow band) of Fig. 4. In the figure, $\mathrm{d}\eta(t;\omega) = \theta(t)\omega$ is the random process that drives the system (solid black line) and the pullback attraction is depicted by the flow of an arbitrary set $\mathcal{B}$ from "pullback times $t = -\tau_2$ and $t = -\tau_1$ onto the attractor (heavy blue arrows).

More explicitly, we show in Fig. 5 four snapshots $\{\mathcal{A}_j(\omega) = \mathcal{A}(\omega; t = t_j) : j = 1, 2, 3, 4\}$ that correspond to the vertical cross-sections (heavy solid) in the attractor of Fig. 4; a short video, from which these snapshots are taken, appears as Supplementary Information in Chekroun *et al.* (2011b). These snapshots were calculated for the random attractor $\mathcal{A}(t;\omega)$ of a stochastically perturbed Lorenz system, given by

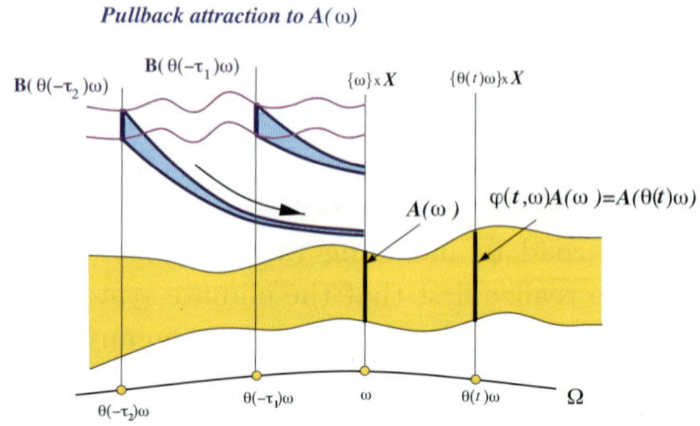

Fig. 4. Schematic diagram of a random attractor $\mathcal{A}(\omega)$ and of the pullback attraction to it; here ω labels the particular realization of the random process $\theta(t)\omega$ that drives the system. We illustrate the evolution in time t of the random process $\theta(t)\omega$ (solid black line at the bottom); the random attractor $\mathcal{A}(\omega)$ itself (yellow band in the middle) with the "snapshots" $\mathcal{A}(\omega) = \mathcal{A}(t = 0;\omega)$ and $\mathcal{A}(t;\omega)$ (the two vertical sections, heavy solid); and the flow of an arbitrary set $\mathcal{B}$ from pullback times $t = -\tau_2$ and $t = -\tau_1$ onto the attractor (heavy blue arrows). [After Ghil *et al.* (2008a) with permission from Elsevier.]

$$\mathrm{d}X = Pr(Y - X)\mathrm{d}t + \sigma X\mathrm{d}\eta, \tag{14a}$$

$$\mathrm{d}X = (rX - Y - XZ)\mathrm{d}t + \sigma Y\mathrm{d}\eta \tag{14b}$$

$$\mathrm{d}X = (-bZ + XY)\mathrm{d}t + \sigma Z\mathrm{d}\eta. \tag{14c}$$

The parameters r, Pr and b in Eqs. (14) have the usual meanings for 2-D thermal convection: $r = R/R_c$ is the Rayleigh number R normalized by its critical value R_c at the onset of convection, Pr is the Prandtl number, and b is a normalized wave number for the most unstable wave at the onset of convection. The noise in this case is multiplicative: its intensity $\sigma = 0.5$ is multiplied in each one of the three coupled, nonlinear equations above by the corresponding variable X, Y or Z.

To be precise, what is plotted in Fig. 5, and in the associated video, is the density

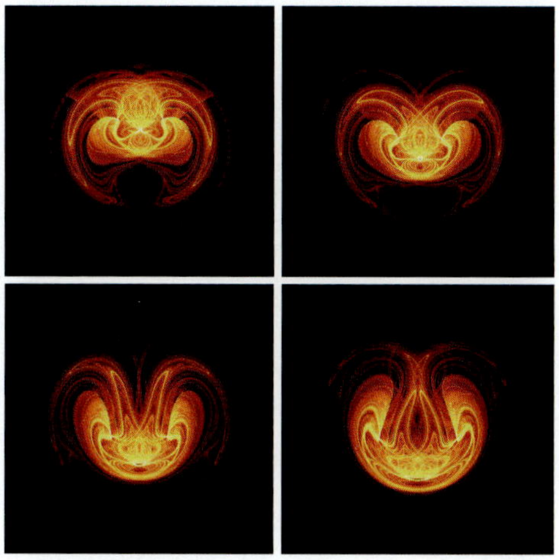

Fig. 5. Four snapshots of the stochastically perturbed Lorenz (1963) model's random attractor $\mathcal{A}(\omega)$ and the invariant measure $\nu(\omega)$ supported on it. The parameter values are the classical ones — normalized Rayleigh number $r = 28$, Prandtl number $Pr = 10$, and normalized wave number $b = 8/3$ — while the noise intensity is $\sigma = 0.5$ and the time step is $\delta t = 5 \cdot 10^3$. The color bar used is on a log-scale and quantifies the probability to end up in a particular region of phase space; shown is a projection of the 3-D phase space (X, Y, Z) onto the (X, Z)-plane. Notice the complex, interlaced filament structures between highly (yellow) and moderately (red) populated regions. The time interval Δt between two successive snapshots — moving from left to right and from top to bottom — is $\Delta t = 0.0875$. Note that the support of the invariant measure $\nu(t;\omega)$ may change quite abruptly, from time t to time $t + \Delta t$; see the related short video in Chekroun *et al.* (2011b). Weakly populated regions cover an important part of the random attractor and are, in turn, entangled with regions that have near-zero probability (black). [After Chekroun *et al.* (2011b) with permission from Elsevier.]

of the invariant measure $\nu(\omega)$ supported on the random attractor of the stochastically perturbed Lorenz system governed by Eq. (14). In the standard literature of random dynamical systems (Arnold, 1998), these time- and realization-dependent measures are called *disintegrated measures*, since it is their integration with respect to the realizations ω over the probability space Ω in which they live that yields — under suitable technical assumptions of existence and uniqueness — the system's time-invariant PDF. We prefer the more intuitive name of *sample measures*, introduced by the pioneering paper of Ledrappier and Young (1988).

Such a sample measure indicates the probability of trajectories winding up in a particular region of phase space and it is very highly concentrated on the attractor, as inferred from the huge range of density values: the color bar in the figure is on a logarithmic scale, and extends over more than 10 orders of magnitude. The situation is thus very different from that expected when studying additive noise — in that case, the noise tends to smear out the fine, Cantor-set–like structure of the deterministic, strange attractor and the associated PDF has nonzero-volume support.

It hardly needs saying that additive noise has been studied in climate dynamics much more extensively than the multiplicative sort, for two reasons: (i) it was easier to do so; and (ii) it was suggested by the simple Brownian motion analogy of "weather $\simeq$ water molecules" and "climate $\simeq$ pollen particle," as proposed by Hasselmann (1976). Across the hierarchy of climate models discussed in the previous two sections of this article, however, it is clear that small-and-fast scales of motion do not enter exclusively in an additive manner: they pop up in many, if not all terms of the governing equations, as summarized in Eq. (13) above. The insights offered, therefore, by Fig. 5 and the video are likely to be of interest across the hierarchy of models, all the way up to and including coupled GCMs and Earth system models.

The sample measure in Fig. 5 exhibits amazing complexity, with fine, very intense filamentation: there is no fuzziness whatsoever in its topological structure, which does evoke the Cantor-set foliation of the deterministic attractor. This fine structure strongly suggests that an object of vanishing volume supports this measure, i.e., that the random attractor $\mathcal{A}(\omega)$ of system (13) has — like the strange attractor of the classical, deterministic version, with $\sigma = 0$ — dimension smaller than 3.

5. A Rigorous Definition of Climate Sensitivity

5.1. *Equilibrium climate sensitivity*

The usual view of climate sensitivity — as reflected in the work of the IPCC (Houghton *et al.*, 1991; Solomon *et al.*, 2007) — is that of climate being in equilibrium, in the absence of external perturbations. In the setting of deterministic, autonomous dynamical systems, this view can be described by the change in the position of a fixed point, $\mathbf{X}_0 = \mathbf{X}_0(\mu)$, as a function of a parameter μ.

Going back to Fig. 1, we can see how the scalar functional $\bar{T} = \bar{T}\{\mathbf{X}_0(\mu)\}$, namely the global-mean temperature, varies as a function of the fractional change μ of insolation at the top of the atmosphere; here the fixed point $\mathbf{X}_0 = \mathbf{X}_0(\mu)$ is the equilibrium solution $T = T(x;\mu)$ of the EBM given by Eq. (2). Climate sensitivity for the present climate is thus simply the partial derivative $\partial\bar{T}/\partial\mu$, i.e., the tangent of the angle γ between the upper branch of Fig. 1 and the abscissa. The sensitivity increases, in general, as we approach the bifurcation point (X_0, μ_0) in Eqs. (3) or (4), and it decreases away from it.

But we have seen in Secs. 3 and 4 here that internal climate variability can be better described by limit cycles and strange attractors than by fixed points. Moreover, the presence of time-dependent forcing, deterministic as well as stochastic, introduces additional complexities into the proper definition of climate sensitivity.

5.2. *Defining climate sensitivity in the presence of variability*

We illustrate in Fig. 6 the difference between the ways that a change in a parameter can affect a climate model's behavior in the case of purely periodic solutions vs. the case of equilibrium solutions. One might still think of the former case as the climate of a simpler world, in which ENSO would be purely periodic, rather than irregular. In this case, climate sensitivity can no longer be defined by a single scalar, like $\partial\bar{T}/\partial\mu$, but needs at least three scalars: the sensitivity of the mean temperature along with that of the limit cycle's frequency (or period) and amplitude.

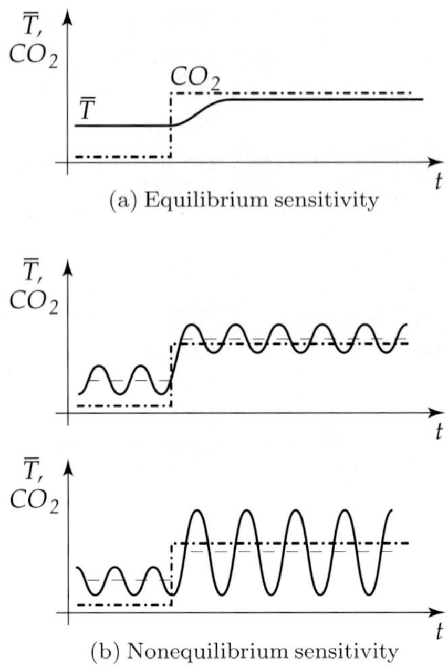

Fig. 6. Climate sensitivity (a) for an equilibrium model; and (b) for a nonequilibrium, oscillatory model. As a forcing (atmospheric CO_2 concentration, say, dash-dotted line) changes suddenly, global temperature (heavy solid) undergoes a transition: in panel (a) only the mean temperature changes; in panel (b) the mean (shown now as light dashed) adjusts as it does in panel (a), but the amplitude of the oscillation can also decrease, increase or stay the same.

More generally, the setting of non-autonomous and of random dynamical systems, as described in Sec. 4, allows one to examine the interaction of internal climate variability with the forcing, whether natural or anthropogenic, and to provide a definition of climate sensitivity that takes into account the climate systems non-equilibrium behavior and its time-dependent forcing. Such a definition is of the essence in studying systematically the sensitivity of global climate models (GCMs) to the uncertainties in tens of semi-empirical parameters; it will be given here in terms of the response of the appropriate probability densities to changes in the parameters. A comparison with numerical results on parameter dependence for a somewhat simplified GCM (Neelin *et al.*, 2010) goes beyond the scope of this brief overview, and will be given elsewhere.

As an illustration of the more general sensitivity definition that we are proposing, we consider here the case of an infinite-dimensional, but still relatively simple ENSO model. The model is due to Galanti and Tziperman (2000), and its two dependent variables are sea surface temperature T in the eastern Tropical Pacific and thermocline depth h there, as a function of time t:

$$\dot{T} = f(T(t), h(t)), \tag{15a}$$

$$h(t) = g(T, h, F)(t, \tau_1, \tau_2), \tag{15b}$$

$$F(t) = 1 + \epsilon cos(\omega t + \phi). \tag{15c}$$

In Eqs. (15), F stands for the seasonal forcing, with period $2\pi/\omega = 12$ months, and all three variables — T, h and F — depend on the time t and the delays τ_1 and τ_2; these delays characterize the traveling times along the Equator of eastward Kelvin and westward Rossby waves. Several authors have studied such delay-differential models of ENSO; see Dijkstra (2005) for a review and Ghil *et al.* (2008b) for further mathematical details on this type of models. Note that this Galanti-Tziperman model is non-autonomous, because of the seasonal forcing, but it is still deterministic.

The solutions of Eqs. (15) exhibit periodic, quasi-periodic and chaotic behavior, as well as frequency locking to the time-dependent, seasonal cycle. Thus, in principle, an infinite number of scalars are required to define the dependence of these solutions on the parameters τ_1 and τ_2; these scalars need to include not just the means of temperature T and depth h, but also their variance and higher-order moments. We have chosen in Fig. 7 to represent this dependence for the zeroth, second and fourth moments

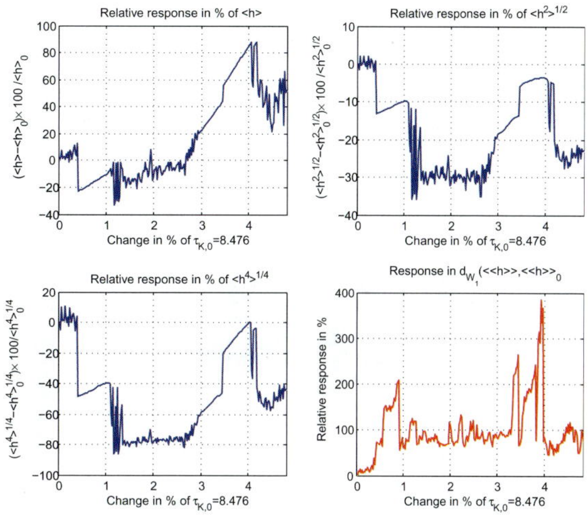

Fig. 7. Relative changes in the statistical properties of the thermocline depth $h(t)$, for the delay-differential ENSO model of Eq. (15) and changes of 05% in the delay parameter τ_K associated with the Kelvin wave transit: (a) mean; (b) second moment; (c) fourth moment; and (d) Wasserstein distance d_W. Note intervals of both smooth and rough dependence of the solution on the parameter. [Courtesy of M. D. Chekroun.]

of $h(t)$; more precisely the plotted quantities are the mean, standard deviation and fourth root of the kurtosis.

The fourth quantity plotted in Fig. 7 is the relative change in *Wasserstein distance* d_W from the same reference solution as for the other three quantities. The Wasserstein distance or "earth mover's distance" d_W is the distance between two measures of equal mass on a metric space, i.e., on a space that has a metric attached to it, like an n-dimensional Euclidean space (Dobrushin, 1970). Roughly speaking, d_W represents the total work needed to move the "dirt" (i.e., the measure) from a trench you are digging to another one you are filling, over the distance between the two trenches.

In fact, the mathematician Gaspard Monge — a founder of the École Polytechnique and an officer in Napoleon's armies — introduced essentially what is now called the Wasserstein distance with precisely this motivation. In general, the shape of the two trenches and the

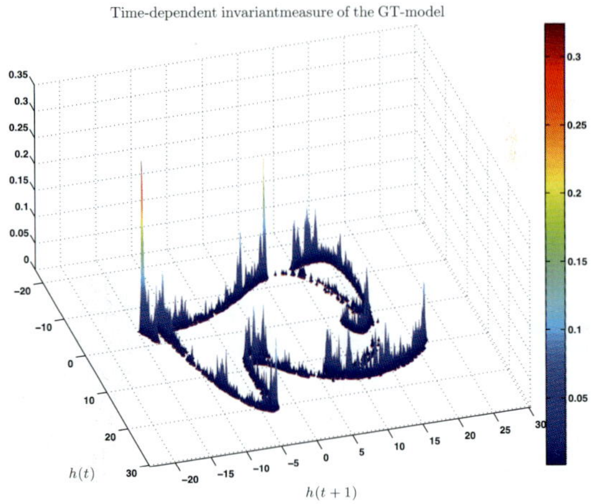

Fig. 8. Time-dependent invariant measure of the Galanti and Tziperman (2000) model, plotted in an isometric projection with the probability density on the perpendicular to the plane spanned by the coordinates $(h(t), h(t+1))$. The time here is in units of years, and the density is highly concentrated on a very "thin" support, as for the stochastically perturbed Lorenz model in Fig. 5. [Courtesy of M. D. Chekroun.]

depth along the trench — i.e., the support of the measure and its density — can differ. In the case at hand, the shape and density of the invariant measure that is being moved are plotted in Fig. 8.

More recently, d_W has become hugely popular in many contexts that involve related problems. Hairer and Mattingly (2006) have used it in proving certain properties of an invariant measure for the Markov semi-group associated with the 2-D Navier-Stokes equations driven by additive noise. Likewise, Hairer and Majda (2010) used it in proving the differentiability of the invariant measure for stochastic differential equations in the presence of a large class of noise processes, i.e., in the so-called hypo elliptic case. The latter authors also did mention climate dynamics as an interesting area of applications of their work.

It is quite clear from Fig. 7 that intervals of smooth dependence of the solution on the parameter alternate with jumps and with intervals of rough dependence. The jump points are the same in all four panels of the figure and there is agreement also in the smooth vs. rough intervals, although the Wasserstein distance in panel (d) shows some roughness even over intervals in which the three statistical moments in panels (a–c) behave smoothly. The latter point is not too surprising, since d_W contains more information than each of the moments.

The very high concentration of probability density in the peak at the extreme left of Fig. 8 might seem surprising but actually agrees with such a peak in the stochastically perturbed ENSO model of Timmermann and Jin (2002). The latter model is based on three ordinary differential equations, for sea surface temperatures in the eastern and western Tropical Pacific and for thermocline depth, and it was analyzed in detail by Chekroun *et al.* (2011b). The corresponding near-singularity in invariant measure on the model's random attractor evolves regularly in time and thus suggests that the interaction of deterministic nonlinearities with time-dependent forcing, including even stochastic perturbations, can help seasonal-to-interannual prediction, rather than hinder it. Chekroun *et al.* (2011a) have taken some interesting steps in exploiting this possibility.

With the results illustrated in Figs. 7 and 8 in hand, it becomes natural to define climate sensitivity in the presence of internal variability and of time-dependent forcing as the partial derivative of the Wasserstein distance d_W with respect to a parameter μ, $\partial d_W/\partial\mu$. Clearly, d_W has to be defined in turn with respect to the particular climate whose sensitivity we wish to evaluate.

At this time, it would be awkward, difficult or even impossible to compute with absolute accuracy the function $d_W = d_W(\mu)$ and its derivative at $\mu = \mu_0$; but reasonable approximations should become available shortly, as is the case already for the calculation of the invariant measures of models of intermediate complexity. We outline a way to proceed in this direction in Sec. 6 below.

The definition given here for a deterministic but non-autonomous ENSO model can be generalized further to the random case, assuming the existence of the suitable invariant measures. The advantage of using the definition of climate sensitivity $\partial d_{\mathrm{W}}/\partial\mu$ proposed herein over the customary one of $\partial\bar{T}/\partial\mu$ is twofold. First, it makes sense for any climate model with natural variability and with time-dependent forcing. Second, it involves the sensitivity of the higher-order moments of the model's PDF, up to and including changes in the model's extreme events. Interest in the latter has been increasing dramatically in recent years [*e.g.*, Yiou and Nogaj (2004); Ghil *et al.* (2011)], along with the realization that changes in mean global temperature $\bar{T}$ are not that interesting in and of themselves.

A complementary approach to climate sensitivity out of equilibrium is the one based on the *fluctuation-dissipation theory (FDT)* of statistical mechanics (Leith, 1975; Ghil *et al.*, 1985; Lucarini and Sarno, 2011; Wouters and Lucarini, 2013). The complementarity arises from D. Ruelle's extending the many-particle, statistical-mechanics ideas to low-order dynamical systems subject to certain mathematical conditions on the latter (Ruelle, 1997, 1999; Chekroun *et al.*, 2011b).

The key concept underlying both approaches is that of *linear response* in a highly nonlinear system. Important results in applying this concept have been obtained across the modeling hierarchy by several authors (Gritsun, 2001; Gritsun and Branstator, 2007; Majda and Wang, 2010).

We thus expect the theory of random dynamical systems, along with FDT, to provide robust tools for studying the parameter dependence of a nonlinear, randomly perturbed system's various "metrics." These metrics — a word used by the climate community in a much broader sense than the standard mathematical term — can include global quantities, like mean temperature or total energy, but also much finer functionals of the state of the system, such as regional temperatures or precipitation. In addition, random dynamical systems theory can help improve prediction of future system properties, by relying on a judicious combination of the history of its slow and fast behavior (Chekroun *et al.*, 2011a).

6. Concluding Remarks

A complete theory of climate variability and climate sensitivity, across the entire range of time scales of interest, is still in the future. We have shown, though, that powerful conceptual and numerical tools exist in order to organize the emerging knowledge so far. The approach described herein relies on applying systematically dynamical systems theory, both deterministic and stochastic, across a hierarchy of models, from the simplest toy models to the most detailed, coupled GCMs.

The dynamical systems approach has progressed from its first modest steps, taken half-a-century ago, to the analysis of the behavior of atmospheric, oceanic and coupled GCMs over the last two decades. We have reviewed, in particular, the robustness of fundamental bifurcations — namely those associated with multiple equilibria and with transition from equilibrium to oscillatory behavior — across the model hierarchy, for models that treat the Earth system's energy balance and for those that treat the oceans' thermohaline circulation. Particularly interesting strides have been taken over the last decade in studying the interaction of the faster time scales with the slower ones, within a genuinely nonlinear framework.

A number of GCMs — particularly some of those used in numerical weather prediction and in seasonal-to-interannual forecasting — combine a deterministic representation of large-scale processes, in particular of the fluid dynamics, with a stochastic parametrization of one or more of the small-scale physical processes, such as cloud feedbacks (Lin and Neelin, 2003;

Palmer and Williams, 2009). Furthermore, the stochastic parametrizations are often of the multiplicative-noise type discussed herein.

Typically — as discussed by Neelin *et al.* (2010), among others — the GCM team in question will verify that a particular stochastic parametrization does achieve a particular improvement aimed at in developing it. But it has been difficult so far to check how one or another such parametrization does affect the GCM's overall performance, in particular its sensitivity to changes in certain parameter values or in other parametrizations. The concepts and tools reviewed herein should help in carrying out more systematically such climate model sensitivity studies.

How to do so in practice raises, of course, the question of how to approximate in a useful way the pullback attractor of a GCM or of high-dimensional models in general? First of all, we remind the reader that the Galanti and Tziperman (2000) model studied in Sec. 5.2 here is an infinite-dimensional model; it falls short of the complexity of a GCM but its evolution in time still needed to be computed by suitable numerical approximations.

More generally, the empirical model reduction (EMR) approach (Kondrashov *et al.*, 2005; Kravtsov *et al.*, 2005, 2009) provides precisely the required kind of reduction from a high-dimensional model, like a GCM, to a low-order model that is both nonlinear and noise-driven. EMR models based on this approach have proven to be particularly skillful at capturing and simulating the power spectrum of higher-order models, as well as their PDF in the phase space of the leading empirical orthogonal functions (Kravtsov *et al.*, 2005). Barnston *et al.* (2012) have shown that real-time ENSO prediction by an EMR model (Kondrashov *et al.*, 2005) is highly competitive among two dozen dynamical and statistical forecasts in the ENSO multi-model prediction plume of the International Research Institute for Climate and Society at Columbia University.

The obvious route to studying the pullback attractor of a GCM is to first approximate the GCM by an EMR model, using a sufficiently long simulation or ensemble of simulations of the GCM. The next step is to study the pullback attractor of the EMR, in the same way as done in Secs. 4.2 and 5.2 here.

Doing so should include a study of the EMR model's sensitivity, as defined herein, to cover changes in its variability, including extreme events, as well as in its mean. This is not to say that such a study will be easy or give the right answers right away. But it should provide an interesting start in the right direction.

Acknowledgments

It is a pleasure to thank Mickael D. Chekroun for Figs. 7 and 8. Discussions with him and with several other colleagues — including but not restricted to D. Kondrashov, J. C. McWilliams, J. D. Neelin, E. Simonnet, S. Wang, and I. Zaliapin — have helped develop the ideas in this chapter, and more particularly those formulated in Sec. 5. Equation (6) and the associated discussion in Sec. 3.2 is based on a talk given by Henk A. Dijkstra at a meeting on "Mathematical Paradigms of Climate Science" in Rome, June 2013. Thoughtful and constructive comments by two anonymous reviewers have helped improve the presentation. Related work is partially supported by NSF grant DMS-0934426 and by ONR-MURI grant N00014-12-1-0911.

References

Andronov, A. A., and L. S. Pontryagin, 1937: Systèmes grossiers. *Dokl. Akad. Nauk SSSR*, **14**(5), 247–250.

Arnold, L., 1998: *Random Dynamical Systems*. Springer-Verlag, New York/Berlin, 602 pp.

Arnol'd, V. I., 1983: *Geometric Methods in the Theory of Ordinary Differential Equations*. Springer-Verlag, New York/Berlin, 334 pp.

Barnston, A. G., M. K. Tippett, M. L. Heureux, S. Li, and D. G. DeWitt, 2012: Skill of real-time seasonal ENSO model predictions during 2002–2011 — is our capability improving?, *Bull. Am. Meteorol. Soc.*, **93**, 631–651.

Bryan, F., 1986: High-latitude salinity effects and interhemispheric thermohaline circulations. *Nature*, **323**, 301–304.

Bryden, H. L., B. A. King, and G. D. McCarthy, 2011: South Atlantic overturning circulation at 24 S. *J. Mar. Res.*, **69**, 38–55, doi:10.1357/002224011798147633.

Budyko, M. I., 1969: The effect of solar radiation variations on the climate of the Earth. *Tellus*, **21**, 611–619.

Charney, J. G., A. Arakawa, D. J. Baker, *et al.*, 1979: *Carbon Dioxide and Climate: A Scientific Assesment.* National Academies Press, Washington, D.C., 34 pp.

Chekroun, M. D., D. Kondrashov, and M. Ghil, 2011a: Predicting stochastic systems by noise sampling, and application to the El Niño-Southern Oscillation. *Proc. Natl. Acad. Sci. USA*, **108**(29), 11,7660–11,771, doi:10.1073/pnas.1015753108.

———, E. Simonnet, and ———, 2011b: Stochastic climate dynamics: Random attractors and time-dependent invariant measures. *Physica D*, **240**(21), 1685–1700, doi:10.1016/j.physd.2011.06.005.

Chen, F., and M. Ghil, 1996: Interdecadal variability in a hybrid coupled ocean-atmosphere model. *J. Phys. Oceanogr.*, **26**, 1561–1578.

Dijkstra, H. A., 2005: *Nonlinear Physical Oceanography: A Dynamical Systems Approach to the Large-Scale Ocean Circulation and El Niño*, 2nd ed., Kluwer Acad. Publishers, Dordrecht/Norwell, Mass., 532 pp.

———, and W. Weijer, 2005: Stability of the global ocean circulation: Basic bifurcation diagrams. *J. Phys. Oceanogr.*, **35**, 933–948.

———, 2007: Characterization of the multiple equilibria regime in a global ocean model. *Tellus*, **59A**, 695–705.

———, and M. Ghil, 2005: Low-frequency variability of the large-scale ocean circulation: A dynamical systems approach. *Rev. Geophys.*, **43**, RG3002, doi:10.1029/2002RG000122.

Dobrushin, R. L., 1970: Prescribing a system of random variables by conditional distributions. *Theor. Prob. Appl.*, **15**, 458–486.

Doob, J. L., 1953: *Stochastic Processes.* Wiley, New York, 654 pp.

Drijfthout, S., S. Weber, and E. van der Swaluw, 2011: The stability of the MOC as diagnosed from model projections for pre-industrial, present and future climates. *Clim. Dyn.*, **37**, 1575–1586, doi:10.1007/s00382-010-0930-z.

Eckmann, J.-P., and D. Ruelle, 1985: Ergodic theory of chaos and strange attractors. *Rev. Mod. Phys.*, **57**, 617–656 (addendum, *Rev. Mod. Phys.*, **57**, 1115).

Einstein, A., 1905: On the movement of small particles suspended in a stationary liquid demanded by the molecular-kinetic theory of heat, *Ann. d. Physik (Leipzig)*, **17**, 549 ff., reprinted in *Investigations on the Theory of the Brownian Movement, five articles by A. Einstein*, Fürth, R. (Ed.) and A. D. Cowper (transl.) (1956, Dover Publ., New York), 122 pp.

Galanti, E., and E. Tziperman, 2000: ENSO's phase locking to the seasonal cycle in the fast-SST, fast-wave, and mixed-mode regimes. *J. Atmos. Sci.*, **57**, 2936–2950.

Ghil, M., 1976: Climate stability for a Sellers-type model. *J. Atmos. Sci.*, **33**, 3–20.

———, 1994: Cryothermodynamics: The chaotic dynamics of paleoclimate. *Physica D*, **77**, 130–159.

———, 2001: Hilbert problems for the geosciences in the 21st century. *Nonlin. Processes Geophys.*, **8**, 211–222.

———, R. Benzi, and G. Parisi, Ed., 1985: *Turbulence and Predictability in Geophysical Fluid Dynamics and Climate Dynamics.* North-Holland, Amsterdam, 449 pp.

———, and S. Childress, 1987: *Topics in Geophysical Fluid Dynamics: Atmospheric Dynamics, Dynamo Theory and Climate Dynamics.* Springer-Verlag, New York/Berlin, 485 pp.

———, and A. W. Robertson, 2000: Solving problems with GCMs: General circulation models and their role in the climate modeling hierarchy. *General Circulation Model Development: Past, Present and Future*, D. Randall, Ed., Academic Press, San Diego, pp. 285–325.

———, A. Mullhaupt, and P. Pestiaux, 1987: Deep water formation and quaternary glaciations. *Climate Dyn.*, **2**, 1–10.

———, M. R. Allen, M. D. Dettinger, *et al.*, 2002: Advanced spectral methods for climatic time series. *Rev. Geophys.*, **40**(1), pp. 3.1–3.41, doi:10.1029/2000RG000092.

———, M. D. Chekroun, and E. Simonnet, 2008a: Climate dynamics and fluid mechanics: Natural variability and related uncertainties. *Physica D*, **237**, 2111–2126, doi:10.1016/j.physd.2008.03.036.

———, I. Zaliapin, and S. Thompson, 2008b: A delay differential model of ENSO variability: Parametric instability and the distribution of extremes. *Nonlin. Processes Geophys.*, **15**, 417–433.

———, P. Yiou, S. Hallegatte, *et al.*, 2011: Extreme events: Dynamics, statistics and prediction. *Nonlin. Processes Geophys.*, **18**, 295–350, doi:10.5194/npg-18-295-2011.

Gritsun, A. S., 2001: Fluctuation-dissipation theorem on attractors of atmospheric models. *Russ. J. Numer. Analysis Math. Modell.*, **16**, 97–190.

Gritsun, A. S., and G. Branstator, 2011: Climate response using a three-dimensional operator based on the fluctuation dissipation theorem. *J. Atmos. Sci.*, **64**, 2558–2575.

Guckenheimer, J., and P. Holmes, 1983: *Nonlinear Oscillations, Dynamical Systems and Bifurcations of Vector Fields.* Springer-Verlag, New York/Berlin, 453 pp.

Hairer, M., and J. C. Mattingly, 2006: Ergodicity of the 2D Navier-Stokes equations with degenerate stochastic forcing. *Annals Math.*, **164**, 993–1032.

Hairer, M., and A. J. Majda, 2010: A simple framework to justify linear response theory. *Nonlinearity*, **23**, 909–922.

Hasselmann, K., 1976: Stochastic climate models, Part I: Theory. *Tellus*, **28**, 473–485.

Hawkins, E., R. S. Smith, L. C. Allison, J. M. Gregory, T. J. Woollings, H. Pohlmann, and B. de Cuevas, 2011: Bistability of the Atlantic overturning circulation in a global climate model and links to ocean freshwater transport. *Geophys. Res. Lett.*, **38**, doi:10.1029/2011GL047208.

Held, I. M., and M. J. Suarez, 1974: Simple albedo feedback models of the ice caps. *Tellus*, **26**, 613–629.

Houghton, J. T., G. J. Jenkins, and J. J. Ephraums, Ed., 1991: *Climate Change, The IPCC Scientific Assessment.* Cambridge Univ. Press, Cambridge, MA, 365 pp.

Kalnay, E., 2003: *Atmospheric Modeling, Data Assimilation and Predictability.* Cambridge Univ. Press, Cambridge/London, UK, 341 pp.

Kennett, R. P., and L. D. Stott, 1991: Abrupt deep-sea warming, paleoceanographic changes and benthic extinctions at the end of the Palaeocene. *Nature*, **353**, 225–229.

Kondrashov, D., S. Kravtsov, A. W. Robertson, and M. Ghil, 2005: A hierarchy of data-based ENSO models. *J. Climate*, **18**, 4425–4444.

Kravtsov, S., D. Kondrashov, and M. Ghil, 2005: Multilevel regression modeling of nonlinear processes: Derivation and applications to climatic variability. *J. Climate*, **18**, 4404–4424.

Kravtsov, S., D. Kondrashov, and M. Ghil, 2009: Empirical model reduction and the modeling hierarchy in climate dynamics, in *Stochastic Physics and Climate Modeling*, edited by T. N. Palmer and P. Williams, pp. 35–72, Cambridge Univ. Press.

Ledrappier, F., and L.-S. Young, 1988: Entropy formula for random transformations. *Prob. Theory Related Fields*, **80**, 217–240.

Leith, C. E., 1975: Climate response and fluctuation dissipation. *J. Atmos. Sci.*, **32**, 2022–2026.

Li, Z.-X., K. Ide, H. Le Treut, *et al.*, 1997: Atmospheric radiative equilibria in a simple column model. *Clim. Dyn.*, **13**, 429–440.

Lin, J. W. B., and J. D. Neelin, 2003: Toward stochastic deep convective parameterization in general circulation models. *Geophys. Res. Lett.* , **30**, 1162, doi:10.1029/2002GL016203.

Lorenz, E. N., 1963: Deterministic nonperiodic flow. *J. Atmos. Sci.*, **20**, 130–141.

Lucarini, V., and S. Sarno, 2011: A statistical mechanical approach for the computation of the climatic response to general forcings. *Nonlin. Processes Geophys.*, **18**, 7–28, doi:10.5194/npg-18-7-2011.

Majda, A. J., and X. Wang, 2010: Linear response theory for statistical ensembles in complex systems with time-periodic forcing. *Commun., Math. Sci.*, **8**, 145172.

Marotzke, J., 2000: Abrupt climate change and thermohaline circulation: Mechanisms and predictability. *Proc. Natl. Acad. Sci. USA*, **97**, 1347–1350.

Neelin, J. D., A. Bracco, H. Luo, *et al.*, 2010: Considerations for parameter optimization and sensitivity in climate models. *Proc. Natl. Acad. Sci. USA*, **107**(50), 21,349–21,354.

North, G. R., 1975: Analytical solution to a simple climate model with diffusive heat transport. *J. Atmos. Sci.*, **32**, 1301–1307.

———, J. G. Mengel and D. A. Short, 1983: Simple energy-balance model resolving the seasons and the continents — application to the astronomical theory of the ice ages. *J. Geophys. Res.*, **88**, 6576–6586.

Palmer, T. N., and P. Williams (Eds.), 2009: *Stochastic Physics and Climate Modelling.* Cambridge University Press.

Ramanathan, V., and J. A. Coakley, 1978: Climate modeling through radiative convective models. *Rev. Geophys. Space Phys.*, **16**, 465–489.

Ruelle, D., 1997: Application of hyperbolic dynamics to physics: Some problems and conjectures. *Bull. Amer. Math. Soc.*, **41**, 275–278.

Ruelle, D., 2006: What is a strange attractor? *Notices Amer. Math. Soc.*, **53**, 764–765.

Ruelle, D., 2009: A review of linear response theory for general differentiable dynamical systems. *Nonlinearity*, **22**, 855–870.

———, and F. Takens, 1971: On the nature of turbulence. *Commun. Math. Phys.*, **20**, 167–192, and **23**, 343–344.

Schneider, S. H., and R. E. Dickinson, 1974: Climate modeling. *Rev. Geophys. Space Phys.*, **25**, 447–493.

Sellers, W. D., 1969: A global climatic model based on the energy balance of the earth-atmosphere system. *J. Appl. Meteor.*, **8**, 392–400.

Solomon, S., D. Qin, M. Manning, *et al.*, Eds., 2007: *Climate Change 2007: The Physical Science Basis. Contribution of Working Group I to the Fourth Assessment Report of the IPCC.* Cambridge Univ. Press, Cambridge/London, UK, 996 pp.

Stommel, H., 1961: Thermohaline convection with two stable regimes of flow. *Tellus*, **13**, 224–230.

Timmermann, A., and F.-F. Jin, 2002: A nonlinear mechanism for decadal El Niño amplitude changes. *Geophys. Res. Lett.*, **29** (1), 1003, doi:10.1029/2001GL013369.

Wetherald, R. T., and S. Manabe, 1975: The effect of changing the solar constant on the climate of a general circulation model. *J. Atmos. Sci.*, **32**, 2044–2059.

Wouters, J., and V. Lucarini, 2013: Parametrization of cross-scale interaction in multi-scale systems. Chapter 4, this volume.

Yiou, P., and N. Nogaj, 2004: Climatic extremes and weather regimes: Where and when? *Geophys. Res. Lett.*, **31**, L07202 10.1029/2003GL019119.

Chapter 3

Fluctuation-dissipation Theorem with Application to Climate Change Studies with Seasonal Impact

Xiaoming Wang

*Department of Mathematics, Florida State University,
Tallahassee, FL 32306, USA*

wxm@math.fsu.edu

We survey a few novel linear response formulas for chaotic (stochastic) complex dynamical systems with time periodic coefficients developed by Andrew Majda and the author recently. Such time periodic systems arise naturally in climate change studies due to the diurnal and seasonal cycles. Several linear response formulas for general time-dependent statistical ensembles that include the effects of deterministic model errors, initial ensembles, and model noise perturbations simultaneously are reviewed.

1. Introduction

The study of climate change is one of the grant challenges in modern sciences. The climate, defined as the long time statistical properties of a given climate model, is intrinsically hard to estimate. For a fixed chaotic climate model, direct approach of estimating the climate would involve the integration of the chaotic climate dynamical system over a very long time interval which usually leads to large errors.[a] Direct approach to climate change study by comparing the climate of the same model under perturbation is even more difficult since the errors incurred at the stage of estimating the climate might overwhelm the true climate change signal. On top of that, the true climate model may not be known or is too expensive in terms of long time simulation although long time observations may be available. Therefore, alternative approaches that rely more on the data and less on the model are highly desirable.

Fortunately, the fluctuation-dissipation theorem (FDT) from statistical physics (Kubo *et al.*, 1985; Risken, 1984; Balescu, 1997) provides us a tool to estimate the averaged means response of appropriate physical system to small external perturbations using suitable data (correlation function) of the unperturbed system. Strictly speaking, the FDT applies to large systems of identical particles. It is Leith (Leith, 1975) who proposed the application of FDT in climate change studies despite the fact that the climate system is a forced dissipative chaotic dynamical system which is physically quite far from the classical physicists' setting for FDT. Leith's proposition has led to a lot of recent activities in generating new theoretical formulations and approximate algorithms for FDT with applications to climate response (Bell, 1980; Gritsoun and Dymnikov, 1999; Gritsoun

[a]See, however, (Wang, 2010) for recent progress on numerical schemes that are able to capture the long time statistical properties.

Climate Change: Multidecadal and Beyond
edited by Chih-Pei Chang et al.

et al., 2002; Gritsoun and Branstator, 2007; Gritsoun *et al.*, 2008; Abramov and Majda, 2007; Abramov and Majda, 2008; Majda *et al.*, 2005; Abramov and Majda, 2009; Abramov, 2010; Abramov and Majda, 2011; Majda *et al.*, 2010; Abramov and Kjerland, 2013). Most of these approximate FDT algorithms are concerned with autonomous climate models of varying complexity and have ignored the important time periodic effect of the diurnal and seasonal cycle (Branstator and Frederiksen, 2003; Frederiksen and Branstator, 2005; Achatz and Opsteegh, 2002). Recently, Andrew Majda and the author (Majda and Wang, 2010) have developed a version of FDT that is applicable to systems with time periodic forcing. Therefore, this new theory has the potential to be applied to climate change studies involving climate models with seasonal impact. The main purpose of this survey manuscript is to review some of the formulas and algorithms developed in the work of Majda and the author (Majda and Wang, 2010). In particular, the new algorithms that takes into account the seasonal impact are recalled in Section 4 below.

The rest of the manuscript is organized as follows. In Section 2, we recall the mathematical set-up of time-periodic dynamical systems and their time periodic climate. In Section 3, we review systematic FDT theorems for the time-periodic equilibrium setting with various important approximations. In Section 4, we utilize the theory developed earlier in Sections 2 and 3 to introduce a suite of new systematic quasi-Gaussian (Leith, 1975; Majda *et al.*, 2005; Gritsoun *et al.*, 2002; Gritsoun and Dymnikov, 1999; Gritsoun and Branstator, 2007; Gritsoun *et al.*, 2008) and adjoint algorithms (Abramov and Majda, 2007; Abramov and Majda, 2008; Abramov and Majda, 2009) for climate response with a seasonal cycle (periodic forcing) developed recently (Majda and Wang, 2010). An application-oriented reader mainly interested in the algorithms for climate response can go to Section 4 directly without difficulty. A few concluding remarks are included at the end. An appendix that illustrate the connection between the FDT and linear response theory is included for the benefit of those who are not familiar with the background.

2. Climate Models with Seasonal Cycle

Many climate models can be formulated as an (Ito) Stochastic Differential Equations (SDE). A generic climate model that takes into account seasonal cycle in the sense that the forcing term may be time periodic (solar heating for instance) may take the form

$$\frac{d\mathbf{X}}{dt} = \mathbf{F}(\mathbf{X},t) + \sigma(\mathbf{X})\dot{W}, \quad \mathbf{X} \in \mathbb{R}^N, \qquad \mathbf{F} = (F_1, \dots, F_N), \quad \mathbf{f} = (f_1, \dots, f_N) \tag{1}$$

where $\dot{W}$ denotes a standard M dimensional white (in time) noise, σ is an $N \times M$ matrix, and $\mathbf{F}(\mathbf{X},t)$ has time periodic coefficients reflecting seasonal impact for instance. A typical example is

$$\mathbf{F}(\mathbf{X},t) = \mathbf{F}(\mathbf{X}) + \mathbf{f}(t) \tag{2}$$

where $\mathbf{f}(t)$ is a time-periodic forcing with period T_0 (representing the seasonality of the solar heating for instance). In the case of zero noise, i.e. $\sigma = 0$, the SDE reduces to an ODE system. The deterministic part could come from spatial discretization of the PDE models, and the stochastic part models the effect of the unresolved small scales on the resolved scales (stochastic back-scattering).

For an ensemble prediction with initial distribution characterized by $\bar{p}_0$, the (unperturbed) statistical solution (future distribution at time t) $\bar{p}(\mathbf{X},t)$ is the solution to the associated (unperturbed time-dependent) **Fokker-Planck equation** (Gardiner, 2004; Risken, 1984)

$$\frac{\partial \bar{p}}{\partial t} = -\nabla \cdot (\bar{p}\mathbf{F}(\mathbf{X},t)) + \frac{1}{2}\nabla \cdot \nabla \cdot (Q\bar{p}) (\stackrel{def}{=} L_{FP}\bar{p}), \tag{3}$$

$$\bar{p}(\mathbf{X},t)|_{t=0} = \bar{p}_0(\mathbf{X}), \tag{4}$$

where

$$Q = \sigma\sigma^T \geq 0 \tag{5}$$

is the $N \times N$ diffusion coefficient matrix,

$$\nabla \cdot \nabla \cdot (Q\bar{p}) = \sum_{i=1}^{N}\sum_{j=1}^{N} \frac{\partial^2 (Q_{ij}\bar{p})}{\partial x_i \partial x_j} \tag{6}$$

is the associated diffusion operator and the time dependent Fokker-Planck operator L_{FT} is defined as

$$L_{FP}(t)p = -\nabla \cdot (p\mathbf{F}(\mathbf{X},t)) + \frac{1}{2}\nabla \cdot \nabla \cdot (Q\,p). \tag{7}$$

The Fokker-Planck equation reduces to the **Liouville equation** in the case of zero noise ($Q \equiv 0$). Our goal here is to present a calculus useful for both climate response and system perturbation/model error in time dependent ensemble prediction that is developed in (Majda and Wang, 2010).

The climate of the model (1) is defined as the long time behavior of the (statistical) solution to the Fokker-Planck equation (3). In the case when the deterministic forcing term $\mathbf{F}$ is time-independent (no seasonal cycle), and the system is dissipative in certain appropriate sense, one can show that there exists a statistical equilibrium characterized by a pdf p_∞ via the classical Bogliubov-Krylov approach (Foias *et al.*, 2001; Vishik and Fursikov, 1988), so that $\bar{p}(\mathbf{X},t)$ converges to $p_\infty(\mathbf{X})$ using the relative entropy as the measure as time approaches infinity. Therefore, p_∞ encodes the statistical information of the climate.

For models with seasonal forcing term, it is impossible to have a climate that is time-independent. We instead look for time-periodic climate. Indeed, for many dissipative systems with time-periodic forcing, one can prove the existence of a time-periodic climate (with the same period as the external forcing) by appropriately applying the Bogliubov-Krylov theory to the (extended) associated skew-product system. These dissipative systems include the following type of special systems (Majda *et al.*, 2005; Majda and Wang, 2006) that are of great importance in geophysical fluid dynamics

$$\frac{d\mathbf{X}}{dt} = B(\mathbf{X},\mathbf{X}) + L\mathbf{X} - \gamma\mathbf{X} + \mathbf{f}(t) + \sigma\dot{W} \tag{8}$$

where B is a bilinear anti-symmetric operator with $\mathbf{X} \cdot B(\mathbf{X},\mathbf{X}) = 0$ (resulting from the quadratic advection term for instance), L is a linear anti-symmetric operator (derived from the Coriolis forcing for example), γ represents dissipation and $\mathbf{f}$ denotes external periodic (seasonal, annual) forcing. Both $L(t)$ and $\gamma(t)$ can also have time periodic components in applications centered about time periodic equilibrium states in a reformulation (Frederiksen and Branstator, 2005; Branstator and Frederiksen, 2003; Achatz and Opsteegh, 2002).

Recall that the relative entropy between two probability density functions p_j, $j = 1, 2$ is defined as

$$H(p_1,p_2) = \int p_1(\mathbf{x}) \ln \frac{p_1(\mathbf{x})}{p_2(\mathbf{x})}\, d\mathbf{x}. \tag{9}$$

An alternative notation is $\mathcal{P}(p_1,p_2)$. The relative entropy can be interpreted as the lack of information of p_1 versus p_2. It enjoys the nice property that $H(p_1,p_2) \geq 0$ for any p_j and $H(p_1,p_2) = 0$ if and only if $p_1 = p_2$. Therefore, it is almost like a distance function although it is not symmetric in its arguments (Majda *et al.*, 2005; Majda and Wang, 2006). Alternative measures, such as the Wasserstein distances (Hairer and Majda, 2010) or Hellinger's distance, can be used to gauge distance between two probability distributions as well.

The following result on the existence and uniqueness of time-periodic climate for climate models with seasonal forcing term was obtained in (Majda and Wang, 2010) utilizing H-theorem type arguments.

Theorem 1 (existence and uniqueness of time-periodic climate). *If the time-dependent system (1) is dissipative in an appropriate sense, then it possesses at least one time-periodic (with the same period as the external forcing term) statistical solution p_{per} (***the seasonal climate***). Moreover, under generic noise assumptions, i.e., rank$(Q) = N$, and appropriate decay property at infinity, the time-periodic climate p_{per} captures all long time statistical properties of the system (1) in the sense that for any statistical solution p*

$$\lim_{t\to\infty} H(p(t), p_{per}(t)) = \lim_{t\to\infty} H(p_{per}(t), p(t)) = 0. \tag{10}$$

Furthermore, we have, for any bounded continuous functional $\varphi(\mathbf{x})$, and t_0,

$$\int \varphi(\mathbf{x}) p_{per}(\mathbf{x}, t_0)\, d\mathbf{x} = \lim_{K\to\infty} \frac{1}{K} \sum_{k=0}^{K} \varphi(\mathbf{X}(t_0 + kT_0)) \tag{11}$$

with the right hand naturally interpreted as the climatological mean.

We will proceed to the generalization of the classical linear response calculation to this case with periodic in time forcing in the next section.

3. The Linear Response Formula

The study of the change of the averaged (with respect to the climate distribution) value of certain functionals as the climate model is perturbed either in terms of the initial condition or model parameters is one of the central themes in climate change studies. Frequently used functionals include, but not limited to, the first and second moments of the state variables. For instance, if the state variable is the temperature, the first moment is the mean temperature.

It is well-known that we frequently encounter various uncertainties both in the deterministic forcing term $\mathbf{F}$, and in the noise term σ. Therefore it is natural to consider **system perturbation** induced by perturbation in the deterministic forcing term and the noise term.

For most applications in climate studies, it is sufficient to consider space-time separable perturbation (Majda *et al.*, 2005). We also assume that the perturbation in noise is of the same order for convenience although more general perturbations can be handled as well. Therefore we are interested in the following perturbed system

$$\frac{d\mathbf{X}}{dt} = \mathbf{F}(\mathbf{X}, t) + \mathbf{a}(\mathbf{X}) \bullet \delta\tilde{\mathbf{F}}(t) + (\sigma(\mathbf{X}) + \delta\tilde{\sigma}(\mathbf{X}))\dot{W}, \tag{12}$$

where $\mathbf{a} \bullet \mathbf{w}$ denotes the Hadamard (or Schur, or entrywise) product between $\mathbf{a}$ and $\mathbf{w}$, i.e., the jth component of it is the product of the jth component of $\mathbf{a}$ and $\mathbf{w}$, i.e.,

$$(\mathbf{a} \bullet \mathbf{w})_j = a_j w_j. \tag{13}$$

The statistical properties of this system is governed by its corresponding Fokker-Planck equation, and its solution denoted p^{δ}.

Suppose that the perturbation to the initial distribution is of the order of δ, i.e.,

$$p_0^{\delta} = \bar{p}_0 + \delta p_0'$$

we then speculate that

$$p^{\delta}(\mathbf{X}, t) = \bar{p}(\mathbf{X}, t) + \delta p'(\mathbf{X}, t) + \mathcal{O}(\delta^2).$$

The mathematical validity of such an expansion for all time has been established in (Hairer and Majda, 2010) for the stationary forcing case. Therefore, the practical challenge is to estimate $\delta p'$, or the statistical average with respect to this perturbative distribution. We will assume that $p_0' \equiv 0$ in the sequel for simplicity.

For a given functional (observable) $A(\mathbf{X})$, the statistics under the perturbed dynamics is given by

$$E^\delta(A)(t) = \iint A(\mathbf{X})p^\delta(\mathbf{X},t)\,d\mathbf{X}$$
$$= E^0(t) + \delta E'(t) + \mathcal{O}(\delta^2) \quad (14)$$

where

$$\delta E'(t) = \delta \int A(\mathbf{X})p'(\mathbf{X},t)\,d\mathbf{X}. \quad (15)$$

A direct approach is to approximate p' (or p^δ and $\bar{p}$) and then compare the spatial averages. This is usually a very costly and unstable (at large time at least) method (Majda *et al.*, 2005). Here we advocate an alternative approach utilizing a generalization of the classical fluctuation-dissipation theorem.

One can show that the leading order perturbation in the statistics can be written as (Majda and Wang, 2010)

$$E'(A)(t) = \int_0^t \vec{R}_{\mathbf{a},A}(t,\tau)\tilde{\mathbf{F}}(\tau)\,d\tau$$
$$+ \int_0^t R_{\sigma,A}(t,\tau)\,d\tau \quad (16)$$

with appropriately defined **linear response operators** $\vec{R}_{\mathbf{a},A}$ and $R_{\sigma,A}$.

In the case when the unperturbed pdf $\bar{p}$ is smooth and non-vanishing, we may formally rewrite the linear response operator in the form of a statistical average which can be replaced by long time average if the unperturbed pdf is assumed to be ergodic (Risken, 1984; Majda *et al.*, 2005). This is in the spirit of the fluctuation-dissipation theory in statistical physics (Risken, 1984; Majda *et al.*, 2005; Balescu, 1997).

Theorem 2 (FDT). *Suppose that the unperturbed climate $\bar{p}(\mathbf{X},t)$ of the original time-dependent system is smooth and positive, then the linear response operators can be represented via the following statistical correlations*

$$\vec{R}^T_{\mathbf{a},A}(t,\tau) = \langle A(\mathbf{X}(t))\mathbf{B_a}(\mathbf{X}(\tau))\rangle$$
$$= \mathbb{E}\int \left[e^{(t-\tau)\hat{L}^T_{FP}}A(\mathbf{X})\right]\Big|_{s=\tau}\mathbf{B_a}(\mathbf{X},\tau)$$
$$\bar{p}(\mathbf{X},\tau)\,d\mathbf{X}, \quad (17)$$
$$R_{\sigma,A}(t,\tau) = \langle A(\mathbf{X}(t))B_\sigma(\mathbf{X}(\tau))\rangle, \quad (18)$$

where $\hat{L}^T_{FP}$ is the transpose of the extended Fokker-Planck operator $\hat{L}_{FP}$

$$\hat{L}_{FP}p = -\nabla\cdot(p\mathbf{F}(\mathbf{X},s)) - \frac{\partial p}{\partial s} + \frac{1}{2}\nabla\cdot\nabla\cdot(Q\,p) \quad (19)$$

for the associated skew-product system (Majda and Wang, 2010), and the $\mathbf{B_a}$ and B_σ are computed through $\bar{p}(\mathbf{X},\tau)$ directly

$$\mathbf{B_a}(\mathbf{X},\tau) = \frac{L_\mathbf{a}\bar{p}(\mathbf{X},\tau)}{\bar{p}(\mathbf{X},\tau)},$$
$$B_\sigma(\mathbf{X},\tau) = \frac{L_\sigma\bar{p}(\mathbf{X},\tau)}{\bar{p}(\mathbf{X},\tau)}. \quad (20)$$

The external operators corresponding to the deterministic uncertainty ($\mathbf{L_a}$) and the uncertainty in noise (L_σ) are defined as

$$\mathbf{L_a}p = -\nabla\bullet(\mathbf{a}p), \quad (21)$$
$$L_\sigma p = \frac{1}{2}\nabla\cdot\nabla\cdot\left((\sigma\tilde{\sigma}^T + \tilde{\sigma}\sigma^T)p\right). \quad (22)$$

The significance of the formulas above lies in the fact that the linear response operators are determined by suitable correlation functions of the unperturbed solution with respect to the unperturbed climate. These correlation functions can be estimated provided we have the necessary information regarding the current unperturbed climate. In another word, as long as we know the unperturbed climate, we can predict climate response. We also like to point out that there are practical issues in applications. In particular, the functional $\mathbf{B}$ involves taking the derivative of the current climate distribution which is usually hard to estimate unless the dimension of the underlying model is small. However, in the case when the climate is Gaussian or near Gaussian, quasi-Gaussian approximations can be used to circumvent this problem. Quasi-Gaussian approximation will be discussed later in this section. It is also worth mentioning that the linear response theory provides answer beyond climate change which is the limiting case of $t \to \infty$ in $E'(t)$. It is easy to see that the climate response only makes sense when the involved correlation functions decay sufficiently fast as time approaches infinity so that the time integral of the correlation functions converge. The decay of the correlation functions is in accordance of the chaotic nature of the dynamics. On the practical side, we may need to have extremely long observational data, depending on the decay rate of the correlation functions (related to the mixing rate), in order to compute the leading order climate response $\delta E'(\infty)$. In the case with stationary external forcing, one can show that systems with generic noise is exponentially mixing (Hairer and Majda, 2010). Unfortunately, the mathematically estimated mixing rate is a function of the noise level and vanishes at the noise vanishes. It is still a challenge to derive mixing rate for deterministic systems except for a few toy models.

The interested reader is referred to (Majda and Wang, 2010) for the derivation of the linear response operators presented here, and their generalizations that include perturbation to initial data, perturbation away from non-climate, and time-dependent observables $A(\mathbf{X}, t)$.

In the special case when the unperturbed pdf is an equilibrium pdf of the unperturbed system, i.e., $\hat{L}_{FP}\bar{\bar{p}} \equiv 0$, and hence $B(\hat{\mathbf{X}}, \tau)$ is independent of τ, the correlation can be written in a simpler form

$$
\begin{aligned}
R_{A,B}(t,\tau) &= \langle A(\hat{\mathbf{X}}(t))B(\hat{\mathbf{X}}(\tau))\rangle \\
&= R_{A,B}(t-\tau, 0) \\
&= \lim_{T\to\infty} \frac{1}{T}\int_{T_*}^{T+T_*} A(\hat{\mathbf{X}}(t'+t-\tau)) \\
&\quad B(\hat{\mathbf{X}}(t'))\, dt' \qquad (23) \\
&= \lim_{T\to\infty} \frac{1}{T}\int_{T_*}^{T+T_*} A(\mathbf{X}(t'+t-\tau)) \\
&\quad B(\mathbf{X}(t'))\, dt' \qquad (24)
\end{aligned}
$$

where we have invoked the ergodicity assumption in the second to the last step, and we have assumed that A, B are independent of s in the last step. This last two formula may be particularly useful in the case of an ergodic statistical equilibrium since they do not involve the potentially unstable computation of tangent map although long time integration of the skew-product system is required. The issue of long time integrator which captures the equilibrium statistics is itself an interesting and challenging issue. See (E and Li, 2008) for the case of finite dimensional Hamiltonian system using Andersen thermostat approach, and (Cheng and Wang, 2008a; Cheng and Wang, 2008b; Wang, 2010) for infinite dimensional dissipative systems.

3.1. *Tangent map approach*

In the case of zero noise, the Fokker-Planck equation reduces to the Liouville equation which is a transport equation that can be solved explicitly via the method of characteristics. In this case, the linear response operator can be expressed in the following alternative form

$$
\begin{aligned}
\vec{R}^T_{\mathbf{a},A}(t,\tau) &= \vec{R}^T_{\mathbf{a},A}(t-\tau, 0) \\
&= \frac{1}{T_0}\int_{\mathbb{R}^N}\left(\int_0^{T_0} \nabla_{\mathbf{x}} A\left(\mathbf{X}\left(\begin{pmatrix}\mathbf{x}\\ s\end{pmatrix}, t\right)\right)\right. \\
&\quad \left.\bullet\ \mathbf{a}(\mathbf{x})\, p_{per}(\mathbf{x}, s)\, ds\right) d\mathbf{x} \qquad (25) \\
&= \lim_{T\to\infty}\frac{1}{T}\int_{T^*}^{T+T^*} \nabla_{\mathbf{x}} A(\mathbf{X}(\hat{\mathbf{x}}, t+\tau)) \\
&\quad \bullet\, \mathbf{a}(\mathbf{X}(\hat{\mathbf{x}}, \tau))\, d\tau \qquad (26)
\end{aligned}
$$

where $\hat{\mathbf{x}} = \binom{\mathbf{x}}{s}$ and $\mathbf{X}(\binom{\mathbf{x}}{s}, t)$ is the solution to the associated skew-product system with initial data $\hat{\mathbf{x}}$ at time t, and we have invoked the ergodicity in the last step.

The formula here was derived in (Majda and Wang, 2010), and it generalizes those for the autonomous case (Majda *et al.*, 2005).

Note that for applications of linear response theory to climate change with seasonal forcing, it is very interesting to have more general functionals, $A(\mathbf{X}, s)$, in applications. See Section 4 below.

3.2. *Quasi-Gaussian approximation*

The tangent map approach presented in the previous subsection requires the solution and its tangent map of the underlying chaotic/turbulent system over a relatively long time which may be difficult to accomplish. As an alternative we may utilize the assumption that the unperturbed climate pdf $\bar{p}$ is close to a Gaussian. This is a very reasonable hypothesis in many applications and hence we may approximate the climate by a Gaussian with the same mean and variance in some appropriate fashion. Depending on the manner on how the equivalent Gaussians are utilized, we may end-up with the so-called quasi-Gaussian approximation (Majda *et al.*, 2005) or the simple direct Gaussian approximation. A crucial advantage of quasi-Gaussian approximation is that *we do not need linear tangent model to assess response behavior approximately.*

For each fixed time t, let $p^G(\mathbf{X}, t)$ be a Gaussian pdf with the same mean and second moments as the unperturbed pdf $\bar{p}(\mathbf{X}, t)$, i.e.,

$$\int \mathbf{X} p^G(\mathbf{X}, t)\, d\mathbf{X} \tag{27}$$

$$= \int \mathbf{X}\bar{p}(\mathbf{X}, t)\, d\mathbf{X} \overset{def}{=} \bar{\mathbf{X}}(t), \tag{28}$$

$$\int (\mathbf{X} - \bar{\mathbf{X}}(t)) \otimes (\mathbf{X} - \bar{\mathbf{X}}(t))\, p^G(\mathbf{X}, t)\, d\mathbf{X} \tag{29}$$

$$= \int (\mathbf{X} - \bar{\mathbf{X}}(t)) \otimes (\mathbf{X} - \bar{\mathbf{X}}(t))\, \bar{p}(\mathbf{X}, t)\, d\mathbf{X} \overset{def}{=} \mathcal{C}, \tag{30}$$

where $\mathcal{C}(t, t)$ is the covariance matrix of $\mathbf{X}(t)$, and hence

$$p^G(\mathbf{X}, t) = \frac{1}{(2\pi)^{\frac{N}{2}} \det \mathcal{C}^{\frac{1}{2}}} \exp\left(-\frac{(\mathbf{X} - \bar{\mathbf{X}}(t)) \cdot \mathcal{C}^{-1}(\mathbf{X} - \bar{\mathbf{X}}(t))}{2}\right). \tag{31}$$

Now we replace $\bar{p}(\mathbf{X}, t)$ by $p^G(\mathbf{X}, t)$ in the calculation of the $\mathbf{B_a}$ and B_σ in the correlation formulation (20) in the previous theorem, i.e., we propose

$$\begin{aligned} \left(\vec{R}^G_{\mathbf{a},A}\right)^T(t, \tau) &= \left\langle A(\hat{\mathbf{X}}(t))\mathbf{B}^G_{\mathbf{a}}(\mathbf{X}(\tau))\right\rangle, \\ R^G_{\sigma,A}(t, \tau) &= \left\langle A(\hat{\mathbf{X}}(t)) B^G_\sigma(\mathbf{X}(\tau))\right\rangle, \end{aligned} \tag{32}$$

with the approximate special (vector) nonlinear functionals

$$\begin{aligned} \mathbf{B}^G_{\mathbf{a}}(\mathbf{X}, \tau) &= \frac{\mathbf{L_a} p^G(\mathbf{X}, \tau)}{p^G(\mathbf{X}, \tau)}, \\ B^G_\sigma(\mathbf{X}, \tau) &= \frac{L_\sigma p^G(\mathbf{X}, \tau)}{p^G(\mathbf{X}, \tau)}. \end{aligned} \tag{33}$$

It is worthwhile to point out that no tangent map is needed here which is a huge advantage for this approximation. This should be contrasted to the tangent map adjoint approach presented in the previous subsection.

In the special case of perturbation in the external forcing only, i.e., $\mathbf{a}(\mathbf{X}) \equiv \mathbf{a}$ is a constant vector, we may assume, without loss of generality,

$$\mathbf{L_a} = -\nabla_{\mathbf{X}}. \tag{34}$$

Hence we have

$$\mathbf{B}^G_{\mathbf{a}}(\mathbf{X}, \tau) = \mathcal{C}^{-1}(t, t)(\mathbf{X} - \bar{\mathbf{X}}(\tau)). \tag{35}$$

Thus if A is a linear function in $\mathbf{X}$, the linear response operator $\vec{R}^G_{\mathbf{a},A}(t,\tau)$ is essentially the *auto-correlation* of $\mathbf{X}$. This is a generalization of classical results on systems of identical particles (Kubo *et al.*, 1985; Risken, 1984; Majda *et al.*, 2005).

Quasi-Gaussian approximation has been successful to many geophysical problems, especially for estimating linear climate response in the mean ($A(\mathbf{X}) = \mathbf{X}$) and variance ($A(\mathbf{X}) = \mathbf{X} \otimes \mathbf{X}$) (Gritsoun and Branstator, 2007; Gritsoun *et al.*, 2008; Majda *et al.*, 2005; Abramov and Majda, 2011). The success of the quasi-Gaussian can be partially explained through the following short time asymptotic expansion.

Theorem 3 (Short time validity of quasi-Gaussian approximation). *Assume that the unperturbed pdf* $\bar{p}(\mathbf{X},t)$ *of the system (1) is smooth and non-vanishing. Furthermore, we assume that the deterministic perturbation is defined by external forcing only, i.e.* $\mathbf{a}(\mathbf{X}) \equiv \mathbf{a}$. *Then for the special linear functionals,* $A(\mathbf{X}) = X_j$, *the quasi-Gaussian approximation defined in (32) satisfies*

$$\vec{R}^G_{\mathbf{a},A}(t,\tau) = \vec{R}_{\mathbf{a},A}(t,\tau) + \mathcal{O}(t-\tau). \quad (36)$$

Instead of taking the quasi-Gaussian approximation, one could consider the Gaussian approximation where the average in the correlation relationship is taken with respect to the approximate Gaussian instead of the original climate distribution. For more details, the interested reader is referred to (Majda *et al.*, 2005) for the stationary case and (Majda and Wang, 2010) for the time-periodic case.

One could also combine the tangent map approach with the quasi-Gaussian approximation. The basic idea behind that is that the quasi-Gaussian approximation (qG) may not be applicable at the beginning when the distribution is far away from Gaussian, while the tangent map approach may not be appropriate after a long time following the accumulation of numerical errors in the computation of the tangent map of a chaotic system. Therefore, a bended approach that combines the two seems reasonable. Such a blended approach has been tested recently and have found considerable success (Abramov and Majda, 2007; Abramov and Majda, 2008; Abramov and Majda, 2009).

4. Numerical Algorithms for Climate Response with Seasonal Impact

The mathematics formulas presented in the previous section can be further approximated to yield numerical algorithms for computing climate response for climate models with time periodic forcing coefficients reflecting the seasonal cycle (Frederiksen and Branstator, 2005; Branstator and Frederiksen, 2003; Achatz and Opsteegh, 2002) with the prototype structure presented in (8). In particular, we recall from (Majda and Wang, 2010) natural generalizations of the quasi-Gaussian FDT algorithms (Leith, 1975; Majda *et al.*, 2005; Gritsoun *et al.*, 2002; Gritsoun and Dymnikov, 1999; Gritsoun and Branstator, 2007; Gritsoun *et al.*, 2008; Abramov and Majda, 2008; Abramov and Majda, 2007; Abramov and Majda, 2009) following 3.2.1, the short time FDT algorithms (Abramov and Majda, 2007; Abramov and Majda, 2008; Abramov and Majda, 2009) and the blended response algorithms (Abramov and Majda, 2008; Abramov and Majda, 2009) to the situation with a seasonal cycle. The interested reader is referred to (Majda and Wang, 2010) for more details.

4.1. *Numerical algorithms*

One of the most useful functionals in climate response has the separable form

$$\hat{A}\begin{pmatrix}\mathbf{x}\\ s\end{pmatrix} = \tilde{A}(\mathbf{x})\chi(|s-s_0|<P) \quad (37)$$

where $\chi(\mathcal{S})$ is the indicator function of the set $\mathcal{S}$. For example, for models with seasonal cycle so that the period T_0 is one year, we might be interested in the change in the low frequency teleconnection patterns (Frederiksen and Branstator, 2005; Branstator and Frederiksen, 2003; Abramov and Majda, 2009) during each month or season so that $\tilde{A}(\mathbf{x})$ is $\mathbf{x}$ (corresponding to the mean) and/or $\mathbf{x}\otimes\mathbf{x}$ (corresponding to the second moments or variance), s_0 is the fifteenth day of each month, and P is fifteen days. Similarly, we might be interested in the mean temperature change of its variance in each month at a specific location rather than just the annual changes (Gritsoun and Branstator, 2007; Gritsoun *et al.*, 2008). We would like to point out that the observable chosen here depends on the time explicitly and hence is more complex than the ones involved in the previous sections. However, the linear response theories in the previous sections can be generalized to the case with time dependent observables (Majda and Wang, 2010).

The first step in our approach is the estimation of the time-periodic climate of the unperturbed system given by p_{per}. According to theorem 1, $p_{per}(\mathbf{x}, s)$ is periodic in s with period T_0 and arises from a long time integration of the periodic dynamical systems in (1) with or without noise assuming ergodicity and strong mixing. It can be estimated from long time series of the system (see Theorem 1). More specifically, take the period interval T_0 and divide it into L equal intervals centered at s_j, $1 \leq j \leq L$ with width $\Delta s = \frac{T_0}{L}$. Then, using the long time series, one can calculate the appropriate statistics of approximate pdf's denoted here by $p_{per}(\mathbf{x}, s_j)$ by doing conditional statistics of the time series of the trajectory to the sets, $\{t, |t - ks_j| < \frac{\Delta s}{2}, k > K_0\}$.

Depending on the size (number of unknowns) of the model, different strategies are called for. In the case of small system, i.e., if $x \in \mathbb{R}^N$ where N is low dimensional (roughly, $N \leq 4$), and (1) is a low order stochastic model (Majda *et al.*, 2008; Achatz and Opsteegh, 2002), then it is reasonable to expect that the entire pdf $p_{per}(\mathbf{x}, s_j)$ be found with reasonably high accuracy (Majda *et al.*, 2008). In this situation, the linear response operators can be approximated using the following straightforward *direct FDT algorithm*

$$A(\hat{\mathbf{x}}) = \hat{A}(\mathbf{x})\hat{B}_{\mathbf{a}}(\mathbf{x}, s_j), \qquad \hat{B}_{\mathbf{a}}(\mathbf{x}, s_j) = \frac{\mathbf{L}_{\mathbf{a}} p_{per}(\mathbf{x}, s_j)}{p_{per}(\mathbf{x}, s_j)}. \tag{38}$$

On the other hand, if $x \in \mathbb{R}^N$ with $N \gg 1$ as occurs in contemporary climate models (Gritsoun and Branstator, 2007; Gritsoun *et al.*, 2008), one can calculate the low order statistics of $p_{per}(\mathbf{x}, s_j)$ involving the mean and covariance with reasonable precision although it is essentially hopeless to find reasonably accurate approximation of the pdf itself. In this case, we can use the following *quasi-Gaussian FDT algorithm* to approximate the linear response operators

$$\hat{B}_{\mathbf{a}}(\mathbf{x}, s_j) = \frac{\mathbf{L}_{\mathbf{a}} p^G(\mathbf{x}, s_j)}{p^G(\mathbf{x}, s_j)}. \tag{39}$$

Both the direct algorithm and the approximated algorithm require evaluation of the response operator, $\vec{R}_{\mathbf{a},A}(t)$, through the correlation with a suitable $\hat{B}$ in either (38) or (39), i.e. the (approximate) response operator, $\vec{R}_{\mathbf{a},A}(t)$, is given by

$$\vec{R}^T_{\mathbf{a},A}(t) \approx \langle A\hat{\mathbf{X}}(t)\hat{B}_{\mathbf{a}}(\hat{\mathbf{X}}(0))\rangle = \frac{1}{T_0}\int_0^{T_0}\int_{\mathbb{R}^N} \mathbb{E}A(\mathbf{X}(\mathbf{x}, t, s), t + s) \hat{B}_{\mathbf{a}}(\mathbf{x}, s)p_{per}(\mathbf{x}, s)\, d\mathbf{x}ds. \tag{40}$$

In (40), $\mathbb{E}$ is the expectation, $\mathbf{X}(\mathbf{x}, t, s)$ is the trajectory of (1) satisfying the phase shifted dynamical systems,

$$\frac{d\mathbf{X}}{dt} = \mathbf{F}(\mathbf{X}, t + s) + \sigma(\mathbf{X})\dot{W}, \quad \mathbf{X}(0) = \mathbf{x}. \tag{41}$$

Using ergodicity with respect to $\mathbf{X}$, and the special form for the functionals in (37), we have

(Majda and Wang, 2010)

$$\vec{R}^T_{\mathbf{a},A}(t) \approx \frac{1}{LK}\sum_{j=1}^{L}\sum_{k=1}^{K}\mathbb{E}\tilde{A}(\mathbf{X}(t+s_j+kT_0,s_j))$$
$$\chi(|s_j+t+kT_0-s_0|\le P)$$
$$\hat{B}(\mathbf{X}(s_j+kT_0,s_j)). \tag{42}$$

With the two approximate formulas for $\hat{B}$ in (38, 39), this leads to the **direct FDT** and **Quasi-Gaussian FDT algorithms** for systems with periodic forcing. Note that there is non-trivial phase averaging for general functionals like those in (37); if $\chi = [0, T_0]$ so that we are interested in only mean averaged statistics, then we can use $L \equiv 1$ in (42) coupled with (38, 39).

For the case of (1) with zero noise, the exact adjoint formula described in Section 3.1 and similar considerations as above in (40, 42) leads to a general response algorithm for functionals of the form in (37)

$$\vec{R}^T_{\mathbf{a},A}(t) \tag{43}$$
$$\approx \frac{1}{LK}\sum_{j=1}^{L}\sum_{k=1}^{K}\nabla_{\mathbf{X}(s_j+kT_0,s_j)}$$
$$\tilde{A}(\mathbf{X}(t+s_j+kT_0,s_j))$$
$$\bullet\mathbf{a}(\mathbf{X}(s_j+kT_0,s_j))$$
$$\chi(|s_j+t+kT_0-s_0|\le P)$$
$$\hat{B}(\mathbf{X}(s_j+kT_0,s_j)) \tag{44}$$

for $K \gg 1$. We call this algorithm, the **short-time FDT algorithm** after (Abramov and Majda, 2007; Abramov and Majda, 2008; Abramov and Majda, 2009); no explicit knowledge of the time-periodic equilibrium measure, p_{per}, is needed. Clearly, these algorithms presented here can be combined in time to create blended response algorithms following (Abramov and Majda, 2007; Abramov and Majda, 2008). The accuracy of these proposed algorithms depends on the sampling width P in (37) and the number, L, of trajectories (the width Δs), as well as the functional $\tilde{A}(\mathbf{X})$, and the length of the time series available.

4.2. *Numerical test*

The algorithms presented in this section has been recently tested by Gershgorin and Majda on the following 3 mode triad nonlinear stochastic model (one real mode u_1, and one complex mode u_2) with exactly solvable first and second moments with Gaussian and non-Gaussian regimes (Gershgorin and Majda, 2010)

$$\frac{du_1}{dt} = -\gamma_1 u_1 + f_1(t) + \sigma_1\dot{W}_1, \tag{45}$$
$$\frac{du_2}{dt} = (-\gamma_2 + i(\omega_0 + a_0u_1))u_2$$
$$+ f_2(t) + \sigma_2\dot{W}_2, \tag{46}$$

where $f_1(t)$ and $f_2(t)$ are periodic functions of time with the same period T_0, which represents the annual cycle, γ_1, γ_2, σ_1, and σ_2 are dissipation and stochastic forcing coefficients that represent the interaction of the triad system with other unresolved modes, ω_0 is deterministic part of the frequency for the complex mode, u_2, and a_0 is the coefficient measuring nonlinearity.

This model mimics the nonlinear interaction of two Rossby waves forced by baroclinic processes with a zonal jet forced by a polar temperature gradient. Periodic forcing naturally introduces the seasonal cycle into the model. The exactly solvable first and second order statistics are utilized to compute both the ideal mean and variance response to the perturbations in forcing or dissipation and the quasi-Gaussian approximation of FDT (qG-FDT) that uses the mean and the covariance in the equilibrium state. The time-averaged mean and variance qG-FDT response to perturbations of forcing or dissipation is compared with the corresponding ideal response utilizing the triad test model in a number of regimes with various dynamical and statistical properties such as weak or strong non-Gaussianity and resonant or non-resonant forcing. It is shown that even in a strongly non-Gaussian regime, qG-FDT has surprisingly high skill for the mean response to the changes in forcing. On the other hand, the performance of qG-FDT for the variance response to the perturbations of dissipation is good in the near-Gaussian regime and deteriorates in the strongly non-Gaussian regime. The results presented in (Gershgorin and Majda, 2010) on the test model may provide useful guidelines for applying the time-periodic FDT to more complex realistic systems such as atmospheric general circulation models.

5. Concluding Remarks

We have surveyed several generalizations of linear response theory (fluctuation-dissipation theorem) as well as a few computational algorithms for systems with periodic external forcing that are developed in (Majda and Wang, 2010). These theories and numerical algorithms may have significant applications to climate change studies for models with seasonal impact. More general theories and additional results can be found in (Majda and Wang, 2010). Preliminary numerical results by Gershgorin and Majda (Gershgorin and Majda 2010) demonstrates the effectiveness and potentials of various algorithms reviewed here. Additional theoretical and numerical works are needed in order to further validate these algorithms.

Appendix: Examples of classical FDT

The purpose of this appendix is to recall two classical examples of FDT which illustrate how the linear response theories are related to the fluctuation-dissipation theorem.

Einstein's relation

Recall that classical Brownian motion with friction takes the form

$$m\frac{du}{dt} = -m\gamma u + \sigma\frac{dW}{dt}, \quad \frac{dx}{dt} = u$$

where x is the position, u is the velocity, m is the mass, γ is the friction coefficient, and W is a standard Brownian motion.

This system enjoys the following equilibrium distribution

$$p(u) = C\exp\left[-\frac{mu^2}{2kT}\right], \quad kT = \frac{\sigma^2}{2m\gamma}$$

where k is the Boltzmann constant, and T is the temperature of the system.

The diffusion coefficient associated with this Brownian motion with friction can be calculated as

$$\begin{aligned} D &\stackrel{def}{=} \lim_{t\to\infty}\frac{1}{2t}\langle|x(t)-x(0)|^2\rangle \\ &= \lim_{t\to\infty}\frac{1}{2t}\int_0^t dt_1\int_0^t dt_2\langle u(t_1)u(t_2)\rangle \\ &= \lim_{t\to\infty}\frac{1}{t}\int_0^t dt_1\int_0^{t-t_1} ds\langle u(t_1)u(t_1+s)\rangle \\ &= \int_0^\infty \langle u(t_0)u(t_0+t)\rangle dt = \frac{\sigma^2}{2m^2\gamma^2} \end{aligned}$$

where we have used the fact that the correlation function decays at infinity.

Therefore, we deduce the following Einstein's relation which can be viewed as a dissipation-fluctuation relation

$$\mu = \frac{1}{m\gamma} = \frac{D}{kT} = \frac{1}{kT}\int_0^\infty \langle u(t_0)u(t_0+t)\rangle dt$$

since it relates the mobility μ to the fluctuation of the system at equilibrium.

Linear response theory

Now we consider a Brownian motion with friction and a periodic forcing

$$m\frac{du}{dt} = -m\gamma u + \sigma\frac{dW}{dt} + K(t), \quad \frac{dx}{dt} = u$$

where $K(t) = K_0\cos\omega t$ represents an external time periodic forcing.

Let $\langle u(t)\rangle$ denote the statistical average of u at large time. The average is zero in the absence of the external deterministic periodic forcing K. Therefore long time linear response of the average with respect to the external forcing K can be determined as

$$\delta\langle u(t)\rangle = \mathcal{R}\mu(\omega)K_0 \exp(i\omega t)$$

where the (complex valued) mobility (or admittance) μ is given by

$$\mu(\omega) = \frac{1}{m}\frac{1}{i\omega+\gamma}.$$

Notice that

$$\begin{aligned}\mu(\omega) &= \frac{1}{m\langle u^2\rangle}\int_0^\infty \langle u(t_0)u(t_0+t)\rangle e^{-i\omega t}dt \\ &= \frac{1}{kT}\int_0^\infty \langle u(t_0)u(t_0+t)\rangle e^{-i\omega t}dt,\end{aligned}$$

$$\int_0^\infty \langle u(t_0)u(t_0+t)\rangle e^{-i\omega t}dt = \frac{\langle u^2\rangle}{i\omega+\gamma}.$$

Hence this linear response relation can be interpreted as the response in the mean is determined by the fluctuation property of the system at equilibrium.

Acknowledgments

This work is supported in part by grants from the National Science Foundation.

References

Abramov, R.V., 2010: Approximate linear response for slow variables of deterministic or stochastic dynamics with time scale separation. *J. Comput. Phys.*, **229**, 7739–7746.

______ and M.P. Kjerland, 2013: The response of reduced models of multiscale dynamics to small external perturbations. Submitted to *Comm. Math. Sci.* http://arxiv.org/abs/1305.0862

______ and A.J. Majda, 2007: Blended response algorithms for linear fluctuation-dissipation for complex nonlinear dynamical systems. *Nonlinearity*, **20**(12), 2793–2821.

______ and ______, 2008: New approximations and tests of linear fluctuation-response for chaotic nonlinear forced-dissipative dynamical systems. *J. Nonlinear Sci.*, **18**, 303–341.

______ and ______, 2009: New algorithms for Low frequency climate response. *J. Atmos. Sci.*, **66**(2), 286–309.

______ and ______, 2011: Low frequency climate response of quasigeostropic wind-driven ocean circulation. *J. Phys. Ocean*, **42**, 243–260, doi: 10.1175/JPO-D-11-052.1.

Achatz, U. and J.D. Opsteegh, 2002: Primitive-equation-based low order models with seasonal cycle. *J. Atmos. Sci.*, **60**, 649–690.

Balescu, R., 1997: *Statistical Dynamics.* Imperial College Press, 330 pp.

Bell, T., 1980: Climate sensitivity from fluctuation dissipation: Some simple model tests. *J. Atmos. Sci.*, **37**, 1700–1708.

Branstator, G. and J. Frederiksen, 2003: The seasonal cycle of Interannual variability and the dynamical imprint of the seasonally varying mean state. *J. Atmos. Sci.*, **60**(13), 1577–1592, doi: 10.1175/3002.1.

Cheng, W. and X. Wang, 2008a: A uniformly dissipative scheme for stationary statistical properties of the infinite Prandtl number model. *Applied Mathematics Letters*, **21**(12), 1281–1285, doi: 10.1016/j.aml.2007.07.036.

______ and ______, 2008b: A semi-implicit scheme for stationary statistical properties of the infinite Prandtl number model. *SIAM J. Num. Anal.*, **47**(1), 250–270, doi: 10.1137/080713501.

Weinaw, E. and D. Li, 2008: The Andersen thermostat in molecular dynamics. *Comm. Pure Appl. Math.*, **61**, 96–136.

Foias, C., O. Manley, R. Rosa and R. Temam, 2001: *Navier-Stokes Equations and Turbulence.* Cambridge University Press, 364 pp.

Frederiksen, J.S. and G. Branstator, 2005: Seasonal variability of teleconnection patterns. *J. Atmos. Sci.*, **62**, 1346–1365.

Gardiner, C.W., 2004: *Handbook of Stochastic Methods*, Springer-Verlag, 464 pp.

Gershgorin, B. and A.J. Majda, 2010: A test model for fluctuation-dissipation theorems with time periodic statistics. *Physica D*, **239**(17), 1741–1757.

Gritsoun, A.S. and G. Branstator, 2007: Climate response using a three-dimensional operator based on the fluctuation-dissipation theorem. *J. Atmos. Sci.*, **64**, 2558–2575, doi: 10.1175/JAS3943.1.

______, ______ and V.P. Dymnikov, 2002: Construction of the linear response operator of an atmospheric general circulation model to small external forcing. *Russ. J Numer. Anal. M*, **17**, 399–416.

———, ——— and A.J. Majda 2008: Climate response of linear and quadratic functionals using the fluctuation-dissipation theorem. *J. Atmos. Sci.*, **65**, 2824–2841.

——— and V.P. Dymnikov, 1999: Barotropic atmosphere response to small external actions: theory and numerical experiments. *Izv. Russ. Acad. Sci. Atmospher. Ocean. Phys.*, **35**(5), 511–525.

Hairer, M. and A.J. Majda, 2010: A simple framework to justify linear response Theory. *Nonlinearity*, **23**(4), 909–922.

Leith, C., 1975: Climate response and fluctuation-dissipation. *J. Atmos. Sci.*, **32**, 2022–2025.

Kubo, R., M. Toda and N. Hashitsume, 1985: *Statistical Physics II: Nonequilibrium Statistical Mechanics.* Springer-Verlag, 279 pp.

Majda, A.J., R. Abramov and M. Grote, 2005: *Information Theory and Stochastics for Multiscale Nonlinear Systems. CRM Monograph Series.* No. 25, American Mathematical Society, 133 pp.

———, C. Franzke and B. Khouider, 2008: An applied mathematics perspective on stochastic modelling for climate. *Phil. Trans. R. Soc. A.*, **366**, 2429–2455.

———, B. Gershgorin and Y. Yuan, 2010: Low-frequency climate response and fluctuation-dissipation theorems: Theory and practice. *J. Atmos. Sci.*, **67**, 1186–1201.

——— and X. Wang, 2006: *Nonlinear Dynamics and Statistical Theories for Basic Geophysical Flows.* Cambridge University Press, 564 pp.

——— and ———, 2010: Linear response theory for statistical ensembles in complex systems with time-periodic forcing. *Comm. Math. Sci.*, **8**(1), 145–172.

Risken, H., 1984: *The Fokker-Planck Equations.* Springer-Verlag, 454 pp.

Vishik, M.I. and A.V. Fursikov, 1988: *Mathematical Problems of Statistical Hydromechanics.* Springer, 576 pp.

Wang, X., 2010: Approximation of stationary statistical properties of dissipative dynamical systems: time discretization. *Math. Comp.*, **79**, 259–280, doi: 10.1090/S0025-5718-09-02256-X.

Chapter 4

Parametrization of Cross-scale Interaction in Multiscale Systems

Jeroen Wouters and Valerio Lucarini*

Klimacampus, Institute of Meteorology, University of Hamburg, Grindelberg 5, 20144 Hamburg, Germany
jeroen.wouters@uni-hamburg.de
http://www.mi.uni-hamburg.de/Theoretische.6.0.html

We consider the problem of deriving approximate autonomous dynamics for a number of variables of a dynamical system, which are weakly coupled to the remaining variables. We show that by using the Ruelle response theory on such a weakly coupled system it is possible to construct a surrogate dynamics, such that the expectation value of any observable agrees, up to second order in the coupling strength, to its expectation evaluated on the full dynamics. We then show that such surrogate dynamics agrees up to second order to an expansion of the Mori-Zwanzig projected dynamics. This implies that the parametrizations of unresolved processes suited for prediction and for the representation of long term statistical properties are closely related, if one takes into account, in addition to the widely adopted stochastic forcing, the often neglected memory effects.

1. Introduction

The investigation of multi-level systems is a central task in many branches of mathematics, natural and social sciences. In multi-level systems it is possible to separate the phase space variables into at least two subsets, such that the variables within each subset are strongly coupled, while variables belonging to different subsets have a much weaker coupling.

In most practical cases the dynamics of each level take place in a distinct range of spatial and temporal scales, so that it is hard to define an optimal resolution for all the variables when we attempt to simulate the system or collect data (Fatkullin and Vanden-Eijnden, 2004). Usually, one is interested in devising ways to account, at least approximately, for the impact of fast processes occurring at small spatial scales on slow variables, often describing large scale features, thus defining an effective autonomous dynamics for the slow variables.

If one assumes a vast time-scale separation between the slow variables X and the fast variables Y, averaging and homogenization methods (Arnold, 2001; Kifer, 2004; Pavliotis and Stuart, 2008) allow for deriving a dynamics for the X variables. A typical example of how this kind of theory can be used in climate science is the setting considered by Hasselmann, where fast "weather" systems influence slow "climate" dynamics. Abramov (Abramov, 2012) has recently presented an application of this method to deriving a simplified dynamics for a system of geophysical relevance. Another rather successful attempt in this direction is given in Majda *et al.* (Majda *et al.*, 2001). Furthermore,

*Also at Department of Mathematics and Statistics, University of Reading, Reading, RG6 6AX, UK.

Climate Change: Multidecadal and Beyond
edited by Chih-Pei Chang et al.

projector operator techniques have been introduced in statistical mechanics with the goal of effectively removing the Y variables. In particular, considerable interest has been raised by the Mori-Zwanzig approach, through which a formal — albeit practically inaccessible — solution for the evolution of the X variables is derived (Zwanzig, 2001). Many other parametrization methods have been proposed from a more heuristic point of view. Particularly relevant with respect to the current study are the results by Kwasniok (Kwasniok, 2012) indicating that non-Markovian parametrizations are superior to Markovian ones when a large time scale separation is absent. A similar conclusion can be reached from the work on empirical model reduction by Chekroun *et al.* (2011), where non-Markovian models are fitted to data without a clear time scale separation.

In a previous paper (Wouters and Lucarini, 2012) we have approached the problem of defining an accurate effective dynamics for the X variable by focusing on the statistical properties of an observable $A(X)$ rather than on its trajectory. The observable A is a smooth function of the variables describing the system, think for example of the kinetic energy of a flow. The starting point was the assumption that the coupling between the X and Y variables is weak, treating such a coupling as a perturbation to the autonomous dynamics of the X and Y variables, regarded as unperturbed system. A possible application of this theory in climate science can be found in the interaction between cloud formation and large scale atmospheric flow, where there is no distinct time scale separation, but instead the coupling could be considered as weak. We have then adopted the response theory developed by Ruelle (1997, 2009) (Lucarini, 2008), which allows to compute explicitly how the long-time averages of Axiom A dynamical systems change as a result of small perturbations to the flow (Reick, 2002; Cessac and Sepulchre, 2007; Abramov and Majda, 2008; Lucarini, 2009; Lucarini and Sarno, 2011). The Axiom A class of dynamical systems, even if mathematically non-generic, is nevertheless assumed to serve as a good model for general physical systems, as is put forward by the chaotic hypothesis (Gallavotti and Cohen, 1995; Gallavotti, 1996), which can be interpreted as an extension of the ergodic hypothesis to non-Hamiltonian systems (Gallavotti and Roma, 2006). A detailed discussion of the relevance of the Axiom A dynamical systems for the description of actual physical processes is given in (Lucarini, 2008; Lucarini and Sarno, 2011). Thus, we have derived explicit formulas to compute the changes in the expectation value of $A(X)$ up to the second order in the coupling strength between the X and Y variables. Moreover, we have been able to derive a surrogate perturbed dynamics for the system X such that up to second order the expectation value of $A(X)$ is the same as that of the fully coupled (X, Y) system, thus deriving an explicit parametrization of the coupling. The correction due to the coupling entail a deterministic contribution to the dynamics, a stochastic forcing expressed as a sum of multiplicative noise terms, and an integral expression which describes a memory term.

A question left open in (Wouters and Lucarini, 2012) was the link between the surrogate dynamics for the X subsystem and the exact dynamics one would obtain making use of the Mori-Zwanzig approach (see Section 2.1). Whereas we have shown that the proposed dynamics generate expectation values that are close to those of the fully coupled system, there may be many possible parametrizations that have this property. It is unclear to what extent the surrogate and full dynamics are related and whether they are close to each other. One must note that in general it is possible to construct a simplified model which reproduces well the statistical properties of the full model without describing well its time evolution. We anticipate that this study gives the comforting answer that, indeed, the surrogate dynamics derived in (Wouters and Lucarini, 2012) and the dynamics projected using the Mori-Zwanzig operator are identical up to second order of perturbation. As we will discuss in the conclusions, the fact that both the invariant measure and the dynamics of the full system are well approximated and the fact that a smoothing stochastic noise term is present, suggests the possibility of applying applying the fluctuation-dissipation theorem (FDT) for relating response and fluctuations of the X variables. This addresses some of the questions raised in (Ruelle, 1999, 2009), (Lucarini, 2008, 2009), (Lucarini and Sarno, 2011), and (Colangeli *et al.*, 2012).

This contribution is structured as follows. In Section 2 we present the Dyson expansion for the evolution of the unperturbed and perturbed ows, and provide a way to treat in a unified way the Mori-Zwanzig and Ruelle approaches. We also explain how to construct all perturbative terms and define simple rules for constructing diagrams describing the various linear and nonlinear contributions to the response. In Section 3 we then use our approach to deal with multi-level systems, thus deriving the explicit expression of the projected dynamics for the X subsystem according to the Mori-Zwanzig and Ruelle formalisms up to the second order of perturbation due to the coupling with the Y subsystem. We show that these two approximate dynamics agree up to second order in the coupling strength. In Section 5 we present our conclusions and perspectives for future work.

2. Perturbations to Deterministic Flows

Given a non-linear dynamical system $\dot{x} = F(x)$ on a manifold $\mathcal{M}$, the analysis of perturbations can be simplified by turning the problem into a linear equation for the evolution of observable functions A: $\mathcal{M} \to \mathbb{R}$. To this end, we define a linear differential operator $L(x) = (F(x).\nabla)$ called the Liouville operator, determining the time derivative of observables, dependent on time through a solution $x(t)$ of the dynamical system. We thus have that by the chain rule $\dot{A}(x(t)) = L(x(t))A(x(t))$. A solution of A over time is then given by $A(t) = \Pi(t)A(0)$ where $\Pi(t) = \exp(Lt)$. Although this equation does not tell us anything about the actual value of the solution, it does allow us to analyze perturbations, where the defining vector field F, and hence also L, are slightly altered.

Both the Mori-Zwanzig projection operator technique (Mori, 1965; Zwanzig, 1961) and the response theory of natural invariant measures (Ruelle, 1997) are cases where use is made of such perturbed evolutions. In the case of Mori-Zwanzig the change in evolution is due to the presence of an "environment" of unresolved variables, while in response theory the perturbation can be general, but instead of looking at the effect on the instantaneous evolution, one is interested in the effect on long term averages.

As described *e.g.* in (Evans and Morriss, 2008), these perturbative relations can be easily derived formally in the resolvent formalism, by taking the Laplace transform of the operator exponentials $\Pi(t)$:

$$\mathcal{L}\{\Pi\}(s) = \int_0^\infty dt \exp(Lt)\exp(-ts) = (s-L)^{-1} \quad (1)$$

If L consists of a perturbation around an operator L_0, i.e. $L = L_0 + L_1$, with L_1 small in an appropriate sense, we can expand the Laplace transform using the equality

$$(A-B)^{-1} = A^{-1} + A^{-1}B(A-B)^{-1}. \quad (2)$$

In the case of the Laplace transform in Eq. (1), we take $A = s - L_0$ and $B = L_1$, so that the A^{-1} and $(A-B)^{-1}$ terms are themselves the Laplace transforms of $\Pi_0(t) = \exp(L_0 t)$ and $\Pi(t)$ respectively. Inverting the Laplace transform, the product of transforms gives a convolution term, resulting in the following expansion of $\Pi(t)$:

$$\Pi(t) = \Pi_0(t) + \int_0^t d\tau \Pi_0(t-\tau)L_1\Pi(\tau) \quad (3)$$

Another expansion can be obtained when making use of the following equality for operator inverses:

$$(A-B)^{-1} = A^{-1} + (A-B)^{-1}BA^{-1}. \quad (4)$$

This gives rise to:

$$\Pi(t) = \Pi_0(t) + \int_0^t d\tau \Pi(t-\tau)L_1\Pi_0(\tau) \quad (5)$$

As the right-hand side of Eqs. (3) and (5) still contain $\Pi(t)$, they can be iterated to expand $\Pi(t) - \Pi_0(t)$ at different orders of the perturbation L_1:

$$\int_0^t d\tau \Pi_0(t-\tau)L_1\Pi_0(\tau) + \int_0^t d\tau_1 \int_0^{\tau_1} d\tau_2 \Pi_0(t-\tau_1) \times L_1\Pi_0(\tau_1-\tau_2)L_1\Pi_0(\tau_2) + O(L_1^3) \quad (6)$$

As Π describes the evolution of observables under the perturbed dynamics $L_0 + L_1$ and Π_0 under the unperturbed L_0, when applied to an observable A, these equations tell us how its evolution is perturbed.

2.1. *Projection operator technique*

In the case of the Mori-Zwanzig approach a projection is carried out on the level of the observables to remove unwanted, irrelevant and usually fast degrees of freedom. Here the expansion is performed around the evolution that involves only the relevant part of the phase space.

If a dynamical system is defined on a manifold $\mathcal{M}$, one defines a projection $\mathcal{P}$ from the space of observable functions on the full phase space $\mathcal{M}$ to a space of observables which are considered to contain only the interesting dynamics. Many different choices are possible; if the manifold $\mathcal{M}$ consists for example of a product of submanifolds $\mathcal{K}$ of relevant and $\mathcal{L}$ of irrelevant variables, one can take a conditional expectation with respect to a measure on $\mathcal{M}$, given the value of the relevant variables $x \in \mathcal{K}$:

$$(\mathcal{P}A)(x) = \frac{\int_{\mathcal{N}} A(x,y)\rho(x,y)dy}{\int_{\mathcal{N}} \rho(x,y)dy}.$$

It is easily verified that this is a projection, i.e. $\mathcal{P}^2 = \mathcal{P}$ and that $\mathcal{P}$ is orthogonal w.r.t the product defined by ρ: $\langle A, B \rangle = \int_{\mathcal{M}} \rho AB$. Another possible choice is a projection onto a set of functions on $\mathcal{M}$, such as linear functions of the coordinates in a Euclidean phase space.

The evolution operator L is now split into its projection $\mathcal{P}L$ onto the relevant space of observables and the complement $\mathcal{Q}L := (1-\mathcal{Q})L$. As described by Zwanzig (Zwanzig, 2001), a generalized Langevin equation can then be derived based on Eq. (5). We write the Liouville equation for an observable A as

$$\frac{dA(t)}{dt} = LA(t) = e^{tL}LA = e^{tL}\mathcal{P}LA + e^{tL}\mathcal{Q}LA$$

The factor $\exp(tL)$ in the second term can be further expanded by making use of Eq. (5) with $L_0 = \mathcal{Q}L$. This gives the following equation

$$\frac{dA(t)}{dt} = e^{tL}\mathcal{P}LA + \left(e^{t\mathcal{Q}L} + \int_0^t ds\, e^{(t-s)L}\mathcal{P}Le^{s\mathcal{Q}L}\right)\mathcal{Q}LA$$

It is then argued (Zwanzig, 2001) that this equation is a generalization of the Langevin equation, where the second term is a correlated noise term dependent on the initial conditions of the irrelevant degrees of freedom and the third term represent the memory of the system due to the presence of irrelevant variables that have interacted with the relevant ones in the past. Note that we have done nothing more than manipulating the original evolution equation $\dot{A} = LA$. Correspondingly, the Mori-Zwanzig equation in itself does not simplify the problem. In order to derive a set of equations that are useful for numerical simulations, assumptions need to be made about the dynamical system. In the limit of an infinite time-scale separation between the relevant and irrelevant variables, the stochastic component of the parametrization can be represented as a white noise term, while the non-Markovian term vanishes, as the irrelevant variables decorrelate quickly. For a discussion of the applicability of Mori-Zwanzig and useful references, see the paper by Gottwald (Gottwald, 2010).

2.2. *Response theory*

While the Mori-Zwanzig projection operator technique results in an expression for the short term behavior of observables in time, the goal of response theory (Ruelle, 1998), on the other hand, is to describe changes in the averages of observables of dynamical systems over a long period of time. We take the average of an observable A over a long trajectory $x(t)$ and assume that the system is ergodic, meaning that the result of this procedure gives the same expectation value $\tilde{\rho}(A)$ for almost any initial point $x(0)$:

$$\tilde{\rho}(A) = \lim_{T\to\infty} \frac{1}{T}\int_0^T dt A(x(t)) \tag{7}$$

$$= \lim_{T\to\infty} \frac{1}{T}\int_0^T dt \Pi(t) A(x(0)) \tag{8}$$

We again assume that the operator L that determines Π consists of a perturbation around an unperturbed evolution: $L = L_0 + L_1$, meaning that we can make use of the expansions (3) and (5) to expand $\Pi(t)$ around $\Pi_0(t)$. The perturbing operator L_1 can derive from an external forcing or, as we will discuss later, from a coupling of internal degrees of freedom. The averages of an observable A for the unperturbed evolution, corresponding to L_0 and Π_0, will be denoted by $\rho(A)$. We now want to derive an expansion of the response of the invariant measure to the perturbation, i.e. expand $\tilde{\rho}(A) - \rho(A)$ in orders of L_1.

To derive the response terms, we can take two different paths. We can start from Eq. (6), expand Π in Eq. (8) in orders of L_1 and then rewrite the different order terms as expectation values under the unperturbed dynamics. Here, instead, we will first derive the equation expressing the full change in expectation value $\tilde{\rho}(A) - \rho(A)$, which can then be expanded in orders of the perturbation.

By inserting Eq. (5) into Eq. (8), we have that

$$\begin{aligned}
\tilde{\rho}(A) - \rho(A) &\\
&= \lim_{T\to\infty} \frac{1}{T}\int_0^T dt \int_0^t d\tau \Pi(\tau)\\
&\quad \times L_1 \Pi_0(t-\tau) A(x(0))\\
&= \lim_{T\to\infty} \frac{1}{T}\int_0^T d\tau \int_0^{T-\tau} dt \Pi(\tau)\\
&\quad \times L_1 \Pi_0(t) A(x(0))\\
&= \tilde{\rho}\left(\int_0^\infty dt L_1 \Pi_0(t) A\right)
\end{aligned}$$

or

$$\tilde{\rho}(A) = \rho\left(\left(1 - \int_0^\infty dt L_1 \Pi_0(t)\right)^{-1} A\right) \tag{9}$$

By expanding the resolvent operator, one gets the response terms at different orders of L_1:

$$\begin{aligned}
\tilde{\rho}(A) &= \rho(A) + \rho\left(\int_0^\infty dt_1 L_1 \Pi_0(t_1) A\right)\\
&\quad + \rho\left(\int_0^\infty dt_1 \int_0^\infty dt_2 L_1 \Pi_0(t_1) L_1 \Pi_0(t_2) A\right)\\
&\quad + \dots
\end{aligned} \tag{10}$$

It is important to note that the response of the expectation values can at all orders be expressed as expectation values under the unperturbed dynamics. This equation shows that in order to determine the first order response, we need to calculate the effect an infinitesimally short forcing L_1 has over a timespan t_1 and integrate over t_1. The higher order contributions follow a similar reasoning, with an increasing number of infinitesimal perturbations acting on the system.

If the measure ρ is sufficiently smooth (absolutely continuous with respect to the Lebesgue measure), we can make use of integration by parts to write the first order term as a cross-correlation function of the equilibrium system, leading to the fluctuation-dissipation theorem (FDT) (Ruelle, 1997, 2009). Although it is a priori not obvious whether the smoothness condition for the derivation of the FDT is satisfied by the invariant measures of geophysical flows, the FDT has been quite successful in applications to forcing experiments in atmospheric circulation models (Langen and Alexeev, 2005; Gritsun and Branstator, 2007; Abramov *et al.*, 2005; Cooper and Haynes, 2011). We will discuss this issue later in the chapter.

An issue that we have not considered here is the convergence of the expressions we have derived. The integrals in the response formula (10) involve a limit to infinity, for which the existence needs to be proven. For dynamical systems this is in general a very complicated task, which Ruelle has been able to accomplish for the class of Axiom A dynamical systems (Ruelle, 1997). For such systems, a split of the tangent space into orthogonal stable, neutral and unstable subspaces exists for almost every point on the attractor. The forcing can hence be split into orthogonal parts along these directions. As the invariant measure is smooth along the unstable direction, this part of the response can be written again as a cross-correlation, which decays rapidly due to mixing. The response to forcings along the stable directions can also be shown to converge as perturbations in these directions get damped exponentially fast by definition.

3. Coupled Multi-scale Systems

The response theory described in Section 2.2 is quite general, in the sense that we have not defined the form of the perturbing operator L_1. The perturbation can be any forcing, where the vector field defining the dynamical system is altered from F to $F + \delta F$. The form of L_1 is then $\delta F.\nabla$. A special case of such a forcing is an internal coupling of degrees of freedom. Here the original vector field F consists of a Cartesian product $(F_X, F_Y)^T$ of the vector fields F_X and F_Y defining the autonomous X and Y dynamics. The perturbing vector field δF is now taken to be a coupling $(\Psi_X, \Psi_Y)^T$ between the two systems. The full dynamical system is then given by

$$\begin{aligned}\frac{dX}{dt} &= F_X(X) + \Psi_X(X, Y)\\ \frac{dY}{dt} &= F_Y(Y) + \Psi_Y(X, Y).\end{aligned} \tag{11}$$

Writing (11) in terms of observables using the chain rule, we have

$$\begin{aligned}\frac{dA(X, Y)}{dt}&\\ &= (L_X(X, Y) + L_Y(X, Y))A(X, Y)\\ &= (F_X(X) + \Psi_X(X, Y)).\nabla_X A(X, Y)\\ &\quad + (F_Y(Y) + \Psi_Y(X, Y)).\nabla_Y A(X, Y)\end{aligned}$$

where ∇_X and ∇_Y denote the gradients with respect to the variables in X and in Y respectively, $L_X = (F_X + \Psi_X)\nabla_X$ and $L_Y = (F_Y + \Psi_Y)\nabla_Y$. For simplicity of presentation, for now we consider the case where $\Psi_X(X, Y) = \Psi_X(Y)$ and $\Psi_Y(X, Y) = \Psi_Y(X)$. We will come back to the general case later.

3.1. *Response theory*

The response of the unperturbed system defined by $(F_X, F_Y)^T$ to the coupling $(\Psi_X, \Psi_Y)^T$ can be calculated by taking $L_0 = F_X(X).\nabla_X + F_Y(Y).\nabla_Y$ and $L_1 = \Psi_X(Y).\nabla_X + \Psi_Y(X).\nabla_Y$ and calculate the different response terms in Eq. (10).

The nth order contribution $\delta^{(n)}\rho(A)$ to the response is calculated by integrating the impact of all n-time couplings over the times between the coupling interactions:

$$\int_0^\infty d\tau_1 \ldots \tau_n \sum_{\substack{i_1,\ldots,i_n\\ \in\{X,Y\}}} \delta^{(n)}\rho(A|i_1, \tau_1; \ldots ; i_n, \tau_n)$$

with

$$\begin{aligned}\delta^{(n)}\rho(A|i_1,\tau_1;\ldots;i_n,\tau_n) \\ &= \int \rho_0(dx)(L_{1,i_1}\Pi_0(\tau_1))\ldots \\ &\quad\times (L_{1,i_n}\Pi_0(\tau_n))A(x)\end{aligned}$$

where $x = (X, Y)^T$ and $L_{1,i}$ represents an interaction affecting the X or Y subsystem, depending on the subscript:

$$\begin{aligned}L_{1,X} &= \Psi_X(Y)\nabla_X \\ L_{1,Y} &= \Psi_Y(X)\nabla_Y.\end{aligned} \tag{12}$$

As we have discussed for Eq. (10), the nth order response can be seen as an integration of the effect infinitesimal perturbations have over a certain timespan. Since the perturbation is now given by the coupling, the response can be seen as a sequence of couplings that are activated subsequently (i_1 to i_n) and the times between the interactions (τ_1 to τ_n).

If one chooses as observable a function A_X that is only dependent on X, all response contributions up to second order are $\delta^{(1)}\rho(A_X|X,\tau)$, $\delta^{(2)}\rho(A_X|Y,\tau_1;X,\tau_2)$ and $\delta^{(2)}\rho(A_X|X,\tau_1;X,\tau_2)$. Processes ending in a coupling from X to Y vanish, as they end in a ∇_Y operator acting on $A_X(X)$. The first order term $\delta^{(1)}\rho(A|X,\tau)$ describes the effect an infinitesimally short coupling $\Psi_X(Y)$ from Y to X has over a time span τ. Such a process is graphically represented in Fig. 1. The full expression for the first order response function

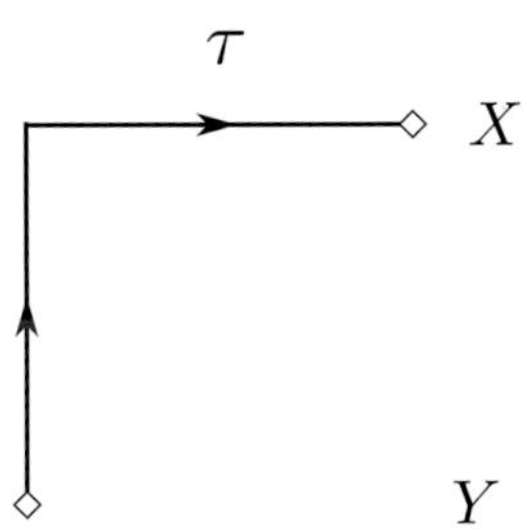

Fig. 1. Diagram for the response term $\delta^{(1)}\rho(A_X|X,\tau)$.

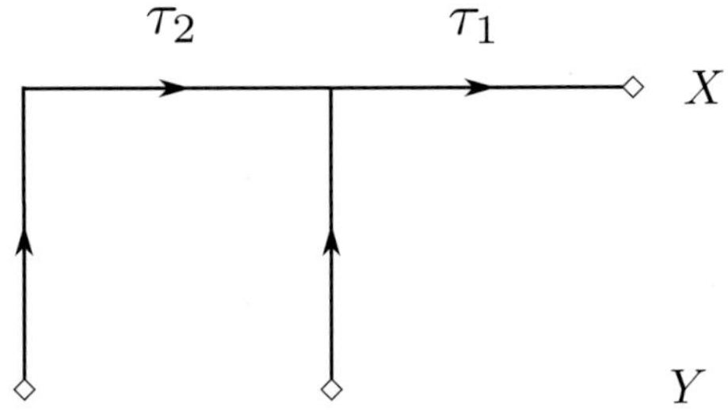

Fig. 2. Diagram for $\delta^{(2)}\rho(A_X|Y,\tau_2;X,\tau_1)$.

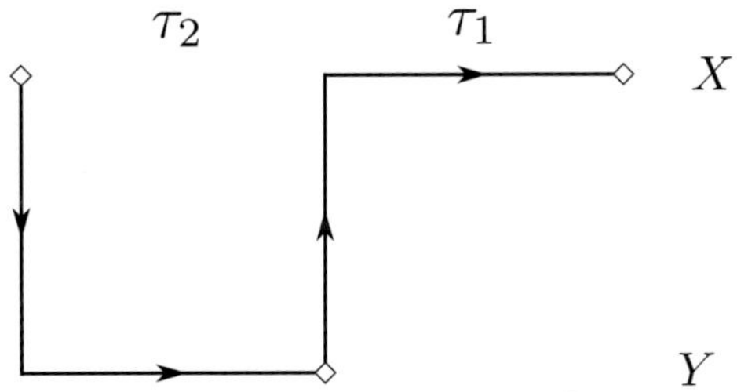

Fig. 3. Diagram for $\delta^{(2)}\rho(A_X|X,\tau_2;X,\tau_1)$.

is given by

$$\begin{aligned}\delta^{(1)}\rho(A|X,\tau) \\ &= \int \rho_0(dx)L_X\Pi(\tau)A_X(x) \\ &= \rho_{0,Y}(\Psi_X(Y))\rho_{0,X}(\nabla_X A_X(f^\tau(X))).\end{aligned}$$

The second order effects result from two distinct processes, one with two $Y \to X$ interactions, and one with a $X \to Y$ interaction followed by a $Y \to X$ interaction. These two processes are represented in Figs. 2 and 3. The corresponding $\delta^{(2)}\rho(A_X|Y,\tau_1;X,\tau_2)$ and $\delta^{(2)}\rho(A_X|X,\tau_1;X,\tau_2)$ terms can be written as

$$\begin{aligned}\delta^{(2)}\rho(A_X|Y,\tau_1;X,\tau_2) \\ &= \rho_{0,Y}\left(\nabla_Y\Psi_X(f^{\tau_1}(Y))\right) \\ &\quad\times \rho_{0,X}\left(\Psi_Y(X)\nabla_X(A_X\circ f^{\tau_2})(f^{\tau_1}(X))\right) \\ \delta^{(2)}\rho(A_X|X,\tau_1;X,\tau_2) \\ &= \rho_{0,Y}\left(\Psi_X(Y)\Psi_X(f^{\tau_1}(Y))\right) \\ &\quad\times \rho_{0,X}\left(\nabla_X(\nabla_X(A_X\circ f_2^\tau)(f^{\tau_1}(X))\right).\end{aligned}$$

Since we perturb around the uncoupled system, the unperturbed measure ρ_0 is a product of invariant measures $\rho_{0,X}$ and $\rho_{0,Y}$ on the X and Y subsystems. For this reason and since the operators in (12) are products of a multiplication and derivation operators that commute whenever the dependence is on different variables, each response term can be written as a product of a $\rho_{0,X}$ and $\rho_{0,Y}$ average.

As we have shown in (Wouters and Lucarini, 2012), if one collects these first and second order responses to the coupling Ψ, an identical change in expectation values from the unperturbed ρ_0 can be obtained by adding a Y-independent perturbing operator $L_{1,p}$ to the uncoupled L_0. We have shown that it is possible to construct explicitly a surrogate non-Markovian dynamics for the X variables able to describe up to second order the effect of coupling:

$$\frac{dX(t)}{dt} = F_X(X(t)) + M + \sigma(t) + \int_0^\infty d\tau h(\tau, X(t-\tau)) \quad (13)$$

where $M = \rho_{0,Y}(\Psi_X)$ is an averaged version of the Y to X coupling, σ is a stochastic term, mimicking the two time correlation properties of the unresolved variables and h is a memory kernel that introduces the non-Markovianity. The memory effect is present due to the finite time scale difference between resolved and unresolved variables. The M term derives from $\delta^{(1)}\rho(A_X|X,\tau)$, the σ term from $\delta^{(2)}\rho(A_X|X,\tau_1;X,\tau_2)$ and the h term from $\delta^{(2)}\rho(A_X|Y,\tau_1;X,\tau_2)$.

Let us first focus on the M term, which is given by $M = \rho_{0,Y}(\Psi(Y))$. By adding this term to the uncoupled dynamical system defined by F_X, we obtain a new model dynamical system, which is a small perturbation around F_X. The expectation values under the dynamics of this first order model system can hence be derived from the response of the system defined by F_X to the perturbation by M. As can be derived from Eq. (10), the first order response of $\rho_{0,X}$ to a constant perturbation M is exactly equal to the first order response to the coupling Ψ.

Similarly, the two second order terms σ and h in the surrogate dynamics Eq. (13) make sure that the expectation values in the model system agree to the expectation values under the fully coupled system (11) agree up to second order in the coupling Ψ. The σ term mimics the correlations present in the fluctuations of the coupling from its uncoupled mean:

$$\langle\sigma_j(t)\sigma_l(t+\tau)\rangle = \rho_Y(\Psi'_X(Y)\Psi'_X(f_Y^\tau(Y))), \quad (14)$$

$$\langle\sigma_j(t)\rangle = 0$$

$$\Psi'_X(Y) = \Psi_X(Y) - \rho_Y(\Psi_X).$$

Here f_Y^τ denotes the uncoupled evolution of Y generated by F_Y. The kernel h represents the memory effected by the presence of unresolved variables. It is given by

$$h(\tau, X) = \Psi_Y(X)\rho_Y(\nabla_Y\Psi_X(f_Y^\tau(Y))). \quad (15)$$

We do not enter into the details of this derivation, which requires an expression for the response of a dynamical system to non-Markovian forcings. This response equation and more details on the derivation of the surrogate dynamics can be found in a previous article (Wouters and Lucarini, 2012).

It should be noted that the choice of parametrization is not unique. Any time-dependent forcing σ with the correct two-point time-correlations will give the right response up to second order. Also for the memory term there is some freedom. One can also use

$$\frac{dX(t)}{dt} = F_X(X(t)) + M + \sigma(t) + \int_0^\infty d\tau h\left(\tau, f_X^{(t-\tau)}(X_0)\right) \quad (16)$$

as $X(t-\tau)$ is to zeroth order in Ψ equal to $f_X^{(t-\tau)}(X_0)$. Hence the difference between the two parametrizations will be of order Ψ^3.

If the coupling functions Ψ_X and Ψ_Y are allowed to be dependent on both X and Y, the above analysis can still be carried out. For the case of separable couplings $\Psi_X(X,Y) = \Psi_{X,1}(X)\Psi_{X,2}(Y)$ and $\Psi_Y(X,Y) = \Psi_{Y,1}(X)\Psi_{Y,2}(Y)$ the average term becomes X dependent and the noise term becomes multiplicative instead of additive. The surrogate dynamics are now given by

$$\dot{X} = F_X(X) + \Psi_{X,1}(X) \times \left(M' + \sigma'(t) + \int_0^\infty d\tau h'(\tau, X(t-\tau))\right)$$

with $M'(X) = \rho_{Y,0}(\Psi_{X,2})$, $\langle \sigma'(t)\sigma'(t+\tau)\rangle = \rho_{0,Y}(\Psi_{X,2}(Y)\Psi_{X,2}(f^\tau(Y)))$ and

$$h' = \Psi_{Y,1}(X(t-\tau))\Psi_{X,1}(X) \times \rho_{0,Y}(\Psi_{Y,2}(Y)\nabla_Y \Psi_{X,2}(f_Y^\tau(Y))).$$

An expression for more general couplings can be derived be decomposing the coupling functions into a basis of separable functions.

3.2. *Mori-Zwanzig projection operator*

We now do a calculation in the style of the Mori-Zwanzig one in Section 2.1 for the dynamical system given in Eq. (11) and for observables A_X that only depend on the relevant variables X. We start the dynamical system at a time $-t$ in the past. The initial condition X_0 of the resolved variables is assumed to be known exactly. The coupled system is assumed to have relaxed into its steady state ρ. The Y variable is hence distributed according to the conditional distribution of ρ given X_0. Expectation values of observables of Y are given by

$$\frac{\int dY B(Y)\rho(X_0,Y)}{\int dY \rho(X_0,Y)}.$$

Since the coupling in the dynamical system corresponding to ρ is small, the invariant measure ρ is a perturbation around the uncoupled product measure $\rho_{0,X} \otimes \rho_{0,Y}$, where $\rho_{0,X}$ and $\rho_{0,Y}$ correspond to the unperturbed vector fields F_X and F_Y. Hence to zeroth order the conditional distribution of Y is simply $\rho_{0,Y}$.

We now try to understand what the expected short term behavior of X is, given these initial conditions. As in Section 2.1, we first perform a projection of the evolution equation of A_X, in order to separate the X and Y variables. As projection, we choose the conditional expectation given by $\rho_{0,Y}$, i.e. $\mathcal{P}A = \int dY A(X,Y)\rho_{0,Y}(Y)$. Applying this projection to the time derivative of A_X decomposes the coupling $\Psi_X(Y)$ into its long term averages and fluctuations around it. We the apply an expansion in terms of Ψ to the evolution of the fluctuations.

The time derivative $(d/dt\, A_X)(X,Y,t)|_{t=0}$ of A_X is given by

$$\begin{aligned}(L_X A_X)(X,Y) &= ((\mathcal{P}L_X + \mathcal{Q}L_X)A_X)(X,Y) \\ &= (F_X(X) + \rho_Y(\Psi_X))\nabla_X A_X(X) \\ &\quad + (\Psi_X(Y) - \rho_Y(\Psi_X))\nabla_X A_X(X). \end{aligned} \tag{17}$$

We now want to find a formal solution for $\Psi_X(Y) - \rho_Y(\Psi_X)$ that we can insert into the previous equation.

The evolution of Ψ_X is given by

$$\begin{aligned}\Psi_X(Y) &= e^{t(L_X+L_Y)}\Psi_X(X_0, Y_0, -t) \\ &= e^{t(L_X+L_Y)}\Psi_X(Y_0).\end{aligned}$$

Making use of the decomposition of the Liouvillian $L = L_X + L_Y$ into $L_0(X_0,Y_0) = F_X(X_0)\nabla_X + F_Y(Y_0)\nabla_Y$ and $L_1(X_0,Y_0) = \Psi_X(Y_0)\nabla_X + \Psi_Y(X_0)\nabla_Y$, we derive by repeated use of Eq. (3) that

$$\begin{aligned}\Psi_X(Y) &= e^{t(L_0+L_1)}\Psi_X(Y_0) \\ &= e^{tL_0}\Psi_X(Y_0) \\ &\quad + \int_0^t d\tau e^{(t-\tau)L_0} L_1 e^{\tau L_0}\Psi_X(Y_0) \\ &\quad + O(L_1^2).\end{aligned} \tag{18}$$

Inserting this equation in (17), we get

$$\left(\frac{d}{dt}A_X\right)(X, X_0, Y_0, t)|_{t=0}$$
$$= (F_X(X) + \rho_Y(\Psi_X) + \tilde{\sigma}(t, Y_0))\, \nabla_X A_X(X)$$
$$+ \left(\int_0^t d\tau \tilde{h}(t, \tau, X_0, Y_0)\right) \nabla_X A_X(X))$$

where

$$\tilde{\sigma}(t, Y_0) = e^{tF_Y(Y_0)\nabla_Y}\Psi_X(Y_0) - \rho_Y(\Psi_X)$$
$$\tilde{h}(t, \tau, X_0, Y_0) = \Pi_0(t-\tau)L_1(X_0, Y_0) \times \Pi_0(\tau)\Psi_X(Y_0)$$

Due to the commutation of $F_X\nabla_X$ and $F_Y\nabla_Y$, we have that

$$\tilde{h} = \left(e^{(t-\tau)F_X\nabla_X}\Psi_Y(X_0)\right) \times e^{(t-\tau)F_X\nabla_X}\nabla_Y e^{\tau F_Y\nabla_Y}\Psi_X(Y_0).$$

Now we can make use of the fact that Y is distributed according to ρ_Y, the invariant measure under the flow generated by F_Y. The average of $\tilde{\sigma}$ is then zero and the auto-correlation is equal to that of σ given before in Eq. 14:

$$\rho_Y(\tilde{\sigma}(t, Y_0)) = 0$$

$$\rho_Y\left(\tilde{\sigma}(t, Y_0)\tilde{\sigma}(t+\tau, Y_0)\right) = \rho_Y\left(\Psi_X(Y_0)e^{\tau F_Y\nabla_Y}\Psi_X(Y_0)\right)$$

and

$$\rho_Y(\tilde{h}) = \Psi_Y(f_X^{t-\tau}(X_0))\rho_Y\left(\nabla_Y e^{\tau F_Y\nabla_Y}\Psi_X(Y_0)\right). \quad (19)$$

This gives us the memory term of Eq. (16).

This derivation shows that the surrogate dynamics that we have derived from the response theory corresponds to the statistical short term behavior of the resolved variables, under the assumption that the system is initially distributed according to the invariant measure. The Mori-Zwanzig projection operator method would have resulted in the exact time derivative $\dot{A}_X$, given the exact initial values X_0 and Y_0. Here we have demonstrated that by expanding this exact time derivative in orders of the perturbation, the time derivative of the parametrized equations equals to second order the exact time derivative, averaged over ρ_Y.

4. Discussion

Our results rely on the presence of a weak coupling, and not on the presence of a large time scale separation. Therefore, they may be of interest for a large number of complex systems. As a specific application, we wish to propose the derivation of parametrizations in geophysical fluid dynamics, where the simulation of long-term properties is the goal of climate modelling and the simulation of short-term behaviour is the problem of interest for weather forecast models. Geophysical fluid systems feature a very large range of time and space scales (Lucarini, 2011; Vallis, 2006), so parametrizations of unresolved small scales are necessary in numerical experiments (Orrell, 2003; Wilks, 2005; Imkeller and Storch, 2001; Hasselmann, 1976; Arnold, 2001; Saltzman, 2001). Many of the approaches used up to now have been inspired by techniques coming from statistical mechanics, where a time scale separation between microscopic and macroscopic processes is often justifiable. In geophysical fluid systems, however, there is no such clear separation between time scales and hence the requirements for disregarding the memory term do not hold. Whereas deterministic and stochastic parametrizations are by now common in geophysical fluid dynamical models (see (Palmer and Williams, 2009) for an overview and (Bony and Emanuel, 2001; Lin and Neelin, 2003) for some specific examples), memory effects are not incorporated. A final motivation for our current study of the connection between the long term statistics and dynamics is the increasing convergence between climate and weather prediction models. On the one hand there is the growing practice of benchmarking climate models by testing their prediction skill for weather forecast (Rodwell and Palmer, 2007), while on the other hand there is an increasing interest in seamless prediction models (Palmer *et al.*, 2008), that have a prediction range from days to years and longer.

While usually the scholarly literature focuses on stochastic parametrizations as crucial tools in this direction (Palmer and Williams, 2009) , the present work underlines that the consideration of — usually neglected — memory effects is as important for achieving this goal. Following some ideas presented in a recent paper by Ruelle (2012) on hydrodynamic turbulence, we speculate that our approach could be useful also for describing systematically the effect of coupling between different scales of motion in a turbulent fluid.

We wish to make a final remark on the main findings of this study. The right-hand side of Eq. (16) contains, as discussed above, a stochastic term. Therefore, the invariant measure of the dynamical system representing the surrogate dynamics is absolutely continuous with respect to Lebesgue. As discussed by Ruelle (1999, 2009), Abramov and Majda (2007), Lucarini (2008, 2009) and Lucarini and Sarno (2011), this implies that the fluctuation-dissipation theorem can be applied, relating response and fluctuations. Since the invariant measure of the surrogate dynamics agrees, up to second order, with the projected measure of the X variables in the full system and, as we have shown in this work, also the dynamics agrees closely with that of the full system, the response of the model should be close to that of the full system. The results therefore suggest the applicability of the fluctuation-dissipation theorem to the projected dynamics, as proposed using a different point of view in (Colangeli *et al.*, 2012). In particular, our results address the concerns expressed in (Lucarini and Sarno, 2011) on the applicability of the FDT in the context of climate dynamics and support the relatively positive findings obtained in this direction by Langen and Alexeev (2005), Gritsun and Branstator (2007), Abramov *et al.* (2005) and Cooper and Haynes (2011).

5. Conclusions

In this study we have first presented a simple formal derivation of the Ruelle response theory (Ruelle, 1997, 2009; Lucarini, 2008) describing how the statistical properties of Axiom A systems are changed when the underlying dynamics is altered. We do not address issues of convergence, but simply present a simple way of deriving the results, under the assumption that the integrals involved converge. We have shown that it is possible to obtain directly the exact expression for the expectation value of an observable A computed according to the invariant probability measure of the altered dynamics as the expectation value of another suitably defined observable computed according to the invariant probability measure of the unperturbed dynamics. Such an expression can be expanded at all orders of perturbations, with the general term of order n corresponding exactly to the nth order term obtained by Ruelle through perturbative expansion. Unsurprisingly, the exact result obtained through direct calculation agrees with what was derived by Ruelle through summation of the perturbative series (Ruelle, 1997).

The Dyson expansion has been instrumental in approaching the problem of studying multilevel systems, by providing a peturbative expansion for the Mori-Zwanzig projection operator (Zwanzig, 2001). Denoting by X the subset of variables we are interested into and by Y the subset of variables we want to project out, we have derived the effective projected dynamics describing the evolution of an observable of the X variables only up to second order of perturbation. Such a dynamics is identical to the surrogate dynamics for the X variables derived in (Wouters and Lucarini, 2012) by imposing, using the Ruelle response theory to describe the impact of the coupling of the X and Y variables, that the expectation value of any observable $A = A(X)$ evaluated on the invariant measure corresponding to the surrogate dynamics agrees up to second order of perturbation to its expectation value evaluated over the complete (X, Y) system. It is important to note that, as discussed in (Wouters and Lucarini, 2012), the surrogate dynamics is not unique, because we require agreement only up to second order. This result provides a connection between the Mori-Zwanzig and Ruelle formalisms, which are seemingly different, the first one pertaining to trajectories, the second one to expectation values: we have that if we are able to follow closely (on average) the individual trajectories, we are also able to model effectively the long-term statistical properties. Such a link between our ability to represent, in some sense, equally well, local and global properties in the phase space strongly relies on the fact that the projection operator leads to introducing a stochastic term and a memory term: the price we have to pay for neglecting the Y degrees of freedom and still retaining a satisfactory representation of the X dynamics on short and long time scales is going from a deterministic representation in terms of ordinary differential equations to an stochastic representation where integro-differential operators are involved. In particular, the consideration of memory effects marks the difference between what is discussed in this contribution and the classic method of averaging (Arnold, 2001; Kifer, 2004), which assumes that there is a vast time-scale separation between the two systems X and Y, so that memory effects are negligible. Our approach, instead, relies on the presence of a relatively weak coupling between the X and Y variables.

So far, we have been able to prove by direct calculation such a correspondence between the Mori-Zwanzig and Ruelle approaches only up to second order, but it is reasonable to conjecture that the same applies at any order n of perturbation. If this is true, and taking the limit of $n \to \infty$, one would get that the exact Mori-Zwanzig projected dynamics provides the unique surrogate dynamics for the X variables which is perfectly statistically compatible with the full (X, Y) system for any observables of the X variables only. Therefore, extending the proof we have given to all orders of perturbations would be extremely relevant.

Acknowledgments

The research leading to these results has received funding from the European Research Council under the European Community's Seventh Framework Programme (FP7/2007-2013) through the Starting Investigator Grant NAMASTE — Thermodynamics of the Climate System/ERC Grant agreement No. 257106. The authors acknowledge the support of CLISAP.

References

Abramov, R. V., 2012: Suppression of chaos at slow variables by rapidly mixing fast dynamics through linear energy-preserving coupling. *Commun. Math. Sci.*, **10**(2), 595–624.

Abramov, R. V. and Majda, A., 2008: New approximations and tests of linear fluctuation-response for chaotic nonlinear forced-dissipative dynamical systems. *J. Nonlinear Sci.*, **18**, 303–341, doi: 10.1007/s00332-007-9011-9.

Abramov, R. V., Majda, A. and Kleeman, R., 2005: Information theory and predictability for low-frequency variability. *J. Atmos. Sci.*, **62**(1), 65–87.

——— and Majda, A. J., 2007: Blended response algorithms for linear fluctuation-dissipation for complex nonlinear dynamical systems. *Nonlinearity*, **20**(12), 2793–2821.

Arnold, L., 2001: Hasselmann's program revisited: The analysis of stochasticity in deterministic climate models. *Stochastic Climate Models*, **49**, 141–158.

Bony, S. and Emanuel, K. A., 2001: A parameterization of the cloudiness associated with cumulus convection: Evaluation using TOGA COARE data. *J. Atmos. Sci.* **58**(21), 3158–3183.

Cessac, B. and Sepulchre, J.-A., 2007: Linear response, susceptibility and resonances in chaotic toy models. *Physica D: Nonlinear Phenomena*, **225**(1), 13–28.

Chekroun, M. D., Kondrashov, D. and Ghil, M., 2011: Predicting stochastic systems by noise sampling, and application to the el nio-southern oscillation. *Proc. Natl. Acad. Sci.*, **108**(29), 11766–11771, doi: 10.1073/pnas.1015753108, PMID: 21730171.

Colangeli, M., Rondoni, L. and Vulpiani, A., 2012: Fluctuation-dissipation relation for chaotic non-hamiltonian systems. *J. Stat. Mech.: Theory and Experiment*, **2012**(04), L04002.

Cooper, F. C. and Haynes, P. H., 2011: Climate sensitivity via a nonparametric fluctuation-dissipation theorem. *J. Atmos. Sci.*, **68**(5), 937–953.

Evans, D. and Morriss, G., 2008: *Statistical Mechanics of Nonequilibrium Liquids*. Cambridge University Press.

Fatkullin, I. and Vanden-Eijnden, E., 2004: A computational strategy for multiscale systems with applications to Lorenz 96 model. *J. Comput. Phys.*, **200**(2), 605–638.

Gallavotti, G., 1996: Chaotic hypothesis: Onsager reciprocity and fluctuation-dissipation theorem. *J. Stat. Phys.*, **84**(5–6), 899–925.

Gallavotti, G. and Cohen, E. G. D., 1995: Dynamical ensembles in stationary states. *J. Stat. Phys.*, **80**(5–6), 931–970.

——— and Roma, F., 2006: Stationary nonequilibrium statistical mechanics. *Encyclopedia of Mathematical Physics*, **3**, 530–539.

Gottwald, G., 2010: On recent trends in climate dynamics. *AMS Gazette*, **37**(5).

Gritsun, A. and Branstator, G., 2007: Climate response using a three-dimensional operator based on the Fluctuation Dissipation theorem. *J. Atmos. Sci.*, **64**(7), 2558–2575.

Hasselmann, K., 1976: Stochastic climate models, part I. theory. *Tellus*, **28**(6), 473–485.

Imkeller, P. and Storch, J. v., 2001: *Stochastic Climate Models (Progress in Probability)*, 1st ed. (Birkhäuser, Basel).

Kifer, Y., 2004: Some recent advances in averaging. *Modern Dynamical Systems and Applications: Dedicated to Anatole Katok on his 60th Birthday*, p. 385.

Kwasniok, F., 2012: Data-based stochastic subgrid-scale parametrization: an approach using cluster-weighted modelling. *Philos. Trans. Roy. Soc. A: Mathematical, Physical and Engineering Sciences*, **370**(1962), 1061–1086.

Langen, P. L. and Alexeev, V. A., 2005: Estimating 2 CO2 warming in an aquaplanet GCM using the fluctuation-dissipation theorem. *Geophys. Res. Lett.*, **32**(23), L23708.

Lin, J. W.-B. and Neelin, J. D., 2003: Toward stochastic deep convective parameterization in general circulation models. *Geophys. Res. Lett.*, **30**(4), doi:10.1029/2002GL016203.

Lucarini, V., 2008: Response theory for equilibrium and non-equilibrium statistical mechanics: Causality and generalized kramers-kronig relations. *J. Stat. Phys.*, **131**, 543–558, doi: 10.1007/s10955-008-9498-y.

Lucarini, V., 2009: Evidence of dispersion relations for the nonlinear response of the Lorenz 63 system. *J. Stat. Phys.*, **134**, 381–400, doi: 10.1007/s10955-008-9675-z.

Lucarini, V., 2011: Modelling complexity: the case of climate science. *arxiv:1106.1265*.

Lucarini, V. and Sarno, S., 2011: A statistical mechanical approach for the computation of the climatic response to general forcings. *Nonlinear Proc. Geophys.*, **18**, 7–28.

Majda, A. J., Timofeyev, I. and Vanden Eijnden, E., 2001: A mathematical framework for stochastic climate models. *Commun. Pure Appl. Math.*, **54**(8), 891–974.

Mori, H., 1965: Transport, collective motion, and Brownian motion. *Prog. Theor. Phys.*, **33**(3), 423–455.

Orrell, D., 2003: Model error and predictability over different timescales in the Lorenz '96 systems. *J. Atmos. Sci.*, **60**(17), 2219–2228.

Palmer, T. N., J., D.-R. F., A., W. and J., R. M., 2008: Toward seamless prediction: calibration of climate change projections using seasonal forecasts. *Bull. Amer. Meteor. Soc.*, **89**, 459–470.

Palmer, T. N. and Williams, P., 2009: *Stochastic Physics and Climate Modelling*. Cambridge University Press.

Pavliotis, G. and Stuart, A., 2008: *Multiscale Methods*, Texts in applied mathematics: TAM, Springer, New York.

Reick, C. H., 2002: Linear response of the Lorenz system. *Phys. Rev. E*, **66**, 036103.

Rodwell, M. J. and Palmer, T. N., 2007: Using numerical weather prediction to assess climate models. *Quarterly J. Roy. Meteor. Soc.*, **133**(622), 129–146.

Ruelle, D., 1997: Differentiation of SRB states. *Commun. Math. Phys.*, **187**(1), 227–241.

Ruelle, D., 1998: Nonequilibrium statistical mechanics near equilibrium: computing higher-order terms. *Nonlinearity*, **11**(1), 5–18.

Ruelle, D., 1999: Smooth dynamics and new theoretical ideas in nonequilibrium statistical mechanics. *J. Stat. Phys.*, **95**(1), 393468.

Ruelle, D., 2009: A review of linear response theory for general differentiable dynamical systems. *Nonlinearity*, **22**(4), 855–870.

Ruelle, D., 2012: Hydrodynamic turbulence as a problem in nonequilibrium statistical mechanics. *Proc. Natl. Acad. Sci.*

Saltzman, B., 2001: *Dynamical Paleoclimatology, Generalized Theory of Global Climate Change*, 1st ed., Academic Press.

Vallis, G. K., 2006: *Atmospheric and Oceanic Fluid Dynamics: Fundamentals and Large-scale Circulation.* Cambridge University Press.

Wilks, D. S., 2005: Effects of stochastic parametrizations in the Lorenz '96 system. *Quart. J. Roy. Meteor. Soc.*, **131**(606), 389–407.

Wouters, J. and Lucarini, V., 2012: Disentangling multi-level systems: averaging, correlations and memory. *J. Stat. Mech.: Theory and Experiment*, **2012**(03), P03003.

Zwanzig, R., 1961: Memory effects in irreversible thermodynamics. *Phys. Rev.*, **124**(4), 983–992.

Zwanzig, R., 2001: *Nonequilibrium Statistical Mechanics.* Oxford University Press.

Chapter 5

Dynamics of Nonlinear Error Growth and the "Spring Predictability Barrier" for El Niño Predictions

Wansuo Duan* and Mu Mu*,†

*LASG, Institute of Atmospheric Physics, Chinese Academy of Sciences, Beijing 100029, China

†Key Laboratory of Ocean Circulation and Wave, Institute of Oceanology, Chinese Academy of Sciences, Qingdao 266071, China

mumu@lasg.iap.ac.cn

In this chapter, we investigate the dynamics of nonlinear error growth in El Niño predictability. The characteristics of the initial errors that cause a significant "spring predictability barrier" (SPB) for El Niño events are described by discussing several previous works associated with applications of the conditional nonlinear optimal perturbation (CNOP) in ENSO predictability studies. From this discussion, it is clear that the occurrence of a significant SPB is closely related to particular initial error patterns. The CNOP is superimposed onto El Niño events and acts as the initial error with the largest effect on the uncertainties of the predictions, ultimately inducing the most significant SPB for El Niño events. In the Zebiak–Cane model, the CNOP-type errors can be classified into two types: one possessing an SSTA pattern with negative anomalies in the equatorial central-western Pacific, positive anomalies in the equatorial eastern Pacific, and a thermocline depth anomaly pattern with positive anomalies along the equator; and another with patterns almost opposite to those of the former type. These two types of CNOP-like initial errors are also found in hindcasts of predetermined model El Niño events and observed El Niño events. The two patterns of initial errors have opposite signs and, consequently, opposite growth behaviors, which may demonstrate two dynamical mechanisms of error growth that are related to the SPB: in one case, the errors grow in a manner similar to El Niño; while in the other, the errors develop with a tendency that is opposite to El Niño. These two types of initial errors have large values concentrated in particular regions, and evidence shows that the particular regions in the CNOP-type errors may represent sensitive areas of ENSO predictions. This result suggests that the CNOP-like errors in realistic ENSO predictions could be filtered by a target observation method, and the skill in forecasting ENSO may subsequently be improved.

1. Introduction

The El Niño–Southern Oscillation (ENSO) cycle, the most well-known coupled ocean–atmosphere phenomenon, is one of the most important seasonal to interannual climate variations (Philander, 1990). Although ENSO originates and develops mainly in the tropical Pacific, its effects are felt worldwide. Knowledge about the ENSO cycle and the ability to forecast its variation supplies valuable information for agriculture, public health and safety, fisheries, forestry, and many other spheres of climate-sensitive human endeavors. In the past three decades, the ENSO phenomenon has received tremendous attention. Models ranging from simple analytical ones, through intermediate coupled numerical models, to sophisticated coupled

Climate Change: Multidecadal and Beyond
edited by Chih-Pei Chang et al.

GCMs, have simulated ENSO with varying degrees of success and have offered different theories for its origin (Wang and Fang, 1996; Jin, 1997a, b; Wang *et al.*, 1999; Wang, 2001; McCreary and Anderson 1991; Suarez and Schopf, 1988; Battisti and Hirst, 1989; Philander, 1990; Kleeman *et al.*, 1995; Picaut *et al.*, 1997; Neelin *et al.*, 1998). Some numerical models, such as that of Zebiak and Cane (Cane *et al.*, 1986; Zebiak and Cane, 1987) and the coupled GCM at the National Centers for Environmental Prediction (NCEP) (Ji *et al.*, 1994), have been used to predict the real ENSO system. Recently, the climate forecast system at the NCEP (Saha *et al.*, 2006), the seasonal forecast systems at the European Center for Medium-Range Weather Forecasts (ECMWF), the fully coupled global ocean–atmosphere general circulation model, SINTEX-F, at the Frontier Research Center for Global Change (Luo *et al.*, 2003, 2005), and the Multimodel Ensemble System (MME) at both the EU (Palmer *et al.*, 2004) and at Asian-Pacific Economic Cooperation (APEC) Climate Center (APCC), have also been developed for seasonal to interannual climate prediction. Even so, there still exist considerable uncertainties in realistic ENSO predictions (Jin *et al.*, 2008; Luo *et al.*, 2008). In particular, if the forecasts are made before and through the spring, the ENSO predictions tend to be much less successful. This low predictability has been related to the so-called "spring predictability barrier" (SPB) of ENSO (Webster and Yang, 1992; Yu and Kao, 2007).

The SPB is a well-known characteristic of ENSO forecasts (Webster and Yang, 1992; Lau and Yang, 1996; McPhaden, 2003). From a statistical perspective, it refers to the fact that most ENSO prediction models often experience an apparent drop in anomaly-persistence correlation skill across April and May (Latif *et al.*, 1994); while in terms of error growth, the SPB is a phenomenon that ENSO forecasting has large prediction error, and, in particular, prominent error growth occurs during spring when the prediction is made before that spring (Mu *et al.*, 2007a, b; Duan *et al.*, 2009; Yu *et al.*, 2009). SPBs exist in coupled and statistical models. In some cases, the SPB is even stronger in statistical models than in general circulation models (GCMs) (van Oldenborgh *et al.*, 2005).

Many works have investigated the SPB phenomenon (e.g., Walker, 1924; Webster and Yang, 1992; Moore and Kleeman, 1996; McPhaden, 2003), but the debate concerning the origin of the SPB continues. One possible cause is the rapid seasonal transition of monsoon circulation during boreal spring, a phenomenon that perturbs the Pacific's basic state when the east–west sea surface temperature (SST) gradient is at its weakest (Webster and Yang, 1992; Lau and Yang, 1996). Another notion is that the SPB is caused by the fact that the ocean–atmosphere coupling is weakest during the spring in the eastern Pacific (Webster, 1995). Other studies have argued that the SST anomalies in boreal spring are relatively small, which makes them difficult to detect and forecast in the presence of atmospheric and oceanic noise (Chen *et al.*, 1995). Samelson and Tziperman (2001) demonstrated that the SPB is an inherent characteristic of ENSO, whereas Chen *et al.* (1995, 2004) suggested that improving a model's initialization could reduce the predictability barrier. McPhaden (2003) showed that subsurface information has a winter persistence barrier and that the predictability of ENSO throughout spring can be greatly enhanced by incorporating this information into the model. In general, the cause of the SPB is still elusive; in addition, there is an urgent need for a better understanding of ENSO in order to further address the problems related to the SPB.

A number of papers have explored the SPB by studying the transient growth of the initial error and emphasizing the role of the initial errors in the SPB. Moore and Kleeman (1996) investigated the SPB by tracing the evolution of the linear singular vector (LSV) (Lorenz, 1965).

Samelson and Tziperman (2001) used the linear stability theory that is related to the LSV to study the SPB induced by initial errors. Blumenthal (1991), Xue *et al.* (1997), and Thompson (1998) also used the LSV approach to study ENSO predictability. As shown in these papers, the linear theory of singular vector is useful in predictability studies and can be used to identify the optimal transient growth of an initial perturbation in a non-self-adjoint system (Farrell and Ioannou, 1996; Farrell, 1988); however, it can only deal with sufficiently small initial errors and is thus limited in describing nonlinear evolution of finite-amplitude initial errors (Oortwijn and Barkmeijer, 1995; Mu *et al.*, 2003; Mu and Zhang, 2006).

Mu *et al.* (2003) proposed a nonlinear technique of conditional nonlinear optimal perturbation (CNOP), which considers the effect of nonlinearity and can be used to explore the evolution of finite-amplitude initial perturbations. Mu *et al.* (2007a, b) used the CNOP approach and investigated the evolution of finite-amplitude initial errors in ENSO events, demonstrating that the SPB may be a result of the combined effect of three factors: the climatological annual cycle, the El Niño event itself, and the initial error pattern. The nonlinearity increases the uncertainties of ENSO forecasting across spring. The former two factors remain in the model and are the origin of the seasonality of error growth. It has been suggested that, even if the seasonality of error growth is determined by the model, particular initial error modes are necessary to bring about the SPB (Mu *et al.*, 2007a). That is, there exists the possibility that some types of initial errors may cause extreme uncertainties in ENSO forecasting through the spring and exhibit a prominent season-dependent evolution that is related to the SPB. This is due to the seasonality of the ocean–atmosphere coupling, while other types tend to yield an unapparent season-dependent evolution. The growth of the initial errors, which result from the nonlinear instability, could play an important role in the occurrence of the SPB for ENSO. But which features display initial error patterns that exhibit a significant season-dependent evolution related to the SPB for El Niño events? What is the growth mechanism of these initial errors? How does the nonlinearity affect such optimal error modes that are associated with the SPB for El Niño events? And what information does the particular pattern of the optimal errors provide for improving ENSO's forecast skill?

In this chapter, we will answer these questions by discussing the works of Mu *et al.* (2007b), Duan *et al.* (2009), Yu *et al.* (2009), Duan and Wei (2012), and Yu *et al.* (2012). In Section 2, the CNOP approach that was used in these papers is introduced. The behaviors of the CNOP-type errors, the LSV-type errors, and other initial errors are discussed in Section 3, demonstrating that the CNOP-type errors are most likely to cause a significant SPB for El Niño events. In Section 4, hindcast experiments are introduced to show the robustness of the CNOP-like errors associated with the SPB, and their dynamical growth mechanism is described in Section 5. In Section 6, the role of the CNOP-like errors in improving ENSO forecast skill is discussed. Based on the discussions in Sections 3–6, we present a summary and a discussion in Section 7.

2. Conditional Nonlinear Optimal Perturbation

The CNOP is an initial perturbation that satisfies a given constraint and has the largest nonlinear evolution at the prediction time (described below). The CNOP approach is a natural generalization of the LSV approach to the nonlinear regime. It has been used to study the nonlinear dynamics of ENSO predictability (Mu and Duan, 2003; Duan *et al.*, 2004; Duan and Mu, 2006; Mu *et al.*, 2007a, b; Duan *et al.*, 2008) and the sensitivity of ocean circulation (Mu *et al.*, 2004; Sun *et al.*, 2005; Wu and

Mu, 2009; Terwisscha van Scheltinga and Dijkstra, 2008). Recently, the CNOP approach has also been used to determine the "sensitive area" in target observations for typhoons (Mu *et al.*, 2009), to generate the initial perturbation fields in an ensemble forecast (Mu and Jiang, 2008), and to estimate the sensitivity of the ecosystem (Mu and Wang, 2007; Sun and Mu, 2009). These studies have shown that the CNOP is a useful tool for studying weather and climate predictability. For the readers' convenience, we briefly review the CNOP approach as follows.

Let $M_{t_0,t}$ be the propagator of a nonlinear model from the initial time t_0 to t. u_0 is an initial perturbation that is superimposed on the basic state $U(t)$, which is a solution to the nonlinear model and satisfies $U(t) = M_t(U_0)$, with U_0 being the initial value of the basic state $U(t)$.

For a chosen norm $\|\cdot\|$, an initial perturbation $u_{0\delta}$ is called the CNOP if and only if

$$J(u_{0\delta}) = \max_{\|u_0\|_L \leq \delta} \|M_{t_0,t}(U_0 + u_0) - M_{t_0,t}(U_0)\| \quad (1)$$

where $\|u_0\|_L \leq \delta$ is the initial constraint defined by the chosen norm $\|\cdot\|_L$. The norm $\|\cdot\|_L$ also measures the evolution of the perturbations. Obviously, we can also investigate a situation in which the initial perturbations belong to other kinds of functional sets. Furthermore, the constraint condition could reflect some physical laws that the initial perturbation should satisfy.

The CNOP is the initial perturbation whose nonlinear evolution attains the maximum value of the cost function J at time τ (Mu *et al.*, 2003; Mu and Zhang, 2006). In predictability studies, the CNOP (rather than the LSV) represents the initial error that has the worst effect on the prediction result at the optimization time (Mu *et al.*, 2003). The CNOP has, therefore, been used to identify the initial error that causes the largest prediction error for the ENSO events (Xu and Duan, 2008) and to determine the initial error that is most likely to cause a significant SPB for the El Niño events in a theoretical model (Mu *et al.*, 2007a). With the Zebiak–Cane (ZC) model, the CNOP approach is also applied to explore the initial error that exhibits season-dependent evolution that is related to the SPB for El Niño events (Mu *et al.*, 2007b; Yu *et al.*, 2009). Based on these studies, in this chapter, we demonstrate the role of initial errors in yielding the SPB for El Niño events and analyze the characteristics of the initial errors that cause a significant SPB.

To compute the CNOP, one needs to solve the nonlinear optimization problem in Eq. (1). There exist some ready solvers to deal with minimization optimization problems. Therefore, we have to transfer the maximization problem in (1) into a minimization problem by considering the negative of the cost function. Then, solvers such as Spectral Projected Gradient 2 (SPG2) (Birgin *et al.*, 2000), Sequential Quadratic Programming (SQP) (Powell, 1982), and Limited memory Broyden–Fletcher–Goldfarb–Shanno (L-BFGS) (Liu and Nocedal, 1989) can be used to compute the CNOP. In these solvers, the gradient of the modified cost function is necessary; furthermore, the adjoint of the corresponding model is usually used to obtain the gradient. With this gradient information, running these solvers with initial guesses can find the minimum of the modified cost function [i.e., the maxima of the cost function in Eq. (1)] along the descendent direction of the gradient. In phase space, the point that corresponds to the minimum of the modified cost function is the CNOP defined by Eq. (1).

3. Dependence of the SPB on the Patterns of Initial Errors

In Mu *et al.* (2007b) and Yu *et al.* (2009), the ZC model was adopted to investigate the SPB for El Niño events. The ZC model was the first coupled ocean–atmosphere model to simulate the interannual variability of observed ENSO and has been a benchmark in the ENSO community for over two decades. The ZC model is a nonlinear anomaly model of intermediate complexity that describes anomalies about a specified seasonally-varying background, which avoids the "climate drift" problem (Zebiak and Cane, 1987). The coupled ZC model describes the essential physics of ENSO and has been widely used in predictability studies and predictions of ENSO (Blumenthal, 1991; Xue *et al.*, 1994; Chen *et al.*, 2004; Tang *et al.*, 2008). The results obtained by the ZC model are therefore instructive for ENSO predictions.

Mu *et al.* (2007b) used the ZC model to demonstrate that the CNOP-type errors cause a significant SPB for El Niño events; however, non-CNOP-type errors neither exhibit an obvious season-dependent evolution nor yield a large prediction error. The authors then showed that the occurrence of the SPB is related to the pattern of initial errors. In Yu *et al.* (2009), El Niño events with different intensities were chosen to illustrate that there are two types of CNOP-type errors that cause a significant SPB to be most likely, which demonstrates the dependence of the SPB on the initial errors. In this section, we summarize these works and emphasize the spatial characteristics of the initial errors that cause a significant SPB to be most probable.

3.1. *The CNOP-type errors*

As mentioned in Section 2, the CNOP represents the initial error that has the largest effect on the predictions, which has the potential for yielding a significant SPB for El Niño events. To reveal the SPB of El Niño events in the ZC model, Mu *et al.* (2007b) and Yu *et al.* (2009) computed the CNOP-type errors for predetermined El Niño events. For the readers' convenience, we follow Mu *et al.* (2007b) and Yu *et al.* (2009) to briefly describe the calculation of the CNOP-type errors.

The CNOP-type error, which is denoted by $u_{0\delta}$, can be obtained by solving the nonlinear optimization problem:

$$J(\vec{u}_{0\delta}) = \max_{\|\vec{u}_0\|_1 \le \delta} \left\|\vec{T}'(\tau)\right\|_2 \tag{2}$$

where $\vec{u}_0 = \left(w_1^{-1}\vec{T}'_0, w_2^{-1}\vec{h}'_0\right)$ is the non-dimensional initial error of the SSTA and the themocline depth anomaly that is superimposed on the initial state of a predetermined reference state El Niño event. $w_1 = 2°\text{C}$ and $w_2 = 50\,\text{m}$ are the characteristic scales of the SST and the thermocline depth. $\|\vec{u}_0\|_1 \le \delta$ is the constraint condition that is defined by a prescribed positive real number δ and the norm $\|\vec{u}_0\|_1 = \sqrt{\sum_{i,j}\left\{\left(w_1^{-1}T'_{0i,j}\right)^2 + \left(w_2^{-1}h'_{0i,j}\right)^2\right\}}$, where $T'_{0i,j}$ represents the dimensional initial error of the SSTA; $h'_{0i,j}$ represents the thermocline depth anomaly at different grid points; (i, j) is the grid point in the domain of the tropical Pacific with latitude and longitude, respectively, from 129.375°E to 84.375°W by 5.625° and from 19°S and 19°N by 2°. The evolution of the initial error is measured by $|\vec{T}'(\tau)\|_2 = \sqrt{\sum_{i,j}\left(T'_{i,j}(\tau)\right)^2}$. $\vec{T}'(\tau)$ represents the prediction error of the SSTA at time τ and is obtained by subtracting the SSTA of the reference state from the predicted SSTA at the prediction time τ.

In order to study the seasonal dependence of the CNOP-type error evolution that is related to the SPB, Mu *et al.* (2007b) and Yu *et al.* (2009) divided a calendar year into four seasons, starting with January to March (JFM), followed by April to June (AMJ), and so forth. Then, they calculated the slope of the curve $\gamma(t) = \|T'(t)\|_2$ for different seasons, where $T'(t)$ represents the SSTA component of the nonlinear evolution of the CNOP-type error. The slope of $\gamma(t)$, which is denoted by κ, indicates the growth tendency of the CNOP-type error for different seasons. A positive (negative) value of κ implies an increase (decrease) in the error; the larger the absolute value of κ, the faster the increase (decrease) in the error.

In Mu *et al.* (2007b) and Yu *et al.* (2009), the CNOP-type errors that cause the largest prediction uncertainty were computed first. Then, their evolutions were traced, where the reference state El Niño events were chosen from a long time series of Niño-3 indices (the SSTA averaged over the Niño-3 region) generated by the ZC model. For example, in Yu *et al.* (2009), they chose two groups of El Niño events: one group consisting of weak events, with Niño-3 indices less than 2.5°C; and the other group including strong events, with Niño-3 indices larger than 2.5°C. Furthermore, by considering that different types of El Niño events exist in nature, the four events in each group were chosen with initial warming times in January, April, July, and October. Figure 1 shows the time-dependent Niño-3 indices for these two groups of El Niño events.

In the numerical experiments of Mu *et al.* (2007b) and Yu *et al.* (2009), each of the predetermined reference state El Niño events were predicted for 12 months (i.e., the lead time was 12 months) with different start months (see Fig. 1). In addition, the CNOP-type errors, i.e., the initial errors that have the largest effect on the prediction uncertainties, were computed by the optimization problem in Eq. (2). By estimating the seasonal growth rates (as measured by the slope κ) of these CNOP-type errors, Mu *et al.* (2007b) and Yu *et al.* (2009) showed that, when the predictions are made before the AMJ season, the CNOP-type errors always exhibit significant season-dependent evolution with their largest growth rates in either the AMJ or JAS seasons and cause the largest prediction uncertainties, which yield a significant SPB (Fig. 2). Nevertheless, when the predictions start in spring, the significant growth of the CNOP-type errors mostly occurs in either the JAS or OND season. That is to say, when the predictions are made directly in the spring, the decrease in the forecast skill is the severest for the El Niño event during either the JAS or the

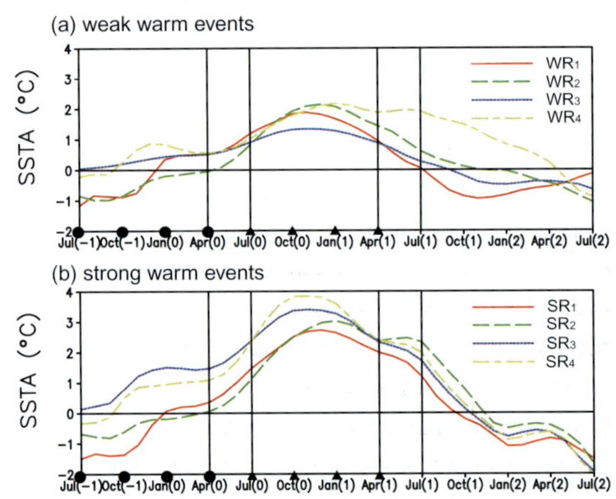

Fig. 1. Reference-state El Niño events: (a) the four weak El Niño events were measured by the Niño-3 index, which are respectively denoted by WRi, i = 1, 2, 3, and 4; and (b) the four strong El Niño events, which are signified by SRi, i = 1, 2, 3, and 4. The start months of the predictions for these El Niño events are marked on the horizontal axis, i.e., July(−1), October(−1), January(0), April(0), July(0), October(0), January(1), and April(1). This figure is from Yu *et al.* (2009).

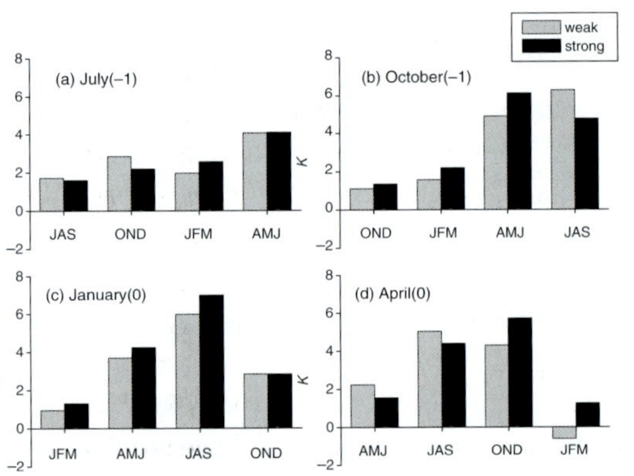

Fig. 2. Ensemble mean of the seasonal growth rate κ for the CNOP-type errors is shown for both weak and strong El Niño events. The start months for the predictions are (a) July(−1), (b) October(−1), (c) January(0), and (d) April(0). For the predictions that started in pre-boreal spring, the largest growth in the CNOP-type errors tends to be in the AMJ or JAS seasons. Although sometimes the largest growth appears in the JAS season, the error growth during the AMJ season became aggressively large and may have also caused the drastic decrease in the El Niño forecast skill across the spring, which yielded the SPB. This figure is from Yu *et al.* (2009).

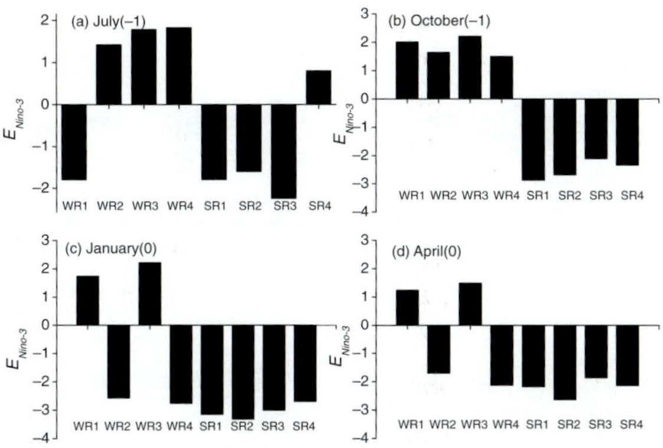

Fig. 3. Prediction errors for the eight El Niño events that are shown in Fig. 2, which are caused by their respective CNOP-type error with start months of (a) July(−1), (b) October(−1), (c) January(0), and (d) April(0). This figure is from Yu *et al.* (2009).

OND season, while the decrease seen during the spring (AMJ) is not nearly as significant. Thus, it is clear that, when the predictions are made before and through the spring, the CNOP-type errors tend to produce a significant SPB for El Niño events.

The prediction errors caused by the CNOP-type errors are shown in Fig. 3, where $E_{Ni\tilde{n}o\text{-}3}$ represents the uncertainty caused by the CNOP-type error of the El Niño prediction with a one-year lead time; this value was obtained by subtracting the Niño-3 index of the reference state from the predicted one. It can clearly be seen that the CNOP-type errors associated with some of the eight El Niño events in Fig. 1 cause a positive error in the Niño-3 SSTA and induce an over-prediction of the corresponding El Niño events; however, those of other El Niño events lead to a negative error in the Niño-3 SSTA and an under-prediction of the El Niño events. This indicates that the CNOP-type errors associated with the SPB for El Niño events can be classified into two types: those that result in positive prediction errors in the Niño-3 SSTA (i.e., the positive values of $E_{Ni\tilde{n}o\text{-}3}$), and those that induce negative prediction errors for the El Niño events (i.e., the positive values of $E_{Ni\tilde{n}o\text{-}3}$). But what characterizes the patterns of these two types of CNOP-type errors?

Yu *et al.* (2009) demonstrated that the SSTA components of the CNOP-type errors always have a localized region in the tropical Pacific and the thermocline depth anomaly components tend to have positive (or negative) anomalies along the equator. In particular, one type of CNOP error ("type-1") possesses the thermocline depth anomaly component with a deepening along the equator and the SSTA pattern of a structure with negative anomalies in the equatorial central-western Pacific and positive anomalies in the equatorial eastern Pacific; while the other type ("type-2") possesses signs almost opposite to the former. In Fig. 4, we plot the two pairs of patterns as in Fig. 2 of Yu *et al.* (2009), except for the chosen norm magnitude of 2.0 [see the optimization problem in Eq. (2)]. Here, we further examine the relationship between the positive or negative prediction errors and the two types of CNOP errors. We find that the CNOP-type errors that cause an over-prediction (i.e., positive value of $E_{Ni\tilde{n}o\text{-}3}$) possess a spatial structure that is similar to the type-1 CNOP

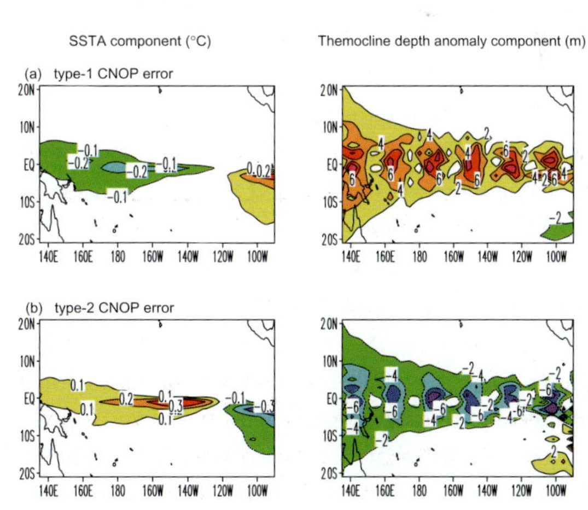

Fig. 4. The patterns for the two categories of CNOP-type errors with a chosen norm magnitude of 2.0. The left column is for the SSTA component, while the right column is for the thermocline depth anomaly. The pattern in (a) is the type-1 CNOP error, while that in (b) is the type-2 CNOP error.

error, while those that cause an under-prediction have a structure of the type-2 CNOP error. In Mu *et al.* (2007b), they also displayed several cases of the CNOP-type errors for some predetermined El Niño events. By observing their patterns, we find that, in the displayed CNOP-type errors, some have the structure of the type-1 CNOP errors, while others resemble the type-2 CNOP errors. Furthermore, the CNOP-type errors in Mu *et al.* (2007b) that are similar to the type-1 CNOP-type errors cause a positive prediction error and yield an over-prediction of El Niño events; however, those similar to the type-2 CNOP errors cause an under-prediction of the corresponding El Niño events.

As mentioned in Section 2, the CNOP is a natural generalization of the LSV in the nonlinear regime and represents the optimal initial perturbation in a nonlinear model, which certainly will yield a larger nonlinear evolution than the LSV. This indicates that, in ENSO prediction, the CNOP-type error will cause a much larger prediction error than the LSV-type error. Therefore, even if the LSV-type errors exhibit a season-dependent evolution, they will cause a less significant SPB than the CNOP-type errors, which reflects the effect of the nonlinearity. This implies that the SPB is dependent on the patterns of the initial errors. Therefore, we naturally ask: what causes the LSV-type errors to have a significantly lower SPB? And what is the difference between the CNOP-type errors and the LSV-type errors?

3.2. *The LSV-type errors*

The LSV is the fastest-growing perturbation in linearized models. Many studies have approximately regarded it as an optimal perturbation of nonlinear models, in which quite a few papers used the LSV to explore the SPB for El Niño events. Here, a comparison between the CNOP-type errors and LSV-type errors is favorable for demonstrating the dependence of the SPB on the patterns of the initial errors.

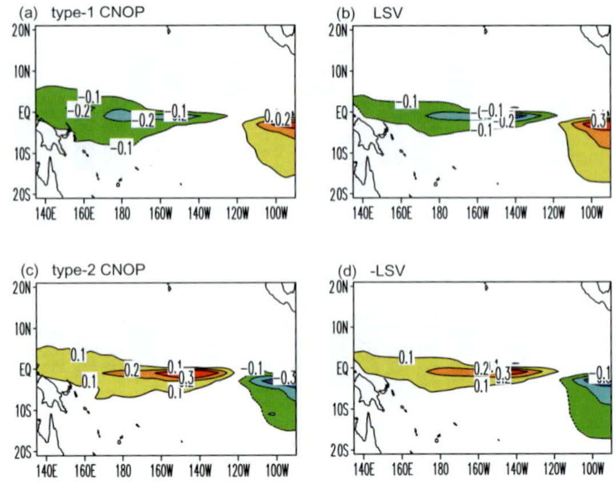

Fig. 5. SSTA components of the composite CNOP-type and LSV-type errors, with a chosen norm magnitude of 2.0. Both the CNOP-type and LSV-type errors possess a large-scale dipole structure along the equator, although the LSV-type errors have a more localized region than the CNOP-type errors. It is this difference that causes the CNOP-type errors to have a much larger prediction error than the LSV-type errors.

The LSV-type errors superimposed on the El Niño events were calculated by replacing the nonlinear model with its linearized version in the optimization problem, Eq. (2). Yu *et al.* (2009) demonstrated that the LSV-type errors with the same magnitude as the CNOP-type errors have a large-scale zonal structure that is similar to the CNOP-type errors; consequently, they can also be classified into two types, but they possess a much more localized spatial region than the CNOP-type errors (see Fig. 5). It is this difference between the CNOP and LSV patterns that causes the LSV-type errors to yield a less significant SPB for the El Niño events than the CNOP-type errors. As a comparison between CNOP and LSV, we illustrate in Fig. 6 the seasonal growth rates of the prediction errors that are caused by CNOP and LSV as in Fig. 5. It is shown that, almost in each season, the growth rates of the LSV-type errors are smaller than those of the CNOP-type errors. The LSV-type errors induce a weaker SPB for El Niño events as compared to the CNOP-type errors.

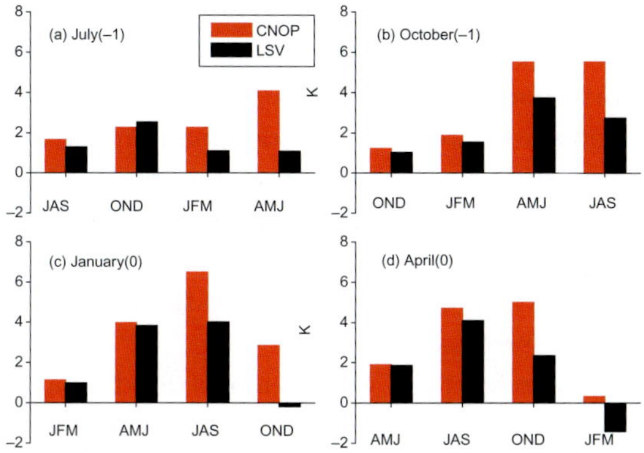

Fig. 6. Ensemble mean of the seasonal growth rate κ for the CNOP-type and LSV-type errors shown in Fig. 2 for the El Niño events. The start months of the predictions are (a) July(−1), (b) October(−1), (c) January(0), and (d) April(0). The growth rates of the LSV-type errors in almost each season are smaller than those of the CNOP-type errors.

In addition, we know that, for a prediction with a lead time and a given magnitude of initial errors, there only exists one CNOP-type error for the El Niño prediction; although the type-1 and type-2 CNOP errors are almost the opposite of each other (but not fully symmetric), the opposite of each of them cannot be regarded as a CNOP-type error for this prediction. However, the two types of LSV errors, i.e., an LSV error called "type-1" and its negative pattern, −LSV, called "type-2" (see Fig. 5), were obtained simultaneously in a prediction and are fully symmetric. The difference between CNOP and LSV reflects the effect of the nonlinearity and indicates that LSV cannot be the initial error that has the largest effect on the prediction in the nonlinear ZC model. Due to this limitation in the singular vector, we instead conclude that the CNOP-type error with respect to nonlinearity captures such an optimal initial error and is therefore more likely to cause a significant SPB.

From the above discussion, it is clear that the initial errors that cause a significant SPB most probably can be identified as two types of CNOP-type errors: one possessing an SSTA pattern with negative anomalies in the equatorial central-western Pacific, positive anomalies in the equatorial eastern Pacific, and a thermocline depth anomaly pattern with positive anomalies along the equator, which yields an over-prediction of El Niño events; and another, with patterns almost opposite to those of the former type, causes an under-prediction of El Niño events. Although the LSV-type errors also exhibit a season-dependent evolution, they have a smaller growth rate than the CNOP-type errors in almost every season and cause a much smaller prediction error. Furthermore, the CNOP-type errors often possess a much broader region than the LSV-type errors. Obviously, the occurrence of the SPB for El Niño events is related to particular patterns of initial errors. To further demonstrate this perspective, we now review the numerical results of Mu *et al.* (2007b) and Yu *et al.* (2009) on the behaviors of some other initial errors.

3.3. *Other types of initial errors*

Mu *et al.* (2007b) constructed some initial errors (which do not present particular patterns associated with CNOP and LSV) by a deterministic approach. These initial errors neither cause trivial prediction uncertainties nor yield apparent season-dependent evolution, which fails to induce an SPB for El Niño events.

Considering that random initial errors in some ENSO prediction studies are often superimposed on El Niño events to explore ENSO predictability, Yu *et al.* (2009) examined whether or not random initial errors with the same magnitude as the CNOP-type errors yield an SPB phenomenon. The random initial errors were composed of the errors at each grid point that satisfy a normal distribution and then scaled to have the same magnitude as the CNOP-type errors. By investigating the seasonal growth rates of such random initial errors, Yu *et al.* (2009) found that the random initial errors neither yield large prediction uncertainties nor exhibit any obvious season-dependent evolution; therefore, they do not cause the SPB phenomenon. Furthermore, the growth rate in each season is fairly small, which indicates that the random initial errors have trivial growth in each season.

It is clear that a particular error pattern is necessary for the occurrence of the SPB for El Niño events, though the seasonality originates from the annual cycle. The CNOP-type errors do cause a significant SPB, while the LSV-type errors yield a less significant SPB and other initial errors either yield much less significant SPB or fail to cause a SPB.

4. The CNOP-like Errors Generated by Hindcast Experiments

As mentioned in the introduction, the SPB often occurs in realistic ENSO predictions; and as discussed in subsequent sections, the initial errors could play an important role in the occurrence of the SPB for ENSO. The CNOP-type errors cause a most significant SPB for El Niño events. However, the CNOPs are generally not computed in ENSO predictions. Therefore, the question arises: are there initial errors that are similar to the CNOP-type errors in the SPB phenomenon of realistic ENSO predictions? The works of Duan and Wei (2012) might answer this question, with the forecast results of hindcast experiments.

Duan and Wei (2012) used predictions for the SST generated by the Flexible Global Ocean–Atmosphere–Land System model of the IAP/LASG (FGOALS-g) (Yan and Yu, 2012; Yu *et al.*, 2002, 2004) to investigate the season-dependent predictability of SST anomalies for El Niño/La Nina events by analyzing the forecast error growth in an imperfect model scenario. They demonstrated that, for the El Niño predictions, the prediction errors induced by both initial errors and model errors tend to have a prominent season-dependent evolution and yield a prominent SPB. Furthermore, they also reported that the initial SST errors that correspond to a significant SPB for El Niño events tend to have the dominant modes in a large-scale dipolar pattern with negative anomalies in the equatorial central-western Pacific and positive anomalies in the eastern Pacific, or vice versa, which are very similar to the two types of CNOP errors.

Therefore, it was demonstrated that, in realistic ENSO predictions generated by the FGOALS-g model, initial SSTA errors exist that are similar to the CNOP-type errors reported by Yu *et al.* (2009) and Duan *et al.* (2009), and they also correspond to a significant SPB for El Niño events. This result encourages us to filter the CNOP-like errors from the initial fields of ENSO predictions to reduce the SPB and improve the ENSO forecast's skill.

5. Dynamical Mechanisms of Error Growth in the SPB for El Niño Events

The results described in Sections 3 and 4 suggest that the initial errors that cause a significant SPB for El Niño events can be classified into two types: type-1 and -2 CNOP errors. Furthermore, they exist in realistic ENSO predictions generated by the FGOALS-g model. Under the perfect model assumption, Duan *et al.* (2009) and Yu *et al.* (2009) demonstrated that the evolutions of the two types of initial errors suggest two kinds of dynamical mechanisms for the error growth that are related to the SPB (Duan *et al.*, 2009; Yu *et al.*, 2009). One mechanism is that the initial errors grow with the same behavior as that of the El Niño events; while the other is that the initial errors evolve with an opposite behavior to the El Niño events. We know that El Niño and La Niña events usually exhibit their fastest growth during spring, since the strongest ocean–atmospheric instability in the climatological annual cycle occurs in that season (Wang and Fang, 1996). Thus, the error growth related to a significant SPB has the same growth mechanism as that of ENSO events, which in essence results from the Bjerknes (Bjerknes, 1969) positive feedback mechanism.

These two dynamical mechanisms of the error growth in the SPB for El Niño events, as shown above, were identified from perfect model predictability experiments. If the two kinds of realistic CNOP-like errors discussed in the last section can be found as having the two dynamical mechanisms, the model errors may have a small effect on the SPB. As such, we should avoid the presence of the initial errors that exhibit these two mechanisms in realistic ENSO predictions, so as to improve ENSO prediction skill. Actually, we only need to destroy the structure of and filter the CNOP-like errors to achieve the desired aim. For example, a target observation approach may be useful in managing to filter the CNOP-like errors.

6. Implications of the Location of CNOP-like Errors on the Target Observation

Given the dipole pattern of SSTA for CNOP-type errors with the location of large anomalies in the equatorial central and eastern Pacific, Yu *et al.* (2012) investigated the contribution of the location of the CNOP-type errors to the error evolution in El Niño prediction experiments and analyzed implications on the target observation.

Yu *et al.* (2012) first divided the tropical Pacific into six regions, of which one is the region that has the large SST anomalies of CNOP-type errors (i.e., the equatorial central and eastern Pacific; also Domain 5 in Fig. 7). For each region, they computed the CNOP-type errors that have the largest effect on El Niño predictions and investigated the resulting prediction errors. They found that the CNOP-type errors in Domian 5 cause aggressively large prediction error compared to those in any other regions. This suggests that the errors in Domain 5 in an initial analysis field may have the largest effect on El Niño predictions compared to those in other regions, and Domain 5 may represent the "sensitive area" of El Niño target observations.

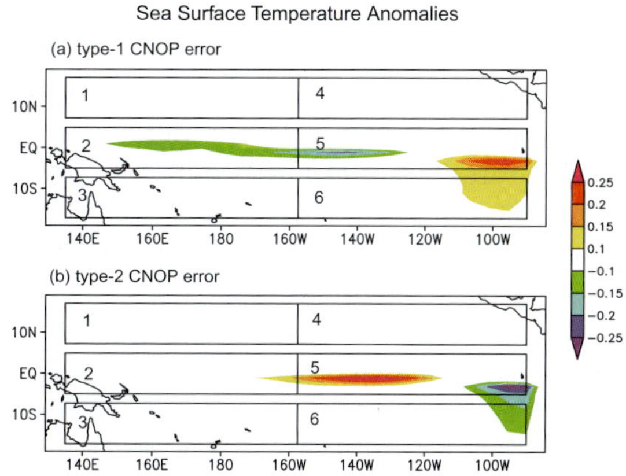

Fig. 7. Two types of CNOP errors shown in six rectangular domains denoted by Domain i (i = 1, ..., 6). Large values of both types of CNOP error occur mainly in Domain 5. This figure is from Yu *et al.* (2012).

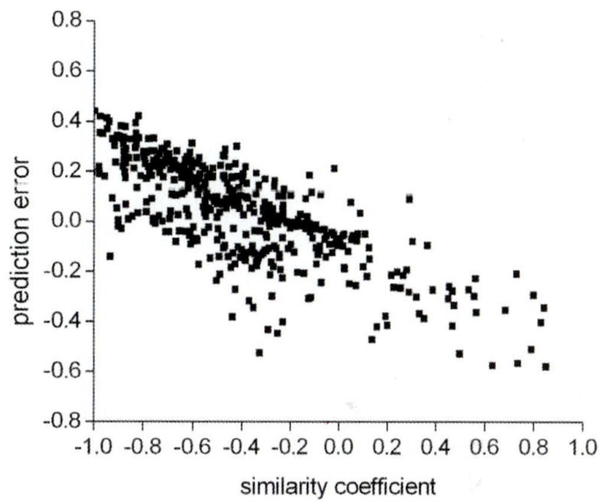

Fig. 8. Prediction errors of Niño-3 indices induced by the uncertain analysis field of a realistic hindcast experiment as a function of the similarity coefficients between the analysis field error and the CNOP-type errors.

To verify the sensitive area identified by the CNOP-type errors, Yu *et al.* (2012) further considered the initial SSTA analysis fields of ENSO realistic hindcast experiments generated by the ZC model. They found that the better the initial errors in the analysis fields correlate to the CNOP-type errors, the larger their resultant prediction errors are (see Fig. 8). It is clear that the CNOP-type errors also exist in realistic hindcast experiments generated by the ZC model. Using these initial analysis fields, Yu *et al.* (2012) investigated the validity of the "sensitive area" identified by CNOP-type errors.

For these analysis fields, Yu *et al.* (2012) replaced the initial anomalies in each of six regions in Fig. 8 with those of the "true" El Niño events (referred to here as model El Niño events) and eliminated the initial errors in each region. They found that, when the initial errors in Domain 5 are removed, the prediction errors of El Niño events are largely reduced; but the elimination of initial errors in other regions cannot aggressively reduce the prediction errors. This indicates that Domain 5, i.e., the equatorial central and eastern Pacific — also the region of large anomalies of CNOP-like errors — does indeed represent the "sensitive area" of El Niño predictions. Therefore, if we increase observations in this area and improve the initial field, the ENSO forecast skill may be greatly enhanced.

In addition, we noticed that the CNOP-type errors possess a spatial structure very similar to the optimal precursors of ENSO provided by Duan *et al.* (2012), which may shed light on by the assertion that the CNOP-type errors have the same dynamical mechanism as El Niño or La Nina. Therefore, if additional observations are increased in the sensitive area identified by the CNOP-type errors, not only can they reduce the initial errors and improve the accuracy of initial fields, but also help detect the signals for ENSO events, ultimately favoring an improvement in ENSO forecast skill.

7. Summary and Discussion

In this chapter, we have summarized the works of Mu *et al.* (2007b), Duan *et al.* (2009), Yu *et al.* (2009), Duan and Wei (2012), and Yu *et al.* (2012) on the SPB for El Niño events and demonstrated the dependence of the SPB on the patterns of the initial errors. A particular pattern of initial errors is necessary for the SPB of El Niño events to occur. More specifically, the two CNOP-type errors found in Mu *et al.* (2007b) and Yu *et al.* (2009) are the initial errors that probably cause the most significant SPB for El Niño events. As shown in Yu *et al.* (2009), one type of CNOP errors has an SSTA pattern with negative anomalies in the equatorial central-western Pacific, positive anomalies in the equatorial east Pacific, and a thermocline depth anomaly pattern with positive anomalies along the equator; while the other type possesses almost the opposite patterns. Such patterns of initial errors that are related to the SPB have also been identified in hindcast experiments, which further emphasizes the robustness of the particular pattern of the initial errors that cause a significant SPB for El Niño events.

Two types of initial errors suggest two dynamical behaviors of error growth in the SPB phenomenon for El Niño events. In fact, the two types of CNOP errors possess almost the same dynamical behavior as the El Niño and La Niña and induce the largest prediction error; thus, they can be regarded as most likely to cause a significant SPB phenomenon. The two types of LSV errors also have the same dynamical behavior as El Niño and La Nina; however, they cause a less significant SPB because of a much more localized spatial region compared to those of CNOP.

Sensitivity experiments reported in Yu *et al.* (2012) demonstrated that the localized region of the CNOP-type errors, which have large anomalies always arising in the equatorial central-eastern Pacific, is more likely to capture the "sensitive area" of ENSO predictions. This result may guide efforts to intensify observations in this area and subsequently improve ENSO prediction.

The works discussed in this chapter investigated the SPB for El Niño events from viewpoint of error growth and emphasized the role of the initial errors in the SPB for El Niño predictions. The localized region identified by the CNOP-type errors provides information on the "sensitive area" of target observations for ENSO, which needs to be further tested in realistic predictions. Of course, there are other questions to be explored. For example, some studies have used the stochastic optimal approach to investigate the predictability of ENSO (Kleeman and Moore, 1997; Moore and Kleeman, 1999). In these studies, the authors explored the role of stochastic noise forcing, which actually emphasized the effect of a kind of model error on ENSO predictability. It is known that both initial error and model error are origins of prediction uncertainties for El Niño, but what is the effect of model error on ENSO predictability? Does initial error or model error play the more important role in yielding uncertainties for ENSO predictions. Among others, these questions serve to illustrate that in-depth ENSO predictability studies are still required.

Acknowledgments

This work was jointly sponsored by the National Natural Science Foundation of China (Nos. 41230420; 41176013).

References

Battisti, D. S., and A. C. Hirst, 1989: Interannual variability in the tropical atmosphere-ocean model: influence of the basic state, ocean geometry and nonlineary. *J. Atmos. Sci.*, **45**, 1687–1712.

Birgin, E. G., J. M. Martínez, and M. Raydan, 2000: Nonmonotone spectral projected gradient methods on convex sets. *SIAM J. Optim.*, **10**, 1196–1211.

Bjerknes, J., 1969: Atmospheric teleconnections from the equatorial Pacific. *Mon. Wea. Rev.*, **97**, 163–172.

Blumenthal, M. B., 1991: Predictability of a Coupled Ocean-Atmosphere Model. *J. Climate*, **4**, 766–784.

Cane, M. A., S. E. Zebiak, and S. C. Dolan, 1986: Experimental forecasts of El Niño. *Nature*, **321**, 827–832.

Chen, D., M. A. Cane, A. Kaplan, S. E. Zebiak, and D. J. Huang, 2004: Predictability of El Niño over the past 148 years. *Nature*, **428**, 733–736.

———, S. E. Zebiak, A. J. Busalacchi, and M. A. Cane, 1995: An improved procedure for El Niño forecasting: Implications for predictability. *Science*, **269**, 1699–1702.

Duan, W. S., X. C. Liu, K. Y. Zhu, and M. Mu, 2009. Exploring the initial errors that cause a significant "spring predictability barrier" for El Niño events. *J. Geophys. Res.*, **114**, C04022, doi: 10.1029/2008JC004925.

———, and M. Mu, 2006: Investigating decadal variability of El Niño–Southern Oscillation asymmetry by conditional nonlinear optimal perturbation. *J. Geophys. Res.*, **111**, C07015, doi: 10.1029/2005JC003458.

———, ———, and B. Wang, 2004: Conditional nonlinear optimal perturbation as the optimal precursors for ENSO events. *J. Geophys. Res.*, **109**, D23105.

———, and C. Wei, 2012: The spring predictability barrier for El Niño events and its possible mechanism results from a fully coupled model. *Inter. J. Climatology*, doi: 10.1002/joc.3513.

———, H. Xu, M. Mu, 2008: Decisive role of nonlinear temperature advection in El Niño and La Niña amplitude asymmetry. *J. Geophys. Res.*, **113**, C01014, doi: 10.1029/2006JC003974.

———, Y. Yu, H. Xu, and P. Zhao, 2012: Behaviors of nonlinearities modulating El Niño events induced by optimal precursory disturbance. *Clim. Dyn.*, doi: 10.1007/s00382-012-1557-z.

Farrell, B., 1988: Optimal excitation of neutral Rossby waves. *J. Atmos. Sci.*, **45**, 163–172.

Farrell, B. F., and P. J. Ioannou, 1996: Generalized stability theory. Part II: Non-autonomous operators. *J. Atoms. Sci.*, **53**, 2041–2053.

Ji, M., A. Kumar, and A. Leetmaa, 1994: An experimental coupled forecast system at the National Meteorological Center: Some early results. *Tellus*, **46A**, 398–418.

Jin, E. K., J. L. Kinter III, B. Wang, C. K. Park, I. S. Kang, B. P. Kirtman, J. S. Kug, A. Kumar, J. J. Luo, J. Schemm, J. Shukla, and T. Yamagata, 2008: Current status of ENSO prediction skill in coupled ocean-atmosphere models. *Clim. Dyn.*, **31**, 647–664.

Jin, F. F., 1997a: An equatorial ocean recharge paradigm for ENSO. Part I: Conceptual model. *J. Atmos. Sci.*, **54**, 811–829.

———, 1997b: An equatorial ocean recharge paradigm for ENSO. Part II: A stripped-down coupled model. *J. Atmos. Sci.*, **54**, 830–847.

Kleeman, R., and A. M. Moore, 1997: A theory for the limitation of ENSO predictability due to stochastic atmospheric transients. *J. Atmos. Sci.*, **54**, 753–767.

———, ———, and N. R. Smith, 1995: Assimilation of subsurface thermal data into a simple ocean model for the initialisation of an intermediate tropical coupled ocean-atmosphere forecast model. *Mon. Wea. Rev.*, **123**, 3103–3113.

Latif, M., T. P. Barnett, M. A. Cane, M. Flu¨gel, N. E. Grahem, H. von Storch, J. S. Xu, and S. E. Zebiak, 1994: A review of ENSO prediction studies. *Clim. Dyn.*, **9**, 167–179.

Lau, K. M., and S. Yang, 1996: The Asian monsoon and predictability of the tropical ocean-atmosphere system. *Quart. J. Roy. Meteor. Soc.*, **122**, 945–957.

Liu, D. C., and J. Nocedal, 1989: On the limited memory method for large scale optimization. *Math. Program. B*, **45**(3), 503–528.

Lorenz, E. N., 1965: A study of the predictability of a 28-variable atmospheric model. *Tellus*, **17**, 321–333.

Luo, J. J., S. Masson, S. Behera, P. Delecluse, S. Gualdi, A. Navarra, and T. Yamagata, 2003: South Pacific origin of the decadal ENSO-like variation as simulated by a coupled GCM. *Geophys. Res. Lett.*, **30**, 2250, doi: 10.1029/2003GL018649.

———, ———, ———, S. Shingu, and T. Yamagata, 2005: Seasonal climate predictability in a coupled OAGCM using a different approach for ensemble forecasts. *J. Climate*, **18**, 4474–4497.

———, ———, ———, and T. Yamagata, 2008: Extended ENSO predictions using a fully coupled ocean-atmosphere model. *J. Climate*, **21**, 84–93.

McCreary, J. P., and D. L. T. Anderson, 1991: An overview of coupled ocean-atmosphere models of El Niño and the Southern Oscillation. *J. Geophys. Res.*, **96**, 3125–3150.

McPhaden, M. J., 2003: Tropical Pacific ocean heat content variations and ENSO persistence barriers. *Geophys. Res. Lett.*, **30**, doi: 10.1029/2003GL016872.

Moore, A. M., and R. Kleeman, 1996: The dynamics of error growth and predictability in a coupled model of ENSO. *Quart. J. Roy. Meteor. Soc.*, **122**, 1405–1446.

———, and ———, 1999: Stochastic forcing of ENSO by the intraseasonal oscillation. *J. Climate*, **12**, 1199–1220.

Mu, M., and W. S. Duan, 2003: A new approach to studying ENSO predictability: Conditional nonlinear optimal perturbation. *Chinese Sci. Bull.*, **48**, 1045–1047.

———, ———, and B. Wang, 2003: Conditional nonlinear optimal perturbation and its applications. *Nonlinear Proc. Geoph.*, **10**, 493–501.

———, ———, and ———, 2007a: Season-dependent dynamics of nonlinear optimal error growth and El Niño-Southern Oscillation predictability in a theoretical model. *J. Geophys. Res.*, **112**, D10113, doi: 10.1029/2005JD006981.

———, and Z. N. Jiang, 2008: A new approach to the generation of initial perturbations for ensemble prediction: Conditional nonlinear optimal perturbation. *Chinese Sci. Bull.*, **53**(13), 2062–2068.

———, L. Sun, and H. A. Dijkstra, 2004: The sensitivity and stability of the ocean's thermocline circulation to finite amplitude freshwater perturbations. *J. Phys. Oceanogr.*, **34**, 2305–2315.

———, and B. Wang, 2007: Nonlinear instability and sensitivity of a theoretical grassland ecosystem to finite-amplitude perturbations. *Nonlinear Proc. Geoph.*, **14**, 409–423.

———, H. Xu, and W. S. Duan, 2007b: A kind of initial errors related to "spring predictability barrier" for El Niño events in Zebiak-Cane model. *Geophys. Res. Lett.*, **34**, L03709, doi: 10.1029/2006GL-27412.

———, and Z. Zhang, 2006: Conditional nonlinear optimal perturbations of a two-dimensional quasigeostrophic model. *J. Atmos. Sci.*, **63**, 1587–1604.

———, F. F. Zhou, and H. L. Wang, 2009: A method for identifying the sensitive areas in targeted observations for tropical cyclone prediction: conditional nonlinear optimal perturbation. *Mon. Wea. Rev.*, **137**, 1623–1639.

Neelin, J. D., D. S. Battisti, A. C. Hirst, F. F. Jin, Y. Wakata, T. Yamagata, and S. E. Zebiak, 1998: ENSO theory. *J. Geophys. Res.*, **103**, 14261–14290.

Oortwijn, J., and J. Barkmeijer, 1995: Perturbations that optimally trigger weather regimes. *J. Atmos. Sci.*, **52**, 3952–3944.

Palmer, T. N., *et al.*, 2004: Development of a European Multimodel Ensemble System for seasonal to inter-annual prediction (DEMETER). *Bull. Amer. Meteor. Soc.*, **85**, 853–872.

Philander, S. G., 1990: *El Niño, La Niña, and the Southern Oscillation.* Academic Press, San Diego, CA, 293 pp.

Picaut, J., F. Masia, and Y. du Penhoat, 1997: An advective-reflective conceptual model for the oscillatory nature of the ENSO. *Science*, **277**, 663–666.

Powell, M. J. D., 1982: VMCWD: A Fortran subroutine for constrained optimization. *DAMTP Report 1982/NA4*, University of Cambridge, England.

Saha, S., S. Nadiga, C. Thiaw, J. Wang, W. Wang, Q. Zhang, H. M. Van den Dool, H. L. Pan, S. Moorthi, D. Behringer, D. Stokes, M. Peña, S. Lord, G. White, W. Ebisuzaki, P. Peng, and P. Xie, 2006: The NCEP climate forecast system. *J. Climate*, **19**(15), 3483–3517.

Samelson, R. G., and E. Tziperman, 2001: Instability of the chaotic ENSO: The growth-phase predictability barrier. *J. Atmos. Sci.*, **58**, 3613–3625.

Suarez, M. J., and P. S. Schopf, 1988: A delayed action oscillator for ENSO. *J. Atmos. Sci.*, **45**, 3283–3287.

Sun, G. D., and M. Mu, 2009: Nonlinear feature of the abrupt transitions between multiple equilibria states of an ecosystem model. *Adv. Atmos. Sci.*, **26**(2), 293–304, doi: 10.1007/s00376-009-0293-8.

Sun, L., M. Mu, D. J. Sun, and X. Y. Yin, 2005: Passive mechanism of decadal variation of thermohaline circulation. *J. Geophys. Res.*, **110**, C07025, doi: 10.1029/2005JC002897.

Tang, Y., Z. Deng, X. Zhou, Y. Cheng, and D. Chen, 2008: Interdecadal variation of ENSO predictability in multiple models. *J. Climate*, **21**, 4811–4833.

Terwisscha van Scheltinga, A. D., and H. A. Dijkstra, 2008: Conditional nonlinear optimal perturbations of the double-gyre ocean circulation. *Nonlinear Proc. Geoph.*, **15**, 727–734.

Thompson, C. J., 1998: Initial Conditions for optimal growth in a coupled ocean-atmospheric model of ENSO. *J. Atmos. Sci.*, **35**, 537–557.

van Oldenborgh, G. J., M. A. Balmaseda, L. Ferranti, T. N. Stockdale, and D. L. T. Anderson, 2005: Evaluation of atmospheric fields from the ECMWF seasonal forecasts over a 15-year period. *J. Climate*, **18**, 3250–3269.

Walker, G. T., 1924: Correlation in seasonal variations of weather IX: A further study of world weather. *Mem. India Meteorol. Dep.*, **24**, 275–332.

Wang, B., A. Barcilon, and Z. Fang, 1999: Stochastic dynamics of El Niño-Southern Oscillation. *J. Atmos. Sci.*, **56**, 5–23.

———, and Z. Fang, 1996: Chaotic oscillation of tropic climate: A dynamic system theory for ENSO. *J. Atmos. Sci.*, **53**, 2786–2802.

Wang, C., 2001: A unified oscillator model for the El Niño-Southern Oscillation. *J. Climate*, **14**, 98–115.

Webster, P. J., 1995: The annual cycle and the predictability of the tropical coupled ocean-atmosphere system. *Meteorol. Atmos. Phys.*, **56**, 33–55.

———, and S. Yang, 1992: Monsoon and ENSO: Selectively interactive systems. *Quart. J. Roy. Meteor. Soc.*, **118**, 877–926.

Wu, X. G., and M. Mu, 2009: Impact of horizontal diffusion on the nonlinear stability of thermohaline circulation in a modified box model. *J. Phys. Oceanogr.*, **39**, 798–805.

Xu, H., and W. S. Duan, 2008: What kind of initial errors cause the severest prediction uncertainty of El Niño in Zebiak-Cane model. *Adv. Atmos. Sci.*, **25**, 577–584.

Xue, Y., M. A. Cane, and S. E. Zebiak, 1997: Predictability of a coupled model of ENSO using singular vector analysis. Part I: Optimal growth in seasonal background and ENSO cycles. *Mon. Wea. Rev.*, **125**, 2043–2056.

———, ———, ———, and M. B. Blumenthal, 1994: On the prediction of ENSO, a study with a low-order Markov model. *Tellus*, **46A**, 512–528.

Yan, L., and Y. Q. Yu, 2012: Spring prediction barrier of ensemble forecast by statistical analysis. *Chin. J. Oceanol. Limn.*, Accepted.

Yu, J. Y., and H. Y. Kao, 2007: Decadal changes of ENSO persistence barrier in SST and ocean heat content indices 1958–2001. *J. Geophys. Res.-Atmos.*, **112**, D13106, doi: 10.1029/2006JD007654.

Yu, Y. Q., R. C. Yu, X. H. Zhang, and H. L. Liu, 2002: A flexible coupled ocean-atmosphere general circulation model. *Adv. Atmos. Sci.*, **19**, 169–190.

———, X. H. Zhang, and Y. F. Guo, 2004: Global coupled ocean-atmosphere general circulation models in LASG/IAP. *Adv. Atmos. Sci.*, **21**, 444–455.

Yu, Y. S., W. S. Duan, H. Xu, and M. Mu, 2009: Dynamics of nonlinear error growth and season-dependent predictability of El Niño events in the Zebiak-Cane model. *Quart. J. Roy. Meteor. Soc.*, in press.

———, M. Mu, W. S. Duan, and T. Gong, 2012: Contribution of the location and spatial pattern of initial error to uncertainties in El Niño predictions. *J. Geophys. Res.-Oceans*, **117**, C06018, doi: 10.1029/2011JC007758.

Zebiak, S. E., and M. A. Cane, 1987: A model El Niño-Southern Oscillation. *Mon. Wea. Rev.*, **115**, 2262–2278.

Chapter 6

An Adaptive Approach for Nonlinear and Nonstationary Data Analysis

Norden E. Huang
Research Center for Adaptive Data Analysis
Center for Dynamical Biomarkers and Translational Medicine
National Central University
Zhongli, Taiwan, ROC
Norden@ncu.edu.tw

Analyzing data from real world is a challenge; we have to face the limitations imposed by reality: nonstationarity, nonlinearity. The traditional methods cannot fully accommodate these restrictions. To alleviate this difficulty, various assumptions and approximations have often been invoked to process and analyze data. Unfortunately, methods based on the most common assumptions, linearity and stationarity, are not as effective as is expected in revealing useful information and the key information of data remains concealed most of the time.

In this chapter, we will focus on adaptive methods that can overcome many limitations of traditional methods. As will be illustrated, many of the difficulties could be traced back to the lack of logical definition of frequency, a critical physical quantity. In fact, once the frequency is properly defined and extracted from the data, many difficult tasks, such as quantification of degree of nonlinearity and nonstationarity and determination of the trend, can be carried out with ease. It is also showed that the accurate determination of the frequency has to depend on the adaptive method and the Hilbert-Huang Transform (HHT) can serve this purpose HHT, consists of Empirical Mode Decomposition (EMD) and the Hilbert Spectral Analysis (HSA), is a new adaptive method that can provide time-frequency-energy expression of data without using a priori basis, such as in wavelet. Related most recent developments, such as the Nonlinear Matching Pursuit method, Ensemble Empirical Mode Decomposition (EEMD), Instantaneous Frequency computations, Trend determination, Time-dependent Intrinsic Correlation (TIDC), density representation of Hilbert Spectrum, and the extension of the time series analysis method to multi-dimensional data, will also be discussed. These latter advances have made the HHT method much more robust and mature, and many applications of HHT are progress and had produced viable results.

1. Introduction

Data analysis is indispensable to every science and engineering endeavor; it is a critical step to convert the cold numbers to physical insights and understandings of the underlying key processes of a system's evolution. For this reason, a data analysis method should focus more on revealing physically meaningful prospective rather than on simply obtaining mathematical parameters. The data from natural phenomena are highly variable and reflect the hidden nonstationary and nonlinear processes, which poses a severe challenge to the existing data analysis methods, especially probability theory and spectral analysis methods developed for idealized conditions that are both linear and stationary. In pursue of the mathematical rigor, we are forced to make idealized assumptions and live in a pseudo-real linear and stationary state;

Climate Change: Multidecadal and Beyond
edited by Chih-Pei Chang et al.

yet, the real world is neither stationary nor linear. For example, spectral analysis is synonymous with the Fourier based analysis. As Fourier spectrum analysis and all its generalizations can only give meaningful interpretation to linear and stationary process, its application to data from nonlinear and nonstationary processes is problematical. And probability distributions can only represent global properties, which imply homogeneity (or stationarity) in the sampling population. As scientific research getting increasingly sophistic, the inadequacy of the traditional data analysis methods is becoming glaringly obvious. Even with some modifications, the available data analysis method could be applied to nonstationary but linear processes such as wavelet and Wigner-Ville distribution (see, for example, Flandrin, 1999), or nonlinear but stationary and deterministic (Kantz and Schreiber, 1997). For both nonlinear and nonstationary processes, our options are limited and our tools are totally inadequate. The only alternative is to break away from these limitations; we should let data speak for themselves so that the results could reveal the full range of consequence of the true underlying driving mechanisms. To do so, we need new paradigm of data analysis methodology without *a priori* basis to fully accommodating the variations of the physical processes. One available alternative is an adaptive data analysis method, based on the Empirical Mode Decomposition (EMD) and Hilbert Spectral Analysis, officially designated by NASA as Hilbert-Huang Transform (HHT) (Huang et al., 1998). The result is presented in a time-frequency-energy representation. In fact, the solution for analysis nonstationary and nonlinear data hinges on the definition of the true frequency. And the true frequency can only be defined through adaptive method, which would lead to quantify nonstationarity and nonlinearity, and to determine the elusive trend of nonstationary data.

Ever since HHT was introduced, it has been used widely in science and engineering research and applications. Over the years, new improvement such as the Ensemble EMD was introduced that greatly improve the robustness of the method. Further extension also made the method extended to multi-dimensional and multi-variant data sets. Based on the very properties of instantaneous frequency, new approaches to define the degree of nonlinearity and nonstationarity were also proposed. With EMD, the elusive trend also got a good definition and a method to extract it objectively. In this short review, we will summarize the method and give examples of its application. The rest of the paper is organized as follows: Section 2 will give a summary of the EMD method; Section 3 gives the time-frequency-energy presentation and Section 4, the new method for trend determination. Finally, a short conclusion is given at the end.

2. Empirical Mode Decomposition Method

The problem of the traditional data analysis method could be simply traced to the poor definition of frequency. The most fundamental definition of frequency, ω, is based on period, $\boldsymbol{T}$, of a wave:

$$\omega = \frac{\mathbf{1}}{\boldsymbol{T}}. \tag{1}$$

This definition is certainly correct dimensionally, but it is crude and could only serve as a mean value over a whole wave. All serious theoretical wave studies define frequency in term of phase function. In general, for any wave motion, there must be a smooth phase function, $\boldsymbol{\theta}$, so that we can define wave number, $\boldsymbol{k}$, and frequency as

$$\boldsymbol{k} = \frac{\partial \boldsymbol{\theta}}{\partial \boldsymbol{x}}, \quad \text{and} \quad \omega = -\frac{\partial \boldsymbol{\theta}}{\partial \boldsymbol{t}}. \tag{2}$$

Therefore, by cross differentiation, we have the wave conservation law:

$$\frac{\partial \boldsymbol{k}}{\partial \boldsymbol{t}} + \frac{\partial \omega}{\partial \boldsymbol{x}} = \mathbf{0}. \tag{3}$$

All wave motions have to follow this fundamental kinematic conservation law. For it to hold, it is obvious that both the wave number and the frequency have to have instantaneous values and also be differentiable. Based on this criterion, the constant wave number and frequency defined through Fourier analysis is problematic. Certainly the Fourier frequency and wave number satisfy the kinematic conservation, but that would be a trivial condition of zero plus zero equal to zero. Only with a well-defined instantaneous frequency can we describe the richness of the dynamical representation of the variation in frequency of the nonlinear and nonstationary waves, where the intra-wave frequency modulation is the rule rather than the exception.

As the instantaneous frequency depends on a phase function, Huang et al. (1998) showed that the necessary condition for a general time series to have physically meaningful phase function and instantaneous frequency is to represent a time series as a collection of Intrinsic Mode Function (IMF) through Empirical Mode Decomposition (EMD).

The EMD is a nonlinear time domain decomposition of a complicated data into a collection of mono-component functions. It is implemented through the following steps: For any data, we first identify all the local extrema, and connect all the local maxima by a cubic spline line as the upper envelope. Then, we repeat the procedure for the local minima to produce the lower envelope. The upper and lower envelopes should cover all the data between them. Their mean is designated as $\boldsymbol{m}_{1,1}$, and the difference between the data and $\boldsymbol{m}_{1,1}$ is the first proto-IMF (PIMF) component, $\boldsymbol{h}_{1,1}$:

$$\boldsymbol{x}(\boldsymbol{t}) - \boldsymbol{m}_{1,1} = \boldsymbol{h}_{1,1} \tag{4}$$

Technically, the mean, $\boldsymbol{m}_{1,1}$, is the medium, for it is the mean of maximum and minimum. As shown by Hou *et al.* (2009), medium is a better reference. By construction, this PIMF, $\boldsymbol{h}_{1,1}$, should satisfy the definition of an IMF, but the change of its reference frame from rectangular coordinate to a curvilinear one can cause anomalies. To eliminate such anomalies, the sifting process has to be repeated as many times as necessary to get rid of the riding waves. In the subsequent sifting process steps, $\boldsymbol{h}_{1,1}$ is treated as the data. After each repetition, we will check the results against a 'stoppage criterion.' Various criteria had been proposed by Huang *et al.* (1998, 1999, 2003), Flandrin *et al.* (2004) and Wu and Huang (2004), for example. Through extensive study of white noise Wu and Huang (2004, 2010) have found that the EMD is a dyadic filter when the iteration is repeated 10 times.

Once the proto-IMF satisfies the stoppage criterion, it is designate as the first IMF component, $\boldsymbol{c}_1(\boldsymbol{t})$. Thus $\boldsymbol{c}_1(\boldsymbol{t})$ is the result of the data minus a sum of spline functions. After we subtract the first IMF, $\boldsymbol{c}_1(\boldsymbol{t})$, from the data, the residual is the sum of spline functions. Mathematically, we have

$$\boldsymbol{x}(\boldsymbol{t}) - \boldsymbol{c}_1(\boldsymbol{t}) = \boldsymbol{r}_1(\boldsymbol{t}). \tag{5}$$

Therefore, the residual, $\boldsymbol{r}_1(\boldsymbol{t})$, contains all the rest of information in the data other than the first IMF, $\boldsymbol{c}_1(\boldsymbol{t})$. We can, therefore, repeat the above operation using $\boldsymbol{r}_1(\boldsymbol{t})$ as the next set of data, and re-iterate the process, we should have

$$\begin{aligned}
&\boldsymbol{r}_1(\boldsymbol{t}) - \boldsymbol{c}_2(\boldsymbol{t}) = \boldsymbol{r}_2(\boldsymbol{t});\\
&\boldsymbol{r}_2(\boldsymbol{t}) - \boldsymbol{c}_3(\boldsymbol{t}) = \boldsymbol{r}_3(\boldsymbol{t});\\
&\quad\cdots\\
&\boldsymbol{r}_{n-1}(\boldsymbol{t}) - \boldsymbol{c}_n(\boldsymbol{t}) = \boldsymbol{r}_n(\boldsymbol{t});\\
&\Rightarrow \boldsymbol{x}(\boldsymbol{t}) - \sum_{j=1}^{n} \boldsymbol{c}_j(\boldsymbol{t}) = \boldsymbol{r}_n(\boldsymbol{t}) \quad \text{or}\\
&\boldsymbol{x}(\boldsymbol{t}) = \sum_{j=1}^{N(=n+1)} \boldsymbol{c}_j(\boldsymbol{t}).
\end{aligned} \tag{6}$$

We finally obtain the last equation in (6), which means that the data could be decomposed into a sum of IMFs and a residue, $\boldsymbol{r}_n(\boldsymbol{t})$, which can be either a constant, a monotonic mean trend or a curve having only one extrema; it could also be counted as a component.

Recent studies by Flandrin *et al.* (2004) and Wu and Huang (2004, 2010) have established that the EMD is a dyadic filter if the proper stoppage criterion is used. Being adaptive, we have avoided the pitfalls of using an *a priori*-defined basis, and also avoided the spurious harmonics that would have resulted had the *a priori* basis is adopted.

The components of the EMD are usually physically meaningful, if there is no scale mixing (defined as mixed characteristic scales in a single IMF component). To avoid the scale mixing, Wu and Huang (2009) had proposed an Ensemble EMD (EEMD), which is essentially the same EMD procedure, except that the procedure will be repeated $\boldsymbol{n}$ times each with a different white noise added to the data. The procedures are these:

a) add a white noise series to the targeted data;
b) decompose the data with added white noise into IMFs;
c) repeat step 1 and step 2 again and again, but with different white noise series each time; and
d) obtain the (ensemble) means of corresponding IMFs of the decompositions as the final result.

After the ensemble of trials, the final result will be the ensemble mean. A large number collection of white noise signals will cancel each other out in a time domain ensemble mean. This new method proposed in EEMD has utilized all these important statistical characteristics of noise. Based on the study of white noise, Wu and Huang (2004) has also proposed a test for the statistical significance of each IMF, which would determine which IMF component is not noise like. The Ensemble EMD represents a major improvement over the original EMD. It has become a more robust and mature tool for nonlinear and nonstationary time series analysis.

EEMD could be easily extended to data of any dimension known as Multi-dimensional Ensemble Empirical Mode Decomposition (MEEMD) proposed in Wu *et al.* (2009). The implementation is based on the applications of EEMD to slices of data in each and every dimension involved. The final re-construction of the corresponding Multi-dimensional IMF is based on a comparable minimal scale combination principle as given in Wu *et al.* (2009).

Still another variation of EMD is for the Multi-Variate EMD (MVEMD) as proposed by Tanaka and Mandic (2006) for complex variables and Rilling *et al.* (2007) for bivariate, then further extended to trivariate and n-variate by Rehman and Mandic (2010a, 2010b, 2011) and Mandic *et al.* (2013). Rehman and Mandic (2011) have shown that MVEMD has a much sharper filter characteristic than EEMD or the original EMD. This property is a two-sided sword: In some applications, just as in EEMD, the sharp filter bank of MVEMD would strip the harmonics off the fundamental. In order to use it for nonlinear signal, rectification is might be needed.

All above methods are empirical. An exciting new implementation of EMD is not the empirical but analytical as proposed by Hou *et al.* (2009) and Hou and Shi (2011), where the decomposition is reduced to the nonlinear optimization processes through the solution of a set of ordinary differential equation by Newton-Raphson method. This method depends on an initial guess of the answer. More recently, Hou and Shi (2012) proposed further improvement of their earlier trial. The basic idea is to look for the sparsest representation of multi-scale data within the largest possible dictionary consisting of intrinsic mode functions of the form $\{\boldsymbol{a(t)cos(\theta(t))}\}$, where $\boldsymbol{a \geq 0}$ is assumed to be smoother than $\boldsymbol{cos(\theta(t))}$ and $\boldsymbol{\theta}$ is a piecewise smooth increasing function. They formulate the problem as a nonlinear $\boldsymbol{L^1}$ optimization problem and solve it recursively. This new approach is based on a solid mathematical formulation;

therefore, it could offer a solid mathematical foundation for the empirical approach that would facilitate further development the adaptive approach for nonlinear and nonstationary data analysis. Another analytic approach is through Synchrosqueezed wavelet analysis (Daubechies *et al.*, 2011). As the method depends on the integral transformation of the wavelet analysis, its ability to resolve nonlinear phenomena with intra-wave frequency modulations would be problematical.

Other methods such as Singular Spectrum Analysis (SSA, Vautard *et al.*, 1992) and Multitaper Analysis (Thomson, 1982, 2000) are not discussed in details here due to the following reasons: SSA depends on Singular Decomposition method (or Empirical Orthogonal Function), which is based on a covariance matrix of the data. To compute covariance, one would have to invoke stationarity to evaluate the mean, an assumption making this method problematic for applications to nonstationary data. The Multitaper analysis is basically multi-window mean of the standard Fourier spectral analysis. Therefore, it will inherit all the problems associated with Fourier analysis.

3. Hilbert Spectral Representation

Once the data is decomposed into IMF components, the true instantaneous frequency can be computed in a number of ways as given by Hou *et al.* (2009) and Huang *et al.* (2009, 2011). Each IMF component can be written as

$$x(t) = \Re\left(\sum_{j=1}^{N} a(t)e^{i\theta(t)}\right) = \Re\left(\sum_{j=1}^{N} a(t)e^{i\int_{\tau}\omega(\tau)d\tau}\right), \quad (7)$$

where θ is the phase function and ω is the instantaneous frequency. Then, the data could be represented in a time-frequency-energy presentation designated as the Hilbert Spectrum

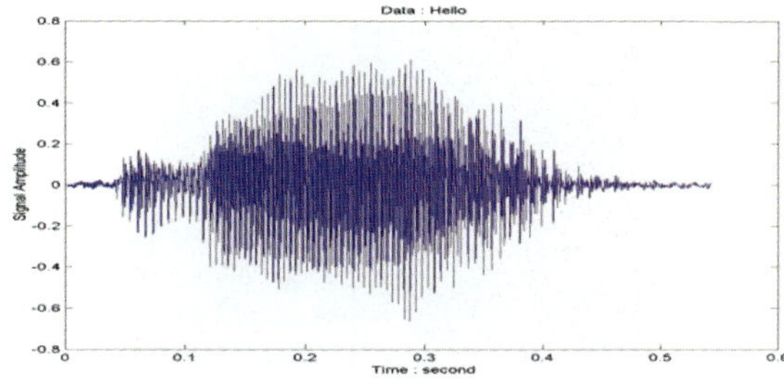

Fig. 1. A speech signal of the sound, ‘Hello.’

(Huang *et al.*, 2011), $H(\omega_j, t_i)$,

$$H(\omega_j, t_i) = \frac{1}{\Delta\omega \cdot \Delta t} a_{i,j}^2, \quad (8)$$

for all bins within $\omega_j = \omega_0 + j\Delta\omega$ and $t_i = t_0 + i\Delta t$. Here the smallest Δt is the sampling rate, but there is no restriction for $\Delta\omega$. In other words, the frequency resolution could be arbitrarily zoomed in any range. As an example, the signal represent to the sound, ‘Hello,’ given in Fig. 1, is analyzed. Comparisons among all the other ‘time-frequency’ representations as in Wavelet, Wigner-Ville (Cohen, 1995; Flandrin, 1999) and spectrogram, the voice signal analysis results are given in Figs. 2a and b.

The effect of uncertainty principle is clearly seen in the top row of Spectrogram results given in Fig. 2. The Hilbert spectrum from the EEMD result, however, gives a physically more satisfactory answer with the steady vocal cord vibration at around 100 Hz and the nonlinear vocal chamber reverberation up to 1,000 Hz.

The crucial differences between the HHT and traditional methods can be summarized in Table 1 as given in Huang (2007). The versatility of the adaptive approach is clearly shown: without the limitation of the uncertainty principle, HHT could give frequency values at any time to any degree of precision limited only by the sampling rate and data quality.

4. The Determination of Trend

Now, let us discuss the problem of determination of the trend. “Trend” and “detrend” are

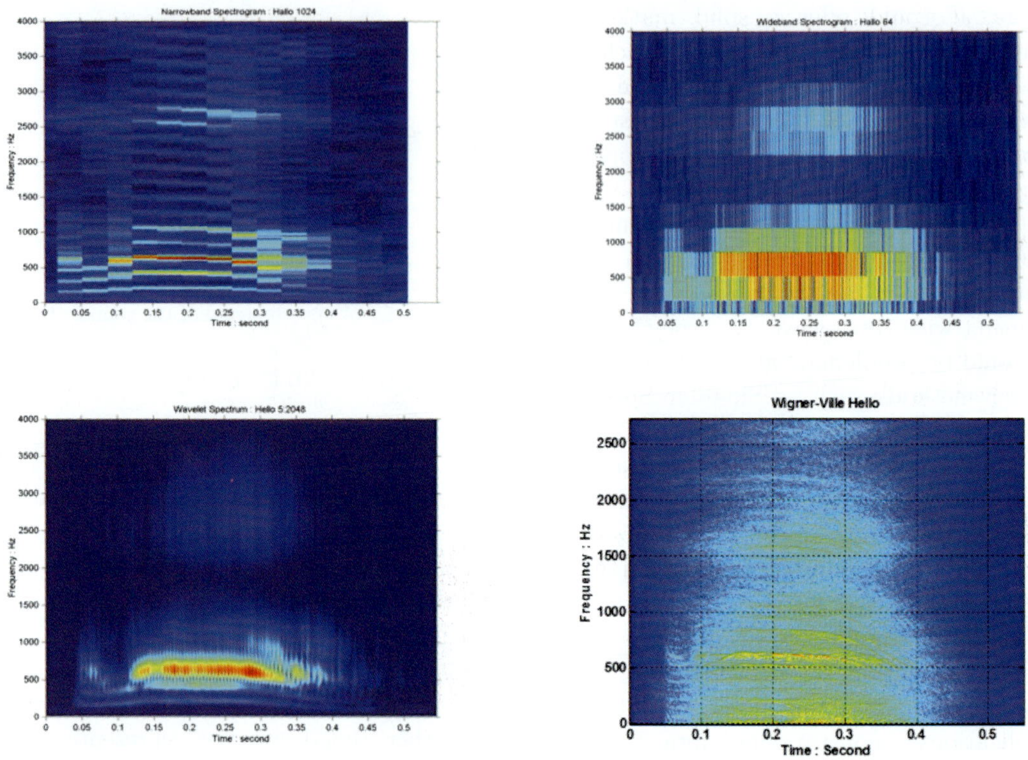

Fig. 2. The sound 'Hello' analyzed by different traditional methods: The top row are the Fourier spectrograms with narrow band (left, based on 1024 data point in each window) and wide band (right, with 64 data points in each window). The effect of uncertainty principle is clear: when frequency is well defined, the time location is smeared, and vice versa. The bottom left is the Morlet wavelet analysis, which give a uniform resolution, yet the resolution is also uniformly poor. The bottom right is the Wigner-Ville analysis, which suffers the similar problem as in the Fourier spectrogram, but it is less localized.

Table 1. Summary comparisons among Different Frequency Analysis Methods.

	Fourier	Wavelet	HHT
Basis	*a priori*	*a priori*	Adaptive
Frequency	Integral Transform: Global	Integral Transform: Regional	Differentiation: local
Presentation	Energy-Frequency	Time-energy-frequency	Time-frequency-energy
Nonlinear	No. Harmonics	No. Harmonics	Yes: Intra-wave frequency modulation
Nonstationary	no	yes	Yes: Inter-wave frequency modulation
Uncertainty	yes	yes	No resolution limit.
Harmonics	yes	yes	no

frequently encountered terms in data analysis. In many applications, such as climatologic and financial data analyses, the trend is one of the most critical quantities sought (Wu *et al.*, 2011). In other applications, such as in computing the correlation function (Chen *et al.*, 2010) and spectral analysis, it is necessary to remove the trend from the data, known as detrending, lest the result would be overwhelmed by the non-zero mean term; therefore, detrending is often a necessary step before meaningful traditional analysis results can be obtained. Yet, there is no precise definition for the trend in a data set. As there is a lack of precise definition for the trend, the determinations of trend and detrend are totally *ad hoc* operations mostly based on regression analysis.

As regression analysis has long been a core topic in the field of statistics and econometrics, it has also been used extensively in scientific research to define causal relationships. Traditionally, there are two approaches for regression analysis: the parametric and non-parametric regression. The idea behind parametric approach is simple and definitive, but the justification for adopting it is rather thin: How could one defend the selection of one fixed functional form over others? As the final result of the regression depends solely on the pre-selected functional form, the *ab initial* determined functional form would not only be the most important step of the analysis, but also totally prejudice the final result. Most critically, the selection of the functional form is totally arbitrary.

With the difficulties of the parametric regression, increasingly the recent developments have concentrated on the non-parametric or semi-parametric approaches advocated by Pearson as given by (Härdle, 1995; Härdle *et al.*, 2004; Fan and Yao, 2005). The techniques used in the non-parametric are much richer: there are the studies of the asymptotic form of the functions assuming infinite many data, as well as smoothing with kernel estimators and spline, for example. Then, there is the crucial problem of deciding which of the numerous spline functions should be adopted.

An alternative definition for trend was proposed by Huang *et al.* (1998) and refined by Wu *et al.* (2007): Within the given data span, the trend is an intrinsically fitted monotonic function, or a function in which there can be at most one extremum. With this more general view, trend of different temporal scale could be found through EMD. If the data $\boldsymbol{x(t)}$ is expanded in terms of IMF through EMD as in Eq. (6), any division of the IMFs into two groups as

$$\begin{aligned} \boldsymbol{x}(\boldsymbol{t}) &= \sum_{i=1}^{N} \boldsymbol{c}_i = \sum_{i=1}^{k-1} \boldsymbol{c}_i + \sum_{i=k}^{N} \boldsymbol{c}_i \\ &= \boldsymbol{\varepsilon}_{\boldsymbol{k}} + \boldsymbol{T}_{\boldsymbol{k}}; \end{aligned} \tag{9}$$

with $\boldsymbol{T_k} = \sum_{i=k}^{N} \boldsymbol{c_i}$ and $\boldsymbol{\varepsilon_k} = \sum_{i=1}^{k-1} \boldsymbol{c_i}$, would give us a trend, $\boldsymbol{T_k}$, and the corresponding error, $\boldsymbol{\varepsilon_k}$, for any $\boldsymbol{k}$. Of course, this trend is only meaningful within a local neighborhood of $\boldsymbol{P_k}$ defined as the mean period of the IMF component $\boldsymbol{c_k}$. If we systematically divide the IMF into different groups, we could get various generalized trend fittings. These generalized trends are only monotonic within each specific time span. The overall trend could have finite number of extrema, which would make them similar to various regression. The central idea here is that the regression should not be determined by parametric or nonparametric method, but by systematically removal of cycles. The natural way of cycle removal is through EMD or its variation EEMD.

Unfortunately, in most applications, the trend is taken as a simple straight line following the parametrical approach. Let us take the global warming phenomenon as an example. This case has been studied by Wu *et al.* (2007, 2011). According to the 4th Assessment Report of the Intergovernmental Panel on Climate Change (IPCC, 2007), the global temperature anomaly is given in Fig. 3, in which various trends are determined from the annual averaged data.

Based on the slopes of the various trends, IPCC reached this conclusion: "*Note that for shorter recent periods, the slope is greater, indicating accelerated warming.*" If we use EMD to analyze the original monthly data down-sampled to yearly by specific month. Thus, we would get 12 different sets of IMFs each representing a specific month of the years such as all January, all February and so on as

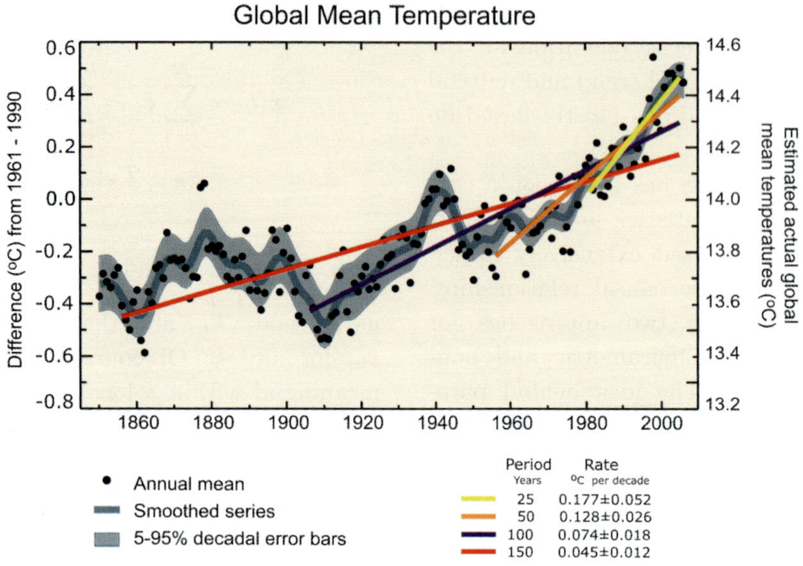

Fig. 3. Instrumental climate data covering the period 1856 to 2007. Various trends and temperature increasing rates are determined for different data lengths: 25 (yellow), 50 (brown), 100 (black) and 150 (red) years.

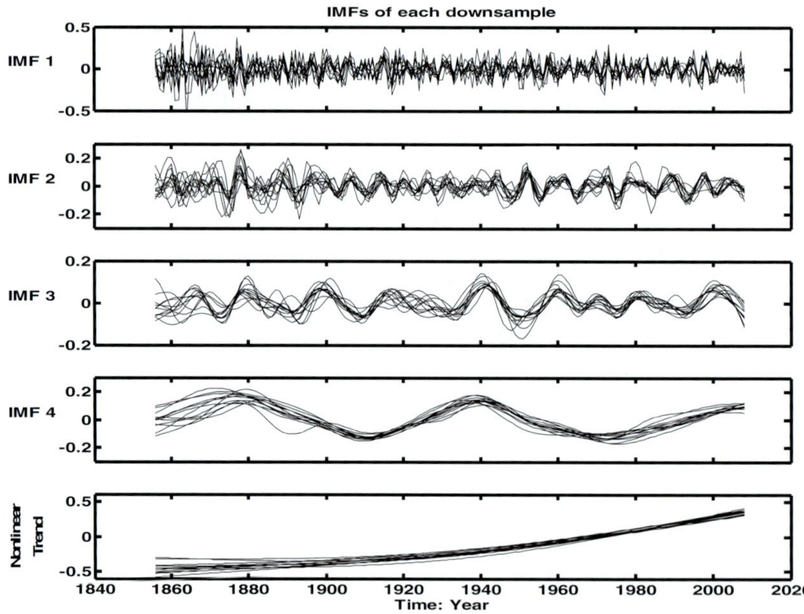

Fig. 4. The various IMFs representing each specific month of the year from the original monthly data. We could get the mean and the scattering bound from these results.

shown in Fig. 4. Clearly the residue after all the oscillation cycles were removed is a trend. From the twelve specific monthly results, we could get an annual mean and the scattering range indicating the confidence limit of this decomposition.

Having obtained this result, we will consider an alternative of the straight line trends given by

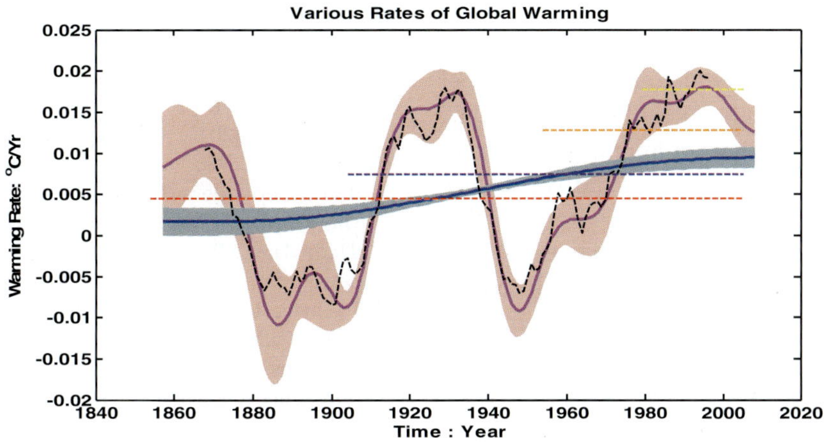

Fig. 5. The 25-year trends determined with any 25 period. The black dashed line gives the running value determined from the annual mean data. The mean of the corresponding slopes determined from the specific monthly data is given as the pink line with the scattering given by the shaded band. The slopes from the various straight line trends are given by the corresponding colored dashed lines (25-year: yellow; 50-year: brown; 100-year: black and 150-year: red), each is roughly the mean values over the time span. The slope from the EMD determined trend is given by the blue line with the scattering band in gray that is only half the value as given by IPCC.

IPCC. Instead of drawing the 25-year trend at the end of the data, we can draw 25-year trend centered at any point. The slope values would be a variable now. The result is plotted in Fig. 5. Clearly, the running 25-years trends showed a fluctuating pattern with the period roughly 65 years. Based on this study, we concluded as in Wu *et al.* (2011) that the true warming rate of the climate is only half the rate as determined by the IPCC. Therefore, the recent global warming is partially contributed by natural fluctuation with a 65 years cycle corresponding to the multi-decadal oscillation as suggested by Schlesinger and Ramankutty (1994). A more detailed analysis of the Global Surface Temperature Anomaly data can be found in Wu *et al.* (2011).

It should be pointed out that there has been controversial concerning the multi-decadal cycle. The main point of contention is on the limited and low degree of freedom of this cycle, for there are only slightly more than two cycles in the available data. Yet the existence and robustness of this cycle have been established in various studies. Zhou and Tung (2013) have used multi-regression method on the sea surface temperature and arrived at the same cycle. Additionally, Tung and Zhou (2013) used 353-year instrumental record in Central England and found this multi-decadal cycles persisted all through the record. Furthermore, the first scientific paper on global warming by Broecker (1975) used more than 800 years of Camp Century ice core data and obtained a mean cool-warm cycle at the average of 70 years. Therefore, the existence of the near 70 years cycle seems to be a robust fact.

It is interesting to point out that the data actually indicate the slight slowing down in the global temperature increasing rate, a fact seems in agreement with the recent hiatus in temperature increase observation by Hawkins (2013) that the global warming data is heading to break the 95% limit of all the model results for Global Climate assessment. Unfortunately, the underlying dynamic processes to drive this multi-decadal cycle are still elusive. Therefore, it behooves us to dedicate ourselves to look for the cause and mechanism of this multi-decadal climate oscillation rather than finding flaws in any particular set of data.

5. Conclusion

In this short review, which is a variation of the papers of the similar titles by Huang (2007, 2010). Additional review for applications in geophysics could be found in Huang and Wu (2008). To be adaptive, the method has to be independent of any a priori basis. The EMD is a time domain scale separation method that is independent of the basis. Yet in mathematical theory, the mainstream approach is still based on a priori basis. The theory of the general approach of adaptive data analysis should be examine seriously. As the technology becoming more sophisticated, data both in quantity and quality would increase. We are overwhelmed by the volume of the data, yet at the same time we are also underserved by the information we could extract from the data. The key here is the strict artificial limitations on data analysis methodology for all the traditional methods. The only way to break through this bottleneck is to use adaptive data analysis. HHT is an adaptive data analysis, which enable us to define instantaneous frequency, the true and physically meaningful way to represent frequency. With the frequency definition, we can also quantify the degree and order of nonlinearity, and also determine the trend. The time for adaptive data analysis seems to have arrived. Our experience indicates that HHT could help us to clarify and provide insight to the underlying physical processes.

Acknowledgments

The author has been supported by a grant for the Center for Dynamical Biomarkers and Translational Medicine, National Central University, Taiwan (NSC 101-2911-I-008-100) from the National Science Council, Taiwan, and a grant from NCU that have made the conclusion of this study possible.

References

Broecker, W. S., 1975: Climatic change: Are we on the brink of a pronounced global warming? *Science*, **189**(4201), 460–463.

Chen, X., Z. Wu, and N. E. Huang, 2010: The Time-Dependent Intrinsic Correlation based on the Empirical Mode Decomposition. *Adv. Adapt. Data Anal.*, **2**, 233–265.

Cohen, L., 1995: *Time-Frequency Analysis*. Prentice-Hall Signal Processing Series, Englewood Cliffs, New Jersey, 299 pp.

Daubechies, I., J. Lu, and H.-T. Wu, 2011: Synchrosqueezed wavelet transforms: An Empirical Mode Decomposition-like tool. *Appl. Comp. Harmon. Anal.*, **30**, 243–261.

Fan, J., and Q. Yao, 2005: *Nonlinear Time Series: Nonparametric and Parametric methods*. Springer, New York, 551 pp.

Flandrin, P. 1999: *Time-Frequency/Time-Scale Analysis*. Academic Press, San Diego, CA, 386 pp.

———, G. Rilling, and P. Gonçalves, 2004: Empirical Mode Decomposition as a filter bank. *IEEE Signal Proc. Let.*, **11**(2), 112–114.

Härdle, W., 1995: *Applied Nonparametric Regression*. Cambridge University Press, Cambridge, 333 pp.

———, M. Müller, S. Sperlich, and A. Werwatz, 2004: *Nonparametric and Semi-parametric Models*. Springer, Berlin, 300 pp.

Hawkins, E., 2013: Climate science: A sensitive matter. *The Economist*, March 30, 2013.

Hou, T. Y., and Z. Shi, 2011: Adaptive data analysis via sparse time-frequency representation. *Adv. Adapt. Data Anal.*, **3**, 1–28, doi: 10.1142/S1793536911000647.

———, and ———, 2012: Data driven time-frequency analysis. *Appl. Comp. Harmon. Anal.*, **35**, 284–308, doi: 10.1016/j.acha.2012.10.001.

———, M. P. Yan, and Z. Wu, 2009: A variant of the EMD method for multi-scale data. *Adv. Adapt. Data Anal.*, **1**, 483–516.

Huang, N. E., 2007: A plea for adaptive data analysis. *Independent Component Analyses, Wavelets, Unsupervised Nano-Biomimetic Sensors, and Neural Networks, V*, H. H. Szu, and J. Agee, Eds., Proc. of SPIE, Society of Photo Optical, doi: 10.1117/12.725199.

———, 2010: An introduction to an adaptive data analysis method. *From Waves in Complex Systems to Dynamics of Generalized Continua: Tributes to Professor Yih-Hsing Pao on His 80th*

Birthday, K. Hutter, T.-T. Wu, and Y.-C. Shu, Eds., World Scientific, Singapore, 137–158.

———, X. Chen, M.-T. Lo, and Z. Wu, 2011: On Hilbert spectral representation: A true time-frequency representation for nonlinear and nonstationary data. *Adv. Adapt. Data Anal.*, **3**, 63–93, doi: 10.1142/S1793536911000659.

———, Z. Shen, R. S. Long, 1999: A new view of nonlinear water waves: The Hilbert spectrum. *Annu. Rev. Fluid Mech.*, **31**, 417–457.

———, ———, ———, M. C. Wu, H. H. Shih, Q. Zheng, N.-C. Yen, C. C. Tung, and H. H. Liu, 1998: The Empirical Mode Decomposition method and the Hilbert spectrum for nonstationary time series analysis. *Proc. R. Soc. Lond. A*, **454**, 903–995.

———, M.-L. C. Wu, ———, S. S. P. Shen, W. Qu, P. Gloersen, and K. L. Fan, 2003: A confidence limit for the Empirical Mode Decomposition and the Hilbert spectral analysis. *Proc. R. Soc. Lond. A*, **459**, 2317–2345.

———, and Z. Wu, 2008: A Review on Hilbert-Huang transform: Method and its applications to geophysical studies. *Rev. Geophys.*, **46**, RG2006, doi: 10.1029/2007RG000228.

———, ———, S. R. Long, K. C. Arnold, X. Chen, and K. Blank, 2009: On instantaneous frequency. *Adv. Adapt. Data Anal.*, **1**, 177–229, doi: 10.1142/S1793536909000096.

Intergovernmental Panel on Climate Change. *Climate Change, 2007: The Physical Science Basis. Contribution of Working Group I to the Fourth Assessment Report of the IPCC*, S. Solomon, *et al.*, Eds., Cambridge University Press, Cambridge, U.K., 996 pp.

Kantz, H., and T. Schreiber, 1997: *Nonlinear Time Series Analysis.* Cambridge University Press, Cambridge, 304 pp.

Mandic, D. P., N. U. Rehman, Z. Wu, and N. E. Huang, 2013: Empirical Mode Decomposition-Based time-frequency analysis of multivariate signal. *IEEE signal Proc. Mag.*, **30**, 74–86.

Rehman, N., and D. P. Mandic, 2010a: Empirical Mode Decomposition for trivariate signals. *IEEE T. Signal Proces.*, **58**, 1059–1068.

———, and ———, 2010b: Multivariate Empirical Mode Decomposition. *Proc. R. Soc. Lond. A*, **466**, 1291–1302, doi: 10.1098/rspa.2009.0502.

———, and ———, 2011: Filter bank property of Multivariate Empirical Mode Decomposition. *IEEE T. Signal Proces.*, **59**, 2421–2426.

Rilling, G., P. Flandrin, P. Gonçalves, and J. M. Lilly, 2007: Bivariate Empirical Mode Decomposition. *IEEE Signal Proc. Let.*, **14**, 936–939.

Schlesinger, M. E., and N. Ramankutty, 1994: An oscillation in the global climate system of period 65–70 years. *Nature*, **367**, 723–726.

Tanaka, T., and D. P. Mandic, 2006: Complex Empirical Mode Decomposition. *IEEE Signal Proc. Let.*, **14**, 101–104.

Thomson, D. J., 1982: Spectrum estimate and harmonic analysis. *P. IEEE*, **70**, 1055–1096.

———, 2000: Multitaper analysis of nonstationary and nonlinear time series data. *Nonlinear and Nonstationary Signal Processing*, W. J. Fitzgerald, Ed., Cambridge University Press, Cambridge, 317–393.

Tung, K.-K., and J. Zhou, 2013: Using data to attribute episodes of warming and cooling in instrumental records. *Proc. Natl. Acad. Sci.*, **110**, 2058–2063.

Vautard, R., P. Yiou, and M. Ghil, 1992: Singular-spectrum analysis: A toolkit for short, noisy chaotic signals. *Physica D: Nonlinear Phenomena*, **58**, 95–126.

Wu, Z., and N. E. Huang, 2004: A study of the characteristics of white noise using the empirical mode decomposition method. *Proc. R. Soc. Lond. A*, **460**, 1597–1611.

———, and ———, 2009: Ensemble Empirical Mode Decomposition: A noise-assisted data analysis method. *Adv. Adapt. Data Anal.*, **1**, 1–42, doi: 10.1142/S1793536909000047.

———, and ———, 2010: On the filtering properties of the Empirical Mode Decomposition. *Adv. Adapt. Data Anal.*, **2**, 397–414, doi: 10.1142/S1793536910000604.

———, ———, and X. Chen, 2009: The Multi-dimensional Ensemble Empirical Mode Decomposition method. *Adv. Adapt. Data Anal.*, **1**, 339–372.

———, ———, S. R. Long, and C.-K. Peng, 2007: On the trend, detrending, and variability of nonlinear and nonstationary time series. *Proc. Natl. Acad. Sci.*, **104**, 14,889–14,894.

———, ———, J. M. Wallace, B. V. Smoliak, and X. Chen, 2011: On the time-varying trend in global-mean surface temperature. *Clim. Dyn.*, **37**, 759–773, doi: 10.1007/s00382-011-1128-8.

Zhou, J., and K.-K. Tung, 2013: Deducing multidecadal anthropogenic global warming trends using multiple regression analysis. *J. Atmos. Sci.*, **70**, 3–8.

Chapter 7

Internal Southern Ocean Centennial Variability: Dynamics, Impacts and Implications for Global Warming

Mojib Latif[*,†], Torge Martin[‡], Wonsun Park[*], and Mohammad H. Bordbar[*]

[*]*GEOMAR Helmholtz Centre for Ocean Research Kiel, Düsternbrooker Weg 20, 24105 Kiel, Germany*

[†]*Kiel University, Kiel, Germany*

[‡]*Polar Science Center, Applied Physics Laboratory, University of Washington, Seattle, WA, USA*

mlatif@geomar.de

www.geomar.de/∼mlatif

It is well established that centennial climate variability can be externally forced by, e.g., quasi-oscillatory fluctuations of the solar constant or slowly varying atmospheric aerosol concentrations in association with changes of volcanic activity. Climate models recently suggested that substantial centennial variability can be also produced internally, and different competing mechanisms have been proposed. This paper deals with the internal centennial variability originating in the Southern Ocean Sector simulated by the Kiel Climate Model (KCM). In that model, the Southern Ocean centennial variability (SOCV) is linked to Weddell Sea deep convection activity and drives regional as well as global climate variations, as witnessed, for example, by coherent changes in Antarctic sea ice extent and globally averaged surface air temperature (SAT). Furthermore, the SOCV is associated with changes in deep Southern Ocean temperature in the KCM. Interestingly, a warming of the abyssal Southern Ocean has been observed during the recent decades, suggesting a contribution from SOCV. Another important impact of the SOCV in the model concerns the Atlantic Meridional Overturning Circulation (AMOC). The AMOC strengthens and deepens after the cessation of Weddell Sea deep convection and Antarctic Bottom Water (AABW) formation with a time delay of several decades to a century. Internal North Atlantic sea level variations can be as large as ± 15cm/century in the model with a strong contribution from the SOCV. Such regional sea level variations are of the same order of magnitude as the observed globally averaged 20th century sea level rise amounting to about 15-20cm. Finally, the KCM simulation suggests that the SOCV may have contributed to the current hiatus in global warming through an enhanced deep ocean heat uptake.

1. Introduction

Climate variability can be either generated internally by interactions within or between the individual climate system components (e.g., atmosphere, ocean, and sea ice) or externally by, for example, volcanic eruptions or variations in the insolation at the top of the atmosphere. Prominent examples of internal variability, also termed unforced variability, are the North Atlantic Oscillation (NAO) originating from the chaotic atmospheric dynamics, the El Niño/Southern Oscillation (ENSO), a coupled ocean-atmosphere mode on interannual time scales, the Pacific Decadal Oscillation (PDO), or the Atlantic Multidecadal Oscillation (AMO)

Climate Change: Multidecadal and Beyond
edited by Chih-Pei Chang et al.

(see e.g. Latif and Keenlyside (2011) for a review of decadal variability and predictability, and several chapters in this book which deal with various aspects of decadal variability and predictability). All these phenomena are interrelated to some extent, cover a wide range of frequencies, and project to a certain degree onto hemispheric and even global surface air temperature (SAT). Of special interest is the decadal to centennial variability which has the potential to mask the regional effects of anthropogenic climate change, as this variability evolves on a similar time scale (see Chapter 1 by Wallace *et al.*). An example is the Atlantic hurricane activity. The latter featured strong multidecadal variability during the 20th century which superimposed a weak long-term upward trend. Some of the multidecadal variability may be predictable. It is therefore of primary importance to understand the mechanisms of the internal decadal to centennial variability in order to facilitate the detection of anthropogenic climate signals and improve climate projections for the 21st century.

The existence of centennial variability is well established from proxy data. For instance, centennial sea surface temperature (SST) variability in the subpolar Atlantic over the last two millennia was documented by Sicre *et al.* (2011). Centennial variability of the SST gradient across the subpolar North Atlantic over the last 2,800 years has been recently described by Miettinen *et al.* (2012). To understand the mechanisms of centennial variability we additionally need to employ climate models. The proxy data is sparse and the instrumental records not sufficiently long to allow a detailed inspection of the underlying dynamics. Vellinga and Wu (2004) described internal centennial variability from the Hadley Centre climate model version 3 that relies on large-scale ocean-atmosphere interactions involving the tropics and extratropics. Menary *et al.* (2012) identified the same mechanism also in two other climate models. Enhanced northward ocean heat transport associated with an anomalously strong Atlantic Meridional Overturning Circulation (AMOC) generates a cross-equatorial SST gradient and causes the Intertropical Convergence Zone (ITCZ) to move to a more northerly position when it is anomalously strong. The extra ITCZ rainfall imposes a freshwater flux and produces a salinity anomaly in the tropical North Atlantic. Such sustained salinity anomalies slowly propagate toward the subpolar North Atlantic at a lag of 5–6 decades. The accumulated low-salinity water lowers the upper ocean density there, which causes the AMOC to slow, thereby providing a delayed negative feedback necessary to maintain an oscillation. Recently, Delworth and Zeng (2012) reported pronounced internal centennial variability of the AMOC in the GFDL climate model version 2, which has a significant impact on Northern Hemisphere average SAT. The advection of upper ocean salinity anomalies from the South Atlantic into the subpolar North Atlantic sinking region was shown to be important in that model. However, the precise mechanism is still under investigation. Finally, Karnaukas *et al.* (2012) describes unforced centennial variability in the western equatorial Pacific SST simulated by three climate models, which is a result of the strong asymmetry of interannual ocean heat content anomalies.

The focus of this book chapter is on the internal Southern Ocean centennial variability which will be referred to as the SOCV in the following. We summarize here the findings from several papers which describe results from different versions of the Kiel Climate Model (KCM) (Park *et al.*, 2009; Park and Latif, 2008; Martin *et al.*, 2013, 2014; Latif *et al.*, 2013). These papers deal with various aspects of the SOCV in the KCM and its regional and global impacts. The SOCV in the KCM was found to drive, for example, regional and global SAT anomalies and also to significantly affect the AMOC and North Atlantic sea level. The main conclusion that can be drawn from the KCM simulations is that internal variability matters even on centennial time scales, rendering projections for the climate throughout the 21st century an initial value problem.

This book chapter is organized as follows. In Section 2, we describe some evidence of Southern Ocean centennial variability from instrumental observations and proxy data. The main characteristics of the SOCV in the KCM are presented in Section 3. The possible implications of the SOCV for the climate during the 21st century are discussed by means of an ensemble of global warming integrations with the KCM in Section 4. The main conclusions are provided in Section 5.

2. Observational Evidence of Southern Ocean Centennial Variability

Although the Earth's globally averaged SAT did strongly warm during the recent decades, the Southern Hemisphere SAT rose only little (Fig. 1). Further, in marked contrast to the Northern Hemisphere sea ice extent, which strongly declined during the recent decades, the Southern Hemisphere sea ice extent slightly increased (e.g. Parkinson and Cavalieri, 2012). These seemingly conflicting trends often cause confusion among the public about the reality of global warming. Moreover, the low-level westerlies over the Southern Ocean intensified during the recent decades (Thompson and Solomon, 2002; Visbeck, 2009); and the abyssal Southern Ocean considerably warmed, especially in its Atlantic Sector (e.g. Purkey and Johnson, 2010).

At this stage of research the origin of these decadal trends in the Southern Hemisphere remain unclear. Both internal and external mechanisms have been proposed. Here we shall put forward a hypothesis that reconciles the different decadal changes by interpreting them collectively in terms of internal centennial variability of the Southern Ocean circulation. It should be mentioned, however, that this topic is a very controversial issue. Our proposal which is based on internal variability is just one possible explanation for the origin of the observed trends in the Southern Hemisphere during the recent decades.

We investigate next the sea surface temperature (SST) during the complete instrumental period. An SST index is defined as the

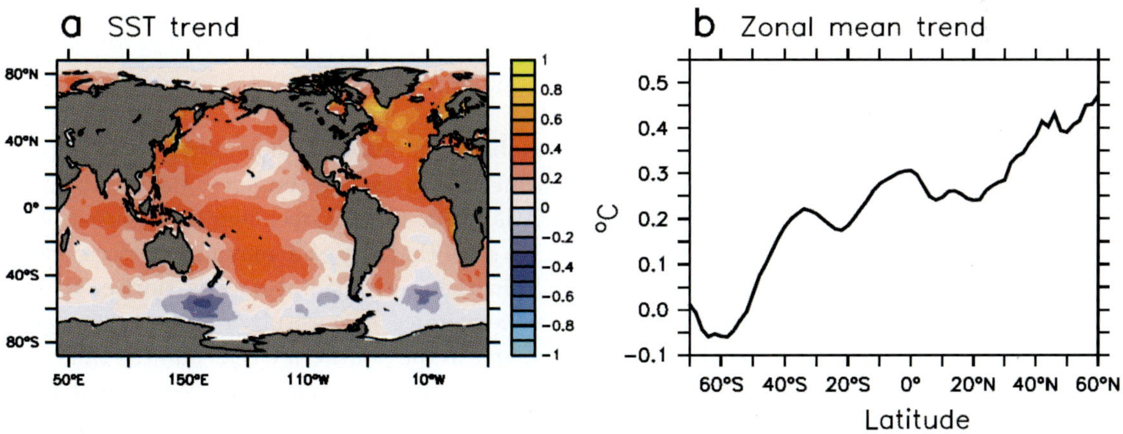

Fig. 1. Changes in sea surface temperature (SST) during the recent decades based on the ERSSTv3b dataset (Smith *et al.*, 2008, http://www.ncdc.noaa.gov/ersst/). (a) Global pattern of annual SSTs averaged over the period 1991–2010 presented as anomalies with respect to the reference period 1971–1990. (b) The corresponding zonal mean temperature anomalies. While the Northern Hemisphere experienced a strong warming during the recent decades, the Southern Hemisphere warmed only little. The globally averaged SST difference between the two time periods amounts to about 0.2°C. From Latif *et al.* (2013).

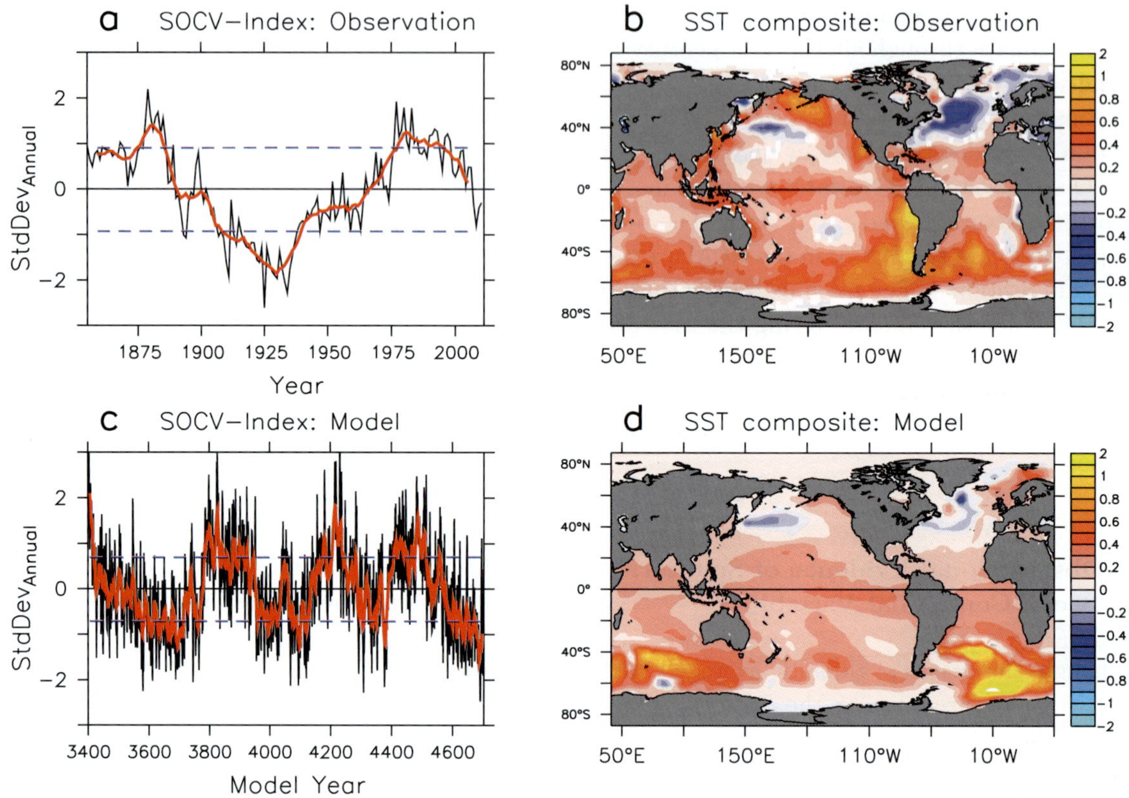

Fig. 2. The Southern Ocean sea surface temperature (SST) depicts strong centennial variability both in the observations (a, b: 1854–2010, ERSSTv3b) and the Kiel Climate Model (KCM) simulation (c, d: 1,300years). (a, c) The SOCV index defined as the zonal average annual SST anomalies over the latitude band 50° S–70° S, de-trended and normalized by the long-term standard deviation. (b, d) SST composite pattern (°C, high minus low) associated to the SOCV index using the thresholds shown as dashed blue lines in (a) and (c). The thresholds are defined as plus and minus one standard deviation of the 11-year running mean-filtered SOCV index which are shown as red bold lines. From Latif *et al.* (2013).

average over the latitude belt 50°S-70°S [hereafter referred to as the Southern Ocean centennial variability index (SOCV index), Fig. 2a]. We note that both typical PDO and AMO indices exhibit a rather different time evolution to that of the SOCV index during the instrumental period (not shown, but see Fig. 15 in Park and Latif, 2012). The SOCV index depicts a pronounced centennial variation, with maxima around the mid-1870s and mid-1970s. The global SST anomaly pattern associated to the SOCV index as obtained from composite analysis is dominated by anomalies of the same sign (Fig. 2b), indicating that the SOCV is of global significance. Anomalies of the opposite sign are seen only in the North Pacific, the North Atlantic and the Greenland and Barents Seas. We note that the SST anomaly pattern derived from the instrumental data is subject to large uncertainties, because observations are generally sparse in the Southern Ocean and the time series covers only one cycle of the centennial mode. This also applies to the time scale of the mode, as it cannot be reliably determined from such a short record.

The existence of pronounced centennial variability in the Southern Ocean Sector is supported by the Tasmanian summer temperatures of the last 3,600 years, as reconstructed from tree rings (Cook *et al.*, 2000) and discussed by Latif *et al.* (2013). The reconstructed temperatures (not shown) depict significantly enhanced variability (above the 99% significance level) at centennial timescales, even though tree ring temperature reconstructions are known to underestimate the level of centennial variability. We cannot rule out the possibility that the multidecadal and multicentennial variability seen in the spectrum of the reconstructed Tasmanian summer temperatures are forced externally, specifically by the known solar cycles which have periodicities in these frequency ranges. The magnitude of the solar radiation changes associated with the solar cycles, however, is rather small. On the other hand, a weak external forcing can resonantly excite a damped internal eigenmode.

Let us return to the instrumental SST record. The rapid increase of the SOCV index during the mid-1970s (Fig. 2a) coincided with the so-called climate shift in the Pacific Ocean (e.g., Trenberth and Hurrell, 1994), which was characterized by an anomalously warm equatorial Pacific and an anomalously cold North Pacific. Similar anomalies are also seen in the SOCV index pattern (Fig. 2b). Although the origin of the climate shift is assumed to have been in the Tropical Pacific, the Southern Ocean also may have contributed, as suggested by the KCM results discussed below. The SOCV index dropped thereafter and was well below-normal during the most recent years, thereby cooling the Earth's surface and retarding global warming.

We would like to highlight one more observation, which will be of large importance in the subsequent discussion about the mechanism behind the model's centennial variability. A rather prominent event in the Southern Ocean during the 20th century was the Weddell Polynya (e.g. Zwally and Gloersen, 1977; Carsey, 1980; Gordon and Comiso, 1988), a very large area (350,000 km^2) of open water within the ice-covered Weddell Sea during the austral winters of 1974–76. There was vigorous air-sea interaction within the polynya that resulted in deep convection and the formation of vast amounts of cold and very dense Antarctic Bottom Water (AABW, Gordon, 1982). The appearance of the Weddell Polynya coincided with the last warm extreme of the SOCV index (Fig. 2a), and it is argued here that the polynya could be possibly understood as part of a longer-term centennial variation, putting the Weddell Polynya into a new perspective (Latif *et al.*, 2013). If this assertion is correct, there may have been another polynya during the previous warm phase around 1880. Unfortunately, little is known about the sea ice in the Weddell Sea during that time.

3. Internal Centennial Variability in the KCM

The dynamics and global impacts of SOCV are investigated here by analyzing a 1,300 year long control integration of the present-day climate conditions (Martin *et al.*, 2013, 2014; Latif *et al.*, 2013) of the KCM (Park *et al.*, 2009). The model version used here differs from that in Park and Latif (2008) with respect to the sea ice parameterization in that one single parameter was changed to allow for a more realistic simulation of Antarctic sea ice. The dynamics of the centennial variability, however, did not fundamentally change relative to the previous version which was integrated over a much longer time period (Martin *et al.* , 2013). This also applies to another version that was used in a large ensemble of global warming simulations which will be described below in Section 4. It should also be noted that the time scale of the mode cannot be reliably determined from the model, as there is a strong sensitivity of the main periodicity to the model formulation (Latif *et al.*, 2013).

Consistent with the instrumental SSTs (Fig. 2a), the model version discussed in this section simulates pronounced centennial Southern Ocean SST variability (Fig. 2c). The KCM integration depicts SOCV with a dominant period of about 300 to 400 years. However, the variability is irregular, and an about 100 year fluctuation similar to that observed during the instrumental record is within the broad range of the model variability. For instance, a 100 year fluctuation is simulated during the model years 4,000–4,100 (Fig. 2c). On the other hand, a longer multicentennial period would not be inconsistent with the Tasmanian summer temperature record discussed above. The spectrum of the reconstructed temperatures exhibits enhanced variability on both multidecadal and multicentennial time scales.

The SST anomaly pattern related to the SOCV index from the model (Fig. 2d) is similar to that derived from the instrumental observations (Fig. 2b) in that most regions depict positive SST anomalies, while negative SST anomalies are seen in the North Pacific and North Atlantic Oceans. There are a number of noticeable differences between the model pattern and that derived from the instrumental data. The KCM is a coarse-resolution climate model and suffers, as many other climate models, from large regional biases. A good correspondence is therefore only expected for rather large spatial and on rather long time scales. Furthermore, the observed variability during the instrumental period is composed of several modes so that one single mode alone cannot account for the full variability. The instrumental SSTs cover only one realization of the centennial variability, and the SST anomaly pattern is therefore subject to large uncertainties. Yet we believe that the correspondence between the model and the observed anomaly pattern is good enough to justify a more detailed investigation of the dynamics behind the SOCV in the KCM in order to obtain some insight into the processes that may operate in the real ocean.

The long-term Southern Ocean SST variability (Fig. 2c) is clearly reflected in globally averaged SAT (Fig. 3a) and associated with rather rapid Antarctic sea ice area changes in the model (Fig. 3b). The magnitude of global SAT changes can be as large as 0.4°C/century (see, for instance, the centennial warming trend starting just before the model year 3,800), a trend which is about half of the 20th century global warming. The magnitude of Antarctic sea ice area changes can amount to about $2 \times 10^6\,\text{km}^2/\text{century}$ in the model. This can be compared to the observed increase of about $1 \times 10^6\,\text{km}^2$ since 1979. Thus, if the model variability is realistic, the Southern Ocean can drive climate changes which clearly have the potential to mask anthropogenic climate change locally, regionally and even globally. The current hiatus in global surface warming, for instance, may partly be the result of an enhanced deep Southern Ocean heat uptake (Fig. 3c).

The SOCV in the KCM is linked to variations in Weddell Sea deep convection (Martin *et al.*, 2013). The latter is characterized by a slow century-long built-up of heat during the non-convective regime (Fig. 3c, gray shading) primarily taking place at mid-depth between 1,000 m and 3,000 m, followed by a rather abrupt cooling after deep convection onset. The variations of the vertically integrated heat content amount to about $\pm 40 \times 10^{21}$ J in the Atlantic-Indian Ocean Basin, defined here as the region 40°S–90°S and 70°W–80°E. The heat accumulation is effectively orchestrated by the Weddell Gyre. During active deep convection a smaller, local gyre enhances upwelling. Convection onset and shutdown additionally depend on the stochastic occurrence of favorable sea surface conditions (e.g. regarding precipitation, wind, and sea ice thickness), which also contributes to the multicentennial period of the mode.

The heat at mid-depth is provided by the lower branch of the AMOC, a mechanism previously described from uncoupled ocean model simulations (Mikolajewicz and Maier-Reimer, 1990; Yin and Sarachik, 1995; Pierce *et al.*, 1995). Those, however, have been criticized for using mixed boundary conditions, which distorts

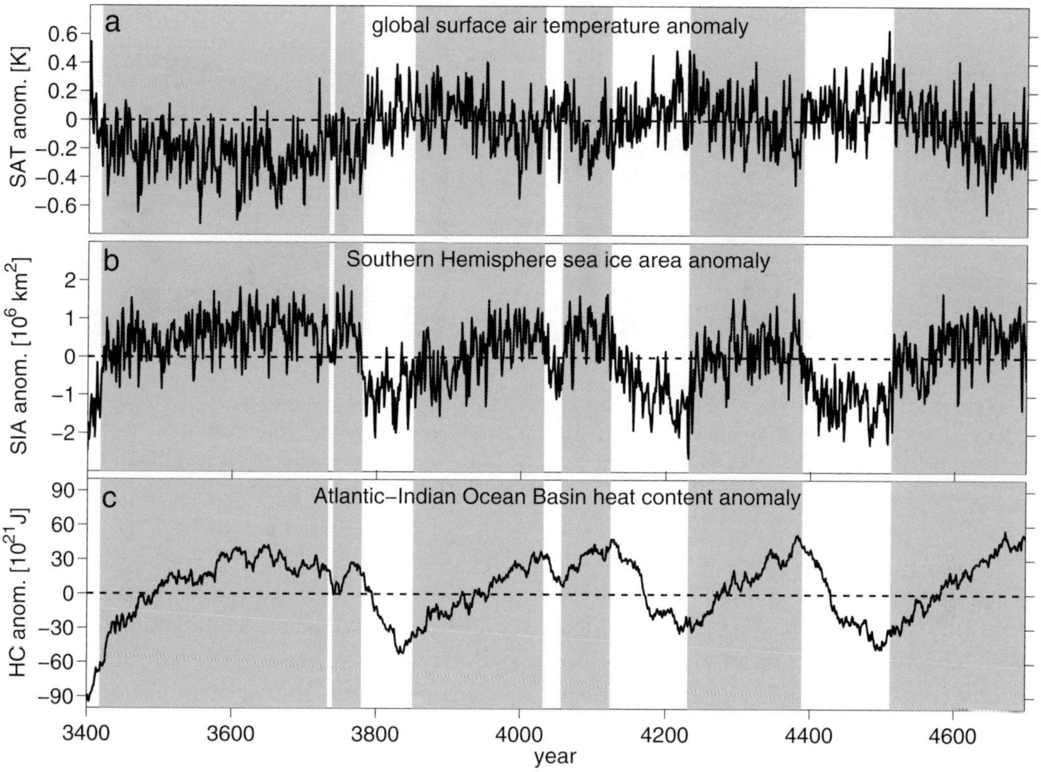

Fig. 3. The KCM features pronounced centennial variability in the Southern Ocean Sector which is even seen in the globally averaged SAT (a). The centennial variability is also seen in Antarctic sea ice area (b). The variability is linked to the oceanic heat content in the Atlantic-Indian Ocean Basin of the Southern Ocean (c). All time series show annual mean values of anomalies that are derived with respect to the model years 3,800–4,100, which approximately cover one full period of the deep convection oscillation. Grey shading denotes phases without deep convection in the Weddell Sea. Redrawn after Latif *et al.* (2013).

the air-sea coupling. Here we recover the mechanism in a fully coupled model. During phases of deep convection, the heat is rapidly released to the atmosphere, similar to what was observed during the Weddell Polynya (Moore *et al.*, 2002). When the oceanic heat reservoir is virtually depleted a coincidental strong freshening event at the sea surface shuts down the Weddell Sea convection (Martin *et al.*, 2013) and the heat accumulation resumes (Fig. 3c).

The heat release to the atmosphere during phases of deep convection drives global teleconnections in the KCM (Latif *et al.*, 2013). We computed a composite SLP anomaly pattern, again using the SOCV index shown in Fig. 2c, and compare the model pattern with the observed (National Center of Environmental Predictions, NCEP) SLP trend pattern of 1975–2010 (Fig. 4). This period also represents a switch from a convecting to a non-convecting state in the Weddell Sea. The switch is associated in the observations with an enhanced SLP gradient across the Southern Ocean, implying intensified Southern Hemisphere mid-latitude low-level westerlies (Fig. 4a). Consistent with the observations during 1975–2010, the KCM simulates an anomalously strong meridional SLP gradient over the mid-latitudes in association with the shift from a convecting to a non-convecting state in the Weddell Sea (Fig. 4b).

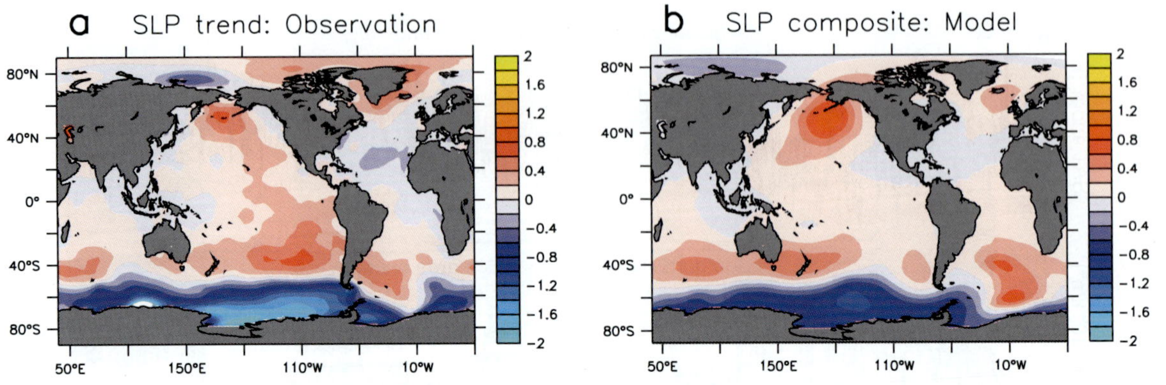

Fig. 4. The KCM simulates global teleconnections in association with the Southern Ocean centennial variability (SOCV). (a) The observed SLP trend 1975–2010 (hPa/decade). (b) The model composite (low-high) SLP anomaly pattern (hPa) using the SOCV index shown in Fig. 2c. The SLP data are from Kalney *et al.* (1996). From Latif *et al.* (2013).

Thus our model results support the conjecture that internal centennial variability originating in the Southern Ocean did contribute to the observed SLP trends during the recent decades. External forcing alone (e.g., Arblaster and Meehl, 2006; Arblaster *et al.*, 2011; Thompson *et al.*, 2011) may not completely explain the recent decadal SLP changes observed in the Southern Ocean Sector. This assertion is further supported by the slow weakening of the meridional SLP gradient from the 1920s to the 1970s, as exemplified by the decline in the Southern Annual Mode (SAM) reported by Visbeck (2009).

The Aleutian low and to a lesser extent the Icelandic low weaken in association with the phase change of the centennial mode in the model (Fig. 4b). This is also consistent with the observations during the last decades (Fig. 4a). The observed and simulated global SLP anomaly patterns exhibit a pattern correlation of 0.74. The observed SLP trend pattern, however, is likely a superposition of several internal modes and an externally forced signal. The good correspondence of the observed pattern with the model pattern may be thus coincidental, especially in the Northern Hemisphere.

Deep water warming in the Southern Ocean and a global-scale contraction of Antarctic Bottom Water were observed during the recent decades, after the occurrence of the Weddell Polynya in the mid-1970s (Robertson *et al.*, 2002; Purkey and Johnson, 2010, 2012). Such a behavior is simulated in the KCM during the decades immediately following the cessation of Weddell Sea deep convection (Fig. 5). While the surface ocean cools in the Weddell Sea (Fig. 5a), the mid-depth (Fig. 5c) and deep ocean (Fig. 5e) warm. The model simulates the strongest deep ocean warming in the Atlantic and Indian Ocean Sectors, in line with the observations which also show the strongest warming in these two regions.

The salinity tendencies in the Southern Ocean after convection shutdown in the Weddell Sea (Fig. 5, right panels) exhibit a pattern similar to that of the temperature tendencies: freshening accompanies the cooling at the surface and enhanced salinity the mid-depth (Fig. 5d) and deep warming (Fig. 5f).

A cautionary note should be made at this point. When making a reference to the observed trends during the recent decades, one has to keep in mind the additional effects of external forcing. In the real world, unforced and forced variability go side by side. For instance, the enhanced

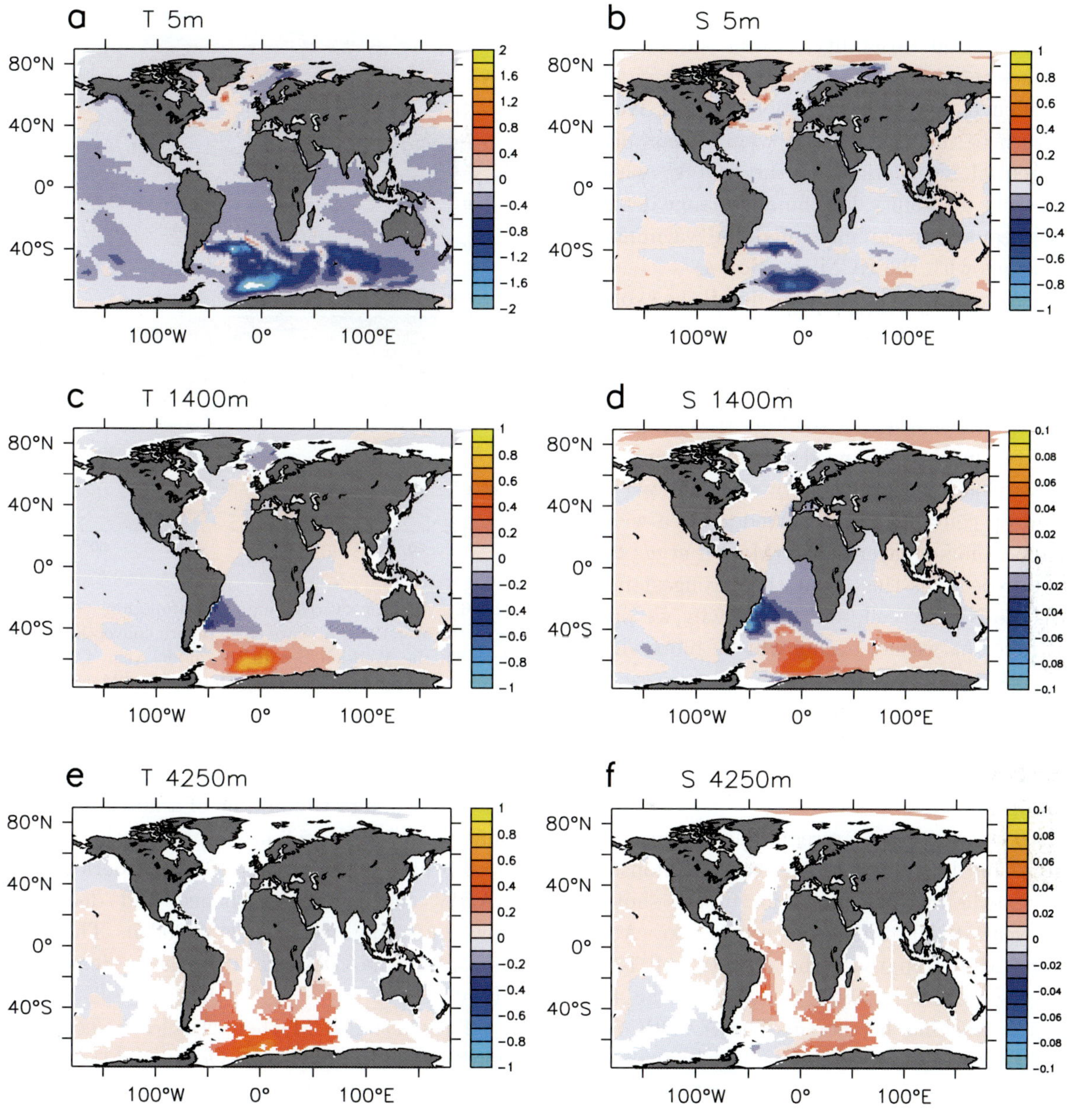

Fig. 5. The Southern Ocean centennial variability (SOCV) is associated with temperature and salinity changes throughout the water column. Composite (low-high) anomaly patterns using the SOCV index shown in Fig. 2c of (a) SST (5 m-ocean temperature), (b) SSS (5 m-salinity), (c) temperature at 1,400 m, (d) salinity at 1,400 m, (e) temperature at 4,250 m, (f) salinity at 4,250 m. Units are (°C) and (psu) for temperature and salinity, respectively. From Latif *et al.* (2013).

oceanic freshwater input in response to anthropogenic warming from ice melt or changes in the hydrological cycle can be rapidly communicated to the abyssal ocean in the high-latitude convection regions. Such forced processes by definition are not incorporated in control integrations of climate models which thus provide only unforced variability. What we would like to point out here is that the patterns of internal and external variability may be similar, which makes the detection of anthropogenic signals a challenge, especially on regional scales. We shall return to this point below when discussing the potential implications of the SOCV, which is internal variability, for climate change detection by means of an ensemble of global warming simulations with the KCM (Section 4).

We now describe in more detail the evolution of the Antarctic Bottom Water (AABW) and North Atlantic Deep Water (NADW) cells after one particular shutdown event, namely that of the model year 4,230 (see Fig. 3). This is one of the two most pronounced shutdown events in the sense that it was preceded by a rather long phase of deep convection in the Weddell Sea with the deepest mixed layer depth (Martin *et al.*, 2013). We expect for this event the strongest impact on the NADW cell, and this is why we present it here. As weaker events will be more affected by other processes, we believe that a case study is better suited to demonstrate the impact of Weddell Sea deep convection cessation on the NADW cell than, for instance, correlations or composites.

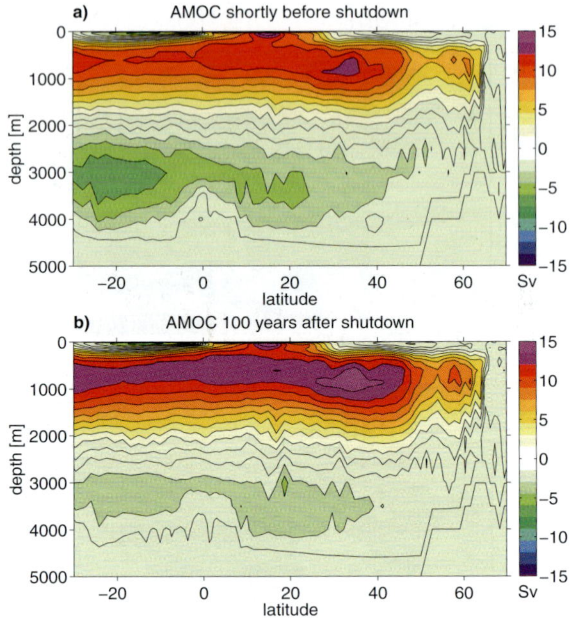

Fig. 6. Atlantic overturning streamfunction (Sv) averaged (a) over the ten years right before convection shutdown in the Weddell Sea in the model year 4,230 and (b) over the ten years a century later. After Martin *et al.* (2014).

The overturning streamfunction at convection shutdown, defined as the conditions averaged over the 10 years prior to the shutdown in year 4,230, is shown in Fig. 6a. At this time, the strength of the AABW cell is close to its maximum, with relatively large transports reaching far north. The NADW cell is relatively shallow extending barely to 2,000 m depth, and the overall cell strength is relatively weak. The AABW cell weakens with an e-folding time of about 50 years, which was estimated from the individual shutdown events. The shutdown of the deep convection in the Weddell Sea, which is associated with a decrease in the AABW formation rate, is followed by a strengthening of the NADW cell during the subsequent decades. This is visualized by another 10-year mean of the overturning streamfunction which was computed 100 years after the shutdown of Weddell Sea deep convection (Fig. 6b). At this time, which is in the middle of a non-convective phase, the AABW cell has almost reached quasi-equilibrium. During this 100-year period, the NADW cell has strengthened and deepened now extending well below 2,000 m. The outflow of the NADW at 30°S has increased by about 3Sv.

Thus in the KCM, there is a clear teleconnection between the Southern Ocean and the North Atlantic such that strong changes of the convective activity in the Weddell Sea are followed by a slow adjustment of the North Atlantic Deep Water cell. The response time of the latter is rather long and of the order of a century. The changes in the KCM are consistent with the conceptual picture presented by Seidov *et al.* (2001, see their Fig. 3) on the basis of forced ocean general circulation model experiments. They suggest that the southern deep water source "can be a more powerful modulator of the meridional

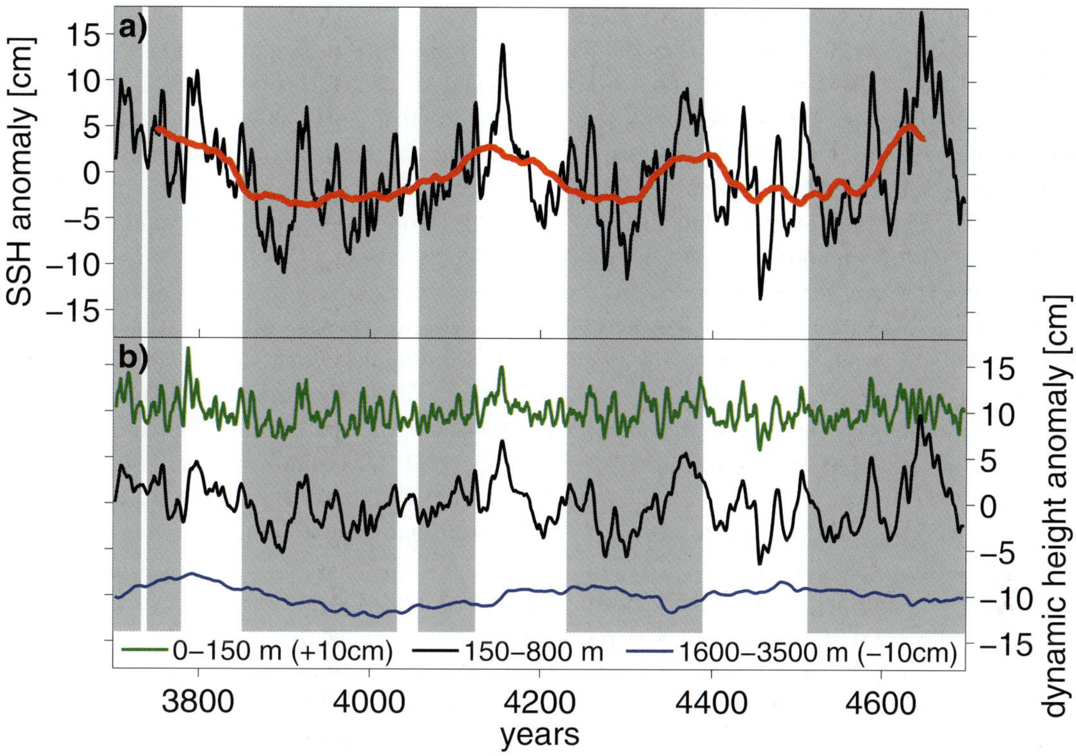

Fig. 7. (a) Annual mean (black) and low-pass filtered (red) SSH anomalies at the center of strongest variability in the North Atlantic. (b) Decomposition of SSH anomalies shown in (a) by deriving the dynamic height from salinity and potential temperature time series for the depth ranges surface to 150 m (green), 150 m–800 m (black), and 1,600 m–3,500 m (blue). A 5-year running mean filter was applied to all time series in (a) and (b) except for the red curve in (a), for which a 101-year running mean was used. Gray shaded areas mark time periods without Weddell Sea deep convection.

deep-ocean conveyor than the northern source". We extend this picture to the unforced (internal) variability at centennial time scales. In the KCM, a prolonged phase of Weddell Sea deep convection drives an anomalously strong AABW cell, which weakens the NADW cell with regard to both strength and vertical extent. After the halt of deep convection, the AABW cell becomes less prominent and the NADW cell starts to strengthen and deepen.

The SOCV considerably impacts the sea surface height (SSH) in several regions of the World Ocean. Strongly enhanced SSH variability at centennial time scales is found, for example, in the North Atlantic in association with the waxing and waning of the AABW formation (Fig. 7). The strongest centennial SSH variability is found in the region 45°N–55°N and 10°W–30°W, the region of the eastern portion of the North Atlantic Current in the model. We note that in the model, the North Atlantic Current has a too zonal path. The response time of the SSH in this index region with respect to the Weddell Sea convection shutdown is similar to that of the NADW cell (Fig. 6b), i.e. an anomalously high SSH follows the cessation of Weddell Sea deep convection with a time delay of about a century.

Peak-to-peak SSH changes amount to about ±15 cm/century in the region of the strongest variability, with a strong contribution from the SOCV (Fig. 7a). Such changes are of the same order of magnitude as the globally averaged sea level rise during the 20th century which amounted to about 15–20 cm. We note, however, that the model time series is rather "noisy" with pronounced decadal-scale variability superimposed on the longer-term centennial variability. Yet the model results suggest that unforced centennial variability cannot be ignored when assessing 20th century sea level trends in the North Atlantic. Moreover, internal centennial variability in sea level, given its long time scale, cannot be neglected either when projecting regional sea level in the North Atlantic for the 21st century, provided nature exhibits a level of centennial variability similar to that simulated by the KCM.

Dynamic height calculations (Fig. 7b) highlight the role of the upper ocean (150 m–800 m) warming for the sea level rise along the path of the model's North Atlantic Current after the cessation of Weddell Sea deep convection. This is plausible as a stronger AMOC will be connected with a stronger upper ocean northward heat transport in the North Atlantic (see Martin *et al.* (2014) for more discussion on this matter).

4. Implications for Global Warming

We analyze next an ensemble of 22 global warming simulations with the KCM. The version used for these experiments is that described in Park *et al.* (2009) where the tropical response in terms of both mean SST and ENSO is presented from a subset of the complete ensemble. Each of the integrations was forced by a 1% yr^{-1} (compound) increase of the atmospheric carbon dioxide (CO_2) concentration, but the runs start from different initial conditions obtained from a millennial control integration of the model including various states of Southern Ocean deep convection. The initial CO_2 concentration amounts to 348 ppmv. The doubling of the initial CO_2 concentration is reached after about 70 years, and the integrations were continued with fixed CO_2 for another 30 years to allow for stabilization. Each of the integrations has thus a total length of 100 years. The transient climate response (TCR), i.e. the global mean surface air temperature change averaged over a 20-yr period centered at the time of CO_2 doubling (model years 61 to 80) amounts to 2.9°C, which is considerably larger than the TCR in most other climate models (Park *et al.*, 2009).

We present in Fig. 8a the global mean SST anomalies. Although there is considerable spread about the long-term warming trend, it takes only a few decades for the mean warming signal (black line) to exceed the range of the internal variability (green lines indicate individual ensemble members). Clearly, the detection of the anthropogenic signal on these very large spatial scales is not an issue, given the simple experimental setup that we employ here, where we changed only one external forcing agent. The time-dependent direct and indirect effects of anthropogenic aerosols or changes in solar radiation, for instance, can considerably complicate the picture.

The warming in the Southern Ocean region (Fig. 8b) is considerably delayed by more than a decade relative to the global mean (Fig. 8a) and features a greater spread over the entire 100 year period. Here we are mostly interested in such regional variations of the ensemble-mean warming pattern and the spread of the centennial SST trends on regional scales. We computed at each grid point the ensemble-mean surface temperature trend over all 100 years of the integrations (Fig. 9a) and the corresponding ensemble spread which is expressed by the standard deviation of the centennial surface temperature trends (Fig. 9b). The surface temperature warms virtually everywhere, with the exception of a small region in the North Atlantic that cools (Fig. 9a). The cooling is due to a slowing of the AMOC, with a reduction of the overturning streamfunction at 30°N by about 40% in the ensemble mean (not shown) by the end of the global warming integrations.

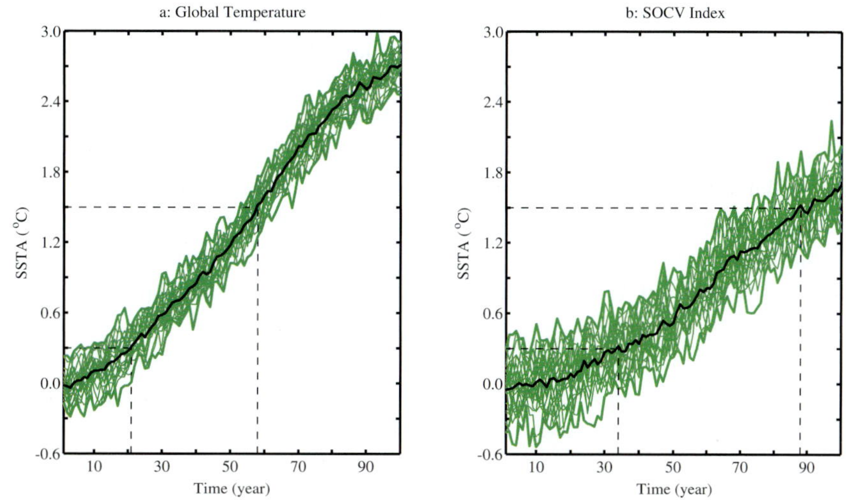

Fig. 8. The SST evolution in an ensemble of greenhouse warming (1% yr^{-1}) simulations with the KCM. (a) Global mean SST anomalies (SSTA, °C) and (b) anomalies of the Southern Ocean Centennial Variability (SOCV) index defined as the SST averaged over the region 50°S to 70°S (Latif *et al.*, 2013). Anomalies are computed as deviations from the mean state of the control simulation. Green lines depict the individual realizations, whereas the black lines show the ensemble means. Thin dashed lines indicate the timing of 0.3°C and 1.4°C warming, which are about 10% and 50% of the global mean warming after 100 years.

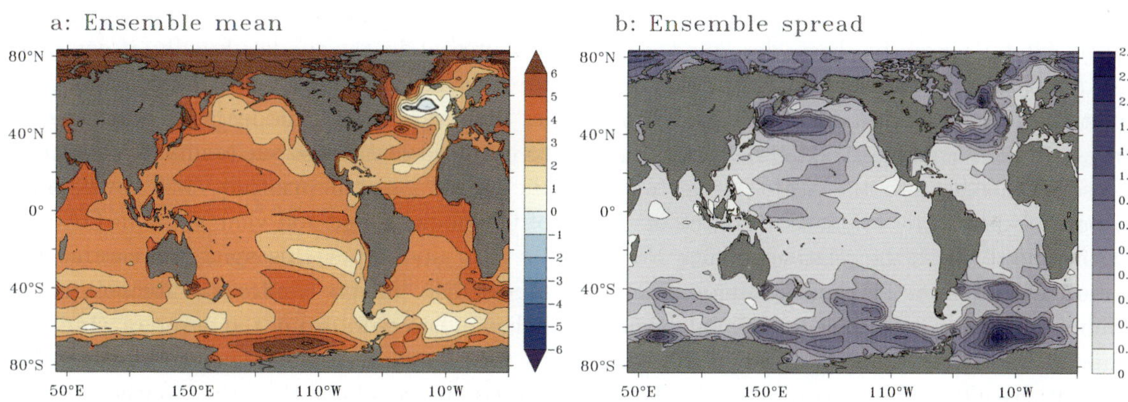

Fig. 9. (a) The ensemble-mean surface temperature trend (°C/century) and (b) the ensemble spread expressed by the standard deviation (°C/century) obtained from an ensemble of 22 global warming simulations with the KCM. Please note the change in contour interval width at 0.6°C/century in (b).

The Southern Ocean warms only little, with the SST in the region between 50° and 70°S exhibiting a warming in the ensemble mean of less than 2°C/century (Fig. 8b). The model thus suggests that this region may continue to stay "cold" even in the presence of a rather strong global sea surface warming of the order of about 3°C/century. Again, our experimental setup is incomplete, as several processes such as the long-term effects of a potential recovery of the Antarctic ozone hole are not considered here.

The ensemble spread of the centennial surface temperature trends is particularly large, and consequently the signal-to-noise ratio (Latif and Park, 2012) small, in parts of the North Pacific and North Atlantic, and in the Southern

Ocean (Fig. 9b). In the Weddell Sea region, which exhibits the strongest internal multicentennial SST variability in the KCM, the spread is as large as the mean warming signal (Fig. 9a). The observed SST trends south of 50°S during the last decades, which exhibited a weak cooling trend in the zonal mean (Fig. 1), may be thus interpreted just as a realization of internal centennial variability to the extent that the model is realistic.

5. Conclusions

A strong multicentennial oscillation of open ocean deep convection between two extreme regimes, one with and the other without convection in the Southern Ocean, is found in a millennial control simulation with the Kiel Climate Model (KCM), a coupled atmosphere-ocean-sea ice general circulation model. The driving mechanism of the Southern Ocean centennial variability (SOCV) can be best described as a recharge oscillator that consists of a slow accumulation of heat at mid-depth during the non-convective regime in the Weddell Sea and a fast release of this heat to the atmosphere during episodes of deep convection. The source of the heat at mid-depth is the relatively warm North Atlantic Deep Water. The multicentennial mode significantly affects regional and global surface temperature, impacts sea level even outside the Southern Ocean, for instance in the North Atlantic, and influences the AMOC, and all this with amplitudes large enough to mask the effects of anthropogenic climate change over intervals as long as several decades.

During phases of deep convection in the Weddell Sea the surface Southern Ocean warms, the abyssal Southern Ocean cools, Antarctic sea ice extent retreats, and the low-level atmospheric circulation over the Southern Ocean weakens. After the cessation of deep convection in the Weddell Sea the surface Southern Ocean cools, the abyssal Southern Ocean warms, Antarctic sea ice expands, and the low-level atmospheric circulation over the Southern Ocean intensifies, consistent with what has been observed during the recent decades. Thus some of the trends observed during the recent decades may be, at least partly, due to the effects of the SOCV. This may also pertain to the current hiatus in global warming, which could be partly due to an enhanced heat uptake of the deep Southern Ocean.

The global three-dimensional ocean circulation and the mechanisms of its variability are not well known from observations. We therefore heavily rely on climate models to understand the mechanisms of climate variability, especially those in the Southern Ocean. The KCM like other climate models suffers from large biases. Processes important to Southern Ocean variability such as mesoscale eddies are not explicitly resolved and instead parameterized. Moreover, shelf convection, the main formation mechanism of the very dense Antarctic Bottom Water is not present in the model. Finally, if compared to other climate models the KCM seems to be a "high variability" model, both with respect to unforced and forced variability. In particular, the KCM's climate sensitivity is at the high end of the model range.

It is unclear at this stage of research how model bias affects centennial variability. We believe, however, from the level of similarity to the observations that the model is realistic enough to conclude that the Southern Ocean plays an important role in the generation of internal centennial variability and that this variability has significant climate impacts, even outside the Southern Ocean sector.

Acknowledgments

This study was supported by the European Union Project NACLIM and the BMBF Project RACE. The model integrations were performed at the DKRZ Hamburg and the Computing Centre at Kiel University.

References

Arblaster, J. M., and G. A. Meehl, 2006: Contributions of external forcings to Southern Annular Mode changes. *J. Climate*, **19**, 2896–2095.

———, ———, and D. J. Karoly, 2011: Future climate change in the Southern Hemisphere: Competing effects of ozone and greenhouse gases. *Geophys. Res. Lett.*, **38**, L02701, doi: 10.1029/2010GL045384.

Carsey, F. D., 1980: Microwave observations of the Weddell Polynya. *Mon. Wea. Rev.*, **108**(5), 2032–2044.

Cook, E. R., B. M. Buckley, R. D. D'Arrigo, and M. J. Peterson, 2000: Warm-season temperatures since 1600 BC reconstructed from Tasmanian tree rings and their relationship to large-scale sea surface temperature anomalies. *Clim. Dyn.*, **16**(2), 79–91, doi: 10.1007/s003820050006.

Delworth, T. L., and F. Zeng, 2012: Multicentennial variability of the Atlantic meridional overturning circulation and its climatic influence in a 4000 year simulation of the GFDL CM2.1 climate model. *Geophys. Res. Lett.*, **39**, L13702, doi: 10.1029/2012GL052107.

Gordon, A. L., 1982: Weddell deep water variability. *J. Mar. Res.*, **40**, 199–217.

———, and J. C. Comiso, 1988: Polynyas in the Southern Ocean. *Sci. Amer.*, **256**, 90–97.

Kalnay, E., *et al.*, 1996: The NCEP/NCAR 40-Year Reanalysis Project. *Bull. Amer. Meteor. Soc.*, **77**, 437–471.

Karnauskas, K. B., J. E. Smerdon, R. Seager, and J. F. González-Rouco, 2012: A Pacific centennial oscillation predicted by coupled GCMs. *J. Climate*, **25**, 5943–5961, doi: 10.1175/JCLID-11-00421.1.

Latif, M., and N. S. Keenlyside, 2011: A perspective on decadal climate variability and predictability. *Deep-Sea Research II*, **58**, 1880–1894, doi: 10.1016/j.dsr2.2010.10.066.

———, T. Martin, and W. Park, 2013: Southern Ocean sector centennial climate variability and recent decadal trends. *J. Climate*, **26**, 7767–7782, doi: 10.1175/JCLI-D-12-00281.1.

———, and W. Park, 2012: Climatic variability on decadal to century time-scales. *The Future of the World's Climate*, A. Henderson-Sellers, and K. McGuffie, Eds., Elsevier Science, Amsterdam, The Netherlands, 167–195.

Martin, T., W. Park, and M. Latif, 2013: Multi-centennial variability controlled by Southern Ocean convection in the Kiel Climate Model. *Clim. Dyn.*, **40**, 2005–2022, doi: 10.1007/s00382-012-1586-7.

———, ———, and ———, 2014: Southern Ocean forcing of the North Atlantic at multi-centennial timescales in the Kiel Climate Model. *Deep Sea Research II, Special issue on "Southern Ocean Dynamics and Biogeochemistry in a Changing Climate"*, doi: 10.1016/j.dsr2.2014.01.018, in press.

Menary, M., W. Park, K. Lohmann, M. Vellinga, M. D. Palmer, M. Latif, and J. H. Jungclauss, 2012: A multimodel comparison of centennial Atlantic meridional overturning circulation variability. *Clim. Dyn.*, **38**, 2377–2388, doi: 10.1007/s00382-011-1172-4.

Miettinen, A., *et al.*, 2012: Multicentennial variability of the sea surface temperature gradient across the Subpolar North Atlantic over the last 2.8 kyr. *J. Climate*, **25**, 4205–4219, doi: 10.1175/JCLI-D-11-00581.1.

Mikolajewicz, U., and E. Maier-Reimer, 1990: Internal secular variability in an ocean general circulation model. *Clim. Dyn.*, **4**, 145–156.

Moore, G. W. K., K. Alverson, and I. A. Renfrew, 2002: A reconstruction of the air-sea interaction associated with the Weddell Polynya. *J. Phys. Oceanogr.*, **32**, 1685–1698.

Park, W., N. S. Keenlyside, M. Latif, A. Ströh, R. Redler, E. Roeckner, and G. Madec, 2009: Tropical Pacific climate and its response to global warming in the Kiel Climate Model. *J. Climate*, **22**, 71–92, doi: 10.1175/2008JCLI2261.1.

———, and M. Latif, 2008: Multidecadal and multicentennial variability of the Meridional Overturning Circulation. *Geophys. Res. Lett.*, **35**, L22703, doi: 10.1029/2008GL035779.

———, and ———, 2012: Atlantic Meridional Overturning Circulation response to idealized external forcing. *Clim. Dyn.*, **39**(7–8), 1709–1726, doi: 10.1007/s00382-011-1212-0.

Parkinson, C. L., and D. J. Cavalieri, 2012: Antarctic sea ice variability and trends, 1979–2010. *The Cryosphere*, **6**, 871–880, doi: 10.5194/tc-6-871-2012.

Pierce, D. W., T. P. Barnett, and U. Mikolajewicz, 1995: Competing roles of heat and freshwater flux in forcing thermohaline oscillations. *J. Phys. Oceanogr.*, **25**(9), 2046–2064.

Purkey, S. G., and G. C. Johnson, 2010: Warming of global abyssal and deep Southern Ocean Waters between the 1990s and 2000s: Contributions to global heat and sea level rise budgets. *J. Climate*, **23**, 6336–6351.

———, and ———, 2012: Global contraction of Antarctic Bottom Water between the 1980s and 2000s. *J. Climate*, in press, doi: 10.1175/JCLI-D-11-00612.1.

Robertson, R., M. Visbeck, A. L. Gordon, and E. Fahrbach, 2002: Long–term temperature trends in the deep waters of the Weddell Sea. *Deep–Sea Research II*, **49**(21), 4791–4806.

Seidov, D., B. J. Haupt, E. J. Barron, and M. Maslin, 2001: Ocean bi-polar seesaw and climate: Southern versus northern meltwater impacts. *The Oceans and Rapid Climate Change: Past, Present, and Future, Vol. 126*, D. Seidov, B. J. Haupt, and M. Maslin, Eds., AGU, Washington, D. C., 147–167, doi: 10.1029/GM126p0147.

Sicre, M.-A., *et al.*, 2011: Sea surface temperature variability in the subpolar Atlantic over the last two millennia. *Paleoceanography*, **26**, PA4218, doi: 10.1029/2011PA002169.

Smith, T.M., R.W. Reynolds, T. C. Peterson, and J. Lawrimore, 2008: Improvements to NOAA's historical merged land-ocean surface temperature analysis (1880–2006). *J. Climate*, **21**, 2283–2296.

Thompson, D. W. J., and S. Solomon, 2002: Interpretation of recent Southern Hemisphere climate change. *Science*, **296**, 895–899.

———, ———, P. J. Kushner, M. H. England, K. M. Grise, and D. J. Karoly, 2011: Signatures of the Antarctic ozone hole in Southern Hemisphere surface climate change. *Nat. Geosci.*, **4**, 741–749, doi: 10.1038/ngeo1296.

Trenberth, K. E., and J. W. Hurrell, 1994: Decadal atmosphere-ocean variations in the Pacific. *Clim. Dyn.*, **9**, 303–319.

Vellinga, M., and P. Wu P, 2004: Low latitude freshwater influence on centennial variability of the Atlantic thermohaline circulation. *J. Climate*, **17**, 4498–4511.

Visbeck, M., 2009: A station-based Southern Annular Mode index from 1884 to 2005. *J. Climate*, **22**, 940–950, doi: 10.1175/2008JCLI2260.1.

Yin, F. L., and E. S. Sarachik, 1995: Interdecadal thermohaline oscillations in a sector ocean general circulation model: Advective and convective processes. *J. Phys. Oceanogr.*, **25**, 2465–2484.

Zwally, H. J., and P. Gloersen, 1977: Passive microwave images of the polar regions and research applications. *Polar Rec.*, **18**, 431–450.

Chapter 8

Atlantic Meridional Overturning Circulation and Climate

Rong Zhang
GFDL, NOAA,
Princeton, NJ 08540, USA
rong.zhang@noaa.gov
www.gfdl.noaa.gov/rong-zhang-homepage

Changes in the Atlantic Meridional Overturning Circulation (AMOC) could have a profound impact on global scale climate, as indicated by both observations and climate model simulations. This review chapter highlights the global and regional scale climate impacts of the AMOC on paleo and modern climate, the tropical and extra-tropical AMOC fingerprints and the origin of the multidecadal North Atlantic sea surface temperature (NASST) variations, the meridional coherence and propagation of AMOC variations, as well as the impacts of the Nordic Sea overflow on the AMOC and large scale North Atlantic ocean circulation. The results reviewed in this chapter are mainly focused on the Geophysical Fluid Dynamics Laboratory (GFDL) coupled climate modeling results and their comparisons with available paleo and modern observations, although modelling results from some other climate models are also discussed.

1. Introduction

The AMOC includes both the wind-driven component (the overturning circulation directly induced by surface wind stress) and the component associated with the formation of the North Atlantic Deep Water (NADW) at northern high latitudes, in particular, in the Labrador Sea and Nordic Sea. In climate modeling studies, the AMOC Index is often defined as a max value of the Eulerian mean overturning streamfunction integrated over the Atlantic basin at certain latitude. The AMOC component associated with the formation of the NADW includes a northward flow of warm water in the upper Atlantic, and a southward flow of cold water in the deep Atlantic, and carries huge amount of heat northward. Changes in the AMOC component associated with the formation of the NADW could have a profound impact on global scale climate, as indicated by both observations and climate model simulations. This chapter reviews the global and regional scale climate impacts of the AMOC during the last ice age and Holocene (Section 2) and during the 20th century (Section 3). It also highlights studies on the tropical and extratropical AMOC fingerprints and the origin of the multidecadal North Atlantic sea surface temperature (NASST) variations in the 20th century (Section 4), the meridional coherence and propagation of low frequency AMOC variations (Section 5), as well as the impacts of the Nordic Sea overflow on the AMOC and large scale North Atlantic ocean circulation (Section 6). The results reviewed in this chapter are mainly focused on the Geophysical Fluid Dynamics Laboratory (GFDL) coupled climate modeling results and their comparisons with available paleo and modern observations,

Climate Change: Multidecadal and Beyond
edited by Chih-Pei Chang et al.

although modelling results from some other climate models are also discussed.

2. AMOC and Abrupt Climate Changes during the Last Ice Age

Early modeling studies show that the AMOC collapse leads to an abrupt climate change over the Atlantic, however, very few responses were found in the tropics outside the Atlantic. Recently, many new paleo records discovered in the tropics outside the Atlantic show coherent abrupt climate changes with the millennial-scale Dansgaard–Oeschger (D–O) cycles recorded in the Greenland ice core during the last ice age. For example, when Greenland was in stadial (cooling) condition, the summer Asian monsoon was reduced as indicated by a record from Hulu Cave in eastern China (Wang *et al.*, 2001), and the Indian summer monsoon was weakened (Altabet *et al.*, 2002). A paleo proxy from the Cariaco basin off Venezuela suggests that the Intertropical Convergence Zone (ITCZ) shifted southward during cooling stadials of the Greenland D-O oscillations (Peterson *et al.*, 2000). Drying conditions in the northeastern tropical Pacific west of Central America were synchronous with the Younger Dryas and the latest Heinrich event — H1 (Benway *et al.*, 2006). Wet periods in northeastern Brazil are synchronous with Heinrich events, cold periods in Greenland, and periods of weak east Asian summer monsoons and decreased river runoff to the Cariaco Basin (Wang *et al.*, 2004). Those remote signals were extremely strong during H1 when the AMOC was almost shutdown (McManus *et al.*, 2004).

The global synchronization of abrupt climate changes as indicated by these paleo records is consistent with simulated response to a substantial weakening of the AMOC in climate models. For example, Zhang and Delworth (2005) simulated climate changes in response to the weakening of the AMOC using the GFDL coupled climate model (CM2.0). In the water hosing experiment using GFDL CM2.0, the AMOC was substantially weakened by an extra freshwater forcing of 0.6 Sv ($1\,\text{Sv} = 10^6\,\text{m}^3\,\text{s}^{-1}$) uniformly distributed over the northern North Atlantic (55°–75°N, 63°W–4°E) for the entire 60-yr duration. This substantial weakening of the AMOC weakens the global ocean northward cross-equatorial heat transport, and the global atmospheric northward cross-equatorial heat transport is strengthened for compensation, resulting in a southward shift of the global ITCZ. The simulated southward ITCZ shift over the tropical Atlantic is similar to that found in many modeling studies (Vellinga and Wood, 2002; Dahl *et al.*, 2005; Stouffer *et al.*, 2006), and consistent with paleo records of drier conditions over the Cariaco basin (Peterson *et al.*, 2000) and wetter conditions over northeastern Brazil during Heinrich events (Wang *et al.*, 2004).

Beyond the typical responses in the Atlantic, this experiment also shows that a substantially weakened AMOC leads to significant and persistent remote responses outside the Atlantic. These include a southward shift of the ITCZ in the tropical Pacific (Fig. 1), consistent with drying conditions over the northeastern tropical

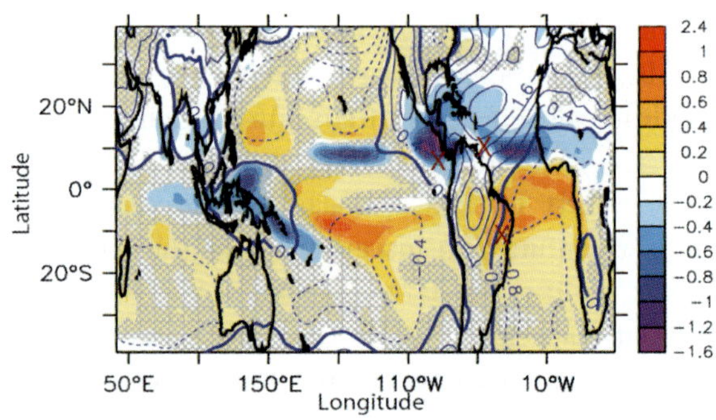

Fig. 1. Annual mean precipitation anomaly (m/yr) from GFDL CM2.0 water hosing experiment (color shade) (adapted from Zhang and Delworth, 2005). The blue contour is the annual mean SLP anomaly with an interval of 0.4 hPa. The brown crosses mark the location of paleo proxies of precipitation at the Cariaco Basin (Peterson *et al.*, 2000), northeastern Brazil (Wang *et al.*, 2004), and northeastern tropical Pacific (Benway *et al.*, 2006). The anomalies are significant at the 95% level in area without shadow with Student's *t* test.

Pacific during the Younger Dryas and Heinrich events (Benway *et al.*, 2006). The weakening of the AMOC also results in an El Niño–like condition and a weakened Walker circulation in the southern tropical Pacific, a La Niña–like condition and a stronger Walker circulation in the northern tropical Pacific, more symmetric western tropical Pacific Ocean circulation. This result is found in the water hosing experiment using the GFDL coupled model, and it is unclear whether such response is common cross other climate models. The simulated weakening of the Indian and East Asian summer monsoon (Fig. 2) in response to the weakening of the AMOC is also consistent with paleo proxies from the Indian Ocean (Altabet *et al.*, 2002) and the Hulu Cave in eastern China (Wang *et al.*, 2001, 2004). Note the interpretation of the isotope record for the East Asia remains controversial and some other climate model does not simulate the response of reduced East Asian summer monsoon to AMOC weakening (Pausata *et al.*, 2011). Ocean dynamics is very important for

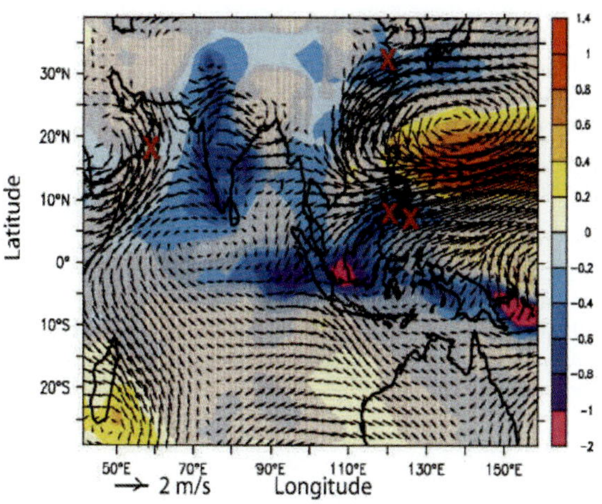

Fig. 2. Anomalous summer precipitation (m/yr) and flow at 925 hPa over the Indian and eastern China regions (adapted from Zhang and Delworth, 2005). Brown crosses mark the location of paleo proxies off the western boundary of the Arabian Sea (Altabet *et al.*, 2002), eastern China (Wang *et al.*, 2001), and around the west Pacific warm pool (Stott *et al.*, 2002; Dannenmann *et al.*, 2003). The precipitation anomalies are significant at the 95% level in area without shadow with Student's *t* test.

amplifying the tropical Pacific responses (Zhang and Delworth, 2005) and this is consistent with another recent modeling study (Wu *et al.*, 2007). When coupled with a slab ocean with no ocean dynamics, the simulated tropical Pacific response is much weaker and there is no statistical significant change of the east Asian summer monsoon. A recent study shows that all four past glacial terminations coincide with a major Heinrich Event and Weak Asian Monsoon Interval (WMI), suggesting the crucial role of AMOC in ice age terminations (Cheng *et al.*, 2009).

The substantial weakening of the AMOC also excites a large scale stationary wave pattern, resulting in the strengthening of the Aleutian Low and large-scale cooling in the central North Pacific, similar to that found in Mikolajewicz *et al.* (1997) and Vellinga and Wood (2002). The teleconnection between North Atlantic and North Pacific is also seen in paleo records, i.e. the cooling stadials of the Greenland D-O oscillations were synchronous with higher oxygen levels off the California coast (indicating reduced upwelling and reduced California Current) (Behl and Kennett, 1996), enhanced North Pacific intermediate-water formation, and the strengthening of the Aleutian Low (Hendy and Kennett, 2000). The global scale teleconnections of abrupt climate changes associated with changes in the AMOC are also found in recent paloeclimate modeling and model-data comparison studies of the last deglaciation (Liu *et al.*, 2009; Shakun *et al.*, 2012).

3. Global Scale Climate Impacts during the 20th Century

Over the 20th century, large scale low frequency variability has been observed in the North Atlantic SST. The low frequency variability in the Atlantic ocean is often called the "Atlantic Multidecadal Oscillation" (AMO), and the AMO index is often defined as the area average over the entire North Atlantic of low-pass filtered (LF) annual mean SST anomalies, after removing any linear trend (Enfield *et al.*, 2001; Sutton and Hodson, 2005; Knight *et al.*, 2005). The observed AMO pattern is similar to simulated SST anomaly pattern induced by AMOC changes (Delworth and Mann, 2000; Knight *et al.*, 2005; Zhang and Delworth, 2005). The AMO is often thought to be induced by AMOC variations and associated ocean heat transport fluctuations.

The observed AMO index is highly correlated with multidecadal variations of the tropical North Atlantic (TNA) SST and Atlantic hurricane activity (Goldenberg *et al.*, 2001; Landsea, 2005). The observed TNA surface warming is correlated with above-normal Atlantic hurricane activity during the 1950–60s and the recent decade since 1995. Linkages between the AMO index and multidecadal fluctuations of Sahel summer rainfall during the 20th century were mainly based on statistical analyses of observed data (Folland *et al.*, 1986), in which an unusually warm North Atlantic is associated with increased summer rainfall over the Sahel. To investigate the causal link between the AMO and other multidecadal variability, Zhang and Delworth (2006) simulated the impact of AMO-like SST variations on climate with a hybrid coupled model modified from GFDL CM2.1. The results show that a warm phase of the AMO leads to a northward shift of the Atlantic ITCZ, and thus an increase in the Sahelian and Indian summer monsoonal rainfall, as well as a reduction in the vertical shear of the zonal wind in the tropical Atlantic region that is important for the development of Atlantic major Hurricanes (Figs. 3, 4, 5). Figure 6 summarized the mechanism for the causal linkage between the AMO and other multidecadal variations. A positive AMO is associated with an enhanced northward ocean heat transport. The enhanced northward ocean heat transport is mainly compensated by a reduced northward atmospheric heat transport, resulting in an anomalous Hadley cell (anticlockwise) and a northward shift of the ITCZ, thus enhanced

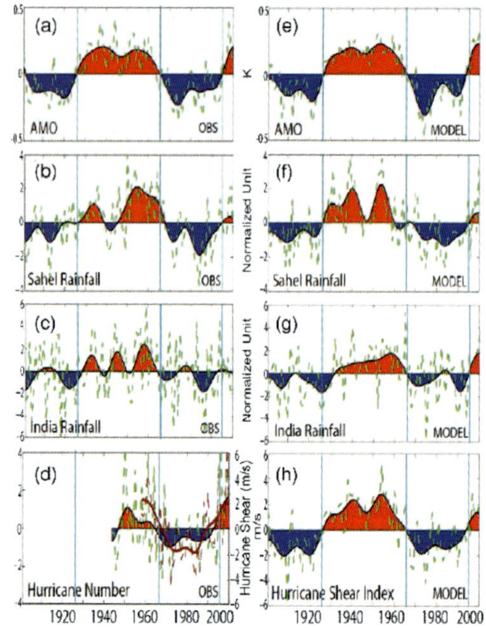

Fig. 3. Observed (OBS, left panels) and modeled (right panels) multidecadal fluctuations association with the AMO over the 20th century (adapted from Zhang and Delworth, 2006). Dashed green lines are unfiltered values, and the red and blue color-shaded values denote low-pass filtered values. The vertical blue lines denote transitions between warm and cold phases of the AMO. (a), (e) AMO Index. (b), (f) Normalized summer rainfall anomalies over the Sahel (20°W–40°E, 10–20°N). (c), (g) Normalized summer rainfall over west-central India (65–80°E, 15–25°N). (d) Number of major Atlantic Hurricanes from the NOAA HURDAT data set (life axis). The brown lines in (d) denote the vertical shear of the zonal wind (multiplied by −1, right axis) derived from the ERA-40 reanalysis, i.e. the difference in the zonal wind between 850 and 200 hPa over the south-central part of the main development region (MDR) for Atlantic Hurricanes (10–14°N, 70–20°W). (h) Vertical shear of the simulated zonal wind (multiplied by −1), calculated as in (d).

summer monsoon rainfall over the Sahel and India. Over tropical North Atlantic, due to the Coriolis Effect, the surface easterly trade winds are weakened and the upper level zonal winds become more westward (i.e. less westerly), resulting in a reduction of the vertical shear of the zonal wind and favoring the development of Atlantic major Hurricanes.

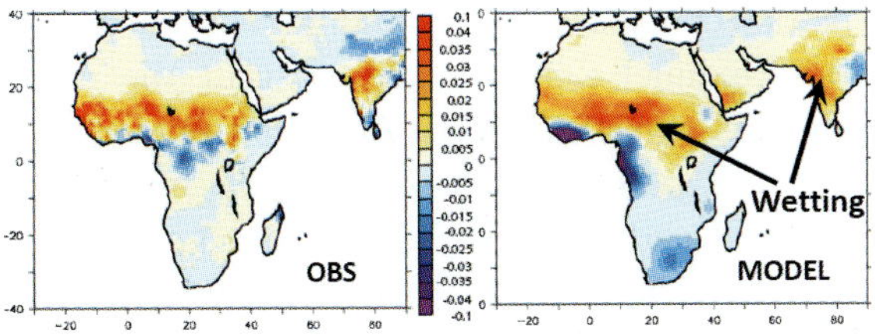

Fig. 4. Regression of low-pass filtered summer (JJAS) rainfall anomalies on the AMO Index (adapted from Zhang and Delworth, 2006). Left panel: Observation, the observed rainfall is from CRU-TS_2.1. Right panel: Model results. The regressions correspond to 1 SD of the AMO index, and are normalized by the SD of observed time series of the leading mode (371 mm/month) for left panel, and normalized by the SD of modeled leading mode (261 mm/month) for right panel.

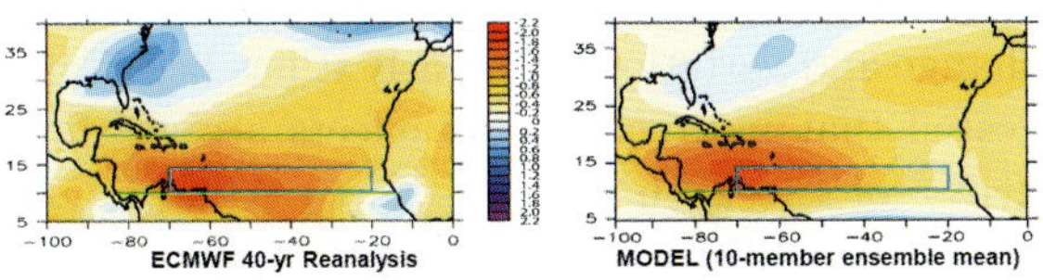

Fig. 5. Regression of low-pass filtered anomalous August-September-October vertical shear of zonal wind on the AMO Index (m/s, 1958–2000), corresponding to 1 SD of the AMO index (adapted from Zhang and Delworth, 2006). Left panel: Using ERA-40 and HADISST data. Right panel: Using model results. Green lines mark the MDR between 10°N and 20°N. The Hurricane Shear Index in Figures 3d,h is averaged over the blue box (70°W–20°W, 10°N–14°N).

Knight *et al.* (2006) also found a clear relationship between AMO-like SST fluctuations and surface air temperature over North America and Eurasia, modulation of the vertical shear of the zonal wind in the tropical Atlantic, and large-scale changes in Sahel and Brazil rainfall using the HadCM3 control simulations. Simulated vertical wind shear over the tropical North Atlantic forced by prescribed observed SST also exhibits similar anticorrelated multidecadal variations with the AMO index (Latif *et al.*, 2007).

The climate impacts of the AMO extend well beyond the tropics. Observational analyses (Enfield *et al.*, 2001) suggest that the AMO has a strong impact on the multidecadal variability

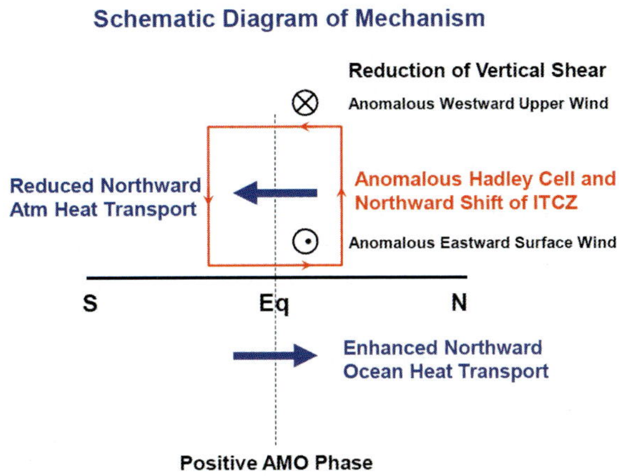

Fig. 6. Schematic diagram of the mechanism for the causal linkage between the AMO and other multidecadal variations. A positive AMO is associated with an enhanced northward ocean heat transport. The enhanced northward ocean heat transport is mainly compensated by a reduced northward atmospheric heat transport, resulting in an anomalous Hadley cell (anticlockwise) and a northward shift of the ITCZ, thus enhanced summer monsoon rainfall over the Sahel and India. Over tropical North Atlantic, due to the coriolis effect, the surface easterly trade winds are weakened and the upper level zonal winds become more westward (i.e. less westerly), resulting in a reduction of the vertical shear of the zonal wind and favoring the development of Atlantic major Hurricanes.

of U.S. rainfall and river flows. McCabe *et al.* (2004) further suggest that there is significant positive correlation between the AMO and the Central U.S. multidecadal drought frequency, and the positive AMO phase contributes to the droughts observed over the continental U.S. in the decade since 1995. Modeling studies (Sutton and Hodson, 2005, 2007) provide a clear assessment of the impact of the AMO over the Atlantic, North America, and Western Europe. In response to a warm phase of the AMO, a broad area of low pressure develops over the Atlantic, extending westward into the Caribbean and Southern United States. The pressure anomaly pattern denotes weakened easterly trade winds, potentially reinforcing the positive SST anomalies in the tropical North Atlantic Ocean by reducing the latent heat flux. Precipitation is generally enhanced over the warmer Atlantic waters and is reduced over a broad expanse of the United States. The summer temperature response is clear, with substantial warming over the United States and Mexico, with weaker warming over Western Europe.

Knight *et al.* (2005) find that variations in the AMOC are correlated with variations in the Northern Hemisphere mean surface temperature on decadal and longer time scales in the HadCM3 control simulations. Zhang *et al.* (2007) demonstrate that AMO-like SST variations can contribute to the Northern Hemispheric mean surface temperature fluctuations, such as the early 20th century warming, the pause in hemispheric-scale warming in the mid-20th century, and the late 20th century rapid warming, in addition to the long-term warming trend induced by increasing greenhouse gases. Zhang and Delworth (2007) suggests that the AMO-like SST variations can contribute to the Pacific Decedal Oscillation (PDO), especially the component of the PDO that is linearly independent of ENSO, and the associated Pacific/North America (PNA) pattern. In addition, a recent study (Timmermann *et al.*, 2007) showed that it is robust feature across climate models that the weakening of the AMOC leads to an enhanced ENSO variability through atmospheric teleconnections.

4. AMOC Fingerprints and the Origin of the Multidecadal North Atlantic SST Variations

The source of the observed Atlantic multidecadal SST variations has not been firmly established. One leading candidate mechanism involves fluctuations of the AMOC (Knight *et al.*, 2005; Latif *et al.*, 2006), and the SST contrast between the North and South Atlantic has been suggested as an AMOC fingerprint to reconstruct AMOC changes (Latif *et al.*, 2004). The AMO Index is often defined as detrended area-averaged North Atlantic SST (NASST), however, the role of direct radiative forcing in the detrended NASST is highly debated. Mann and Emanuel (2006) suggested that the main changes in tropical North Atlantic summer SST over the past century are mainly influenced by direct radiative forcing. Various approaches have been applied to separate non-detrended NASST changes into a radiatively forced part and a part arising from AMOC variability (Ting *et al.*, 2009; Zhang and Delworth, 2009; Delsole *et al.*, 2011; Wu *et al.*, 2011; Ting *et al.*, 2011). These studies find that both radiative forcing and ocean internal variability have contributed to non-detrended NASST changes.

Identifying the prime drivers of the 20th century multidecadal variability in the Atlantic Ocean is crucial for predicting future climate change and variations. Recently it has been suggested that indirect aerosol effects are a prime driver of the 20th century North Atlantic climate variability using the HadGEM2-ES climate model simulations (Booth *et al.*, 2012). However, Zhang *et al.* (2013) shows that there are major discrepancies between the HadGEM2-ES simulations and observations in the North Atlantic upper ocean heat content, in the spatial pattern of multidecadal SST changes within and outside the North Atlantic, and in the subpolar North Atlantic sea surface salinity. These discrepancies are strongly influenced by aerosol effects. For example, the HadGEM2-ES simulations, which have strong aerosol effects, show no net radiatively forced warming in the North Atlantic upper ocean over the 2nd half of the 20th century, in contrast to the observations (Gleckler *et al.*, 2012). This discrepancy of the North Atlantic upper ocean heat content with observations is mainly due to the strongly overestimated aerosol effects in HadGEM2-ES. It is also shown that the aerosol effects simulated in HadGEM2-ES cannot account for the observed anti-correlation between detrended multidecadal surface and subsurface temperature variations in the tropical North Atlantic (Zhang *et al.*, 2013). These discrepancies cast considerable doubt on the claim that aerosol forcing drives the bulk of this multidecadal variability.

Because there were no reliable observations of AMOC variability before 2004, it is crucial to reconstruct AMOC variability using fingerprints for understanding the origin of the NASST. The fingerprints need to be variables that can be derived from both climate models and observations. The fingerprints would link the AMOC with variables that are observed extensively. Identification of such fingerprints will contribute to the monitoring of AMOC variations, and improve assessments of the impacts of AMOC variability on global climate change. Zhang (2007) shows that the multidecadal variations of observed TNA SST are strongly anticorrelated with those of the observed TNA subsurface ocean temperature, with long-term trends removed (Fig. 7). Zhang (2007) further shows that the anticorrelated change between the TNA surface and subsurface temperature is a distinctive signature of the AMOC variations, using water-hosing experiments with the GFDL coupled climate model (CM2.1) (Fig. 7). The weakening of the AMOC leads to a southward shift of the Atlantic ITCZ and TNA surface cooling, as well as the thermocline deepening and TNA subsurface warming through the propagation of oceanic waves (Fig. 8). External radiative forced simulations with the same model do not provide a significant relationship between the TNA surface and subsurface temperature variations (Fig. 7). The observed detrended multidecadal TNA subsurface temperature anomaly may be taken as a proxy for the AMOC variability. Contrary to the rapid surface warming, there is a rapid subsurface cooling in the TNA in the recent decade, with long-term trends removed, indicating that the AMOC might have been strengthened since the mid 70s. This is consistent with the switch of the observed AMO index to a positive phase

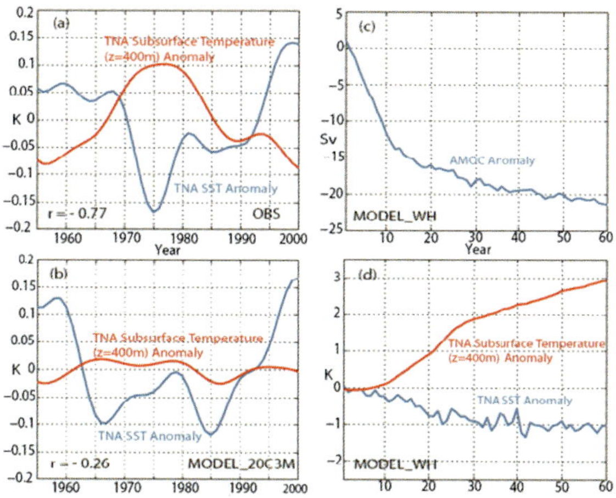

Fig. 7. Observed and modeled time series (adapted from Zhang, 2007). (a) Observeddetrended surface (blue) and subsurface (red) ocean temperature anomalies averaged over the TNA (0–14°N and 70°W–10°E). (b) Modeled detrended TNA surface (blue) and subsurface (red) ocean temperature anomalies from the ensemble of the 20th century historical simulations using GFDL CM2.1. A low-pass filter is applied with a 10-year cutoff period in (a, b). (c) Modeled AMOC anomaly (Sv) from the ensemble of water hosing experiments using GFDL CM2.1. (d) Modeled TNA surface (blue) and subsurface (ref) ocean temperature anomalies from the ensemble of water hosing experiments using GFDL CM2.1.

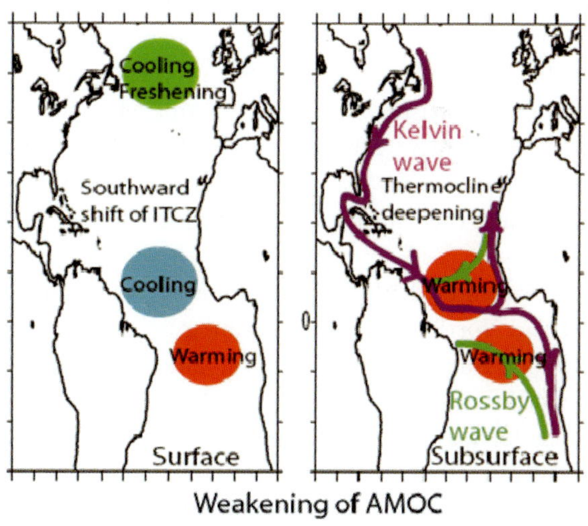

Fig. 8. Schematic diagrams of the mechanism for anticorrelated TNA surface and subsurface temperature anomalies (adapted from Zhang, 2007). The weakening of the AMOC leads to a southward shift of the Atlantic ITCZ and TNA surface cooling, as well as the thermocline deepening and TNA subsurface warming through the propagation of oceanic waves.

around the mid-1990s (Enfield *et al.*, 2001; Goldenberg *et al.*, 2001). Due to the observed anticorrelation between the TNA surface and subsurface temperature anomaly, the AMOC variations inferred from the observed detrended TNA subsurface temperature anomaly independently are in phase with the TNA SST anomaly, indicating that the AMOC variations have played a role in the phases of observed multidecadal TNA SST variations. The detrended NASST anomaly has highly debated origins. The detrended TNA subsurface temperature anomaly does not suffer from this debate, and may serve as an independent fingerprint of AMOC variations.

Similar AMOC-induced anticorrelated surface and subsurface TNA variations have also been found in NCAR CCSM3 coupled model simulations (Chiang *et al.*, 2008). The high-resolution temperature records of the last deglacial transition from a southern Caribbean sediment core also show that warmer subsurface temperatures correspond to colder surface temperature and weaker AMOC during the Younger Dryas (Schmidt *et al.*, 2012). The anticorrelated surface and subsurface temperature variations is also consistent with a recent model-data comparison study for oceanic changes in the last ice age (Marcott *et al.*, 2011). The subsurface warming is also important for understanding the cause of deglaciation (He *et al.*, 2013).

Satellite altimeter data shows a weakening of the North Atlantic subpolar gyre during the 1990s (Häkkinen and Rhines, 2004). However, whether the recent slowing subpolar gyre is a decadal variation or a long-term trend remains unclear. Zhang (2008) shows that the altimeter data is highly correlated with instrumental subsurface ocean temperature data in the North Atlantic, and both show opposite signs between the subpolar gyre and the Gulf Stream path (Figs. 9a, b). Such a dipole pattern is a distinctive fingerprint of AMOC variability, as shown for the first time by a 1000-year coupled ocean-atmosphere model (GFDL CM2.1) simulation (Figs. 9c, d; Fig. 10).

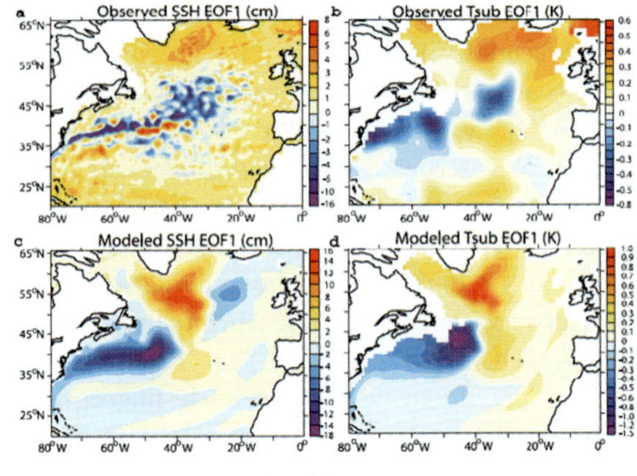

Fig. 9. Observed and modeled annual mean anomalies of SSH and Tsub (adapted from Zhang, 2008). (a) Observed SSH EOF1 (cm) and (b) observed Tsub EOF1 (K) for 1993–2003. (c) Modeled SSH EOF1 (cm) and (d) modeled Tsub EOF1 (K) from GFDL CM2.1 1000-year control integration.

The results suggest that the recent slowdown of the subpolar gyre is part of a multidecadal variation and suggests a strengthening of the AMOC. The historical subsurface temperature data could be used for reconstructing AMOC variations in the past. The ongoing satellite and subsurface temperature measurements could be used to monitor future AMOC variations sensitively. The AMOC variations inferred from the observed leading mode of subsurface temperature are consistent with those inferred from the observed AMO index. The results bring significant new evidence that the observed multidecadal NASST variations are linked to AMOC variations rather than merely a 20th century artifact of changes in radiative forcing.

5. Meridional Coherence and Propagation of AMOC Variations

The pioneering RAPID program was established recently to monitor AMOC variations at 26.5°N since 2004 (Cunningham *et al.*, 2007). This observational program excited two challenging

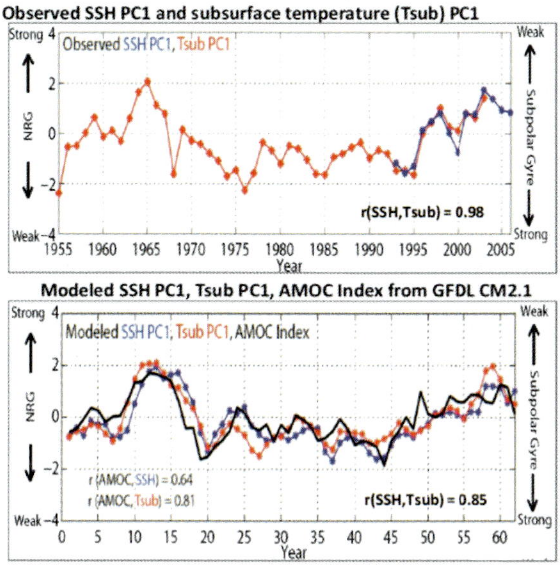

Fig. 10. Observed and modeled leading principle component (PC1) of SSH and Tsub (adapted from Zhang, 2008). Upper panel: Observed SSH PC1 (blue) and Tsub PC1 (red). Lower panel: Modeled SSH PC1 (blue) and Tsub PC1 (red), along with the AMOC index (black) from a segment of the GFDL CM2.1 control integration. The correlation in the upper panel is for the overlapping period. The correlations in the lower panel are from GFDL CM2.1 1000-year control integration.

questions: 1, Is there any meridional coherence of AMOC anomalies associated with changes in the North Atlantic Deep Water (NADW)? 2, How do AMOC anomalies associated with changes in NADW propagate and adjust with latitudes? Earlier studies (Kawase, 1987; Johnson and Marshall, 2002; Getzlaff *et al.*, 2005) suggest that AMOC anomalies propagate with Kelvin wave speed because the NADW outflow only moves along the western boundary as deep western boundary current (DWBC), resulting in a very short time lead (a few months) between subpolar and subtropical AMOC changes. Bingham *et al.* (2007) found a very low correlation between subpolar and subtropical AMOC variations at zero time lag in depth space and suggested a low meridional coherence of AMOC variations based on the above classic picture of AMOC propagation. However, recent RAFOS float observation in the western North Atlantic shows that a large fraction of the NADW outflow moves away from the western boundary and enters into the interior ocean near Flemish Cap and the Grand Banks along interior pathways (Bower *et al.*, 2009). This observational result leads to a fundamental question: given a partial detachment of the DWBC and the interior pathways of NADW outflow in Regime I, will AMOC variations still propagate with the Kelvin wave speed?

Using a coupled climate model (GFDL CM2.1), Zhang (2010) shows that due to the existence of interior pathways of NADW from Flemish Cap to Cape Hatteras as that observed, AMOC variations associated with the NADW formation propagate with the advection speed in this region, resulting in a much longer time lead (several years) between subpolar and subtropical AMOC variations (Figs. 11, 12). The much longer time lead provides a more useful predictability. Zhang (2010) also suggests that the typical estimation of AMOC in depth space

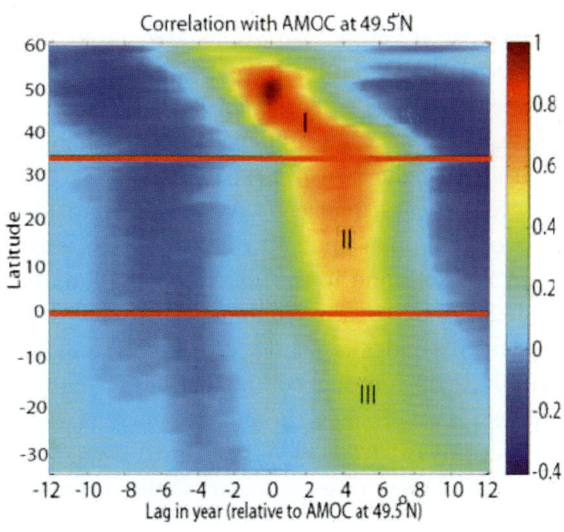

Fig. 11. Correlation between AMOC variations at 49.5°N and AMOC variations at all latitudes from the GFDL CM2.1 1000-year control integration (adapted from Zhang, 2010). The horizontal red lines mark boundaries of the three different regimes (I, II, III) of the southward propagation of AMOC variations.

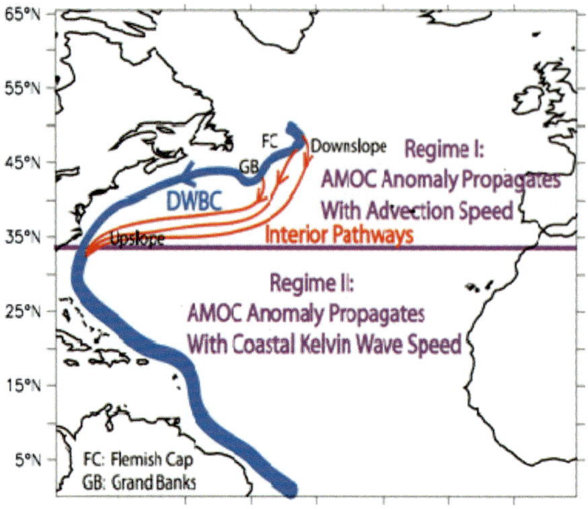

Fig. 12. Schematic diagram of the propagation of a positive AMOC anomaly induced by enhanced NADW formation (adapted from Zhang, 2010). In Regime II, there are no interior pathways. In Regime I, part of the DWBC moves downslope around Flemish Cap as well as the Grand banks, thus is partially detached from the western boundary and moves along interior pathways as observed. Hence the AMOC anomaly propagates with the slow advection speed in Regime I.

at mid and high latitudes is not a good measure of the part associated with the NADW formation, natural AMOC variations associated with the NADW formation have significant meridional coherence in density space, and monitoring AMOC variations in density space at higher latitudes might reveal a stronger signal with a several year time lead. A similar time lead between subpolar and subtropical AMOC variations is also found in the NCAR CCSM3 model (Y. Kwon, personal communication).

One caveat of Zhang (2010) is that the GFDL CM2.1 ocean component lacks explicit eddies, has broaden boundary currents and weaker DWBC from Flemish Cap to Cape Hatteras than that observed, and slows down the Kelvin wave artificially. The results need to be reinvestigated with higher resolution models. Using a high resolution eddy-permitting global coupled ocean-atmosphere model GFDL CM2.5, Zhang *et al.* (2011) shows that in the region north of 34°N, due to the existence of interior pathways, AMOC changes induced by an abrupt change in the Nordic Sea overflow propagate on the slow tracer advection timescale, instead of the fast Kelvin wave timescale. The slow advection timescale results in a time lead of several years between subpolar and subtropical AMOC changes, consistent with that found in the coarse resolution model (Zhang, 2010). The consistency between the high and low resolution coupled models suggests that the simulated AMOC propagation and adjustment process are robust.

6. Nordic Sea Overflow and AMOC

The Nordic Sea overflow, which enters into the deep North Atlantic through the Greenland-Iceland-Scotland (GIS) ridge, is one of the major sources for the NADW and contributes significantly to the deep branch of the AMOC. In many climate models, the simulated Nordic Sea overflow entering into the deep North Atlantic through the GIS ridge is unrealistically weak, compared to the observation (Dickson *et al.*, 1990). This bias is mainly caused by excessive convective entrainment of the overflow over staircase bathymetry in climate models, leading to much lighter and shallower overflow waters (Winton *et al.*, 1998; Danabasoglu *et al.*, 2010). Danabasoglu *et al.* (2010) proposed a new physical parameterization for the Nordic Sea overflow, and studied the impact of the Nordic Sea overflow parameterizations on the North Atlantic ocean circulation in a coupled coarse resolution climate model (NCAR CCSM4) as well as in its uncoupled ocean component. The previous coarse resolution model simulations lack explicit oceanic eddy activities and have broader western boundary currents than observed, and the previous ocean-only models do not have realistic air-sea boundary conditions. Hence it is important to reinvestigate the linkage between the Nordic Sea overflow and the North Atlantic ocean circulation using higher resolution coupled models.

Zhang *et al.* (2011) studied the sensitivity of the North Atlantic ocean circulation to an abrupt change in the Nordic Sea overflow for the first time using a high resolution eddy-permitting global coupled ocean–atmosphere model CM2.5. The model is recently developed at Geophysical Fluid Dynamics Laboratory (GFDL) (Delworth *et al.*, 2011), which has explicit eddy activities and narrow boundary currents. This sensitivity study points to the important role of the Nordic Sea overflow in the large scale North Atlantic ocean circulation. For example, a stronger and deeper penetrating Nordic Sea overflow will lead to stronger and deeper AMOC, stronger northward ocean heat transport, reduced Labrador Sea deep convection, westward shift of the North Atlantic Current (NAC) and southward shift of the Gulf Stream, warmer SST east of Newfoundland and colder SST south of the Grand Banks, stronger and deeper NAC and Gulf Stream, and stronger eddy activities along the NAC and the Gulf Stream paths. A stronger/weaker Nordic Sea overflow (Fig. 13) also leads to a contracted/expanded subpolar gyre (SPG) (Fig. 14), consistent with the relationship indicated by sediment core records of the last millennium from Iceland Basin (Moffa Sanchez *et al.*, 2011). It is crucial for climate models to have a correct representation of the Nordic Sea overflow, so as to have a realistic simulation of the AMOC and the large scale North Atlantic ocean circulation.

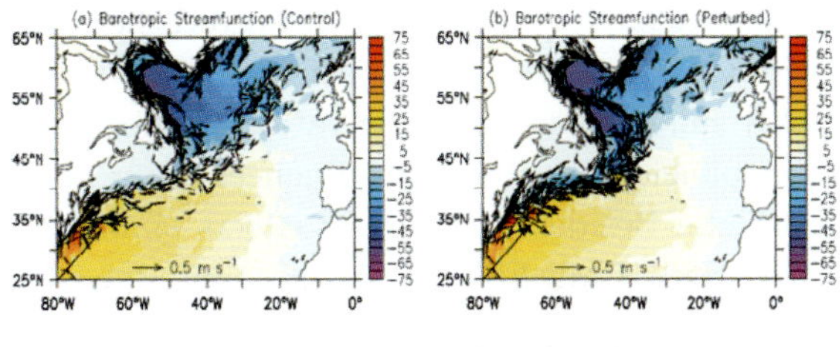

Fig. 14. Annual mean North Atlantic barotropic stream function (Sv) and upper ocean (100m) horizontal velocities (u, v, m/s) averaged from year 6 to 10 of (a) the control and (b) the perturbed experiments in GFDL CM2.5 (adapted from Zhang *et al.*, 2011). Vectors smaller than 8 cm/s are not shown.

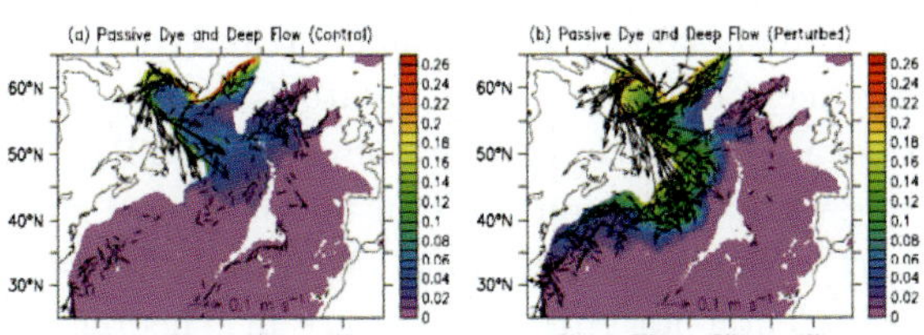

Fig. 13. Annual mean passive dye tracer released from the Denmark Strait, and horizontal velocities (u, v, m/s) in the North Atlantic deep ocean (average of 2500 m to 4000 m) averaged from year 6 to 10 of (a) the control and (b) the perturbed experiments in GFDL CM2.5 (from Zhang *et al.*, 2011). Vectors smaller than 1.5 cm/s are not shown.

7. Summary

The main conclusions reviewed in this chapter are highlighted as follows:

- Both paleo records and climate modeling suggest that the substantial weakening or collapse of the AMOC can result in a global synchronization of abrupt climate change. The response to a substantial weakening or collapse of the AMOC includes a southward shift of the ITCZ in both the Atlantic and the Pacific, and weakening of the Indian and East Asian summer monsoons. Ocean dynamics is important for amplifying the tropical Pacific response.
- Observational analyses and modeling results suggest that the AMO/AMOC has played an important role in many global and regional scale multidecadal climate fluctuations during the 20th century. In particular, a central impact of AMO is to alter the position of tropical ITCZ.
- The origin of the observed 20th century multidecadal North Atlantic SST variations is highly debated, due to the complication of anthropogenic forcings. Recently it has been suggested that aerosols are a prime driver of the 20th century North Atlantic climate variability using the HadGEM2-ES

climate model simulations. However, there are major discrepancies between the HadGEM2-ES simulations and observations. These discrepancies are strongly influenced by aerosol effects. Besides, the aerosol effects simulated in HadGEM2-ES cannot account for the observed anti-correlation between detrended multidecadal surface and subsurface temperature variations in the tropical North Atlantic. These discrepancies cast considerable doubt on the claim that aerosol forcing drives the bulk of this multidecadal variability.

- Independent fingerprints of AMOC variability bring new evidence that the observed AMO is linked to AMOC variations rather than merely a 20th century artifact of changes in radiative forcing. The AMOC fingerprints (SSH/Tsub) could be used for reconstructing AMOC variations in the past and monitoring AMOC variations in the future. Observed fingerprints of AMOC variability show that the AMOC was stronger during the 60's and the recent decade, and weaker during the 70's and 80's.
- Due to the existence of interior pathways, AMOC variations propagate with the slow advection speed, resulting in a much longer time lead (several years) between subpolar and subtropical AMOC variations. The longer time lead provides a more useful predictability. The results are robust in both coarse and high resolution climate models.
- The high resolution global coupled model shows that a stronger and deeper-penetrating Nordic Sea overflow will lead to a stronger and deeper AMOC, westward shift of the North Atlantic Current (NAC), southward shift of the Gulf Stream, warmer SST east of Newfoundland, and cooler SST south of the Grand Banks. A stronger/weaker Nordic Sea overflow also leads to a contracted/expanded subpolar gyre (SPG).

The results reviewed in this chapter are mainly focused on the GFDL coupled climate modeling results and their comparisons with available paleo and modern observations, although modelling results from some other climate models are also discussed. Future studies are needed for a multi-model comparison on subjects such as the AMOC fingerprints, the meridional coherence and propagations of AMOC variations, etc. The climate models used for the multi-model comparison need to be selected carefully based on their simulated climatological mean states. Climate models with a realistic Nordic Sea overflow and realistic North Atlantic deep water formation (location and strength) in the climatological mean states should be given a priority to be selected for the comparison.

Many challenging subjects in this field remain to be pursued in the future, such the mechanisms for the low frequency AMOC variability, the quantitative contribution of the AMOC variability to the observed North Atlantic SST variations, and the connection between various forcing (anthropogenic raditive forcing or natural solar irradiance) and AMOC changes. Strong research efforts in both observations and climate modeling are necessary to improve our understanding and prediction of future AMOC changes and associated climate impacts.

Acknowledgments

I am thankful for the very helpful discussions with many colleagues at the NTU international Science Conference on Climate Change: Multidecadal and Beyond in the Fall of 2012 for this work, and I am grateful for the travel support provided by National Taiwan University and Center for Advanced Study in Theoretical Sciences (CASTS) to attend the meeting.

References

Altabet, M. A., M. J. Higginson, and D. W. Murray, 2002: The effect of millennial-scale changes in

Arabian Sea denitrification on atmospheric CO2. *Nature*, **414**, 159–162.

Behl, R. J., and J. P. Kennett, 1996: Brief interstadial events in the Santa Barbara basin, NE Pacific, during the past 60 kyr. *Nature*, **379**, 243–246.

Benway, H. M., A. C. Mix, B. A. Haley, and G. P. Klinkhammer, 2006: Eastern Pacific Warm Pool paleosalinity and climate variability: 0-30 kyr. *Paleoceanography*, **21**, PA3008, doi: 10.1029/2005PA001208.

Bingham, R. J., C. W. Hughes, V. Roussenov, and R. G. Williams, 2007: Meridional coherence of the North Atlantic meridional overturning circulation. *Geophys. Res. Lett.*, **34**, L23606, doi: 10.1029/2007GL031731.

Booth, B. B. B., N. J. Dunstone, P. R. Halloran, T. Andrews, and N. Bellouin, 2012: Aerosols implicated as a prime driver of twentieth-century North Atlantic climate variability. *Nature*, **484**, 228–232.

Bower, A. S., M. S. Lozier, S. F. Gary, and C. W. Böning, 2009: Interior pathways of the North Atlantic meridional overturning circulation. *Nature*, **459**, 243–248, doi: 10.1038/nature07979.

Cheng H., R. L. Edwards, W. S. Broecker, G. H. Denton, X. Kong, Y. Wang, R. Zhang, and X. Wang, 2009: Ice age terminations. *Science*, **326**, doi: 10.1126/science.1177840.

Chiang, J. C. H., W. Cheng, and C. M. Bitz, 2008: Fast teleconnections to the tropical Atlantic sector from Atlantic thermohaline adjustment. *Geophys. Res. Lett.*, **35**, L07704, doi: 10.1029/2008GL033292.

Cunningham, S. A., *et al.*, 2007: Temporal variability of the Atlantic meridional overturning circulation at 26.5°N. *Science*, **317**, 935–938.

Dahl, K. A., A. J. Broccoli, and R. J. Stouffer, 2005: Assessing the role of North Atlantic freshwater forcing in millennial scale climate variability: A tropical Atlantic perspective. *Clim. Dyn.*, **24**, 325–346.

Danabasoglu, G., W. G. Large, and B. P. Briegleb, 2010: Climate impacts of parameterized Nordic Sea overflows. *J. Geophys. Res.*, **115**, C11005, doi: 10.1029/2010JC006243.

Dannenmann, S., B. K. Linsley, D. W. Oppo, Y. Rosenthal, and L. Beaufort, 2003: East Asian Monsoon forcing of suborbital variability in the Sulu Sea during Marine Isotope Stage 3: Link to Northern Hemisphere climate. *Geochem. Geophys. Geosyst.*, **4**, 1001, doi: 10.1029/2002GC000390.

DelSole, T., M. K. Tippett, and J. Shukla, 2011: A significant component of unforced multidecadal variability in the recent acceleration of global warming. *J. Climate*, **24**, 909–926.

Delworth, T. L., and M. E. Mann, 2000: Observed and simulated multidecadal variability in the Northern Hemisphere. *Clim. Dyn.*, **16**, 661–676.

———, *et al.*, 2011: Simulated climate and climate change in the GFDL CM2.5 high-resolution coupled climate model. *J. Climate*, **25**, 2755–2781, doi: 10.1175/JCLI-D-11-00316.1.

Dickson, R. R., E. M. Gmitrowicz, and A. J. Watson, 1990: Deep water renewal in the North Atlantic. *Nature*, **344**, 848–850.

Enfield, D. B., A. M. Mestas-Nuñez, and P. J. Trimble, 2001: The Atlantic multidecadal oscillation and its relation to rainfall and river flows in the continental U.S. *Geophys. Res. Lett.*, **28**, 2077–2080.

Folland, C. K., T. N. Palmer, and D. E. Parker, 1986: Sahel rainfall and worldwide sea temperatures. *Nature*, **320**, 602–607.

Getzlaff, J., C. W. Böning, C. Eden, and A. Biastoch, 2005: Signal propagation related to the North Atlantic overturning. *Geophys. Res. Lett.*, **32**, L09602, doi: 10.1029/2004GL021002.

Gleckler, P. J., *et al.*, 2012: Human-induced global ocean warming on multidecadal timescales. *Nature Climate Change*, **2**, 524–529, doi: 10.1038/nclimate1553.

Goldenberg, S. B., C. W. Landsea, A. M. Mestas-Nuñez, and W. M. Gray, 2001: The recent increase in Atlantic hurricane activity: Causes and implications. *Science*, **293**, 474–479.

Häkkinen, S., and P. B. Rhines, 2004: Decline of subpolar North Atlantic circulation during the 1990s. *Science*, **304**, 555–559.

He, F., J. Shakun, P. Clark, A. Carlson, Z. Liu, B. Otto-Bliesner, J. Kutzbach, 2013: Northern Hemisphere forcing of Southern Hemisphere climate during the last deglaciation. *Nature*, **494**, 81–85.

Hendy, I. L., and J. P. Kennett, 2000: Dansgaard-Oeschger cycles and the California Current System: Planktonic foraminiferal response to rapid climate change in Santa Barbara basin, Ocean Drilling Program hole 893A. *Paleoceanography*, **15**, 30–42.

Johnson, H. L., and D. P. Marshall, 2002: A theory for the surface Atlantic response to thermohaline variability. *J. Phys. Oceanogr.*, **32**, 1121–1132.

Kawase, M. 1987: Establishment of deep ocean circulation driven by deep water production. *J. Phys. Oceanogr.*, **17**, 2294–2317.

Knight, J. R., R. J. Allan, C. K. Folland, M. Vellinga, and M. E. Mann. 2005: A signature of persistent natural thermohaline circulation cycles in observed climate. *Geophys. Res. Lett.*, **32**, doi: 10.1029/2005GL024233.

———, C. K. Folland, and A. A. Scaife, 2006: Climate impacts of the Atlantic Multidecadal Oscillation. *Geophys. Res. Lett.*, **33**, doi: 10.1029/2006GL026242.

Landsea, C. W., 2005: Hurricanes and global warming. *Nature*, **438**, 11–13.

Latif, M., C. Böning, J. Willebrand, A. Biastoch, J. Dengg, N. Keenlyside, and U. Schweckendiek, 2006: Is the thermohaline circulation changing? *J. Climate*, **19**, 4631–4637.

———, N. Keenlyside, and J. Bader, 2007: Tropical sea surface temperature, vertical wind shear, and hurricane development. *Geophys. Res. Lett.*, **34**, L01710, doi: 10.1029/2006GL027969.

———, *et al.*, 2004: Reconstructing, monitoring, and predicting decadal-scale changes in the North Atlantic thermohaline circulation with sea surface temperature. *J. Climate*, **17**, 1605–1614.

Liu, Z., B. L. Otto-Bliesner, F. He, E. C. Brady, R. Tomas, P. U. Clark, A. E. Carlson, J. Lynch-Steiglitz, W. Curry, E. Brook, D. Erickson, R. Jacob, J. Kutzbach, J. Cheng, 2009: Transient simulation of last deglaciation with a new mechanism for Bølling-Allerød warming. *Science*, **325**, 310–314.

Mann, M. E., and K. A. Emanuel, 2006: Atlantic hurricane trends linked to climate change. *EOS*, **87**, 233–241.

Marcott, S. A., P. U. Clark, L. Padman, G. P. Klinkhammer, S. Springer, Z. Liu, B. Otto-Bliesner, A. E. Carlson, A. Ungerer, J. Padman, F. He, J. cheng, and A. Schmittner, 2011: Ice-shelf collapse from subsurface warming as a trigger for Heinrich events. *PNAS*, doi: 10.1073/pnas.1104772108.

McCabe, G. J., M. A. Palecki, and J. L. Betancourt, 2004: Pacific and Atlantic Ocean influences on multidecadal drought frequency in the United States. *PNAS*, **101**, 4136–4141.

McManus, J. F., R. Francois, J. M. Gherardi, L. D. Keigwin, and S. Brown-Leger, 2004: Collapse and rapid resumption of Atlantic meridional circulation linked to deglacial climate changes. *Nature*, **428**, 834–837.

Mikolajewicz, U., T. J. Crowley, A. Schiller, R. Voss, 1997: Modelling teleconnections between the North Atlantic and North Pacific during the Younger Dryas. *Nature*, **387**, 384–387.

Moffa Sanchez, P., I. R. Hall, S. Barker, and D. J. R. Thornalley, 2011: Centennial to decadal-scale ocean changes in the North Atlantic during the last Holocene. *RAPID-US AMOC International Science Meeting*, July, 2011.

Pausata, F. S. R., D. S. Battisti, K. H. Nisancioglu, and C. M. Bitz., 2011: Chinese stalagmite $\delta^{18}O$ controlled by changes in the Indian monsoon during a simulated Heinrich event. *Nat. Geosci.*, **4**, 474–480.

Peterson, L. C., G. H. Haug, K. A. Hughen, and U. Rohl, 2000: Rapid changes in the hydrologic cycle of the tropical Atlantic during the last glacial. *Science*, **290**, 1947–1951.

Schmidt, M. W., P. Chang, J. E. Hertzberg, T. R. Them II, L. J, and B. L. Otto-Bliesner, 2012: Impact of abrupt deglacial climate change on tropical Atlantic subsurface temperatures. *PNAS*, **109**, doi: 10.1073/pnas.1207806109.

Shakun, J., P. Clark, F. He, Z. Liu, B. Otto-Bliesner, S. A. Marcott, A. C. Mix, A. Schmittner, and E. Bard, 2012: Global warming preceded by increasing CO_2 during the last deglaciation. *Nature*, **484**, 49–54, doi: 10.1038/nature10915.

Stott, L., C. Poulsen, S. Lund, and R. Thunell, 2002: Super ENSO and global climate oscillations at millennial time scales. *Science*, **297**, 222–226.

Stouffer, R. J., *et al.*, 2006: Investigating the causes of the response of the thermohaline circulation to past and future climate changes. *J. Climate*, **19**, 1365–1387.

Sutton, R. T., and D. L. R. Hodson, 2005: Atlantic Ocean forcing of North American and European summer climate. *Science*, **309**, 115–118.

———, and ———, 2007: Climate response to basin-scale warming and cooling of the North Atlantic Ocean. *J. Climate*, **20**, 891–907.

Timmermann, A., Y. Okumura, S. I. An, A. Clement, B. Dong, E. Guiyardi, A. Hu, J. H. Jungclaus, M. Renold, T. F. Stocker, R. J. Stouffer, R. Sutton, S.-P. Xie, and J. Yin, 2007: The influence of a weakening of the Atlantic Meridional Overturning Circulation on ENSO. *J. Climate*, **20**, 4899–4919.

Ting, M., Y. Kushnir, R. Seager, and C. Li, 2009: Forced and internal twentieth-century SST trends in the North Atlantic. *J. Climate*, **22**, 1469–1481, doi: 10.1175/2008JCLI2561.1.

———, ————————, and ———, 2011: Atlantic multi-decadal variability and its climate impacts in CMIP3 and CMIP5 models. *WCRP Open Science Conference: Climate Research in Service to Society*, October, 2011.

Vellinga, M., and R. A. Wood, 2002: Global climatic impacts of a collapse of the Atlantic thermohaline circulation. *Climatic Change*, **54**, 251–267.

Wang, X., A. S. Auler, R. L. Edwards, H. Cheng, P. S. Cristalli, P. L. Smart, D. A. Richards, and C.-C. Shen, 2004: Wet periods in northeastern Brazil over the past 210 kyr linked to distant climate anomalies. *Nature*, **432**, 740–743.

Wang, Y. J., H. Cheng, R. L. Edwards, Z. S. An, J. Y. Wu, C.-C. Shen, and J. A. Dorale, 2001: A high-resolution absolute-dated late Pleistocene monsoon record from Hulu Cave, China. *Science*, **294**, 2345–2348.

Winton, M., R. W. Hallberg, and A. Gnanadesikan, 1998: Simulation of density-driven frictional downslope flow in z-coordinate ocean models. *J. Phys. Oceanogr.*, **28**, 2163–2174.

Wu, L., F. He, Z. Liu, and C. Li, 2007: Atmospheric teleconnections of tropical Atlantic variability: Interhemispheric, tropical-extratropical and cross-basin interactions. *J. Climate*, **20**, 856–870.

Wu, Z., N. E. Huang, J. M. Wallace, B. Smoliak, and X. Chen, 2011: On the time-varying trend in global-mean surface temperature. *Clim. Dyn.*, **37**, 759–773, doi: 10.1007/s00382-011-1128-8.

Zhang, R., 2007: Anticorrelated multidecadal variations between surface and subsurface tropical North Atlantic. *Geophys. Res. Lett.*, **34**, L12713, doi: 10.1029/2007GL030225.

———, 2008: Coherent surface-subsurface fingerprint of the Atlantic Meridional Overturning Circulation. *Geophys. Res. Lett.*, **35**, L20705, doi: 10.1029/2008GL035463.

———, 2010: Latitudinal dependence of Atlantic Meridional Overturning Circulation (AMOC) variations. *Geophys. Res. Lett.*, **37**, L16703, doi: 10.1029/2010GL044474.

———, and T. L. Delworth, 2005: Simulated tropical response to a substantial weakening of the Atlantic thermohaline circulation. *J. Climate*, **18**, 1853–1860.

———, and ———, 2006: Impact of Atlantic multidecadal oscillations on India/Sahel rainfall and Atlantic hurricanes. *Geophys. Res. Lett.*, **33**(5), doi: 10.1029/2006GL026267.

———, and ———, 2007: Impact of the Atlantic multidecadal oscillation on North Pacific climate variability. *Geophys. Res. Lett.*, **34**, L23708, doi: 10.1029/2007GL031601.

———, and ———, 2009: A new method for attributing climate variations over the Atlantic Hurricane Basin's main development region. *Geophys. Res. Lett.*, **36**, L06701, doi: 10.1029/2009GL037260.

———, ———, and I. M. Held, 2007: Can the Atlantic Ocean drive the observed multidecadal variability in Northern Hemisphere mean temperature? *Geophys. Res. Lett.*, **34**, L02709, doi: 10.1029/2006GL028683.

———, ———, A. Rosati, W. G. Anderson, K. W. Dixon, H.-C. Lee, and F. Zeng, 2011: Sensitivity of the North Atlantic Ocean Circulation to an abrupt change in the Nordic Sea overflow in a high resolution global coupled climate model. *J. Geophys. Res.–Oceans*, **116**, C12024, doi: 10.1029/2011JC007240.

———, ———, R. Sutton, D. L. R. Hodson, K. W. Dixon, I. M. Held, Y. Kushnir, J. Marshall, Y. Ming, R. Msadek, J. Robson, A. J. Rosati, M. Ting, and G. A. Vecchi, 2013: Have aerosols caused the observed Atlantic multidecadal variability? *J. Atmos. Sci.*, **70**, 1135–1144, doi: 10.1175/JAS-D-12-0331.1.

Chapter 9

North Atlantic Multi-Decadal Variability — Mechanisms and Predictability

Noel S. Keenlyside
Geophysical Institute, University of Bergen and Bjerknes Centre for Climate Research, Allégaten 70, 5007 Bergen, Norway
noel.keenlyside@gfi.uib.no
www.uib.no/gfi/en

Jin Ba, Jennifer Mecking, Nour-Eddine Omrani, Mojib Latif
GEOMAR Helmholtz-Zentrum für Ozeanforschung Kiel
Düsternbrooker Weg 20, 24105 Kiel, Germany

Rong Zhang and Rym Msadek
GFDL, NOAA
Princeton NJ, 08540, USA

The North Atlantic Ocean undergoes pronounced basin-wide, multi-decadal variations. The corresponding fluctuations in sea surface temperature (SST) have become known as the Atlantic Multidecadal Oscillation (AMO) or Atlantic multi-decadal variability (AMV). AMV is receiving increasing attention for three key reasons: (1) it has been linked to climate impacts of major socio-economic importance, such as Sahel rainfall; (2) it may temporarily mask anthropogenic global warming not only in the North Atlantic Sector, but over the Northern Hemisphere (NH); and (3) it appears to be predictable on decadal timescales. This chapter provides an overview of current understanding of AMV, summarizing proposed mechanisms, our ability to simulate and predict it, as well as challenges for future research.

1. Introduction

The existence of AMV has been long known (Bjerknes, 1964; Deser and Blackmon, 1993; Kushnir, 1994). During the instrumental record, AMV exhibited a periodicity of 70–80 years (Fig. 1a). There were two cold periods (1900–1920 and 1970–1990), and three warm periods: one at the start of the record, another in the middle of last century (1930–1960), and an ongoing one that began around 1990. These variations had major socio-economic impacts, including large ecosystem shifts and the collapse of major fish stocks in the North Atlantic, such as herring in the English Channel (Edwards *et al.*, 2013). AMV was also linked to precipitation and temperature changes, over Northern Africa, North America, Europe, and India, as well as to North Atlantic Hurricane activity (see Chs. 8 & 18 of this book and references therein). The large multi-decadal fluctuations in summertime Sahel rainfall had profound consequence for people living in the region. For example, the 1970–80s drought caused the death of at least 100,000 people, and displaced many more (Ch. 2, UNEP, 2002).

Climate Change: Multidecadal and Beyond
edited by Chih-Pei Chang et al.

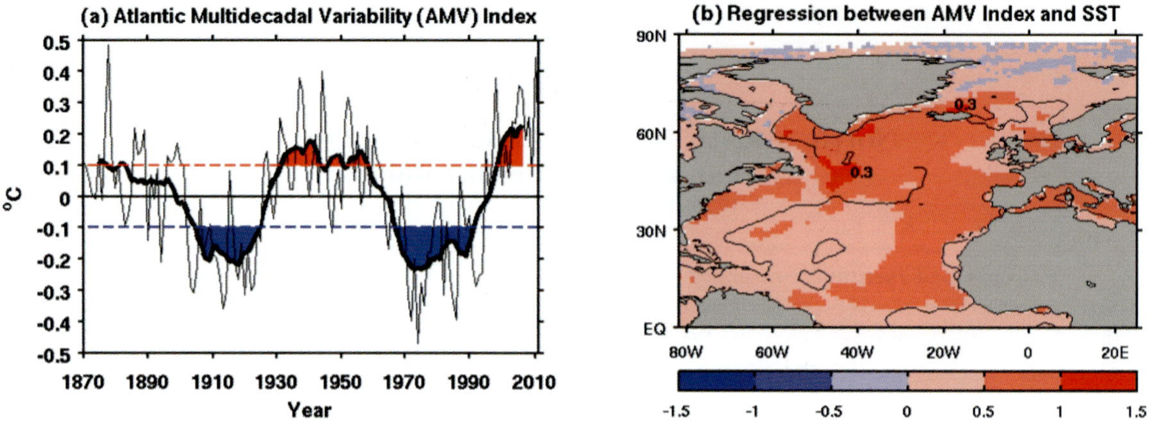

Fig. 1. Observed AMV (a) index and (b) pattern, computed by regressing the index on to SST in North Atlantic (Rayner *et al.*, 2003). Explained variance is contoured (Contour Interval = 0.3). See Section 2 for definition of index.

AMV was also linked to multi-decadal fluctuations in observed global mean surface air temperature (SAT), which showed enhanced warming from 1920–1940 and 1970–2000, and weak cooling from 1950–1970s. (Schlesinger and Ramankutty, 1994). These SAT variations superimposed on centennial scale global warming (Keenlyside and Ba, 2010) and could have resulted from either external factors or internal unforced climate dynamics (Ch. 9, IPCC, 2007). AMV may play a role in the second case, as models show that it may have driven NH multi-decadal SAT changes (Zhang *et al.*, 2007) and likely results from internal dynamics (Ting *et al.*, 2009). However, the possible contribution of external factors in AMV is also debated (Booth *et al.*, 2012; Zhang *et al.*, 2013).

This chapter reviews the understanding of AMV. We describe the observed AMV (Sec. 2) and its potential causes (Sec. 3). Relevant studies from climate models, which are an essential tool given limited observations, are summarized (Sec. 4), along with the leading mechanisms for AMV (Sec. 5). Section 6 reviews the predictability of AMV and discusses future research challenges. A summary concludes the chapter.

2. Observed Atlantic Multi-Decadal Variability

A common AMV index is the low-pass (*e.g.*, 11 year moving average) filtered, North Atlantic (0°–60°N, 7.5°–75°W) average of linearly detrended SST (Sutton and Hodson, 2005) (Fig. 1a). The linear trend is removed because of the persistent warming that was observed in the north Tropical Atlantic and may be related to global warming. While other methods may better exclude global warming, the basic characteristics of AMV remain the same (Ting *et al.*, 2009). In particular, North Atlantic SST differ by around 0.4°C on average between warm and cold phases (Fig. 1a). The changes are of same sign across the entire basin, but are largest over the subpolar North Atlantic (40°–60°N), Gulf Stream region, and the eastern subtropics/tropical North Atlantic (Fig. 1b). Furthermore, the SST variations persist throughout the year.

Boreal winter is a season of particular relevance for AMV as several processes are active that may be key to ocean-atmosphere interaction; these include formation of deep water and stratosphere-troposphere interaction (Sec. 6). Monte Carlo Singular Spectrum

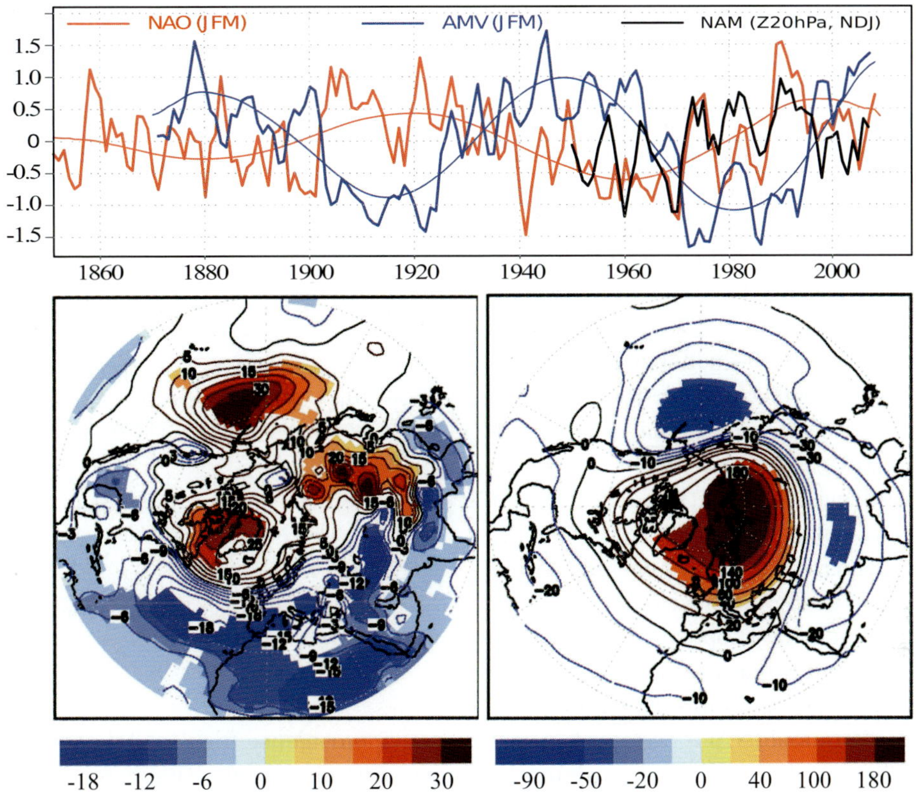

Fig. 2. Observed multi-decadal atmospheric variability associated with AMV. (Upper) The 3-year running mean winter (JFM) NAO (red) and AMV (blue) indices, and their multi-decadal MC-SSA reconstructions (thin blue and red) and the 3-year running mean early winter (NDJ) Northern Annular Mode (NAM) index (black) at 20hPa level. A typical warm phase is depicted by anomalies for 1951–1960 relative to 1961–1990 for (lower left) winter (JFM) 1000hPa geopotential height (gpm), and (lower right) early winter (NDJ) 20hPa geopotential height; only significant values at the 90% level are shaded. Figure adapted from Omrani *et al.* (2013).

Analysis (MC-SSA) (Ghil *et al.*, 2002) identifies the leading mode of the winter AMV index as multi-decadal. This variability differs significantly ($P = 0.05$) from a first order auto-regressive (AR-1) process and explains 48% (69%) of the raw (3-year running mean) variance (Fig. 2a). The strong multi-decadal signal is clearly visible in both winter and annual mean SST variations (Fig. 1a & 2a, blue curve).

The large-scale atmospheric circulation, and in particular, the North Atlantic Oscillation (NAO) also exhibited multi-decadal variations (Fig. 2a). The NAO is the dominant large-scale pattern of NH winter atmospheric variability, explaining up to 31% of the NH winter-mean SAT variability (Hurrell *et al.*, 2003). A negative (positive) NAO phase is associated with anomalous weak (strong) high latitude westerlies and a southward (northward) shift of subpolar jet, storm track and weather systems in the North Atlantic Sector (Hurrell *et al.*, 2003). Although many alternative NAO indices exist, they are all designed to capture sea level pressure (SLP) difference between the Azores High and Icelandic Low.

The multi-decadal variability of the winter NAO index explains 14% (36%) of the raw (3-year running mean) variance (Omrani *et al.*, 2013) and gives the NAO a weakly red spectrum (Wunsch, 1999). (In a red spectrum amplitude decreases with frequency.) The multi-decadal variations in the winter NAO and AMV indices tend to occur out of phase: a warm (cold) AMV phase is associated with a negative (positive) NAO. Composite analysis using historical SLP observations confirm this relation (Omrani *et al.*, 2013), which we illustrate with the winter surface geopotential height anomalies for the 1950–60s warm period (Fig. 2b). Atmospheric reanalysis (Kalnay *et al.*, 1996) suggest that associated changes in the zonal winds extend deep into the stratosphere, where they occur in early winter (Fig. 2a). In particular, the stratospheric polar vortex was anomalously weak in early winter during 1950–60s (Fig. 2c).

Oceanic circulation is thought key to AMV. The North Atlantic horizontal oceanic circulation consists of the anti-cyclonic subtropical gyre, the cyclonic subpolar oceanic gyre (SPG), and the Gulf Stream and North Atlantic Current (NAC) that separates the two gyres. The Gulf Stream and NAC transport warm and salty subtropical water northward in the upper ocean. Through surface cooling and vertical mixing the water becomes denser and eventually sinks in the high-latitudes, returning southward at intermediate (~1000–2500 m) levels (Kuhlbrodt *et al.*, 2007). Much of the sinking occurs in the Greenland, Iceland and Norwegian Seas (GIN), and in the Labrador and Irminger Seas (Marshall and Schott, 1999). The upper-level northward and intermediate-level southward flows form the Atlantic Meridional Overturning Circulation (AMOC), which transports vast amounts of heat poleward (Kuhlbrodt *et al.*, 2007).

Although salinity, subsurface temperature, and ocean circulation data are more limited than SST, they also show evidence for multi-decadal changes. In particular, sea surface salinity (SSS), for which observations extend back till 1895, show multi-decadal variations over the subpolar North Atlantic that slightly lagged the SST variations there (Reverdin, 2010). Hydrographic observations during the last 100 years

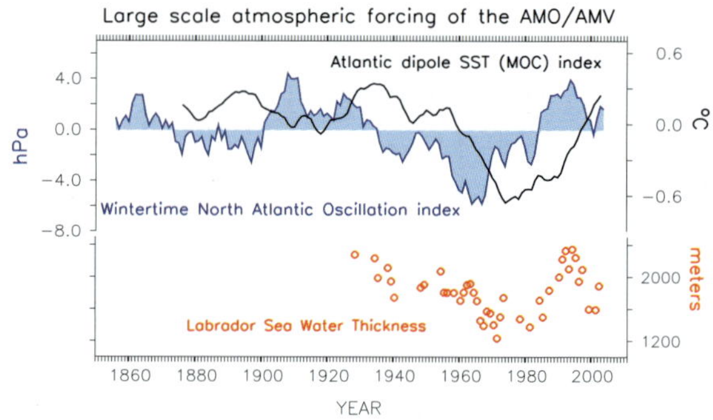

Fig. 3. Time series of winter NAO index (shading blue curve), the Atlantic dipole-SST anomaly index (black curve) and annual data of LSW thickness (red). The dipole index is a measure of the AMOC strength and defined as the difference in SST between the North (40°–60° N, 50°–10° W) and South (40°–60° S, 50° W–0°) Atlantic. The NAO and dipole indices are smoothed with an 11-year running mean. Multi-decadal changes of AMOC as indicated by the dipole index lag those of the NAO by about one decade. Figure from Latif and Keenlyside (2011).

provide evidence for decadal variations in temperature and salinity that extend over the upper 3000 m (Polyakov *et al.*, 2005). The subsurface changes have been related to variations in deep convection over the Labrador Sea, driven by changes in the NAO (Curry *et al.*, 1998; Dickson *et al.*, 1996): Labrador Sea Water (LSW) thickness decreased from the 1930s to the 1970s, then increased until the mid-1990s, and then decreased sharply; these changes lagged the winter NAO by a few years (Fig. 3).

LSW is one of the main water masses of the AMOC return flow. Thus, the observed LSW thickness changes suggest the AMOC may have undergone pronounced multi-decadal changes. It is not possible to confirm the AMOC changes, as direct observations of the AMOC only exist since 2004 at 26.5°N (Cunningham *et al.*, 2007). However, fingerprint techniques, which link observable patterns to AMOC variability, suggest the AMOC underwent multi-decadal changes (Latif *et al.*, 2004)(Ch. 9). For example, the SST difference between the North and South Atlantic is linked to AMOC driven changes in northward heat transport in models (Latif *et al.*, 2004); it suggests multi-decadal AMOC changes lagged those in the NAO and LSW by around a decade (Fig. 3). In addition, hydrographic data and satellite measurements of sea surface height show that the SPG, Gulf Stream and NAC strengthened from the 1960s to the 1990s, and then weakened, likely related to the NAO changes (Häkkinen and Rhines, 2004). The link to AMOC, however, remains unclear.

3. What Drove the Observed AMV?

On short (interannual) timescales extra-tropical SST variations result primarily from anomalous turbulent heat flux, particularly in winter (Cayan, 1992). Extra-tropical atmospheric variability, including the NAO, is often considered as a stochastic, white noise process (i.e., spectral amplitude is frequency independent) (Wunsch, 1999). The direct thermodynamic response of the ocean mixed-layer to such variability describes a red noise AR-1 process — the simplest stochastic climate model in the Hasselmann (1976) paradigm. AMV, however, is not consistent with an AR-1 process (Sec. 2) and thus not with this simplest model (Sec. 5).

There is evidence that the observed AMV was driven by the multi-decadal variations in the winter NAO. On these timescales the two phenomena tended to vary out of phase with each other (Fig. 2a). The turbulent fluxes associated with the negative (positive) NAO drive a tripolar SST pattern: warming (cooling) over higher latitudes and in the tropics, and cooling (warming) around the Gulf Stream and NAC (Visbeck *et al.*, 2003). Thus, the direct thermodynamic forcing associated with the NAO may explain changes in the high latitudes and subtropics (Fig. 1b). It cannot explain the temperature changes in the Gulf Stream and NAC (Bjerknes, 1964; Gulev *et al.*, 2013; Kushnir, 1994), suggesting a role for ocean dynamics there.

Observations are presently insufficient to assess the upper ocean heat budget on these timescales. Forced ocean model experiments, however, indicate that ocean circulation changes, associated with the AMOC and SPG, drove the recent decadal changes in upper ocean heat content over the SPG (Robson *et al.*, 2012; Yeager *et al.*, 2012). The winter NAO variations were shown to play a dominant role in the ocean circulation changes, which in turn were related to ocean deep convection and LSW formation with a few years lag (Böning *et al.*, 2006; Eden and Willebrand, 2001; Hatun *et al.*, 2005). Furthermore, idealized experiments show that the NAO variations can explain the observed AMV, through thermodynamic forcing and advection of heat (Eden and Jung, 2001; Klöwer *et al.*, 2014; Visbeck *et al.*, 1998).

The NAO forced picture, however, has problems explaining several aspects of AMV: Firstly, the coherent tropical SST anomalies are not fully explained in all seasons, and may result from local ocean circulation changes (Wang and Zhang, 2013). Climate models also largely reproduce the pattern, despite simulating little relation between the NAO and AMV (Sec. 4). Secondly, decadal variations in ocean circulation may also be influenced by non-linearity (Lohmann *et al.*, 2009) other large-scale patterns of atmospheric circulation (Langehaug *et al.*, 2012; Msadek and Frankignoul, 2009), and salinity anomalies (Zhang and Vallis, 2006). Thirdly, the origin of the multi-decadal variations in the winter NAO is unclear, and they may be partly forced by SST fluctuations. Lastly, external factors may also be important.

4. AMV in Climate Models

Due to the lack of instrumental data on these timescales, climate models become a useful tool to investigate AMV. In these models the North Atlantic is also a region of pronounced decadal variability (Boer and Lambert, 2008). Furthermore, this variability can be simulated in the absence of time varying external forcing (i.e., it is a mode of unforced, internal climate variability) and shares many similarities to observations (Delworth *et al.*, 1993; Knight *et al.*, 2005). Here the key features of unforced AMV will be described; Ba *et al.* (2014) present a similar but more detailed analysis. Section 7 discusses the issue of external forcing.

The analysis here is based on eight climate model preindustrial control simulations from the third coupled model intercomparison project (CMIP3). The simulation lengths vary from 340 to 500 years, and horizontal model resolutions from 2–4 degree (see Ch. 2, Ba, 2013). We note that caution is required when interpreting the model results as they suffer from large biases, with SST errors of several degrees in the North Atlantic. The AMV index is defined as in Sec. 2, and the AMOC index as the maximum overturning streamfunction at 30°N. In all correlation (regression) analysis we filter the indices with an 11-year running mean. A Student's T-test that accounts for serial correlation is used to assess significance.

The power spectra of simulated AMV indices in the eight CMIP3 preindustrial control runs show a red noise character (Fig. 4). The models all simulate some multi-decadal variability, but provide little evidence for a preferred 70–80 year

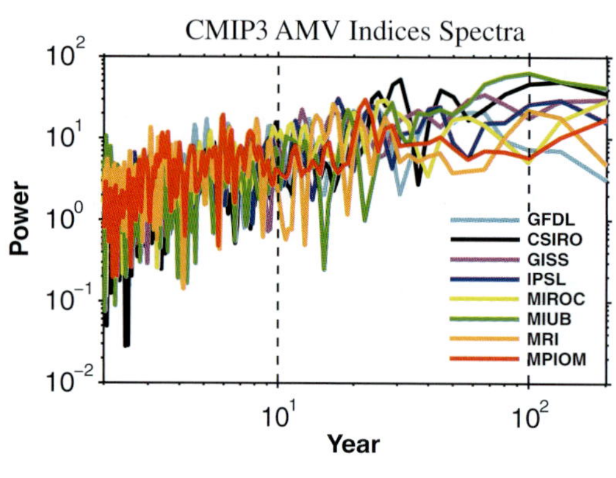

Fig. 4. Spectra of AMV indices in eight CMIP3 CGCMs. Due to the varying autocorrelation for the models, the individual AR-1 red-noise spectra are not shown.

periodicity suggested by observations. The spectra are consistent with past studies that have shown many models simulate variability in the North Atlantic on multi-decadal time scales, but with little agreement on preferred time scales (Ba *et al.*, 2013a; Delworth *et al.*, 1993; Jungclaus *et al.*, 2005; Knight *et al.*, 2005).

The AMV SST anomaly patterns from the eight CMIP3 models (Fig. 5) show the largest loadings in mid-latitudes, as observed (Fig. 1b). To varying degrees, all models show a positive relation in the tropical and subtropical North Atlantic and a tendency for weaker values to the west as observed (Fig 1b). The relation with SST north of 60°N differs significantly among models.

As indicated above, AMOC fluctuations are thought to drive AMV. This hypothesis is based mostly on ocean model and CGCM experiments that have shown a close relation between variations in the AMOC and northward oceanic heat transport (Delworth *et al.*, 1993; Knight *et al.*, 2005; Latif *et al.*, 2004). Consistently, the lead-lag relationship between the AMV and AMOC indices in the eight CMIP3 models strongly suggests that the AMOC impacts SST variability on multi-decadal time scales (Fig. 6a). In most models, AMOC variations tend to lead AMV changes by 4–6 years, with correlations above 0.5. Cross-correlation analysis of AMV indices and North Atlantic SST averaged over 0–30°N and over 30–60°N further indicates that the decadal AMOC changes mainly impact SST variability in the mid-latitude North Atlantic (not shown). This is consistent with the larger loadings in mid-latitudes seen in the AMV SST anomaly patterns (Fig. 5).

The power spectra of the AMOC indices have a red character, with all models exhibiting enhanced power on multi-decadal timescales, but with no preferred periodicity (Fig. 6b). Some models show quite strong variability with 20–30 year periodicity (GFDL, CSIRO), some with 40–50 years (MIUB-ECHO-G, ECHAM5/MPIOM, GISS), and others depict even longer-timescale

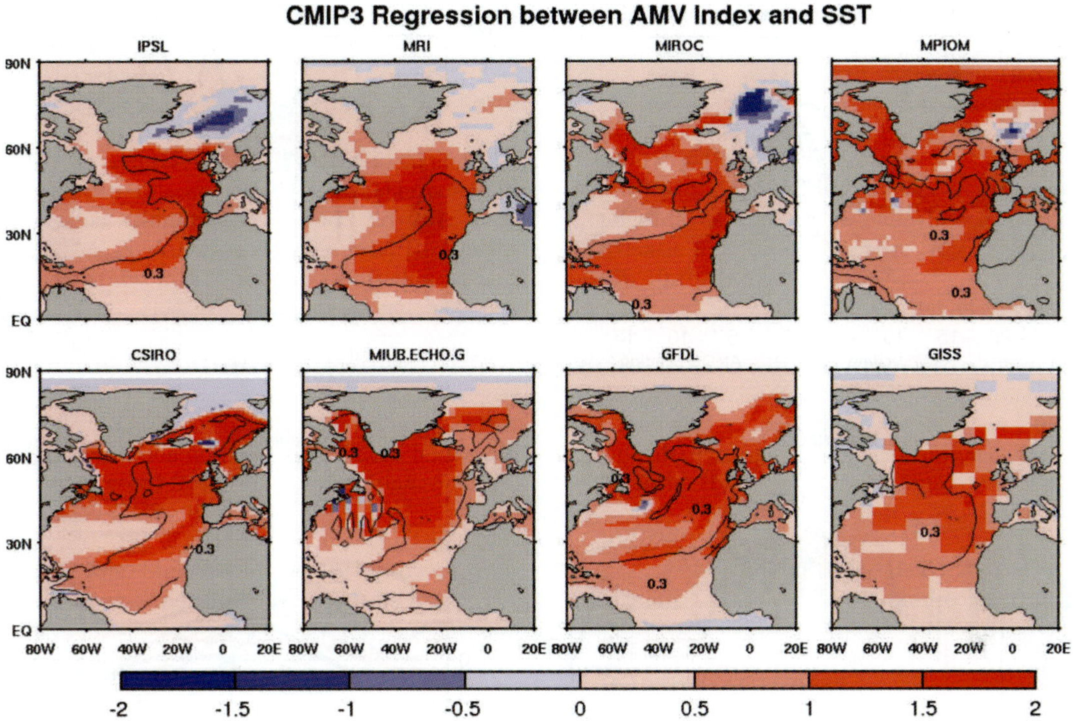

Fig. 5. The regression (shade) of AMV Indices on the SST in 8 CMIP3 models. Explained variance is shown by contours (Contour Interval = 0.3).

variability (MIROC, MIUB-ECHO-G). For the models with a strong AMOC and AMV relation (*e.g.*, CSIRO, MPIOM, MIUB), the AMV and AMOC spectra show correspondence in the respective decadal band (Figs. 4 & 6b).

A number of studies based on observations and models have argued that the NAO may drive multi-decadal variations in the AMOC (Secs. 2 & 3). The NAO simulated by the eight CMIP3 models all exhibit spectra consistent with a white noise, stochastic process (Fig. 6c). This is largely in agreement with the observed NAO spectrum, which is weakly red (Wunsch, 1999). The reddening may simply represent a single realisation of a white noise process in the short instrumental record, and/or a weak response to either external forcing or surface variability.

Cross-correlation between decadal variations in NAO and AMOC indices shows little agreement among models (Fig. 6d). There are indications in two models (MIUB, GFDL) that the NAO may drive decadal variations in AMOC, and indications in another model (GISS) that the AMOC may lead to a negative NAO phase 10–15 years later. In contrast, Gastineau and Frankignoul (2012) analyse six CGCMs using more complex statistical methods and show that in five of the models, positive NAO variations precede a stronger AMOC by 2–3 years. However, the explained variances are generally low, similar to the cross-correlation analysis results here. Thus, the climate models seem to provide little support for the hypothesis that NAO variations precede those in the AMOC, by about a decade. This may reflect the importance of other processes, such as other modes of atmospheric variability (Medhaug *et al.*, 2012; Msadek and Frankignoul, 2009). However, it could be also due to model error

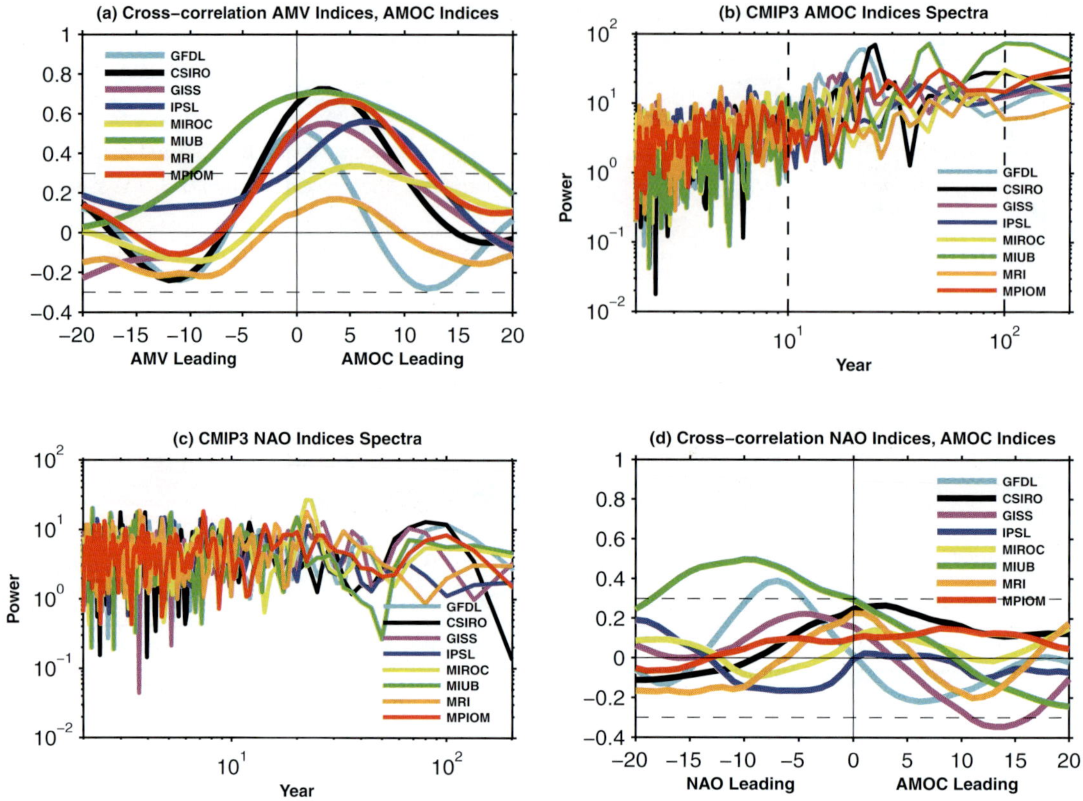

Fig. 6. (a) Cross-correlation among AMV and AMOC indices, power spectra of both (b) AMOC and (c) NAO indices, and (d) cross-correlation among NAO and AMOC indices, as simulated in 8 CMIP3 models. For the correlations an 11-year running mean is applied to the indices, and the 95% confidence level is around 0.3 (dashed line) in each model.

(Secs. 5 & 6). In particular, partially coupled model experiments support a strong NAO influence on the AMOC, while the relation is absent in the fully coupled model (Klöwer *et al.*, 2014).

5. Mechanisms for Simulated AMV

The stochastic climate model paradigm is a powerful framework for understanding and interpreting climate variability (Hasselmann, 1976). It may explain much of the observed AMV, and provides a null hypothesis for more complex mechanisms, such as coupled ocean-atmosphere dynamics and externally forced variability. Relevant aspects of the paradigm are summarised here; see Latif and Keenlyside (2011) for more discussion.

The paradigm recognises that climate is composed of phenomena with very different timescales. In particular, weather has typical lifetimes of hours to days, while the deep ocean has timescales of many centuries. The paradigm separates these timescales. Weather is considered a fast and chaotic process, and modelled as a stochastic variable that drives the slow, deterministic part of the climate system. The resulting models for the slow variables are collectively known as stochastic climate models.

A hierarchy of such models exists depending on the complexity of the equations used to represent the slow variables. The simplest consists of a slab ocean model for upper-ocean temperature that is forced thermodynamically by stochastic atmospheric variability.

The slab ocean integrates the weather noise, which is treated as a white noise process (implicitly including decadal timescales). The resulting SST variations exhibit an AR-1 red spectrum (Hasselmann, 1976). Observed SST variability is consistent with this model over parts of the mid-latitudes, away from coasts, fronts, and other regions where ocean dynamics are important (Frankignoul and Hasselmann, 1977).

As discussed above, while observed and simulated NAO spectra are to first order white, the slab ocean response to such forcing cannot entirely explain the observed AMV pattern. Furthermore, there is evidence that ocean circulation changes drive multi-decadal variations in extra-tropical SST (Secs. 3 & 4). Stochastic climate models that account for mean oceanic advection (Saravanan and McWilliams, 1997) and long baroclinic Rossby waves (Schneider *et al.*, 2002) may explain variations in the North Atlantic on the order of a decade. Model studies have also shown that stochastic atmospheric variability can excite decadal (Weisse *et al.*, 1994), multi-decadal (Delworth and Greatbatch, 2000; Frankcombe *et al.*, 2009), and centennial (Mikolajewicz and Maier-Reimer, 1990) AMOC variability.

To what extent can AMV be described as the oceanic response to stochastic NAO variability? Mecking *et al.* (2013) address this question by driving an ocean general circulation model (OGCM) with a 2000 year long white noise forcing associated with the NAO (Fig. 7a). The forcing patterns are computed by linearly regressing

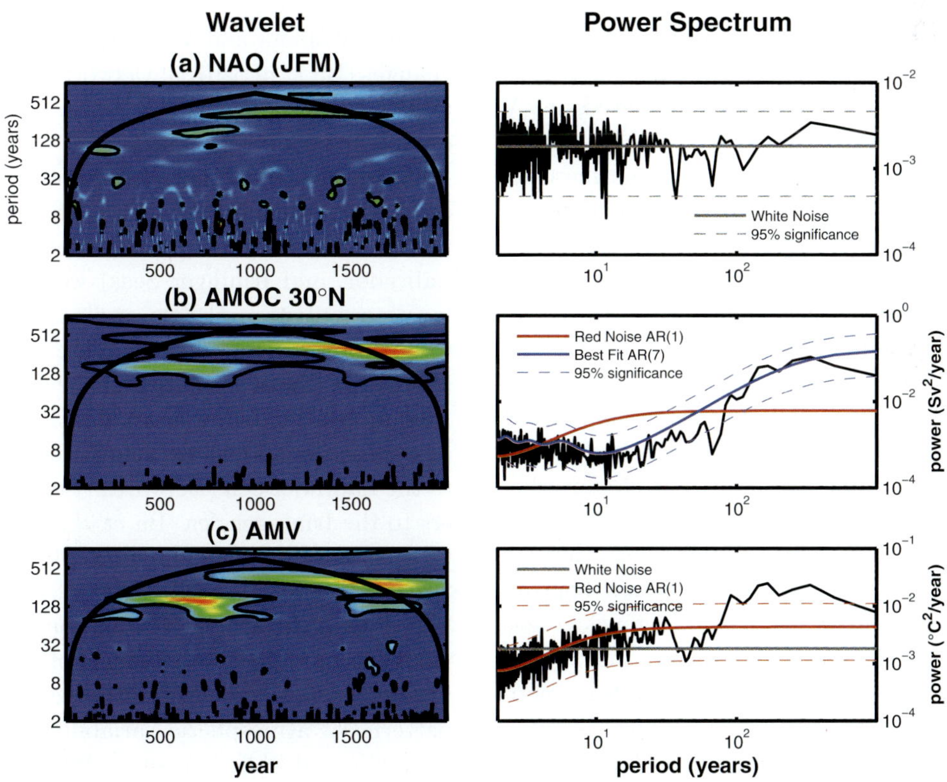

Fig. 7. Results from ocean model experiments forced with idealized stochastic NAO forcing. (left) Wavelet and (right) power spectra of (a) the stochastic NAO index used to drive the model, and simulated (b) AMOC at 30°N, and (c) AMV SST indices. Solid line demarcates the cone of influence on the wavelet spectra, and contour lines regions significant at the 95% level. Confidence intervals for theoretical spectra are shown on right panels.

observed NAO index against ocean model forcing fields. The AMOC at 30°N (Fig. 7b) and SPG (not shown) strength both have enhanced power at low frequencies. However, neither provides evidence for an oscillatory mode of ocean-only variability. Rather, both indices respond linearly to the NAO forcing, as is seen by comparing AMOC and NAO wavelet spectra (Fig. 7a,b). Thus, in this picture the observed AMV may result from a particular realization of stochastic NAO variability.

The variability of the AMOC at 30°N is strongly enhanced on timescales longer than 90 years (Fig. 7b), while that of the SPG on timescales longer than 15 years (not shown). These different response times are linked to the different "memory" of these two indices to past atmospheric forcing. Furthermore, neither index follows the AR-1 spectra of the simplest stochastic climate model; rather the AMOC and SPG spectra are consistent with AR-7 and AR-5 processes, respectively. The simulated AMV SST index exhibits characteristics (Fig. 7c) of both AMOC and SPG variability. On timescales longer than 90 years it shows correspondence to the AMOC at 30°N, while on decadal timescales to the SPG strength. The spectrum of simulated AMV is also not consistent with an AR-1 process. These results show that even though AMV may not be consistent with an AR-1 process, it may still fit the stochastic model paradigm. While the experiments here are limited to the NAO pattern, similar results are likely for other extra-tropical atmospheric circulation patterns.

The response time of the ocean is dictated by the exact set of processes controlling the variability. The inconsistencies seen among CGCMs (Sec. 4) indicate that different processes control variability in these models. Oceanic deep convection, which is often related to deep-water formation, is likely a major source of uncertainty in the mechanisms for AMV. Although atmospheric variability dominates heat loss and drives oceanic deep convection, a cap of surface freshwater anomalies can inhibit wintertime convection (Marshall and Schott, 1999). The relative importance of these processes on longer timescales is not clear, and differs among CGCMs (Ba *et al.*, 2014); in some models salinity variations are key to AMV (Ba *et al.*, 2013a), while others suggest AMV is purely temperature driven (Dijkstra *et al.*, 2006). Furthermore, the impact of salinity is likely timescale dependent (Deshayes *et al.*, 2013). This sensitivity may arise because temperature and salinity anomalies affect density in opposite ways — salinity increasing density, temperature decreasing it — with temperature controlling their relative importance. In particular, as the freezing point of water is approached, salinity anomalies dominate variations in density. Thus, differences in model mean states and source waters to convection sites may lead to different timescales. Salinity variations arising from the tropics (Vellinga and Wu, 2004), SPG strength changes (Ba *et al.*, 2013a; Delworth *et al.*, 1993), and the high-latitudes (Jungclaus *et al.*, 2005) could all play a role.

The locations of deep-water formation introduce more uncertainty. CGCMs often do not simulate the three sites of observed convection (GIN, Labrador, and Irminger Seas) well in terms of location and strength (Ba *et al.*, 2014); convection is commonly used as an index of deep water formation. The contribution of the different convection sites to AMOC decadal variability differs among models, some giving greater importance to the Labrador Sea (Delworth *et al.*, 1993), others to the Irminger Sea (Ba *et al.*, 2013a), while the role of GIN Seas is less clear (Medhaug *et al.*, 2012). A related issue is the different contribution of interior and western boundary transport feeding the lower branch of the AMOC (Zhang, 2010). Furthermore, different large-scale patterns of atmospheric variability may excite convection at different sites. While the NAO may be important for Labrador Sea convection (Delworth *et al.*, 1993), the East Atlantic Pattern may be important for Irminger Sea convection (Msadek and Frankignoul, 2009).

Lastly, multiple timescales may exist in a single model, excited by different processes. For example, detailed analysis of the CMIP3 GFDL model simulation (Sec. 4) identifies two timescales: a 20–30-yr periodicity that results from an internal mode of AMOC, and a 50–70-yr periodicity that results from interactions with high-latitudes (Frankcombe *et al.*, 2010). Also as discussed next, coupled ocean-atmosphere interaction (Timmermann *et al.*, 1998; Vellinga and Wu, 2004) and external factors could also contribute to AMV.

6. Decadal Prediction and Future Challenges

There is growing interest in predicting AMV and its climatic impacts. This is part of the greater goal of predicting climate on decadal timescales (Smith *et al.*, 2007), an area of research that features in the Intergovernmental Panel on Climate Change (IPCC) fifth assessment report (AR5) (Keenlyside and Ba, 2010; Meehl *et al.*, 2013). External forcing and internal dynamics contribute to variability on these timescales, with internal dynamics contributing more at regional scales (Hawkins and Sutton, 2009). Thus, decadal (or near-term) prediction must account for both (Branstator and Teng, 2010).

Extensive experimentation during the last five years indicate that there are two regions where decadal prediction is skillful: the Indian Ocean, where external forcing dominates (Guemas *et al.*, 2012) and the extra-tropical North Atlantic, where internal dynamics dominate and initialization is critical (Keenlyside *et al.*, 2008; Matei *et al.*, 2012; Msadek *et al.*, 2013; Yeager *et al.*, 2012).

Skill in predicting AMV has been shown to be of the order 5–10 years (Doblas-Reyes *et al.*, 2013; Kim *et al.*, 2012), which is short in comparison to the observed periodicity. Detailed analysis shows that skill in predicting extra-tropical North Atlantic SST arises from ocean dynamics (Robson *et al.*, 2012; Yeager *et al.*, 2012). Despite this initial success significant challenges exist. One is the development of techniques to initialize predictions from observations (Meehl *et al.*, 2013). This topic is beyond the scope of this chapter. Instead we cover several issues related to understanding and simulating AMV.

A key issue is the poor representation of the mean climate by CGCMs. In addition to errors in high-latitude processes, models exhibit significant SST and SSS biases in the extra-tropics (Ba *et al.*, 2014). These are mostly associated with an incorrect Gulf Stream separation and too zonal NAC, both of which may be related to insufficient oceanic resolution (Scaife *et al.*, 2011). The errors in the ocean circulation influence the exchange of water masses between tropical and subtropical gyres, and with the high-latitudes; while errors in SST can impact the atmospheric circulation and variability (Scaife *et al.*, 2011). Poor representation of the mean state can also lead to significant forecast drift (Yeager *et al.*, 2012), which is found in most decadal prediction systems and is of major concern. The method of Klöwer *et al.* (2014) avoids such a "coupling shock" and could serve as an intermediate approach to decadal AMV forecasting.

Secondly, the view that the extra-tropical ocean only weakly influences the atmosphere (Kushnir *et al.*, 2002) is being challenged. High-resolution observations and models indicate that the sharp SST front associated with the Gulf Stream drives a deep atmospheric circulation (Minobe *et al.*, 2008). However, whether SST variations in this region are of climatic significance remains unclear (Hand *et al.*, 2013). In addition, recent studies show that changes in the stratospheric circulation may drive multi-decadal changes in the winter NAO (Scaife *et al.*, 2005). Atmospheric models show observed warm SST conditions associated with AMV can drive a weakening of the polar vortex in the early winter that propagates back into the troposphere, leading to a negative NAO in late

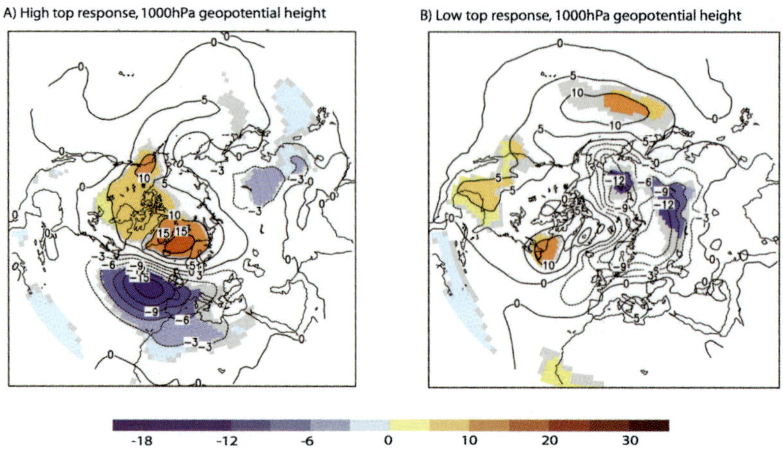

Fig. 8. The winter (JFM) 1000 hPa geopotential height simulated in response to the 1951–1960 warm conditions in the Atlantic with a model (a) including the whole stratosphere (high-top, until ~80 km) and (b) the one only partly resolving it (low-top, until ~30 km). Values significant at the 90% (95%) level are shaded in grey (colour). Results can be compared to Fig. 2b. Figure from Omrani *et al.* (2013).

winter as observed (Fig. 8) (Omrani *et al.*, 2013). Atmospheric models that poorly resolve the stratosphere cannot reproduce this link (Hodson *et al.*, 2010; Omrani *et al.*, 2013) (Fig. 8). Thus, the ocean may have driven part of the observed multi-decadal NAO variations. Better resolving the stratosphere improves simulated variability of the extra-tropics (Charlton-Perez *et al.*, 2013), and this in turn may improve simulation (Manzini *et al.*, 2012; Schimanke *et al.*, 2011) and prediction of AMV and its impacts.

A third issue is the role of external forcing. Anthropogenic and natural (volcanic) aerosol loading and solar forcing also showed decadal variations (IPCC, 2007). Statistical analysis (Mann and Emanuel, 2006) and coupled model experiments (Booth *et al.*, 2012) suggest cooling from tropospheric aerosols together with anthropogenic driven global warming can explain AMV during the 20th century. However, aerosol forcing has large uncertainty (IPCC, 2007). Furthermore, while Booth *et al.* (2012) argue that most climate models underestimate indirect aerosol effects, their simulations fail to reproduce other observed variations, including Atlantic ocean heat content (Zhang *et al.*, 2013). Alternatively, weaker external forcing could synchronize internal dynamics to produce observed AMV (Cheng *et al.*, 2013; Otterå *et al.*, 2010). Thus, while most models indicate AMV is primarily internal in origin (Ting *et al.*, 2009), external factors could have a large contribution.

New data is key to advance understanding, and paleo-proxy data provides an alternative to the short instrumental record. Studies based on marine and terrestrial proxy data have shown the existence of AMV beyond the instrumental record (*e.g.*, Gray *et al.*, 2004). However, disagreement exists among records. Thus more data, particularly from the marine environment are required, and methods to combine multiple proxy records should be explored (Svendsen *et al.*, 2014).

7. Summary

Historical data indicate pronounced multidecadal variations in North Atlantic SST with a 70–80 year periodicity. These basin-wide changes in SST were largest in the subpolar region. They were associated with multi-decadal variations in the atmosphere that extended deep into the stratosphere: the warm (cold) phase of the AMV is associated with a negative (positive) NAO structure in the late winter, and weakening (strengthening) of the stratospheric polar vortex in the early winter. Hydrographic data indicate that the AMV encompassed the upper 3000 m of North Atlantic Ocean, and there is evidence that the NAO drove multi-decadal changes in the formation of LSW and in turn the AMOC. Forced ocean model simulations support these findings, and indicate ocean dynamics drove the observed SST changes in the SPG.

CGCMs simulate AMV in the absence of varying external forcing. Simulated indices of North Atlantic oceanic variability exhibit red spectra, but provide little evidence for a preferred timescale. The simulated SST patterns resemble those observed, and are generally consistent with being driven by variations in the AMOC. There is, however, little evidence from CGCMs of a consistent NAO and AMOC relationship; and simulated NAO indices exhibit white spectra, similar to observations. Despite this climate models largely reproduce the observed AMV pattern, and thus the role of the NAO in AMV may not be as important as observations may suggest.

The leading hypothesis among CGCMs is that AMV represents the response of the ocean to stochastic atmospheric variability. Idealized ocean model experiments demonstrate the applicability of the stochastic climate model: simulated spectra are consistent with high-order AR processes, indicative of ocean dynamics. In this picture, the observed AMV may result from a particular realization of stochastic atmospheric variability. There is, however, disagreement among CGCMs on the exact mechanisms for the oceanic response to atmospheric variability. Some recent model results indicate external factors may largely explain AMV, although these results remain controversial. Mean state errors, which are particularly large in the North Atlantic, could explain much of this uncertainty and need consideration when interpreting model results.

A key issue raised here is that while the comparatively short observations suggest an anti-phase relation between AMV and NAO indices, no consistent relation exists in CGCMs. This implies that models may underestimate the role of ocean-atmosphere interaction for AMV. Experiments showing stratosphere-troposphere interaction is key to capturing the wintertime atmospheric response to AMV support this notion, but further research is required.

CMIP3 preindustrial control experiments were used here to illustrate performance in simulating AMV. Several recent studies based on the newer generation of CGCMs used for the IPCC AR5 present similar findings: the key aspects of the SST pattern and its relation to the AMOC are reproduced; the associated winter atmospheric circulation is poorly simulated and large uncertainties exist in the role of salinity variations; and in addition simulated variability generally has a timescale less than 70–80 years (Kavvada *et al.*, 2013; Marini and Frankignoul, 2013; Zhang and Wang, 2013). No major improvement was found in simulating AMV in the newer models (Ruiz-Barradas *et al.*, 2013).

Despite our limited understanding, models demonstrate skill in predicting AMV on 5–10 year timescales, and operational prediction systems are being developed (Smith *et al.*, 2012). This is generating a new research area that aims to predict the socio-economic impacts of decadal climate variations.

Acknowledgments

The Deutsche Forschungsgemeinschaft (KE 1471/2-1) provided supported this work, with contributions from the EU SUMO (No. 266722) and STEPS (PCIG10-GA-2011-304243) projects. The National Taiwan University and Center for Advanced Study in Theoretical Sciences (CASTS) provided travel support to attend the *NTU International Science Conference on Climate Change: Multidecadal and Beyond*. Discussion at the conference helped shape this chapter.

References

Ba, J., 2013: North Atlantic Multi-decadal Variability Simulated in Coupled General Circulation Models (CGCMs). *Christian-Albrechts-Universität Kiel*, Germany, 81 pp.

———, N. S. Keenlyside, W. Park, M. Latif, E. Hawkins, and H. Ding, 2013a: A mechanism for Atlantic multidecadal variability in the Kiel Climate Model. *Clim. Dyn.*, **41**, 2133–2144.

———, *et al.*, 2014: A multi-model comparison of Atlantic multidecadal variability. *Clim. Dyn.*, **43**, 2333–2348.

Bjerknes, J., 1964: Atlantic air-sea interaction. *Adv. Geophys.*, **10**, 1–82.

Boer, G. J., and S. J. Lambert, 2008: Multimodel decadal potential predictability of precipitation and temperature. *Geophys. Res. Lett.*, **35**, L05706.

Böning, C. W., M. Scheinert, J. Dengg, A. Biastoch, and A. Funk, 2006: Decadal variability of subpolar gyre transport and its reverberation in the North Atlantic overturning. *Geophys. Res. Lett.*, **33**, L21S01.

Booth, B. B. B., N. J. Dunstone, P. R. Halloran, T. Andrews, and N. Bellouin, 2012: Aerosols implicated as a prime driver of twentieth-century North Atlantic climate variability. *Nature*, **484**, 228–232.

Branstator, G., and H. Y. Teng, 2010: Two limits of initial-value decadal predictability in a CGCM. *J. Climate*, **23**, 6292–6311.

Cayan, D. R., 1992: Latent and sensible heat flux anomalies over the Northern Oceans: Driving the sea surface temperature. *J. Phys. Oceanogr.*, **22**, 859–881.

Charlton-Perez, A. J., *et al.*, 2013: On the lack of stratospheric dynamical variability in low-top versions of the CMIP5 models. *J. Geophys. Res. Atmos.*, **118**, 2494–2505.

Cheng, W., J. C. H. Chiang, and D. Zhang, 2013: Atlantic meridional overturning circulation (AMOC) in CMIP5 Models: RCP and historical simulations. *J. Climate*, **26**, 7187–7197.

Cunningham, S. A., *et al.*, 2007: Temporal variability of the Atlantic Meridional Overturning Circulation at 26.5°N. *Science*, **317**, 935–938.

Curry, R. G., M. S. McCartney, and T. M. Joyce, 1998: Oceanic transport of subpolar climate signals to mid-depth subtropical waters. *Nature*, **391**, 575–577.

Delworth, T., S. Manabe, and R. J. Stouffer, 1993: Interdecadal variations of the thermohaline circulation in a coupled ocean-atmosphere model. *J. Climate*, **6**, 1993–2011.

Delworth, T. L., and R. J. Greatbatch, 2000: Multidecadal thermohaline circulation variability driven by atmospheric surface flux forcing. *J. Climate*, **13**, 1481–1495.

Deser, C., and M. L. Blackmon, 1993: Surface Climate Variations over the North-Atlantic Ocean during Winter — 1900–1989. *J. Climate*, **6**, 1743–1753.

Deshayes, J., R. Curry, and R. Msadek, 2014: CMIP5 model inter-comparison of freshwater budget and circulation in the North Atlantic. *J. Clim.*, **27**, 3298–3317. doi: 10.1175/JCLI-D-12-00700.1.

Dickson, R., J. Lazier, J. Meincke, P. Rhines, and J. Swift, 1996: Long-term coordinated changes in the convective activity of the North Atlantic. *Prog. Oceanogr.*, **38**, 241–295.

Dijkstra, H., L. Raa, M. Schmeits, and J. Gerrits, 2006: On the physics of the Atlantic Multidecadal Oscillation. *Ocean Dyn.*, **56**, 36–50.

Doblas-Reyes, F. J., and Coauthors, 2013: Initialized near-term regional climate change prediction. *Nat. Commun.*, **4**, 1715.

Eden, C., and T. Jung, 2001: North Atlantic interdecadal variability: Oceanic response to the North Atlantic Oscillation (1865–1997). *J. Climate*, **14**, 676–691.

———, and J. Willebrand, 2001: Mechanism of interannual to decadal variability of the North Atlantic circulation. *J. Climate*, **14**, 2266–2280.

Edwards, M., G. Beaugrand, P. Helaouët, J. Alheit, and S. Coombs, 2013: Marine Ecosystem Response to the Atlantic Multidecadal Oscillation. *PLOS ONE*, **8**, e57212.

Frankcombe, L. M., H. A. Dijkstra, and A. von der Heydt, 2009: Noise-induced multidecadal variability in the North Atlantic: excitation of Normal Modes. *J. Phys. Oceanogr.*, **39**, 220–233.

———, A. von der Heydt, and H. A. Dijkstra, 2010: North Atlantic multidecadal climate variability: An Investigation of Dominant Time Scales and Processes. *J. Climate*, **23**, 3626–3638.

Frankignoul, C., and K. Hasselmann, 1977: Stochastic climate models, Part II. Application to sea-surface temperature anomalies and thermocline variability. *Tellus*, **29**, 289–305.

Gastineau, G., and C. Frankignoul, 2012: Cold-season atmospheric response to the natural variability of the Atlantic meridional overturning circulation. *Clim. Dyn.*, **39**, 37–57.

Ghil, M., *et al.*, 2002: Advanced spectral methods for climatic time series. *Rev. Geophys.*, **40**, 3-1-3-41.

Gray, S. T., L. J. Graumlich, J. L. Betancourt, and G. T. Pederson, 2004: A tree-ring based reconstruction of the Atlantic Multidecadal Oscillation since 1567 AD. *Geophys. Res. Lett.*, **31**, L12205.

Guemas, V., S. Corti, J. García-Serrano, F. J. Doblas-Reyes, M. Balmaseda, and L. Magnusson, 2012: The Indian Ocean: The region of highest skill worldwide in decadal climate prediction*. *J. Climate*, **26**, 726–739.

Gulev, S. K., M. Latif, N. Keenlyside, W. Park, and K. P. Koltermann, 2013: North Atlantic Ocean control on surface heat flux on multidecadal timescales. *Nature*, **499**, 464–467.

Häkkinen, S., and P. B. Rhines, 2004: Decline of subpolar North Atlantic circulation during the 1990s. *Science*, **304**, 555–559.

Hand, R., N. Keenlyside, N.-E. Omrani, and M. Latif, 2014: Simulated response to interannual SST variations in the Gulf Stream region. *Clim. Dyn.*, **42**, 715–731.

Hasselmann, K., 1976: Stochastic climate models Part 1. Theory. *Tellus*, **28**, 473–485.

Hatun, H., A. B. Sando, H. Drange, B. Hansen, and H. Valdimarsson, 2005: Influence of the Atlantic subpolar gyre on the thermohaline circulation. *Science*, **309**, 1841–1844.

Hawkins, E., and R. Sutton, 2009: The potential to narrow uncertainty in regional climate predictions. *Bull. Amer. Meteor. Soc.*, **90**, 1095–1107.

Hodson, D. L. R., R. T. Sutton, C. Cassou, N. Keenlyside, Y. Okumura, and T. J. Zhou, 2010: Climate impacts of recent multidecadal changes in Atlantic Ocean Sea Surface Temperature: A multimodel comparison. *Clim. Dyn.*, **34**, 1041–1058.

Hurrell, J. W., Y. Kushnir, G. Ottersen, and M. Visbeck, 2003: An Overview of the North Atlantic Oscillation. *The North Atlantic Oscillation: Climate Significance and Environmental Impact*, J. W. Hurrell, Y. Kushnir, G. Ottersen, and M. Visbeck, Eds., American Geophysical Union, 1–35.

IPCC, 2007: *Climate Change 2007: The Physical Science Basis.* Contribution of Working Group I to the Fourth Assessment Report of the Intergovernmental Panel on Climate Change, Cambridge University Press, 996 pp.

Jungclaus, J. H., H. Haak, M. Latif, and U. Mikolajewicz, 2005: Arctic North Atlantic Interactions and Multidecadal Variability of the Meridional Overturning Circulation. *J. Climate*, **18**, 4013–4031.

Kalnay, E., *et al.*, 1996: The NCEP/NCAR 40-Year Reanalysis Project. *Bull. Amer. Meteor. Soc.*, **77**, 437–471.

Kavvada, A., A. Ruiz-Barradas, and S. Nigam, 2013: AMO's structure and climate footprint in observations and IPCC AR5 climate simulations. *Clim. Dyn.*, **41**, 1345–1364.

Keenlyside, N. S., and J. Ba, 2010: Prospects for decadal climate prediction. *WIREs: Climate Change*, **1**, 627–635.

———, M. Latif, J. Jungclaus, L. Kornblueh, and E. Roeckner, 2008: Advancing decadal-scale climate prediction in the North Atlantic sector. *Nature*, **453**, 84–88.

Kim, H. M., P. J. Webster, and J. A. Curry, 2012: Evaluation of short-term climate change prediction in multi-model CMIP5 decadal hindcasts. *Geophys. Res. Lett.*, **39**, L10701.

Klöwer, M., M. Latif, H. Ding, R. J. Greatbatch, and W. Park, 2014: Atlantic Meridional Overturning Circulation and prediction of North Atlantic sea surface temperature. doi: 10.1016/j.lpsl.2014.09.001.

Knight, J. R., R. J. Allan, C. K. Folland, M. Vellinga, and M. E. Mann, 2005: A signature of persistent natural thermohaline circulation cycles in observed climate. *Geophys. Res. Lett.*, **32**, L20708.

Kuhlbrodt, T., A. Griesel, M. Montoya, A. Levermann, M. Hofmann, and S. Rahmstorf, 2007: On the driving processes of the Atlantic meridional overturning circulation. *Rev. Geophys.*, **45**, RG2001.

Kushnir, Y., 1994: Interdecadal variations in North Atlantic sea surface temperature and associated atmospheric conditions. *J. Climate*, **7**, 141–157.

———, W. A. Robinson, I. Blade, N. M. J. Hall, S. Peng, and R. Sutton, 2002: Atmospheric GCM response to extratropical SST anomalies: Synthesis and evaluation. *J. Climate*, **15**, 2233–2256.

Langehaug, H. R., I. Medhaug, T. Eldevik, and O. H. Otterå, 2012: Arctic/Atlantic Exchanges via the Subpolar Gyre. *J. Climate*, **25**, 2421–2439.

Latif, M., and N. S. Keenlyside, 2011: A perspective on decadal climate variability and predictability. *Deep Sea Research Part II: Topical Studies in Oceanography*, **58**, 1880–1894.

———, *et al.*, 2004: Reconstructing, monitoring, and predicting multidecadal-scale changes in the North Atlantic thermohaline circulation with sea surface temperature. *J. Climate*, **17**, 1605–1614.

Lohmann, K., H. Drange, and M. Bentsen, 2009: A possible mechanism for the strong weakening of the North Atlantic subpolar gyre in the mid-1990s. *Geophys. Res. Lett.*, **36**, L15602.

Mann, M. E., and K. A. Emanuel, 2006: Atlantic hurricane trends linked to climate change. *Eos Trans. AGU*, **87**, 233–241.

Manzini, E., C. Cagnazzo, P. G. Fogli, A. Bellucci, and W. A. Müller, 2012: Stratosphere-troposphere coupling at inter-decadal time scales: Implications for the North Atlantic Ocean. *Geophys. Res. Lett.*, **39**, L05801.

Marini, C., and C. Frankignoul, 2013: An attempt to deconstruct the Atlantic Multidecadal Oscillation. *Clim. Dyn.*

Marshall, J., and F. Schott, 1999: Open-ocean convection: Observations, theory, and models. *Rev. Geophys.*, **37**, 98RG0273.

Matei, D., H. Pohlmann, J. Jungclaus, W. Müller, H. Haak, and J. Marotzke, 2012: Two tales of initializing decadal climate prediction experiments with the ECHAM5/MPI-OM model. *J. Climate*, **25**, 8502–8523.

Mecking, J. V., N. S. Keenlyside, and R. J. Greatbatch, 2014: Stochastically-forced multidecadal variability in the North Atlantic: A model study. *Clim. Dyn.*, **43**, 271–288.

Medhaug, I., H. R. Langehaug, T. Eldevik, T. Furevik, and M. Bentsen, 2012: Mechanisms for decadal scale variability in a simulated Atlantic meridional overturning circulation. *Clim. Dyn.*, **39**, 77–93.

Meehl, G. A., *et al.*, 2013: Decadal climate prediction: An update from the trenches. *B. Am. Meteorol. Soc.*

Mikolajewicz, U., and E. Maier-Reimer, 1990: Internal secular variability in an ocean general circulation model. *Clim. Dyn.*, **4**, 145–156.

Minobe, S., A. Kuwano-Yoshida, N. Komori, S.-P. Xie, and R. J. Small, 2008: Influence of the Gulf Stream on the troposphere. *Nature*, **452**, 206–209.

Msadek, R., and C. Frankignoul, 2009: Atlantic multidecadal oceanic variability and its influence on the atmosphere in a climate model. *Clim. Dyn.*, **33**, 45–62.

———, *et al.*, 2014: Predicting a decadal shift in North Atlantic climate variability using the GFDL forecast system. *J. Climate*, **27**, 6472–6496.

Omrani, N. E., N. S. Keenlyside, J. Bader, and E. Manzini, 2014: Stratosphere key for wintertime atmospheric response to warm Atlantic decadal conditions. *Clim. Dyn.*, **42**, 649–663.

Otterå, O. H., M. Bentsen, H. Drange, and L. L. Suo, 2010: External forcing as a metronome for Atlantic multidecadal variability. *Nat. Geosci.*, **3**, 688–694.

Polyakov, I., U. Bhatt, H. Simmons, D. Walsh, J. Walsh, and X. Zhang, 2005: Multidecadal variability of North Atlantic temperature and salinity during the twentieth century. *J. Climate*, **18**, 4562–4581.

Rayner, N. A., *et al.*, 2003: Global analyses of sea surface temperature, sea ice, and night marine air temperature since the late nineteenth century. *J. Geophys. Res.*, **108**, 4407.

Reverdin, G., 2010: North Atlantic subpolar gyre surface variability (1895–2009). *J. Climate*, **23**, 4571–4584.

Robson, J., R. Sutton, K. Lohmann, D. Smith, and M. D. Palmer, 2012: Causes of the Rapid Warming of the North Atlantic Ocean in the Mid-1990s. *J. Climate*, **25**, 4116–4134.

Ruiz-Barradas, A., S. Nigam, and A. Kavvada, 2013: The Atlantic Multidecadal Oscillation in twentieth century climate simulations: Uneven progress from CMIP3 to CMIP5. *Clim. Dyn.*, **41**, 3301–3315.

Saravanan, R., and J. C. McWilliams, 1997: Stochasticity and spatial resonance in interdecadal climate fluctuations. *J. Climate*, **10**, 2299–2320.

Scaife, A. A., J. R. Knight, G. K. Vallis, and C. K. Folland, 2005: A stratospheric influence on the winter NAO and North Atlantic surface climate. *Geophys. Res. Lett.*, **32**, L18715.

———, *et al.*, 2011: Improved Atlantic winter blocking in a climate model. *Geophys. Res. Lett.*, **38**, L23703.

Schimanke, S., J. Korper, T. Spangehl, and U. Cubasch, 2011: Multi-decadal variability of sudden stratospheric warmings in an AOGCM. *Geophys. Res. Lett.*, **38**, L01801.

Schlesinger, M. E., and N. Ramankutty, 1994: An oscillation in the global climate system of period 65–70 years. *Nature*, **367**, 723–726.

Schneider, N., A. J. Miller, and D. W. Pierce, 2002: Anatomy of North Pacific decadal variability. *J. Climate*, **15**, 586–605.

Smith, D. M., S. Cusack, A. W. Colman, C. K. Folland, G. R. Harris, and J. M. Murphy, 2007: Improved surface temperature prediction for the coming decade from a global climate model. *Science*, **317**, 796–799.

———, *et al.*, 2013: Real-time multi-model decadal climate predictions. *Clim. Dyn.*, **41**, 2875–2888.

Sutton, R. T., and D. L. R. Hodson, 2005: Atlantic Ocean forcing of North American and European summer climate. *Science*, **309**, 115–118.

Svendsen, L., S. Hetzinger, N. S. Keenlyside, and Y. Gao, 2014: Marine-based multi-proxy reconstruction of Atlantic multi-decadal variability *Geophys. Res. Lett.*, **41**, 1295–1300, 2013GL059076.

Timmermann, A., M. Latif, R. Voss, and A. Grotzner, 1998: Northern Hemispheric interdecadal variability: A coupled air-sea mode. *J. Climate*, **11**, 1906–1931.

Ting, M. F., Y. Kushnir, R. Seager, and C. H. Li, 2009: Forced and internal twentieth-century SST trends in the North Atlantic. *J. Climate*, **22**, 1469–1481.

United Nations Environment Programme (UNEP)/ GRID-Arendal, 2002: *Africa Environment Outlook — Past, Present and Future Perspectives.* United Nations Publications, 448 pp.

Vellinga, M., and P. Wu, 2004: Low-latitude freshwater influence on centennial variability of the Atlantic Thermohaline Circulation. *J. Climate*, **17**, 4498–4511.

Visbeck, M., E. Chassignet, R. Curry, T. Delworth, B. Dickson, and G. Krahmann, 2003: The Ocean's Response to North Atlantic Oscillation Variability, in "The North Atlantic Oscillation". *The North Atlantic Oscillation: Climate Significance and Environmental Impact*, J. W. Hurrell, Y. Kushnir, G. Ottersen, and M. Visbeck, Eds., American Geophysical Union, 113–146.

———, H. Cullen, G. Krahmann, and N. Naik, 1998: An ocean model's response to North Atlantic Oscillation-like wind forcing. *Geophys. Res. Lett.*, **25**, 4521–4524.

Wang, C., and L. Zhang, 2013: Multidecadal ocean temperature and salinity variability in the tropical North Atlantic: Linking with the AMO, AMOC, and subtropical cell. *J. Climate*, **26**, 6137–6162.

Weisse, R., U. Mikolajewicz, and E. Maier-Reimer, 1994: Decadal variability of the North Atlantic in an ocean general circulation model. *J. Geophys. Res.-Oceans*, **99**, 12411–12421.

Wunsch, C., 1999: The interpretation of short climate records, with comments on the North Atlantic and Southern Oscillations. *B. Am. Meteorol. Soc.*, **80**, 245–255.

Yeager, S., A. Karspeck, G. Danabasoglu, J. Tribbia, and H. Y. Teng, 2012: A decadal prediction case study: Late twentieth-century North Atlantic Ocean heat content. *J. Climate*, **25**, 5173–5189.

Zhang, L., and C. Wang, 2013: Multidecadal North Atlantic sea surface temperature and Atlantic meridional overturning circulation variability in CMIP5 historical simulations. *J. Geophys. Res.-Oceans*, **118**, 5772–5791.

Zhang, R., 2010: Latitudinal dependence of Atlantic meridional overturning circulation (AMOC) variations. *Geophys. Res. Lett.*, **37**, L16703.

———, T. L. Delworth, and I. M. Held, 2007: Can the Atlantic Ocean drive the observed multidecadal variability in Northern Hemisphere mean temperature? *Geophys. Res. Lett.*, **34**, L02709.

———, and G. K. Vallis, 2006: Impact of great salinity anomalies on the low-frequency variability of the North Atlantic climate. *J. Climate*, **19**, 470–482.

———, *et al.*, 2013: Have aerosols caused the observed Atlantic multidecadal variability? *J. Atmos. Sci.*, **70**, 1135–1144.

Chapter 10

A Review of the Dynamics of Pacific Interdecadal Climate Variability

Zhengyu Liu

Department of Atmospheric and Oceanic Sciences
University of Wisconsin-Madison
Madison, WI 53706, USA
zliu3@wisc.edu

This paper is a review of Pacific interdecadal variability with the focus on the Pacific Decadal Oscillation (PDO), the North Pacific Gyre Oscillation (NPGO) and the Pacific Multidecadal Variability (PMV). All the interdecadal variability modes exhibit coherent variability in the tropical and North Pacific. This led to two groups of hypotheses: one attributing the origin of the interdecadal variability to the tropics and the other to the extratropics. Although it still remains difficult to pinpoint the origin of these modes in the real world, it seems robust to conclude that stochastic weather noise in the mid-latitude provides the key driving force for all the interdecadal variability modes, and oceanic dynamics in the extratropical Pacific is critical in determining the time scales of these modes. The PDO and PMV are forced by the atmospheric variability mode of the Aleutian Low, while the NPGO by the atmospheric variability mode of the North Pacific Oscillation. The PDO and NPGO depend on Rossby wave propagation in the mid-latitude, while the PMV is likely determined by Rossby waves in the subpolar region.

1. Observations

Much progress has been made towards the understanding of interdecadal climate variability over the globe (see a recent review by Liu, 2012). Here, we will focus on the Pacific. Significant interdecadal variability has been observed over the Pacific (*e.g.*, Trenberth and Hurrell, 1994; Zhang *et al.* 1997; Minobe, 1997; Folland *et al.*, 2002; Power *et al.*, 1999). Decadal variability (10–20 years) in the North Pacific is dominated by two modes, the Pacific Decadal Oscillation (Mantua *et al.*, 1997; Zhang *et al.*, 1997) and the North Pacific Gyre Oscillation (NPGO) (Di Lorenzo *et al.*, 2008). The PDO can be identified as the EOF1 of sea surface temperature (SST) or sea level height (SSH) variability in the North Pacific (Figs. 1b, d; Figs. 2b, c). The NPGO can be derived as the 2nd EOF of the SST or SSH in the North Pacific (Figs. 2e, f) (Di Lorenzo *et al.*, 2008). The PDO is associated with decadal change of biological activities in the North Pacific (Mantua *et al.*, 1997), while the NPGO is critical for salinity and nutrient changes in the central and eastern North Pacific. In addition to these decadal modes, there is a significant multi-decadal variability on a 50–70-yr timescale, as seen in the 1st PC of the annual mean sea level pressure (SLP) over the North Pacific (NPI index, Figs. 3a, b). This mode shows a coherent pattern throughout the Pacific, and will be referred to as the Pacific

Climate Change: Multidecadal and Beyond
edited by Chih-Pei Chang et al.

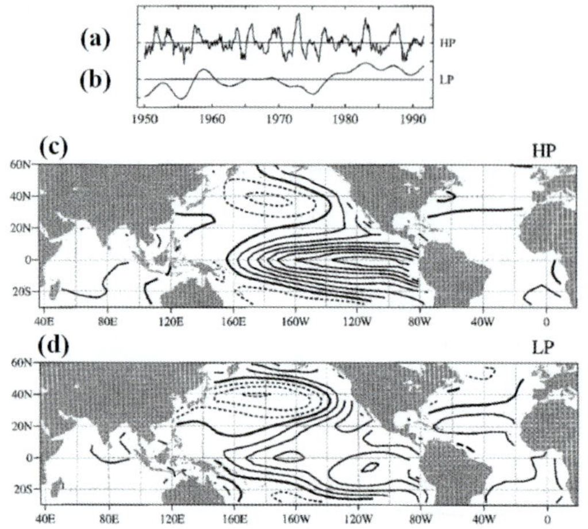

Fig. 1. The leading (normalized) PCs of 6-yr highpass- and lowpass-filtered SST over the Pacific domain shown together with the associated regression patterns for global SST. The interval between tick-marks on the vertical axis of the top panel corresponds to 1.0 standard deviation, and the spacing between the curves is arbitrary. Contour interval 0.1 K per standard deviation of the expansion coefficient time series. Negative contours are dashed; the zero contour is thickened. (Adapted from Zhang *et al.*, 1997).

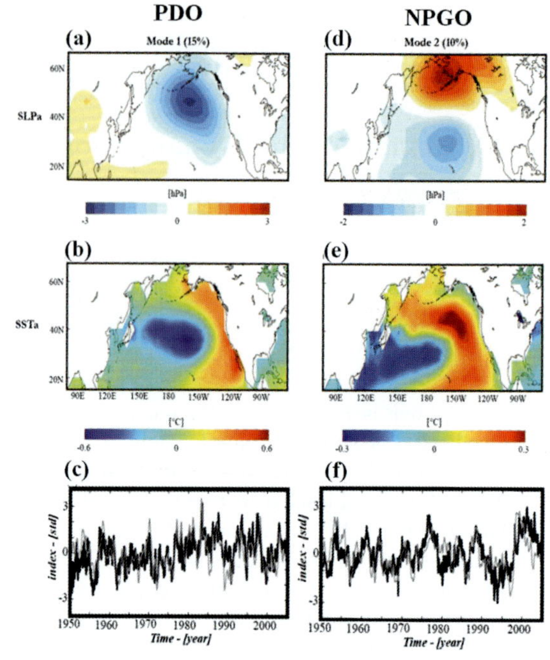

Fig. 2. (a, b) PDO and (d, e) NPGO in the observation as derived as the regression of the (a, d) SLPa (hPa) and (b, e) SSTa on the standard (c) PC1 (PDO index, black) and (f) PC2 (NPGO index, black) time series of the combined EOF analysis of North Pacific SLPa and SSTa. The grey lines are the reconstructed (c) PDO and (f) NPGO index from the AR-1 model forced by the temporal variability of the AL and NPO, respectively (Adapted from Furtado *et al.*, 2010).

Multidecadal Variability (PMV) here. The coexistence of decadal and multidecadal variability in the Pacific can be seen, for example, through wavelet analysis (Minobe, 1999).

The PDO and PMV have different time scales, but share a similar pattern, both being similar to the pattern of El Nino, or the ENSO mode, which is derived as the 1st EOF of highpass (1–7 years) SST variability in the tropical Pacific (Figs. 1a, c). These patterns are characterized by coherent variability over the tropical Pacific with coherent variability of the opposite sign in the mid-latitude North Pacific (Fig. 1d, Fig. 2b). The PDO also exhibits a symmetric pattern with respect to the equator, with significant variability in the South Pacific (Garreaud and Battisti, 1999; Power *et al.*, 1999). So much is the similarity to ENSO that the PDO and PMV are sometimes referred to as the ENSO-like Pacific interdecadal variability (Zhang *et al.*, 1997). Nevertheless, Pacific interdecadal variability has a large amplitude in the mid-latitude (Fig. 1d, Fig. 2b) while ENSO is dominated by the maximum variability in the tropics (Fig. 1c). The region of tropical coherence also extends further poleward for the PDO and PMV than it does for ENSO.

The NPGO and PDO have a similar decadal time scale, but different spatial patterns. The NPGO is dominated by a northeast-southwest tilted SST tri-pole (Fig. 2e), while the PDO is dominated by a maximum loading in the central-western Pacific encircled by the opposite loading (Fig. 2b, Fig. 1b). Their associated SSH and upper ocean circulation pattern also differ: the PDO is dominated by a single gyre over the entire North Pacific (north of 20°N), while

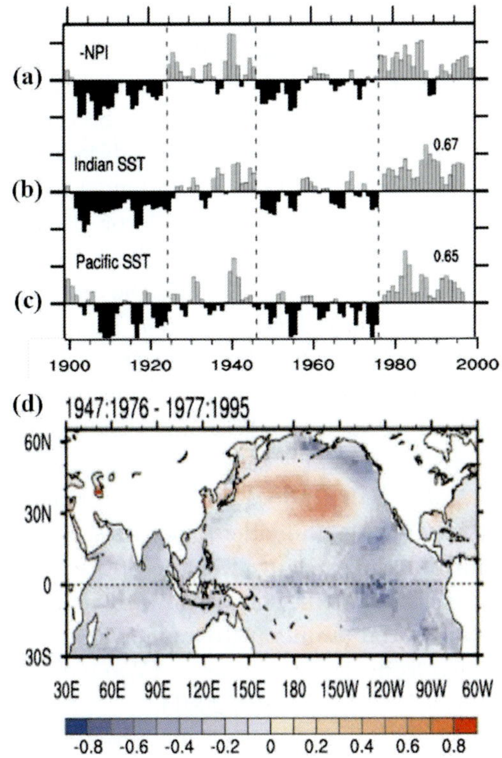

Fig. 3. Selected climate anomaly records for boreal winter during 1900–97: (a) the inverted North Pacific Index (NPI, area-averaged sea level pressure for the region 30°–65° N, 160°–140° W), (b) tropical Indian Ocean SST and (c) southeast tropical Pacific SST. All records are normalized and smoothed with a three-point binomial filter. Each tick mark on the ordinate represents one standard deviation. The numeral to the right of each tropical index represents its correlation coefficient with the inverted NPI. (d) Epoch difference map of boreal winter SST (1977-95–1947-76) from the Kaplan *et al.* (1998) dataset (Adapted from Deser *et al.*, 2004).

the NPGO is characterized by a pair of counter-rotating gyres with the nodal line along 40°N following the Kurosion Extension and the North Pacific Current (Di Lorenzo *et al.*, 2008). As a result, in the positive phase of PDO, the Alaska current is reduced and the California Current is intensified, while in the positive phase of the NPGO, both the Alaska Current and California Current are intensified. The PDO and NPGO therefore reflect, respectively, the variability in the latitudinal position and the strength of the subtropical and subpolar gyres in the North Pacific. The NPGO and PDO also differ in their associated atmospheric variability. The PDO is associated with the variability of the Aleutian Low (AL) while the NPGO is associated with the North Pacific Oscillation (NPO) (Walker and Bliss, 1932), where the AL and NPO can be derived as the 1st and 2nd EOF of sea level pressure anomaly (SLP), respectively. This coherent ocean-atmosphere variability for the two modes can be seen in the combined SST-SLP EOF analysis, which recovers the PDO in SST and the AL in SLP as the first EOF mode (Figs. 2a–c) and the NPGO in SST and NPO in SLP (Figs. 2d–f).

Like the PDO (Power *et al.*, 1999; Folland *et al.*, 2002), the NPGO is also associated with an equatorially symmetric SST and SSH variability in the entire Pacific (Di Lorenzo *et al.*, 2008). Indeed, they can be derived as the 1st and 2nd EOF of the low-pass SST variability in the Pacific, both exhibiting a pattern symmetric about the equator. However, the two modes differ in their associated SST variability in the tropics. The PDO is associated predominantly with the SST variability in the eastern equatorial Pacific, as in a classical El Nino, while the NPGO is associated with the SST variability in the central equatorial Pacific (Di Lorenzo *et al.*, 2010), as in the so called El Nino-Modoki (Ashok and Yamagata, 2009).

2. Mechanisms

2.1. *Tropical Origin*

One important feature of the Pacific interdecadal variability is a strong covariance between the tropics and extratropics. As shown in Fig. 3, major episodes of PDO and PMV in the North Pacific can also be identified in the tropics. The pattern of interdecadal SST variability resembles that of the interannual variability of ENSO, characterized by two centers of opposite signs in the tropical and North Pacific.

The strong negative correlation between tropical and North Pacific SSTs is caused by an effective atmospheric teleconnection from the tropics into the North Pacific through the Pacific-North America (PNA) teleconnection (Alexander, 1990; 1992). The presence of strong ENSO variability in the tropical Pacific and the subsequent atmospheric teleconnection towards the extratropics have led to the hypothesis that Pacific interdecadal variability originates from the tropical Pacific (*e.g.*, Trenberth and Hurrell, 1994; Newman *et al.*, 2003; Deser *et al.*, 2004). Newman *et al.* (2003) showed that the North Pacific interdecadal SST variability can be simulated as an AR-1 process after including the remote forcing of ENSO. Note, however, that such models are linear, and any interdecadal variability in the North Pacific linked to ENSO requires decadal variability forcing in the tropical Pacific. Nevertheless, one is still left with the question how interdecadal variability is generated in the tropical Pacific itself, and ultimately, what is the origin of the interdecadal variability? The tropical origin of decadal variability has also been proposed to be induced by the nonlinear modulation of ENSO (Munnich *et al.*, 1991; Vimont, 2005; Power and Colman, 2006), the nonlinear asymmetry between El Nino and La Nino (Burgers and Stephenson, 1999), and higher baroclinic modes of equatorial waves (Wu *et al.*, 2003).

Gu and Philander (1997) proposed a tropical-origin mechanism for multi-decadal variability, in which the slow timescale is derived from the equatorward ventilation from the mid-latitude North Pacific. Warm North Pacific surface water reaches the equator in ~20 years through the equatorward subduction (Liu *et al.*, 1994; McCreary and Lu, 1994), upwelling along the equator and warming the surface equatorial ocean; this surface warming reduces local easterly, further warming the surface through the positive Bjerkness feedback. The warm equatorial Pacific forces deep atmospheric convection, which then propagates into the North Pacific through the PNA teleconnection, intensifying the Aleutian Low, mid-latitude surface westerly and, in turn, turbulent heat flux loss, cooling the North Pacific. This forms a delayed negative feedback on the North Pacific SST, which eventually leads to a Pacific multi-decadal variability of ~ 50 years. This Gu-Philander hypothesis is basically a delayed oscillator paradigm as in (3), with the positive feedback provided by the Bjerkness feedback in the equatorial Pacific and the delayed negative feedback by the equatoward ventilation. This tropical-origin mechanism, however, was later questioned by Schneider *et al.* (1999) because the equatorward ventilation of North Pacific warm water (Deser *et al.*, 1996) does not seem to reach the equator. The Gu-Philander hypothesis, however, has not been disapproved in the South Pacific, because the possibility of equatoward ventilation of interdecadal temperature anomaly cannot be excluded from the South Pacific (Wang and Liu, 2000; Luo and Yamagata, 2001).

In short, in spite of various hypotheses for the tropical origin of interdecadal variability, the proposed mechanisms are difficult to confirm with observations. Current climate models tend to underestimate the atmospheric teleconnection from the tropical Pacific into the North Pacific (Newman, 2007; Furtado *et al.*, 2010). Therefore, a study based on model may underestimate the role of the tropical origin.

2.2. *Extratropical Origin for Pacific Decadal Variability*

Some early studies of PDV followed that of the delayed oscillator paradigm of ENSO, except for replacing equatorial Rossby waves with mid-latitude Rossby waves, and replacing the Bjerkness feedback with mid-latitude ocean-atmosphere feedbacks. Latif and Barnett (1994) first proposed a mechanism that attributed the origin of the North Pacific decadal variability to the extratropics. Their CGCM simulated a quasi-20-year variability in the Pacific (Figs. 4a, b), with some resemblance to the

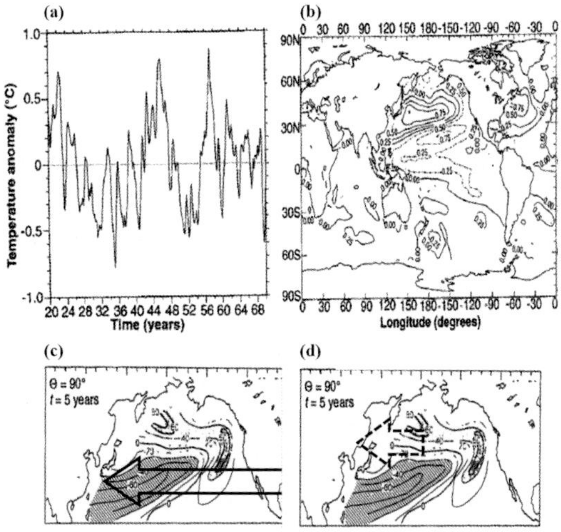

Fig. 4. (a) Time series of the coupled model's anomalous SST (°C) averaged over the region from 150°E to 180°E and 25°N to 35°N. The time series was smoothed with a 9-month running mean filter. (b) Spatial distribution of linear regression coefficients between the index time series in (a) and SST values. The pattern was scaled so that the maximum SST anomalies were to 1°C. (c) Reconstruction of anomalous heat content ($°\mathrm{C\,m^{-1}}$) from the leading CEOF mode at a particular phase of the decadal cycle. (The phase angle measures the phase of cycle, with the full cycle = 360°). The solid double arrow represents schematically the Rossby wave propagation in the subtropics. (d) Same as (c) except for the revised Latif-Barnett model, where the dashed double arrow represents the Rossby wave in the mid-latitude forced by stochastic forcing. [Adapted from Latif and Barnett (1994)].

observed PDV (Zhang *et al.*, 1997). The decadal variability shows a clear westward propagation in the subtropical thermocline, reminiscent of planetary waves (Fig. 4c). Furthermore, a warm KOE SST seems to force a significant ridge response over the Aleutian Low. These two features led to the original Latif-Barnet mode hypothesis: a warm KOE SST induces a high pressure anomaly in the North Pacific, weakening the Aleutian Low and in turn the surface westerly in the mid-latitude. The reduced wind reduces the turbulent heat flux loss, and further warms the North Pacific, leading to a positive ocean-atmosphere feedback. In the mean time, the reduced wind also spins down the subtropical gyre, and in turn, Kuroshio Current, after a decadal timescale of planetary wave propagation across the basin (Anderson and Gill, 1975). The weaker Kuroshio reduces the northward transport of warm water, leading to a delayed cooling in the KOE region, and eventually reversing the SST anomaly there into the cold phase. This Latif-Barnett hypothesis is basically a delayed oscillator paradigm, with the positive feedback provided by the mid-latitude air-sea feedback and the delayed negative feedback by the spin-down of the subtropical gyre associated with the planetary wave propagation.

This original Latif-Barnett mode was, however, later found to be largely invalid. Schneider *et al.* (2002) re-examined the decadal variability simulation of Latif and Barnett (1994) and found that the KOE SST variability lags, rather than leads, the wind variability in the central Pacific by ~5 years. In the CGCM and observation, in contrast to the eastward advection mechanism by KOE as hypothesized in the original Latif-Barnett mode, the KOE SST variability is induced by westward propagating planetary waves in the mid-latitude, forced by the wind-stress curl in the central Pacific (Deser *et al.*, 1999; Schneider and Miller, 2001). Reduced westerly wind-stresses in mid-latitudes then shift the intergyre boundary northward (Seager *et al.*, 2001), inducing a downward Ekman pumping along the intergyre boundary, deepening the thermocline in the central Pacific. The thermocline in the KOE region is then subsequently deepened through the westward propagation of planetary wave. The warmer thermocline water surfaces late winter due to the deepening of the mixed layer, or the reemergence mechanism (Alexander *et al.*, 1999), increasing the KOE SST. Further studies also find that the atmospheric response to typical KOE SST anomalies is not robust (Kushnir *et al.*, 2002), leaving the positive feedback in the original Latif-Barnett mode also questionable.

Schneider *et al.* (2002) modified the Latif-Barnett mode using the paradigm of the stochastic climate model (Hasselmann, 1976). In the stochastic climate model, the origin of the variability is the stochastic climate noise (weather noise), instead of the deterministic paradigm of a delayed oscillator, in which the positive ocean-atmosphere feedback is critical. In this revised model, the reduced surface wind still warms the central Pacific by reducing the turbulent heat flux loss, but this warming is caused by stochastic atmospheric variability associated with the Aleutian Low, instead of the atmospheric response to the warming in the KOE region. This modified Latif-Barnett mode essentially describes a stochastically driven oceanic mode, rather than a coupled mode as described in the original hypothesis. However, both modes share a common feature: decadal variability originates from the extratropical North Pacific (Figs. 4c, d). This modified Latif-Barnett mode seems to be largely consistent with the analysis of variability in the NCAR CCSM2 (Kwon and Deser, 2007), although the role of ocean-atmosphere feedback is thought to be more significant in the latter. The stochastic-driven nature of the Pacific decadal variability is also demonstrated in a CGCM study with interactive ensemble coupling scheme (Yeh and Kirtman, 2006). There, as the internal atmospheric variability is reduced by the ensemble mean of the atmosphere model, the coupled variability is reduced proportionally.

This stochastic climate model for PDO is also applied to the NPGO, with the time scale also determined by the westward propagation of the mid-latitude Rossby waves (Ceballo *et al.*, 2009). The difference is in the spatial pattern of the stochastic atmospheric variability: the PDO is forced by the AL (Fig. 2a) while the NPGO by the NPO (Fig. 2d).

2.3. *Extratropical Origin for Pacific Multi-Decadal Variability*

In contrast to the intensive studies on the decadal (10–20 year) variability in the Pacific, much less attention has been paid to the multi-decadal variability PMV (over ~50 years) in the North Pacific (Fig. 3). It is difficult to explain the multi-decadal variability directly using the Latif-Barnett mechanism, because the latter involves planetary waves in the subtropics to mid-latitude, which cross the basin in less than 10 years. This time scale, although sufficient for decadal variability, is too short to account for the multi-decadal variability.

To identify the origin of the North Pacific multi-decadal variability, sensitivity experiments have been performed in two different CGCMs, with consistent results (Liu *et al.*, 2002; Wu *et al.*, 2003; Zhong *et al.*, 2008). Here, we only show the results in CCSM3. First, to isolate the key region of ocean-atmosphere feedback, a partial-coupling (PC) sensitivity experiment is performed in which ocean-atmosphere coupling is suppressed in the tropics. The distinctive North Pacific multi-decadal variability in the control run (Fig. 5a) remains (Fig. 5b) although the SST variability has been suppressed in the tropics. In contrast, in a complementary PC experiment in which ocean-atmosphere interaction is suppressed in the extratropics, the North Pacific multi-decadal mode disappears completely although tropical variability, including ENSO, remains largely unchanged (not shown). These two PC experiments demonstrate that tropical ocean-atmosphere interaction, albeit essential to ENSO, is unnecessary for the multi-decadal variability in the North Pacific. It is interesting to note that when ENSO is suppressed, the multi-decadal spectral peak of PMV becomes more distinct. This feature has been also seen in another independent model (Wu *et al.*, 2003). One interpretation is that the tropical ENSO acts as a "noise" forcing on PMV. When ENSO is suppressed, the removal of stochastic forcing leaves the PMV more distinct.

To further isolate the role of ocean dynamics in decadal variability, a partial-blocking sensitivity experiment (PB) is performed in which

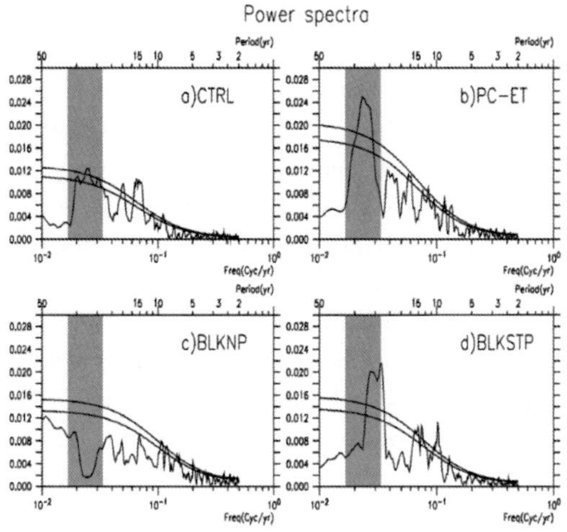

Fig. 5. Power spectra of monthly SST averaged over the western-central North Pacific in four NCAR-CCSM3 modeling surgery experiments (each of 400-year). These experiments are designed to identify the origin of the North Pacific multi-decadal variability. (a) Control simulation, which shows a distinct multi-decadal spectral peak, (b) PC_ET, in which ocean-atmosphere coupling is suppressed in the tropics (<20°N), (c) BLKNP, the same as PC_ET except for an additional sponger layer in the entire North Pacific along the dateline and (d) BLKSTP, the same as BLKNP except that the sponge layer is confined in the subtropical North Pacific (<40°N). Vertical shading bar indicates the range of 30–40 years. The multi-decadal peak persists even if ocean-coupling is disabled in the tropics (in (b)) and wave propagation is blocked in the subtropics (in (d)), but disappears when the subpolar ocean is also blocked (in (c)). Similar results can be obtained if the PC1 of the North Pacific SST is plotted (Adapted from Zhong *et al.*, 2008 and Zhong and Liu, 2009).

ocean wave propagation between the tropics and extratropics is blocked using a vertical sponge wall in the ocean. Despite of the presence of this sponge wall, Pacific multi-decadal variability continued (not shown). These experiments demonstrate unambiguously that the origin of the PMV, that is the process that generates the multi-decadal time scale of the variability, in this model is confined in the North Pacific coupled ocean-atmosphere system.

To further understand the role of planetary waves on multi-decadal timescales, two additional PB experiments were performed with a north-south oceanic sponge wall across the central North Pacific to block the westward propagation of planetary waves. The North Pacific multi-decadal variability disappears over the entire basin when the sponge wall blocks the North Pacific (Fig. 5c). Note that, the multi-decadal variability disappears not only west of the sponge wall, but also east of the wall. This suggests the active role of the planetary wave in the formation of the multi-decadal variability. In contrast, the variability remains similar to the variability in the control run when the sponge wall is confined in the subtropics south of 40°N (Fig. 5d). These two PB experiments demonstrate that planetary wave propagation in the subpolar gyre is an important and necessary process in the PMV. Planetary waves are slow in the subpolar gyre due to both a large Coriolis parameter and weak stratification. In fact, such waves take more than 20 years to cross the basin (Zhong and Liu, 2009). This provides a sufficiently long timescale for a role in driving multi-decadal variability. Further analysis found that a warm KOE SST induces a ridge response and, in turn, downward Ekman pumping in the subpolar gyre. The downward Ekman pumping forces a southward Sverdrup flow and in turn a cold anomaly in the thermocline. This cold anomaly appears to migrate westward and is eventually transported southward by the Oyashio towards the KOE region, generating a delayed negative feedback. This analysis shows evidence of a positive ocean-atmosphere feedback associated with the atmospheric response to the KOE SST, although it remains unclear if the net ocean-atmosphere feedback in the mid-latitude is positive or not. Therefore, this mode is likely to be stochastic-driven, with ocean — atmosphere feedback only modifying the variability.

3. Summary and Discussions

Pacific climate exhibits multiple modes of interdecadal variability, notably the PDO and NPGO at decadal time scales and the PMV at multi-decadal time scales. All the interdecadal variability modes exhibit coherent variability in the tropical and North Pacific, although, relatively, the decadal variability tends to be more prominent in the tropical-subtropical region while the multi-decadal variability tends to be more prominent in the North Pacific. The fact that variability in the tropical Pacific and North Pacific is coherent has led to two groups of hypotheses: one attributes the origin of the variability to the tropics and the other to the extratropics. These hypothesis and related mechanisms proposed are difficult to verify using the analysis of the observation alone. In spite of early studies that emphasizes the role of positive ocean-atmosphere interaction, current studies seems to suggest that all the interdecadal variability modes are mainly forced by the stochastic weather noise, with the time scale determined by ocean dynamics in the North Pacific. The PDO and PMV are forced by the AL atmospheric variability, while the NPGO by the NPO atmospheric variability. The PDO and NPGO depends on Rossby wave propagation in the mid-latitude, while the PMV is likely determined by Rossby waves in the subpolar region.

In spite of significant advances, many questions remain wide open. Perhaps, the most important question concerns the mechanism that determines the time scale of each interdecadal mode? Recent studies suggest that most climate models are able to simulate the SST pattern of the observed Pacific interdecadal variability modes well; however, these models diverge significantly in the time scale of these modes (Overland and Wang, 2007, Furtado *et al.*, 2010). Since the SST pattern of these modes are largely determined by the overlying atmospheric variability, while the time scale depends critically on ocean dynamics, the poorly simulated time scale suggests the lack of our understanding of some fundamental mechanisms in the coupled system, especially on the role of ocean in interdecadal variability.

One issue important for Pacific interdecadal variability is the response of the atmosphere to extratropical oceanic variability (see Liu, 2012 for a more detailed review). Observational and modeling studies show some evidences of an atmospheric response to extratropical SST (and sea ice) variability. The magnitude of the response appears to be modest at monthly to seasonal time scales, but significant at decadal and longer timescales. Therefore, the atmospheric response to extratropical SST forcing can be important for interdecadal climate variability. Even if the extratropical coupling is not critical for the genesis of coupled interdecadal modes, the atmospheric response to extratropical SST can still be of practical importance for the understanding of the climate impact of interdecadal variability in the atmosphere.

We have so far focused on the tropical impact on the North Pacific. It should be pointed out that extratropical climate variability can also impact the tropics through teleconnections via atmospheric, oceanic and the coupled processes (Liu and Alexander, 2007 and references there in). Extratropical climate variability can affect tropical climate through the atmospheric transport of latent and sensible heat (*e.g.* Liu and Yang, 2003; Chiang and Bitz, 2005). Boundary layer ocean-atmosphere coupling through the Wind Speed-Evaporation-SST (WES) Feedback can generate a rapid equatorward propagation in the trade wind regime (Liu and Xie, 1994), enabling extratropical climate variability to affect the tropics in the so called seasonal footprinting mechanism (Vimont *et al.*, 2003). These fast atmospheric and coupled teleconnections apply not only to multi-decadal variability, but also to decadal, interannual (Vimont *et al.*, 2003) and even seasonal (Liu and Xie, 1994) variability in the Pacific. At interdecdal timescale, tropical variability can be further affected by the extratropical climate changes through the equatorward oceanic ventilation

(Liu and Yang, 2003; Wu *et al.*, 2007) and coastal trapped waves (McGregor *et al.*, 2008). Therefore, decadal variability could involve complex interactions between the tropics and extratropics.

Acknowledgments

This work is supported by 2012CB955201, NSFC 41130105 and NSF.

References

Alexander, M. A., 1990: Simulation of the response of the North Pacific Ocean to the anomalous atmospheric circulation associated with El Niño. *Clim. Dyn.*, **5**, 53–65.

———, 1992: Midlatitude atmosphere-ocean interaction during El Niño. Part I: The North Pacific Ocean. *J. Climate*, **5**, 944–958.

Alexander, M., C. Deser, and M. Timlin, 1999: The reemergence of SST anomalies in the North Pacific Ocean. *J. Climate*, **12**, 2419–2433.

Anderson, D. L. T., and A. E. Gill, 1975: Spin-up of a stratified ocean, with application to upwelling. *Deep-Sea Res.*, **22**, 583–596.

Ashok, K., and T. Yamagata. 2009: Climate change: The El Niño with a difference. *Nature*, **461**, 481–484.

Burgers, G., and D. Stephenson, 1999: The "Normality" of El Nino, *Geophys. Res. Lett.*, **26**, 1027–1030.

Ceballos, L., E. Di Lorenzo, N. Schneider, and B. Taguchi, 2009: North Pacific gyre oscillation synchronizes climate fluctuations in the eastern and western North Pacific. *J. Climate*, **22**, 5163–5174.

Chiang, J., and C. Bitz, 2005: Influence of high-latitude ice cover on the marine Intertropical Convergence Zone. *Clim. Dyn.*, **25**, 477–496.

Deser, C., M. A. Alexander, and M. S. Timlin, 1996: Upper-ocean thermal variations in the North Pacific during 1970–1991. *J. Climate*, **9**, 1840–1855.

———, ———, and ———, 1999: Evidence for a wind-driven intensification of the Kuroshio Current Extension from the 1970s to the 1980s. *J. Climate*, **12**, 1697–1706.

———, A. Phillips, and J. Hurrell, 2004: Pacific interdecadal climate variability: Linkages between the tropics and North Pacific during boreal winter since 1900. *J. Climate*, **17**, 3109–3124.

Di Lorenzo, E., K. Cobb, J. C. Furtado, N. Schneider, B. T. Anderson, A. Bracco, M. A. Alexander, and D. J. Vimont, 2010: Central Pacific El Nino and decadal climate change in the North Pacific. *Nat. Geosci.*, **3**, 762–765.

———, N. Schneider, K. Cobb, P. Franks, K. Chhak, A. Miller, J. McWillams, S. Bograd, H. Arango, E. Curchitser, T. Power and P. Rivierex, 2008: North Pacific Gyre Oscillation links ocean climate and ecosystem change. *Geophys. Res., Lett.*, **35**, L08607, doi: 10.1029/2007GL032838.

Folland C. K., J. A. Renwick, M. J. Salinger, and A. B. Mullan, 2002: Relative influences of the interdecadal Pacific Oscillation and ENSO on the South Pacific Convergence Zone. *Geophys. Res. Lett.*, **29**, 21-1-21-4, doi: 10.1029/2001GL014201.

Furtado, J., E. Di Lorenzo, N. Schneider, and N. A. Bond, 2010: North Pacific decadal variability and climate change in the IPCC AR4 models. *J. Climate*, **24**, 3049–3067.

Garreaud R., and D. Battisti, 1999: Interannual (ENSO) and interdecadal (ENSO-like) variability in the Southern Hemisphere tropospheric circulation. *J. Climate*, **12**, 2113–2123.

Gu, D., and S. G. H. Philander, 1997: Interdecadal Climate Fluctuations that depend on Exchanges between the Tropics and Extratropics. *Science*, **275**, 805–807.

Hasselmann, K., 1976: Stochastic climate models. Part I: Theory. *Tellus*, **28**, 473–485.

Kushnir, Y., W. A. Robinson, I. Bladé, N. M. J. Hall, S. Peng, and R. Sutton, 2002: Atmospheric GCM response to extratropical SST anomalies: Synthesis and evaluation. *J. Climate*, **15**, 2233–2256.

Kwon, Y.-O., and C. Deser, 2007: North Pacific decadal variability in the Community Climate System model version 2. *J. Climate*, **20**, 2416–2433.

Latif, M., and T. P. Barnett, 1994: Causes of decadal climate variability over the North Pacific and North America. *Science*, **266**, 634–637.

Liu, Z., 2012: Dynamics of interdecadal climate variability: A historical perspective. *J. Climate*, **25**, 1963–1995.

———, and M. Alexander, 2007: Atmospheric bridge, oceanic tunnel, and global climatic teleconnections. *Rev. Geophys.*, **45**, RG2005, doi: 10.1029/2005RG000172.

———, S. G. H. Philander, and R. Pacanowski, 1994: A GCM study of tropical -subtropical upper ocean mass exchange. *J. Phys. Oceanogr.*, **24**, 2606–2623.

———, W. Wu, R. Gallimore, and R. Jacob, 2002: Search for the origins of Pacific decadal climate variability. *Geophys. Res. Lett.*, **29**, doi: 10.1029/2001GL013735.

———, and S. P. Xie, 1994: Equatorward propagation of coupled air-sea disturbances with application to the annual cycle of the eastern tropical Pacific. *J. Atmos. Sci.*, **51**, 3807–3822.

———, and H. Yang, 2003: Extratropical control on tropical climate: atmospheric bridge and oceanic tunnel. *Geophys. Res. Lett.*, **30**, doi: 10.1029/2002GL016492.

Luo, J.-J., and T. Yamagata, 2001: Long-term El Nino-Southern Oscillation (ENSO)-like Variation with Special Emphasis on the South Pacific. *J. Geophys. Res.*, **106**(C10), 22211–22227.

Mantua, J. S. Hare, Y. Zhang, J. Wallace and R. Francis, 1997: A Pacific interdecadal climate oscillation with impacts on salmon production. *Bull. Amer. Meteor. Soc.*, **76**, 1069–1079.

McCreary, J. P., and P. Lu, 1994: On the interaction between the subtropical and equatorial ocean circulation: The Subtropical Cell. *J. Phys. Oceanogr.*, **24**, 466–497.

McGregor, S., N. J. Holbrook, and S. Power, 2008: Interdecadal sea surface temperature variability in the equatorial Pacific Ocean. Part II: The role of equatorial/off-equatorial wind stresses in a hybrid coupled model. *J. Climate*, **21**, 4242–4256.

Minobe, S., 1997: A 50–70 year climatic oscillation over the North Pacific and North America. *Geophys. Res. Lett.*, **24**, 683–686.

———, 1999: Resonance in bidecadal and pentadecadal climate oscillations over the North Pacific: Role in climatic regime shifts. *Geophys. Res. Lett.*, **26**, 855–858.

Munnich, M., M. A. Cane and S. E. Zebiak, 1991: A study of self-excited oscillations of the tropical ocean-atmosphere system. Part II: Nonlinear cases. *J. Atmos. Sci.*, **48**(10), 1238–1248.

Newman, M., 2007: Interannual to decadal predictability of tropical and North Pacific sea surface temperatures. *J. Climate*, **20**, 2333–2356.

———, G. P. Compo, and M. A. Alexander, 2003: ENSO-forced variability of the Pacific decadal oscillation. *J. Climate*, **16**, 3853–3857.

Overland, J., and M. Wang, 2007: Future climate of the North Pacific Ocean. *EOS*, **88**, 178–182.

Power, S., T. Casey, C. Folland, A. Colman, and V. Mehta, 1999: Interdecadal modulation of the impact of ENSO on Australia. *Clim. Dyn.*, **15**, 319–324.

———, and R. Colman, 2006: Multi-year predictability in a coupled general circulation model. *Clim. Dyn.*, **26**, 247–272.

Schneider, N., and A. J. Miller, 2001: Predicting western North Pacific ocean climate. *J. Climate*, **14**, 3997–4002.

———, ———, M. Alexander, and C. Deser, 1999: Subduction of decadal North Pacific temperature anomaly: Observation and dynamics. *J. Phys. Oceanogr.*, **29**, 1056–1070.

———, ———, and D. W. Pierce, 2002: Anatomy of North Pacific decadal variability. *J. Climate*, **15**, 586–605.

Seager, R., Y. Kushnir, N. H. Naik, M. A. Cane, and J. Miller, 2001: Wind-driven shifts in the latitude of the Kuroshio-Oyashio extension and generation of SST anomalies on decadal timescales. *J. Climate*, **14**, 4149–4165.

Trenberth, K., and J. W. Hurrell, 1994: Decadal atmospheric–ocean variations in the Pacific. *Clim. Dyn.*, **9**, 303–319.

Vimont, D., 2005: The contribution of the interannual ENSO cycle to the spatial pattern of decadal ENSO-like variability. *J. Climate*, **18**, 2080–2092.

———, M. Wallace, and D. Battisti, 2003: The seasonal footprinting mechanism in the Pacific: Implications for ENSO. *J. Climate*, **16**, 2668–2675.

Walker, G. T., and E. W. Bliss, 1932: World weather V. *Mem. R. Meteorol. Soc.*, **4**, 53–84.

Wang, D. X., and Z. Liu, 2000: The pathway of interdecadal variability in the Pacific Ocean. *Chinese Sci. Bull.*, **45**, 1555–1561.

Wu. L, Z. Liu, R. Gallimore, R. Jacob, D. Lee, and Y. Zhong, 2003: A coupled modeling study of Pacific decadal variability: The tropical mode and the North Pacific mode. *J. Climate*, **16**, 1101–1120.

———, ———, C. Li, and Y. Sun, 2007: Extratroical control of recent tropical Pacific decadal climate variability: A relay teleconnection. *Clim. Dyn.*, **28**, 99–112, doi: 10.1007/s00382-006-0198-5.

Yeh S.-W., and B. P. Kirtman , 2006: Origin of decadal El Nino — Southern Oscillation — like variability in a coupled general circulation model. *J. Geophys. Res.*, **111**, C01009, doi: 10.1029/2005JC002985.

Zhang, Y., J. M. Wallace, and D. S. Battisti, 1997: ENSO-like interdecadal variability: 1900–93. *J. Climate*, **10**, 1004–1020.

Zhong Y., and Z. Liu, 2009: On the mechanism of Pacific multidecadal climate variability in CCSM3: The role of subpolar North Pacific Ocean. *J. Phys. Oceanogr.*, **38**, 2052–2076.

———, ———, and R. Jacob, 2008: The origin of Pacific decadal variability in the NCAR-CCSM3. *J. Climate*, **21**, 114–133.

Chapter 11

Global-Scale Decadal Hyper Modes

Dietmar Dommenget
School of Mathematical Sciences, Monash University,
Clayton, Victoria 3800, Australia
dietmar.dommenget@monash.edu
www.monash.edu

This article discusses the idea of stochastic global-scale multi-decadal Hyper Modes. The chapter will first illustrate the spatial pattern and time-scales of the leading mode of global-scale multi decadal climate variability in observations and model simulations. It is further shown that the main characteristic of the leading mode of global-scale multi-decadal variability can be simulated in the absence of any ocean dynamics and by atmospheric teleconnections only. In a simple stochastic model for the Hyper Modes, local (basin-wide) modes of variability act mostly on interannual time-scales forced by atmospheric white noise. Most of these modes have a weak link to the tropical regions via atmospheric teleconnections, which on longer (decadal) time-scales become more dominant. The tropical link provides a synchronization of the local modes to near global multi-decadal Hyper Modes. The persistence of higher latitudes SST variability with an exponentially decreasing mixing into the deeper oceans leads to a power spectrum with a low-frequency tail and different slope compared to a red noise power spectrum. The main elements of this stochastic model do not involve any ocean dynamics beyond local vertical mixing.

1. Introduction

The climate varies on all time and spatial scales. The processes that cause the variability on different time and spatial scales can be either caused by external forcings (either natural such as volcano or variations in the solar radiation, or anthropogenic) or by chaotic process internal to the climate system. The aim of this study is to explore some simple mechanisms that lead to internal climate variability on multi-decadal time scales and on global spatial scales.

The basic idea of Hyper Modes was first mentioned by Tim Barnett in the 1990s (personal communications; not published), proposing that on longer time-scales some large-scale (near global) teleconnections may be hidden in the climate variability that connects most of the local modes, which act on shorter time-scales. The name 'Hyper' is somewhat in analog to the "Hyper Space" used in space science, but it should not be seen as indication of a "Hyper Space" in climate variability.

The study presented here is based on Dommenget and Latif (2008) in which a simple mechanism for global-scale Hyper Modes was presented. The starting point for the analysis is the assumption that internal climate variability follows from the simple stochastic climate model after Hasselmann (1976). Here the chaotic weather fluctuations are integrated by the slow systems of the climate (oceans) leading to reduced short time climate fluctuations and increased low-frequency (decadal) variability.

Climate Change: Multidecadal and Beyond
edited by Chih-Pei Chang et al.

The Hasselmann climate model thus describes mostly a time scale behavior, but gives little to no insight into the spatial organization of low-frequency variability.

Dommenget and Latif (2008) extended Hasselmann's red-noise hypothesis by two aspects: First atmospheric teleconnections will link remote regions on longer time-scales to synchronize the variability globally. Second the stratification of the ocean leads to an effective heat capacity of the ocean that is different on different time and spatial scales. This alters the power spectrum to be slightly different from the red noise power spectrum by adding a low-frequency tail to the total variance.

In the study presented here we will further explore the Hyper Mode idea and discuss some of its aspects in more detail. The analysis starts with exploring the characteristics of the global teleconnections on longer time-scales. We will in particular focus on the tropical link in the teleconnections. In the next section we will explore the characteristics of the power spectrum and discuss how different aspects of the climate dynamics alter the power spectrum. Finally, we will summarize the Hyper Modes idea with a simplified sketch and discuss current limitations in the understanding.

2. The Spatial Pattern of the Leading Hyper Mode

An objective way to determine the spatial structure or teleconnections of global-scale multi-decadal climate variability is by means of principal component analysis (PCA). The leading empirical orthogonal functions (EOFs) are a very efficient way of presenting the most dominant spatial patterns or teleconnections. In Fig. 1a the EOF-1 mode is shown in terms of the correlation of the PC-1 with global SST based on the HADISST data set from 1870–2004 (Rayner *et al.*, 2003). The SST for this analysis has been filtered by a 10-year running mean and by an exponential detrending to focus on internal natural variability on multi-decadal time-scales. A few aspects are remarkable about this pattern:

- The pattern is a near global multi-pole structure with large correlation values on both hemispheres and in all three ocean basins. Here it needs to be noted that PCA does have some constraints that do enforce global patterns (*e.g.*, Dommenget, 2007; Monahan *et al.*, 2009). In particular a spatial red noise processes, such as isotropic diffusion,

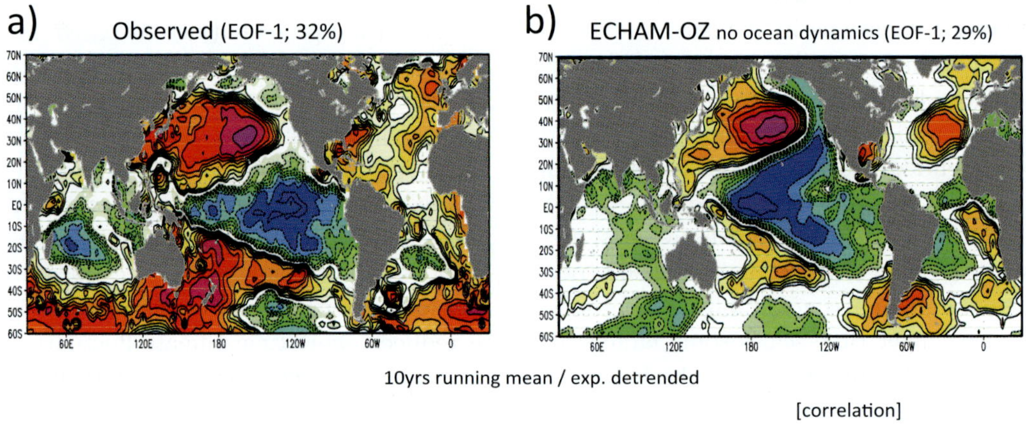

Fig. 1. (a) The leading EOF-mode of observed 10 yrs running mean exponentially detrended SST. (b) the leading EOF-mode in the ECHAM-OZ simulation of 40 yrs running mean SST. The percentage values in the headings are the explained variance for each mode.

would result into a monopole as the leading EOF-mode followed by dipoles (Dommenget, 2007). The fact that we have a quite significant multi-pole structure is a strong support for a non-trivial global-scale teleconnections on the multi-decadal time-scale that is significantly different from a simple spatial red noise process or the geometrical constraints of PCA.

- It resembles the El Nino pattern in many aspects. The pattern in the whole of the Pacific and in the Indian Ocean is quite similar to the El Nino pattern. However, the equatorial Pacific node is wider in its lateral extend and the structures in the Atlantic do not fit to the typical El Nino teleconnections.
- It resembles more a central Pacific El Nino pattern than a east Pacific El Nino. This may indicate that the central El Nino type of variability may have more of a low frequency nature than the east Pacific type of El Nino, which was also discussed in other studies (Yu and Kim, 2011).
- The EOF-1 mode does explain about 1/3 of the total variance on the multi-decadal time-scales suggesting a single mode does explain a large fraction of the variability. The effective number of spatial degrees of freedom (Bretherton *et al.*, 1999), N_{spatial}, is about 6, again supporting the idea that the global-scale multi-decadal climate variability has very few degrees of freedom.

Again, a note of caution has to be given here. The relatively short time period of the observations can artificially force low N_{spatial}. Further, the limited number of observations in the earlier periods of the data set lead to large gaps in the data, which can also artificially force low N_{spatial}., even though these gaps are interpolated in the HADISST data.

Given the interesting observational findings and knowing about the limitations of these observations we would like to confirm the existence of these global-scale modes from model simulations to back-up the observational findings and to explore the physical mechanism causing these structures.

We therefore repeat the analysis for a coupled general circulation model (CGCM) simulation with fixed external boundary conditions (*e.g.*, no greenhouse gas forcings and no change in the incoming solar radiation).

The simulation we use is based on the ECHAM5 atmospheric general circulation model in low (T31; 3.75° × 3.75°) resolution coupled to the simple single column ocean mixed layer model OZ, which does not simulate any lateral interaction in the ocean (Dommenget and Latif, 2008). Each column of the OZ model has 19 levels and goes to 500 m depth. The model is driven at the surface by heat fluxes and mechanical wind mixing following Niiler and Kraus (1977). The vertical mixing is thus a function of the wind input and decreases with depth. The density is a function of temperature and salinity, but only temperature is variable in this model, whereas salinity is kept at climatological profiles. The temperature of the lowest layer at a depth of 500 m is restored to the observed climatology.

The overall structure of this model is similar to other mixed layer models (*e.g.* Alexander and Penland, 1996), however it seems no other study has used such a model for studying longer (multi-decadal) time-scales.

The model is integrated over 2000 yrs and we refer to this model simulation as ECHAM-OZ.

The ECHAM-OZ simulation has some important characteristics in comparison to the observations: First, in the model world we do not have any data sampling problems. Second, the model has 2000-yr of data, which is 20-times more than we have from observations, which allows us to make stronger statements about the variability in this data set. Third and most importantly for the following discussion, the model makes an important simplification: ocean points do not interact laterally and subsequently all spatial coherence or teleconnections are entirely a result of the atmospheric dynamics. Thus ocean dynamics are essentially neglected.

If we compare the leading mode in the ECHAM-OZ simulation (Fig. 1b) with the observations (Fig. 1a), we can find several important results:

- The two patterns of the leading EOF-modes match very well (pattern correlation of 0.9). This holds for the structure within the whole of the Pacific, but also holds for the teleconnections to the Indian and Atlantic Oceans.
- The explained variance of the EOF-1 in the ECHAM-OZ simulation is roughly the same as in the observations. The ECHAM-OZ simulation has also a relatively low $N_{\text{spatial}} = 9$, but still larger than the observations, which could suggest that the much shorter observations period does have some sampling problems. Note that $N_{\text{spatial}} = 20$ in the ECHAM-OZ simulation for annual mean SST without any low-pass filtering.
- Most remarkable is that the ECHAM-OZ simulation has the same El Nino-like pattern in the tropical Pacific, despite the fact that the ocean has no lateral dynamics and all spatial structure is entirely forced by the atmospheric teleconnections. It illustrates that El Nino like SST patterns can exist without any ocean dynamics, but forced by atmospheric feedbacks. This is supported by other studies as well (Dommenget, 2010; Clement *et al.*, 2011).

In summary, we find that the ECHAM-OZ simulation gives a good first order representation of the leading mode of observed multi-decadal global-scale climate variability. Therefore, it appears to be a good tool to study the physical mechanism of this type of climate variability.

3. Limitations of Coupled General Circulation Model (CGCM) Simulations

In Fig. 1 we illustrated that the observed leading mode of global multi-decadal SST variability is

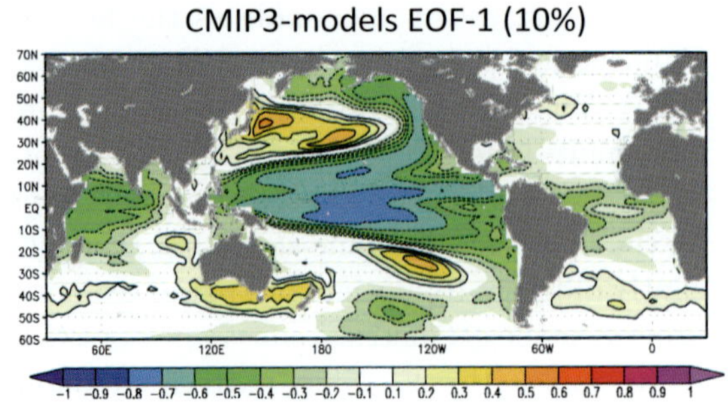

Fig. 2. Correlation maps of the leading EOF mode of detrended multi-decadal global sea surface temperatures (SSTs) as in Fig. 1, but for the CMIP3 multi-model ensemble. From Dommenget and Latif (2008).

well simulated in its spatial structure by the simple ECHAM-OZ model. It therefore seems reasonable to assume that state of the art CGCM simulations should also be capable of simulating the leading mode of global multi-decadal SST variability. Figure 2 shows the leading EOF-1 of the combined multi-model SST variability of seven CMIP3 preindustrial control simulations with each being 340-yr long (2380-yr in total). Although some of the features of the observed pattern are present in this leading EOF-1 mode, the pattern is not as close to the observed as in the ECHAM5-OZ simulation. It is mostly focused on the Indo-Pacific sector, it is missing most of the Atlantic pattern and it is much smaller in its explained variance.

This opens up the question: Why do the more sophisticated CGCM simulations have less skill in simulating the leading mode global multi-decadal SST variability? To some extent this is related to the fact that fully coupled climate models, CGCMs, have much more degrees of freedom than the simplified ECHAM5-OZ simulation. The ECHAM5-OZ simulation is flux corrected and therefore the SST climatology stays close to the observed. In the fully coupled CGCMs the small imbalance in the coupled ocean-atmosphere-land-ice interactions lead to substantial model errors that lead to substantially (up to 5°K) different SST climatologies. Furthermore, the CGCMs have ocean dynamics

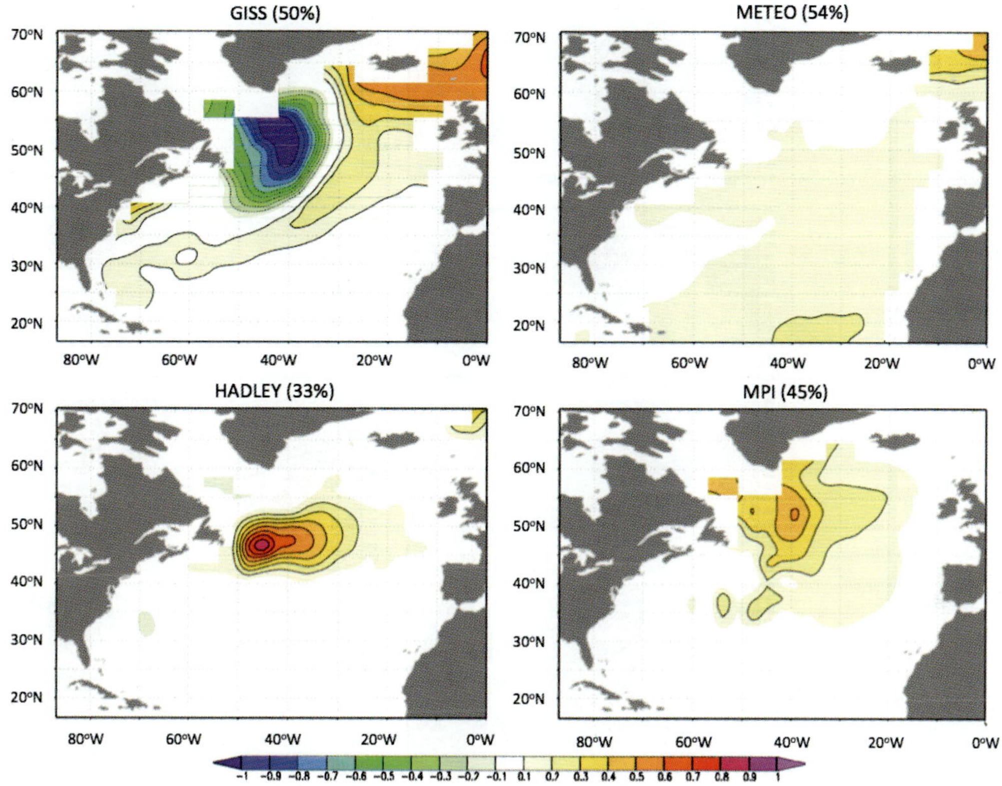

Fig. 3. Correlation maps of the leading EOF modes of detrended multi-decadal global sea surface temperatures (SSTs) as in Fig. 1, but for the CMIP3 multi-model ensemble.

interacting with the atmospheric dynamics, whereas the ECHAM5-OZ simulations do not have any lateral ocean dynamics. Thus the ocean dynamics in CGCM with substantially different SST climatologies can cause quite different modes of SST variability.

This can be illustrated by looking at the leading EOF-modes in different CGCM simulations. In Fig. 3 we show the EOF-1 mode in the North Atlantic for 4 different CGCM preindustrial (no-external forcings) for 5-yr running mean SST variability as an example.

It can be noted that they are quite different from each other and from the observed (not shown). It is also notable that they have quite distinct amplitudes at some locations, but these locations are different in each model. So the models do have quite significant decadal models of variability that are quite local, but different in each model. These local modes are most likely caused by ocean dynamics, as atmospheric dynamics tend to be on larger spatial scales and the AGCM coupled to slab ocean or the ECHAM5-OZ simulations do not capture such local modes. The results are similar for the other extra tropical oceans (North Pacific and Southern Ocean).

In summary, this suggests that the CGCMs simulations strongly disagree on the structure of global Hyper Modes mainly, because they all have different ocean dynamics forcing different SST modes. The global EOF-1 mode (Fig. 2) is only the common structure in all the models and the low explained variance value and the missing Atlantic pattern indicate strong disagreement between the models.

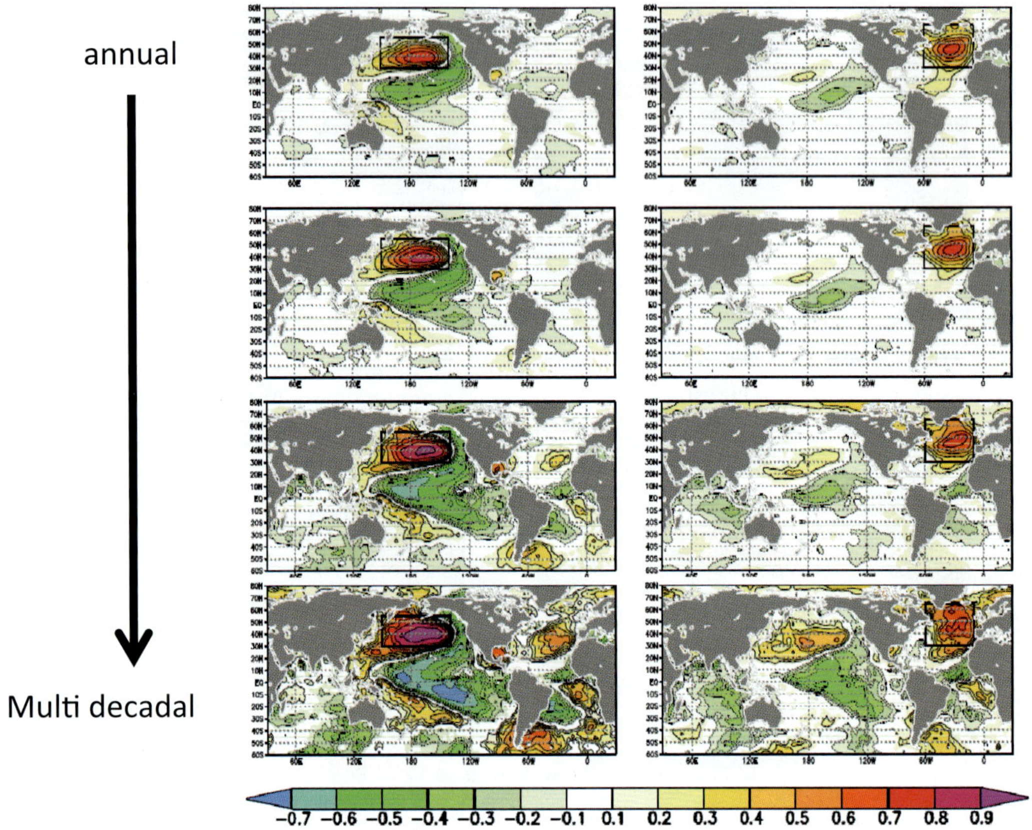

Fig. 4. Left column: The correlation of the SST in the 2000 yrs ECHAM-OZ simulation averaged over the North Pacific box with the SST in all regions of the world. Upper panel for high-pass 5 yrs filter annual mean SST, second from the top for 5 yrs to 10 yrs filtered SST, third from the top for 10 yrs to 40 yrs filtered SST and lower panel for low-pass 40 yrs SST. Right column the same as left, but for the SST-box in the North Atlantic.

4. The Tropical Link of the Inter-Hemispheric Teleconnections

In the previous section we already indicated that the global modes of variability on the shorter, inter-annual, time-scales are much more complex ($N_{\text{spatial}} = 20$). It suggests that on the longer-times scales the more localized smaller scale modes of variability are somehow unified or linked by some teleconnections. To illustrate how this link may operate, we analyze how the global correlation patterns change with time-scales. We therefore defined two reference regions in the North Pacific and North Atlantic (see boxes in Fig. 4) and correlate the mean SST in these boxes with the SST in all parts of the oceans. Figure 4 illustrates how the teleconnections of these boxes changes with time-scales: On the shorter time-scales (annual) both SST-boxes correlate mostly to nearby regions. As the time scales increase to longer periods, the correlation of the SST-boxes spread out to near global teleconnections. Most interestingly, the teleconnections of the SST-box in the North Pacific and North Atlantic become more similar. It is also important to note, that even at the shortest time scales both SST-boxes have a similar link to the subtropical to equatorial north Pacific, suggesting that this region may be a central element to create the global link to remote regions. Several other studies have also suggested that the North Pacific is linked or forcing the tropical Pacific by the subtropical region (Vimont *et al.*, 2001; Yu and Kim, 2011).

To illustrate, that multi-decadal atmospheric variability in the northern and southern hemisphere are indeed linked we can analysis the global correlations of the 500 hPa geopotential heights in the north Pacific with the rest of the world, see Fig. 5a. We find remarkable high correlations all over the world and in particular in the southern extra-tropical Pacific (at around 30°S). With the help of cross-spectral analysis we can illustrate that this inter-hemispheric link becomes dominant only on the very long time-scales of decades and longer, see Fig. 5b.

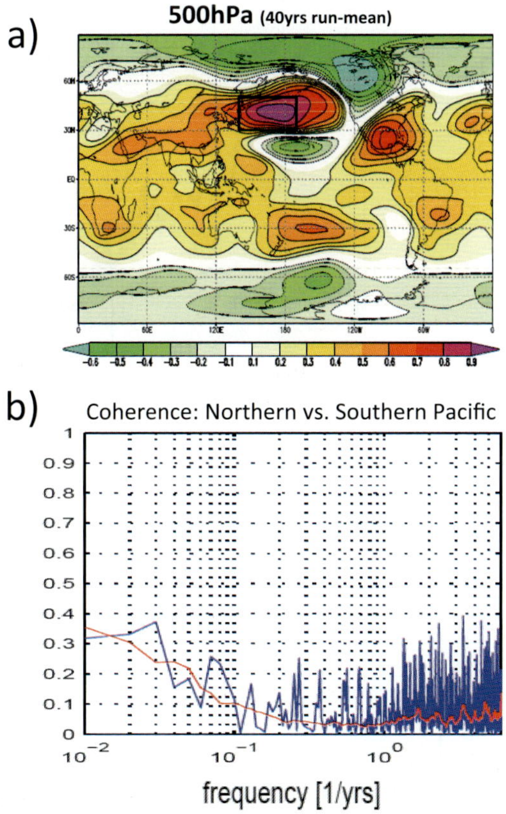

Fig. 5. (a) Correlation of the 500 hPa geopotential height in the ECHAM-OZ simulation in the box over the western North Pacific with the geopotential height globally for 40 yrs low-pass filter data. (b) cross-spectral coherence between the 500 hPa geopotential height in the North Pacific box with the mid-latitudes South Pacific (20°S–40°S; 170°E–230°E). The red line is a smoothed version of the blue line to reduce the noise.

The question arises, whether the tropical Pacific SSTs cause this link, or does the link exist independent of tropical Pacific SSTs. We can address this question by performing a sensitivity experiment with the ECHAM-OZ model in which we hold the SST in the tropical Pacific (from 20°S to 20°N) fixed to climatology (no SST variability allowed). The simulation was integrated for 500-yr and is referred to as ECHAM-OZ$_{\text{noeq}}$.

Figure 6a shows the cross-spectral coherence between the 500 hPa geopotential heights in the North Pacific box and in the South Pacific box. In comparison to the ECHAM-OZ simulation (Fig. 5b) we do not see any increase in coherence at any time-scale. The coherence is low on all time scales and gives no indication of coherent variability between the two regions. Thus we can conclude that the tropical SST variability is a necessary link for the inter-hemispheric interactions.

We can go one step further and compare the results to a simulation without any SST variability (climatological SST everywhere; atmosphere only simulation). This sensitivity simulation is also integrated for 500-yr and is refereed to as ECHAM$_{\text{FIXSST}}$. Figure 6b shows the power spectrum of 500 hPa geopotential heights in the North Pacific box for the three different simulations. Only the simulation that does have tropical SST variability shows an increase in spectral variance for the longer, multi-decadal time-scales. The simulation with extra-tropical SST variability and the one without any SST variability do not differ in the 500 hPa geopotential heights power spectrum at all. This suggests that the tropical SST variability does not only cause the link between the hemispheres, but also does cause most of the variance on the multi-decadal time-scales in the atmospheric variability in the extra-tropics.

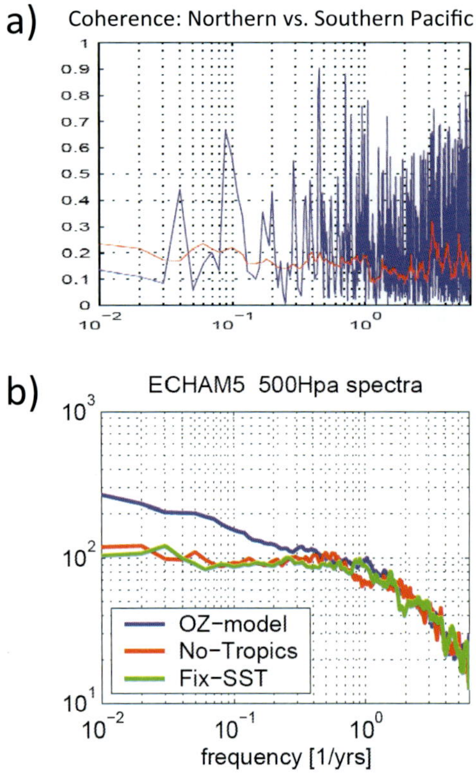

Fig. 6. (a) Cross-spectral coherence between the 500 hPa geopotential height in the North Pacific box with the mid-latitudes South Pacific as in Fig. 5b, but for the ECHAM-OZ_{noeq} simulation. (b) power spectrum of the 500 hPa geopotential height in the North Pacific box in the ECHAM-OZ (blue), ECHAM-OZ_{noeq}(red) and the $ECHAM_{FIXSST}$ (green) simulations.

5. The Time-Scale Characteristics of Stochastic Climate Variability

We now like to analyze the time-scale behavior of global-scale climate variability. The aim of this analysis is to understand what causes the multi-decadal time-scale climate variability.

Stochastic climate models will often have a quite clear signature in the power spectrum of the variability. By comparing the observed power spectrum of SST variability with theoretical stochastic models we can gain some insight into the processes that case the SST variability.

We start with a look at the average power spectrum in the mid-latitudes SST variability of the northern North Pacific and Atlantic, as these regions are known to have relatively strong multi-decadal SST variability. Figure 7a shows the mean power spectrum of the northern North

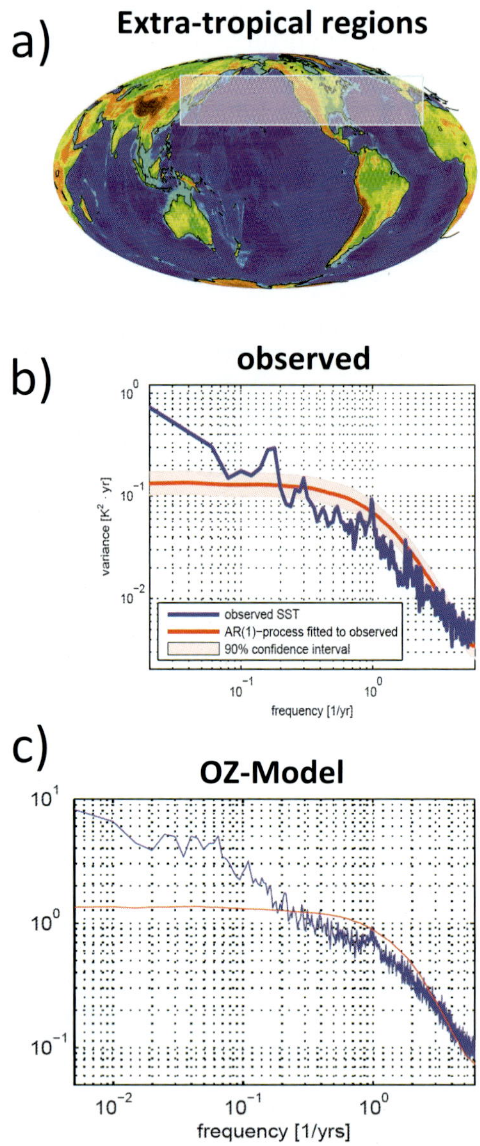

Fig. 7. (a) Region of the mid-latitudes box (b) the average power spectrum of the observed SST (blue line) for all points in the mid-latitudes of the North Pacific and Atlantic (see (a)). The red line is the fitted red noise power spectrum with its 90% confidence interval (light red shading). (c) as (b), but for the 2000 yrs ECHAM-OZ simulation.

Pacific and Atlantic against a fitted red noise power spectrum. The red noise power spectrum represents the simple stochastic climate model that assumes that random weather fluctuations (white noise) are integrated by a fixed heat capacity, such as the ocean surface mixed layer.

We can first of all recognize that the observed mean SST power spectrum increases in variance for increasing periods (smaller frequencies). This increase (the slope of the power spectrum) is almost constant through out all time-scales and it does not stop at long periods.

This time-scale signature is quite different from the fitted red noise spectrum. The fitted red noise spectrum has a faster increase in variance for decreasing frequencies on time-scales from months to about a year. It then saturates at some variance level and does not increase in variance any more for periods longer than a decade.

The fitted red noise process assumes, by construction, the same lag-1 (month) autocorrelation as in the observed SSTs. Thus we can conclude that the time scale behavior of the observed SSTs does not fit to a red noise process. The signature of the observed SST power spectrum is a continuous increase in variance with a slope that is flatter than that of a red noise process, but continues with out any sign of saturation.

The SST power spectrum of the ECHAM-OZ simulation for the North Pacific and Atlantic is quite similar in this respect to the observed power spectrum. The much longer time series of the simulation allows us to extend the power spectrum to longer periods. On the longer periods the ECHAM-OZ spectrum becomes even more clearly separated from the fitted red noise power spectrum. What is also quite remarkable is that the ECHAM-OZ SST variability still increases with no sign of saturation until time scales of 100-yr, despite the fact that this model simulation does not include any ocean dynamics beyond the single column vertical mixing.

The dynamics of the ECHAM-OZ model can be simplified to a simple stochastic model of the SST anomalies, T:

$$c\frac{dT}{dt} = -\gamma_{surf} \cdot T + \kappa_z \cdot \nabla_z^2 T + \xi_{surf} \qquad (1)$$

The heat capacity of the surface layer, c, is assumed to be a constant. The damping of T is proportional to the constant, γ_{surf} and the vertical mixing below the mixed layer depth is proportional to κ_z (as a function of the depth underneath the mixed layer) and the vertical temperature curvature $\nabla_z^2 T$. The system is forced at the surface by white noise weather heat fluxes, ξ_{surf}. So the extension to the Hasselmann red noise hypothesis is the inclusion of the vertical mixing term.

The power spectrum of this simple model, for realistic parameters, is similar to that observed or in the ECHAM-OZ simulation, see Fig. 8a. We can discuss how the time-scale behavior of this model depends on the model parameters. The heat capacity, c, is damping the high-frequency variability, but does not change the amount of variability on the longer time-scales. It essentially shifts the point in the power spectrum, at which the variance saturates (see Fig. 8b). The damping γ_{surf} shifts the variance at frequency zero (infinitively long periods) down (see Fig. 8c). This also shifts the point in the power spectrum, at which the variance saturates. A zero damping would have infinite power at frequency zero. Note, that the strength of the damping is mostly an atmospheric characteristic and will depend on the spatial extend of the SST anomalies: a small-scale SST anomaly is more strongly damped than a global scale SST anomaly, because the small-scale SST anomaly is damped by exchanging heat with nearby regions. Thus, the damping γ_{surf} will be spatial scale dependent and it will be weakest for global-scale SST pattern.

Finally, the mixing profile underneath the mixed layer κ_z alters the slope of the power spectrum. In Eq. (1) the mixing parameter κ_z

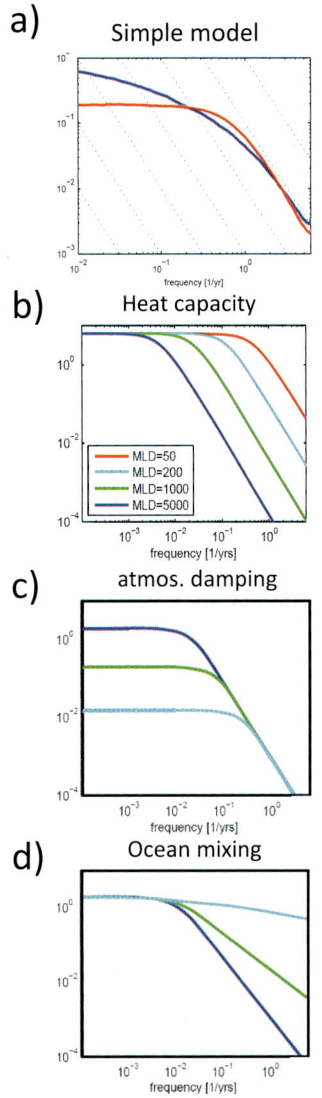

Fig. 8. (a) Power spectrum of the SST from Eq. (1) with realistic parameters (blue line) compared against a fitted red noise spectrum (red line). (b) red noise power spectra for different heat capacities c. (c) red noise power spectra for different damping parameter γ_{surf} (dark blue weakest, light blue strongest). (d) the power spectra of the SST from Eq. (1) with different mixing profiles for κ_z (dark blue fully mixed ocean, green exponentially decreasing mixing, light blue no mixing thin skin layer ocean).

is a function of depth. It is largest at the surface due to the wind input and decreases with depth. In the extreme case of a perfectly mixed ocean (κ_z is very large in all depth; mixed layer goes to the bottom of the ocean) the power spectrum reduces to that of a red noise power spectrum. In the other extreme case of no mixing (κ_z is very small) and a mixed layer of a thin ocean surface skin the power spectrum becomes a white noise power spectrum. The in between realistic cases of an observed mixed layer depth with an exponential decreasing κ_z mixing profile underneath the mixed layer gets the power spectrum with the observed slope (Fig. 8a), that is shallower than red noise, but continues to increase through-out all observed time scales.

The question arises: On what time-scales will the power spectrum of this simple stochastic climate model saturate? For the realistic parameters we have chosen in Fig. 8a it seems to saturate in the 100-yr to 1000-yr time-scales. In the ECHAM-OZ simulation with a 800 m deep ocean it seems to be similar, but it is yet not clear, as the 10,000-yr long simulations needed to test this do not yet exist.

In the observations it is unclear, but it will very much depend on how much deep ocean mixing and ocean internal or atmosphere-ocean coupled modes, such as ENSO, will affect this stochastic mechanism.

6. Summary and Discussion

We can summarize the global-scale multi-decadal Hyper Modes idea by the sketch shown in Fig. 9.

I. On the shorter, seasonal to interannual, time-scales most climate variability is essentially regional (*e.g.*, ocean basin, or near continental) with only weak teleconnections to remote regions (Figs. 4 and 5b).

II. In the higher latitudes of the northern hemisphere and potentially also on the southern hemisphere the SST variability can persist on multi-decadal time-scales due to the deeper mixing in the oceans at higher latitudes (Fig. 7). This SST power spectrum has more variance on multi-decadal and longer time scales and a different slope than a simple red noise process (Fig. 7).

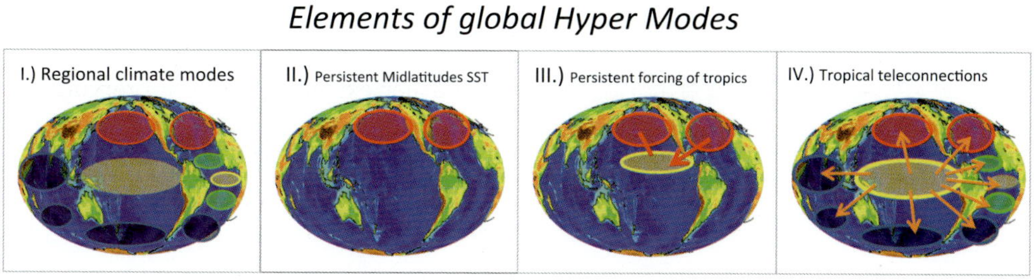

Fig. 9. Sketch summarizing the elements of the global-scale multi-decadal hyper modes.

III. These high latitudes SST modes seem to interact with the subtropical Pacific (Fig. 4).
IV. The tropical Pacific or the tropical SST in general appear to be the central link that connects global SST modes across both hemispheres and across ocean basins (Fig. 6).
V. On the multi-decadal time scales the variability in all ocean basins is linked via the tropics to have synchronized variability leading to the global scale hyper modes (Fig. 1).

Thus the idea of Hyper Modes can be summarized as near global-scale climate modes that connect local mode of variability on longer time scales via atmospheric teleconnections. They can therefor be considered as being a teleconnection in the climate system that is on top of or hidden from the dominant local and shorter time-scale mode, but are probably the dominant modes of internal natural climate variability on time scales longer than decades.

Acknowledgments

I like to thank Claudia Frauen for comments and proof reading. The work was supported by the ARC Centre of Excellence in Climate System Science (CE110001028).

References

Alexander, M. A., and C. Penland, 1996: Variability in a mixed layer ocean model driven by stochastic atmospheric forcing. *J. Climate*, **9**, 2424–2442.

Bretherton, C. S., M. Widmann, V. P. Dymnikov, J. M. Wallace, and I. Blade, 1999: The effective number of spatial degrees of freedom of a time-varying field. *J. Climate*, **12**, 1990–2009.

Clement, A., P. DiNezio, and C. Deser, 2011: Rethinking the ocean's role in the Southern Oscillation. *J. Climate*, **24**, 4056–4072.

Dommenget, D., 2007: Evaluating EOF modes against a stochastic null hypothesis. *Clim. Dyn.*, **28**, 517–531.

———, 2010: The slab ocean El Niño. *Geophys. Res. Lett.*, **37**, L20701.

———, and M. Latif, 2008: Generation of hyper climate modes. *Geophys. Res. Lett.*, **35**, L02706.

Hasselmann, K., 1976: Stochastic climate models Part I. Theory. *Tellus*, **28**, 473–485.

Monahan, A. H., J. C. Fyfe, M. H. P. Ambaum, D. B. Stephenson, and G. R. North, 2009: Empirical orthogonal functions: The medium is the message. *J. Climate*, **22**, 6501–6514.

Niiler, P. P., and E. B. Kraus, 1977: One dimension models of the upper ocean. *Modeling and Prediction of the Upper Layers of the Ocean*, E. B. Kraus, Ed., Pergamon, New York, 143–172.

Rayner, N. A., D. E. Parker, E. B. Horton, C. K. Folland, L. V. Alexander, D. P. Rowell, E. C. Kent, and A. Kaplan, 2003: Global analyses of sea surface temperature, sea ice, and night marine air temperature since the late nineteenth century. *J. Geophys. Res. Atmos.*, **108**, 4407.

Vimont, D. J., D. S. Battisti, and A. C. Hirst, 2001: Footprinting: A seasonal connection between the tropics and mid-latitudes. *Geophys. Res. Lett.*, **28**, 3923–3926.

Yu, J. Y., and S. T. Kim, 2011: Relationships between extratropical sea level pressure variations and the central Pacific and eastern Pacific types of ENSO. *J. Climate*, **24**, 708–720.

Chapter 12

Evidence for a Recurrent Multi-Decadal Oscillation in Global Temperature and Possible Impacts on 21st Century Climate Projections

Ka-Kit Tung and Jiansong Zhou

Department of Applied Mathematics,
University of Washington, Seattle, WA, USA
ktung@uw.edu

We summarize various pieces of evidence in support of interpreting the two-and-a half cycles of multi-decadal variation seen in the observed global surface temperature as an internal variability related to the Atlantic Multidecadal Oscillation (AMO). These include evidence from the pre-industrial era, modes of coupled- model internal variability in the absence of anthropogenic forcing and the reasonableness of the dynamical mechanism for the negative feedbacks necessary for a quasi-periodic behavior of this internal oscillation of ∼70 year period. The impact of this interpretation on the projection of future warming in the 21st century is then explored. Models participation in the Intergovernmental Panel on Climate Change (IPCC), Fourth Assessment Report (AR4) simulated the observed global warming in the last two and half decades of the 20th century as almost entirely anthropogenically forced. If, as we argue here, roughly half of the observed warming was possibly due to an upswing in the internal variability, the modeled forced response could be seen during that period too high, and the projected future warming under various forcing scenarios will likely be too high when the same models are used to project into the future. The discrepancy is further exacerbated when the predicted cooling phase of the AMO is taken into account to mid-21st century. Current IPCC methodology of using ensemble means or models of intermediate complexity in projections excludes or greatly reduces the effect of internal variability. It is suggested that the range of projections of future global warming should include an additional scenario with a multi-decadal internal variability, instead of having all scenarios being entirely responses to anthropogenic forcing. Incorporating such an internal variability in the scenarios yields a possible delay of the next accelerating warming episode by two to three decades.

1. Introduction

Our work was inspired by the paper of Wu *et al.* (2011), who showed, using the novel method of Ensemble Empirical Mode Decomposition (Wu and Huang, 2009; Huang *et al.*, 1998), that there exists, in the 150-year global-mean surface temperature record, a multidecadal oscillation. With an estimated period of 65 years, 2.5 cycles of such an oscillation was found in that global record (Fig. 1, top panel). They further argued that it is related to the Atlantic Multi-decadal Oscillation (AMO) (with spatial structure shown in Fig. 1, bottom panel) and, if this oscillation is separated out, a monotonic trend emerges in the global mean temperature (red line, Fig. 1 top panel), with a smaller warming rate of 0.08°C/decade compared to the linear trend of the raw data since 1980. Given the importance of this last implication on the recent anthropogenic global warming, it is quite natural that the scientific community is demanding more evidence that this oscillation is real, recurrent and natural, and in particular that it is not instead a response to time-varying anthropogenic forcing that happens to look like an oscillation during the 20th century.

Climate Change: Multidecadal and Beyond
edited by Chih-Pei Chang et al.

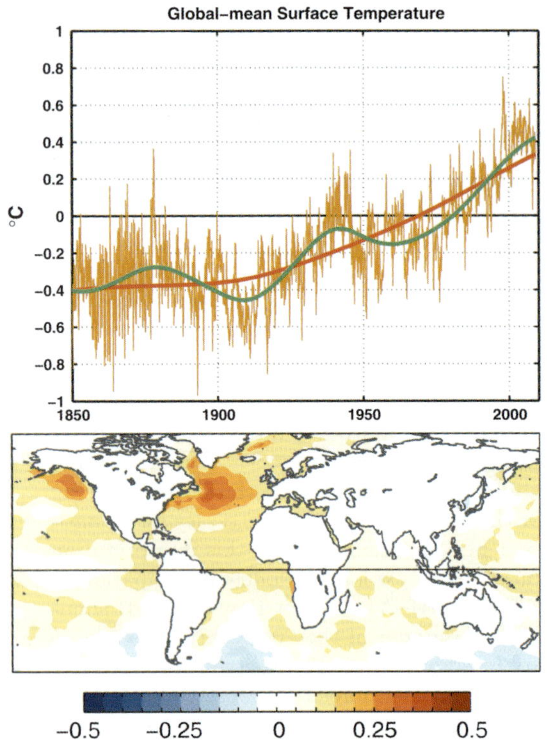

Fig. 1. Taken from Wu *et al.* (2011). Top panel: Raw global surface temperature in brown. The secular trend in red. The low-frequency portion of the data constructed using the secular trend plus the gravest multi-decadal variability, in green. Bottom panel: the global sea-surface temperature regressed onto the gravest multi-decadal mode.

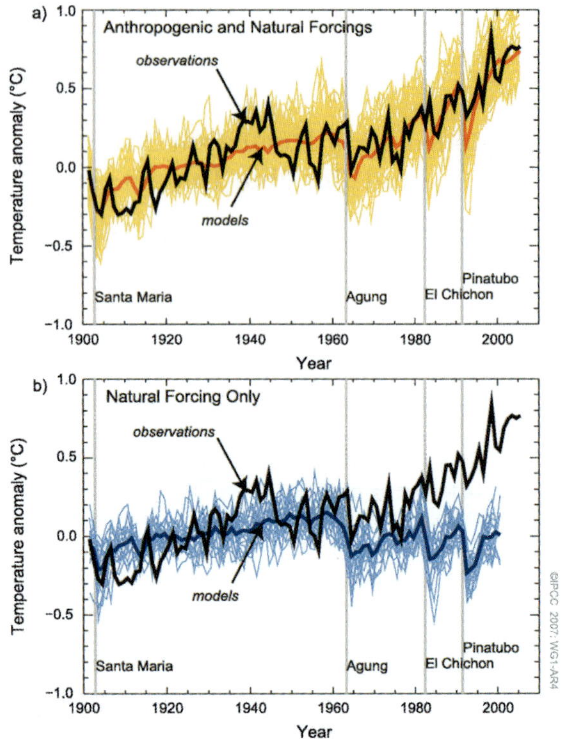

Fig. 2. Taken from IPCC AR4: top panel: global mean surface temperature anomaly relative to the period 1901–1950, as observed (black) and as simulated with both anthropogenic and natural forcings. The thick red curve shows the multi-model ensemble mean and the thin lighter curves show the individual simulations. Bottom panel: same as the top panel except that the simulations are for natural forcing only. The thick blue curve shows the multi-model ensemble mean and the thin lighter curves show individual simulation.

The Intergovernmental Panel on Climate Change (IPCC) Fourth Assessment Report (AR4) (Solomon *et al.*, 2007) presented results on the simulation of 20th century climate using a large number of coupled atmosphere-ocean general circulation models, and compared the model results with the observed global mean temperature. It appears that the observed 20th century climate change can be well simulated as responses to the combined anthropogenic and natural forcing. The former includes greenhouse-gas radiative forcing and tropospheric sulfate aerosols due to man-made pollution, and the latter includes solar irradiance variability and stratospheric aerosols due to explosive volcanic eruptions.

Figure 2, taken from AR4, shows that without anthropogenic forcing these models do not produce any warming trend in the second half of the 20th century (bottom panel). The implication is that the observed warming during that period can be attributed entirely to anthropogenic forcing, although AR4 was more conservative and concluded only that “most” of the observed warming during the second half of the 20th century is due to anthropogenic forcing. This point is worth keeping in mind when we discuss later how these same models are used for future climate projections. Internal variability in models tends to be greatly reduced with ensemble averaging. Due to the difficulty in accurately initializing the subsurface ocean, model internal variability will likely be of different phases and amplitudes compared with the observation, and these vary between each ensemble member and among models.

In an innovative study, DelSole *et al.* (2011) circumvented this problem by looking at the spatial pattern of the dominant internal variability in the AR4 models. That pattern (Fig. 3, top panel) resembles the AMO, with warming centered in the North Atlantic but also in the Pacific and generally over the Northern Hemisphere (Delworth and Mann, 2000). When the *observed* temperature is projected onto this model spatial pattern, the time series (in Fig. 3 bottom panel) that results varies like the AMO Index (Enfield *et al.*, 2001), even though individual models do not necessarily have an oscillation that behaves like the AMO Index (Fig. 3, middle panel). This "unforced" mode contributes 0.08°C/decade to the recent observed 30-year warming trend. When this internal multidecadal mode is removed from the observation, DelSole *et al.* (2011) found a warming rate of ~0.1°C/decade remains as forced response.

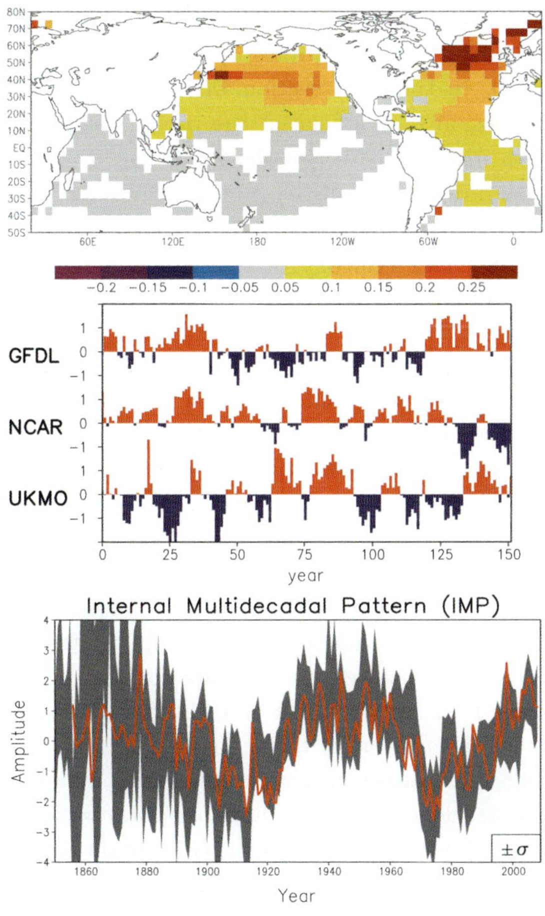

Fig. 3. Taken from DelSole *et al.* (2011). Top panel: the spatial pattern that maximizes the average predictability time of sea-surface temperature in 14 climate models run with fixed forcing (i.e., "control runs"). Middle panel: the time series of this component in three representative control runs. Bottom panel: time series obtained by projecting the observed data onto the model spatial pattern from the top panel. The red curve in the bottom panel is the annual average AMO index after scaling.

Models participating in IPCC AR4 are archived under the Phase 3 of Coupled Model Intercomparison Project (CMIP3) (Meehl *et al.*, 2007). Figure 2 shows that while most models have succeeded in producing the observed warming rates in the global mean in the second half of the 20th century, they failed to simulated the early 20th century warming that is also in the observed data. A careful examination of Fig. 2 reveals that the rates of warming (the slope of the model curves) for this period in most models are too small, compared to the positive rates in the observation (as previously pointed out by Semenov *et al.*, 2010). In fact the observed rate of warming in the early 20th century is not any less than that during the second half (see Fig. 1). Furthermore, a more detailed examination of the modeled vs observed warming shows some discrepancy in the spatial pattern of warming (Semenov *et al.*, 2010), with models producing less warming in the North Atlantic and more in the South Atlantic, implying that the spatial pattern of AMO, which warms in the North Atlantic but cools in the South Atlantic, contributed to the observed warming. Recently, Wigley and Santer (2012) performed a simulation of the 20th century warming using a model of intermediate complexity (without ocean internal variability), and concluded that the observed early 20th century warming is too large to be explained by anthropogenic forcing and/or current estimates of solar forcing, irrespective of a selected combination of climate sensitivities and aerosol forcing. They suggested the discrepancy may be explainable by the change in the AMOC, previously explored by Wigley and Raper (1987).

For the forthcoming IPCC assessment (AR5), the model results are available under the Phase 5 Coupled Model Inter-comparison Project (CMIP5) (Taylor *et al.*, 2012). Of the 10 CMIP5 models examined by Cheng *et al.* (2013) the early 20th century warming in the North Atlantic SST was significantly weaker in the ensemble means of each model as compared to the observed, ERSST (see their Fig. 2(g)). An exception (not consider by Cheng *et al.*, 2013) is HadGEM-ES2, which simulated the observed AMO quite well. The HadGEM-ES2 result is the subject of a recent paper by Booth *et al.*(2012). The newly incorporated indirect aerosol effects from a time-varying aerosol forcing are apparently responsible for driving the multi-decadal variability in the model *ensemble-mean* global mean temperature variation. Because of the stable vertical stratification of the upper ocean, aerosol cooling penetrates to the subsurface layer more easily than surface heating. Consequently the signature of aerosol cooling can be detected more easily in the subsurface. Zhang *et al.* (2013) showed evidence that the indirect aerosol effects in HadGEM-ES2 have been overestimated by showing that while the real upper 700 m of the ocean is observed to warm secularly from 1955, there is no warming in the model ocean, but the model becomes in agreement with observation when the aerosol forcing were held constant. More importantly, while this model has succeeded in simulating the time behavior of the area-mean North Atlantic sea-surface temperature variation in the 20th century, the vertical distribution of temperature in the subsurface ocean and the patterns in other ocean basins are inconsistent with the observation. Wang and Zhang (2013) showed that some single ensemble members of CMIP5 models, including HadGEM-ES2, can simulate the characteristic AMO subsurface pattern, even though the latter's ensemble mean does not. The implication is that the multidecadal variability associated with the AMO is an internal variability, at least in these models.

A strong piece of evidence that the multidecadal variability is natural could be if the oscillation is found to extend into the pre-industrial era, when anthropogenic forcing was small. In the next section we will precede the discussion of evidence in the pre-industrial era with a brief history of the AMO and follow it with a discussion of the proposed mechanisms.

2. A Brief History of the AMO Observations

The acronym "AMO" for Atlantic Multi-decadal Oscillation was coined by the *Science* writer Richard Kerr (Kerr, 2000), who attributed the discovery to Delworth and Mann (2000) in *Climate Dynamics*. There were two earlier papers: Schlesinger and Ramankutty (1994) in *Nature*, who found two cycles with period of 65–70 years, and Folland *et al.* (1984), also in *Nature*, who found a worldwide temperature fluctuation of 0.6 K with power at 83 years for the period 1856–1981. In Delworth and Mann's study, 330 years of multi-proxy data, involving mostly tree rings, but also some ice cores, corals and sediments, were used and they found 4.5 cycles of the AMO. We shall follow the convention of Kerr and still keep the "O" in "AMO".

There are longer records of ice cores in Greenland and tree rings in other locations, but establishing statistical significance of the oscillation has been difficult in the longer proxy records. Gray *et al.* (2004) considered tree-ring based reconstruction of the AMO back to 1567, one cycle longer, but the early cycle with very large amplitudes was probably contaminated by the severe cooling of the Little Ice Age, caused probably by volcanic aerosols. Chylek *et al.* (2011) examined five ice-cores over the period 1303–1961 and reported two time scales for the AMO: A 20-year and a 45–85-year oscillation. The 20-year oscillation was statistically significant, but the longer oscillation was not. The use of the Hamming spectral filter may have significantly reduced the 50–60 year spectral peak. Broecker (1975) extracted two cycles for the period 1800–1975 from the 800-year Camp Century ice core in Greenland. He used a combination of two Fourier harmonics, of 80-year and 180-year periods, to reconstruct the multi-decadal temperature variation in the past 176 years. Neither of the two harmonics used is statistically significant in the 800-year record.

Tung and Zhou (2013) analyzed the longest instrumental temperature record at Central England (Manley, 1974; Parker *et al.*, 1992) going back to 1659, in the middle of the Little Ice Age. The use of instrumental record has a few advantages. It does not have the time uncertainty in some proxy records. It also does not have the divergence issue related to tree-ring reconstructions, which do not agree with the instrumental record in the industrial era when the latter becomes available (Briffa *et al.*, 1998). Although the multi-proxy reconstruction of Delworth and Mann (2000) involved more proxies than tree rings, it also diverges from the instrumental record after 1980. There is no divergence issue with the Central England Temperature (CET), and the 354-year record is slightly longer than that of Delworth and Mann (2000). Using Wavelet spectrum Tung and Zhou (2013) demonstrated that the oscillation with spectral peak between 50 and 80 years is statistically significant at 95% confidence level and extends throughout the 354-year record. The authors argued that this is the AMO mode. Figure 4

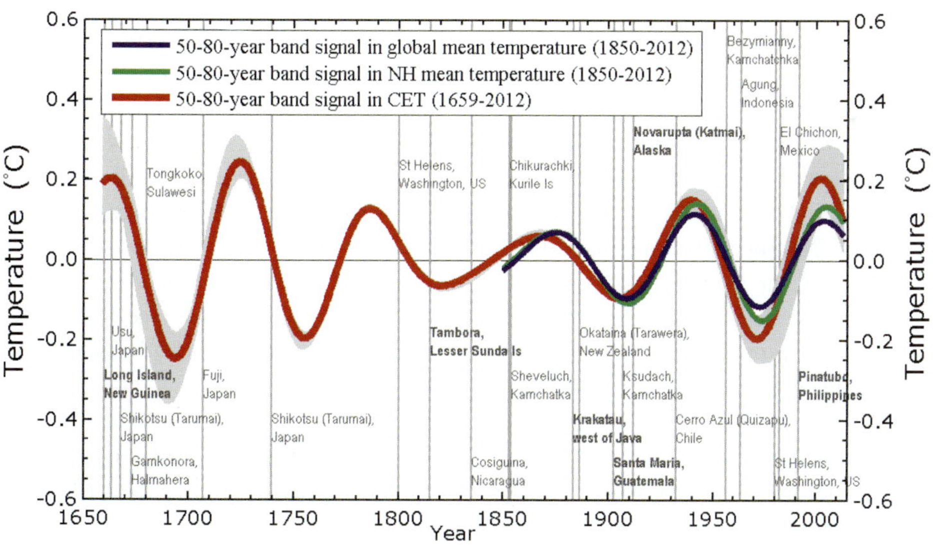

Fig. 4. The multi-decadal variability (50–80 year band-passed in the continuous Wavelet analysis) in CET (in red), in global-mean temperature (in blue) and Northern Hemisphere mean (in green) using HadCRUT4 are compared. The light grey curves are the corresponding wavelet band pass of 3,000 synthetic data obtained by random resampling 200-year blocks of CET data to extend the record before 1659 and after 2012 to as long as needed. The spread of the grey curves gives a quantification of the errors due to the fact that the time series is not infinitely long. To show that the troughs of the AMO are not systematically aligned with large volcano eruptions, years of large volcano eruptions (with Volcano Explosivity Index 5 and higher) are indicated in light grey vertical lines. Bold letters are used for the names of major volcano eruptions with VEI of 6 and 7.

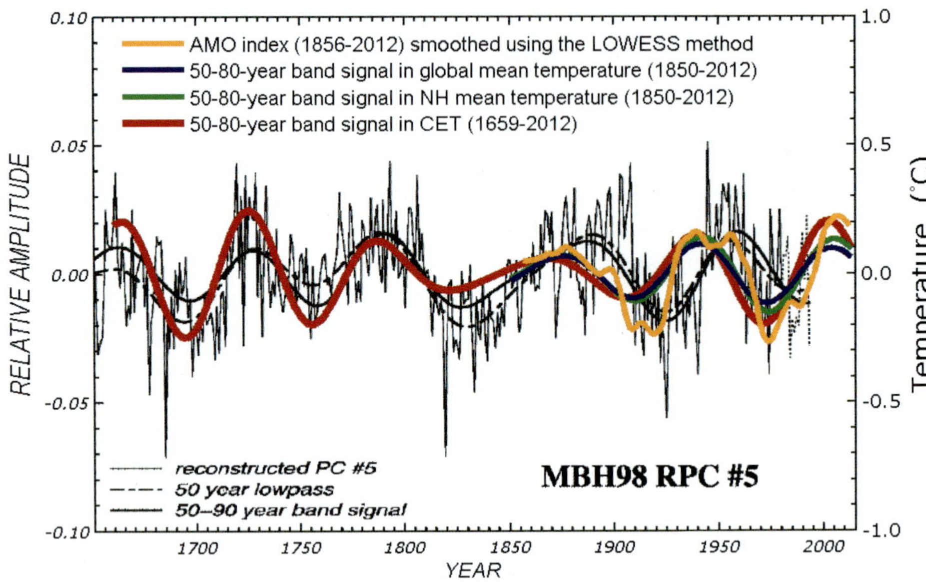

Fig. 5. Comparison of the AMO mode in CET (red) and in global mean HadCRUT4 (blue), obtained from Wavelet analysis in Fig. 4, with the multi-proxy AMO of Delworth and Mann (2000) (in thin black line). The amplitude of multi-proxy data is only relative (left axis). The orange curve is a smoothed version of the AMO index originally available in monthly form.

shows this mode with the Northern Hemisphere and global means superimposed, updated from Tung and Zhou (2013) to 2012, and a narrower bandwidth (50–80 years, vs 50–90 years) is now used — with approximately the same results. As can be seen in the lower panel of Fig. 1, the AMO is out of phase in the South Atlantic subpolar region and so the Northern Hemisphere mean should be, and is, larger than the global mean (except for the early decades in the "global" record, when there were not much Southern ocean measurements). Based on the recent cycle the global oscillation is about 0.3°C from minimum to maximum. During the 163-year overlap with the global record, this mode in the Central England Temperature agrees with the oscillation that is present in the global mean (when scaled to account for the difference in amplitude between a midlatitude measurement and the global mean), and with the AMO Index (Enfield *et al.*, 2001), which is defined as the North Atlantic Sea-Surface Temperature mean, linearly detrended. In the preindustrial era, it agrees (Fig. 5) with the multi-proxy data of Delworth and Mann (2000). The use of the multi-proxy data is intended here to only show that the AMO is "near global", with a synchronous spatial structure (Principle Component #5 in that paper) similar to the AMO.

Central England Temperature record reveals 5 cycles of the AMO with an average period of 70 years. The cycles are not aligned with the eruptions of major volcanos (see Fig. 4 background). It also exists prior to the Industrial Revolution, and so is likely natural.

3. Spatial Pattern, Mechanisms and Predictability of the AMO

The North Atlantic dominates the AMO spatial pattern. Over the Northern Hemisphere the temperature anomalies are mostly in phase (see Fig. 1). Older data, which has almost no observation south of 60S and sparse observation of the Southern Oceans, tended to show anomalies of the same sign for the AMO, hence often the description of the AMO spatial pattern as a "monopole". Our Fig. 4 shows that the Southern Hemisphere subtracts from the Northern Hemisphere to give a global mean that is somewhat smaller than the Northern Hemisphere mean, but still of the same sign.

It is likely that the AMO is a surface manifestation of the slight variability of the thermohaline circulation called the Atlantic Meridional Overturning Circulation (AMOC). Theories for its quasi-periodic variations are many but unsettled. Climatologically, AMOC itself upwells in the Antarctic Circumpolar Current region driven by wind stress, as most of the wind energy of the atmosphere is transferred to the ocean there (Wunsch, 1998). The upwelled water is transported near the surface to the Equator and then towards to the North Atlantic, warmed along the way by the absorption of solar heating.

The anomaly that is the AMO is thought to be a result of a slight speeding up or slowing down of the AMOC. According to Dima and Lohmann (2007), if for some reason the AMOC speeds up a little, caused by some stochastic forcing (*e.g.*, wind stress variation, or the North Atlantic Oscillation). After a delay of a few years, it leads to more warmed surface water advected into the North Atlantic. Ice melt is increased in response to the warmer surface temperature. The resulting less saline water reduced the deep water formation for the thermohaline circulation, which after a delay of about 15–20 years leads to a slowdown of the AMOC. A slower AMOC transports less warm water to the Arctic, more Arctic ice forms, which in turn leads to more saline deep-water formation and finally an acceleration of the AMOC. The cycle then continues. The energy input comes from the absorption of solar energy as the northward surface current transits the tropics in the Atlantic, and is not merely a conservative transport of heat from the Southern Atlantic to the Northern Atlantic. The negative feedback processes outlined above are what makes the AMO an oscillation, albeit only quasi-periodic. Although there is still much that is not understood about this phenomenon this conclusion is tentatively supported by ocean model simulations (*e.g.*, Knight *et al.*, 2005), and by observational records.

Since anthropogenically forced global warming also projects into the North Atlantic sea-surface temperature, the question arises as to how to separate that forced component in the 20th century. Since greenhouse forcing tends to be monotonic, the classical definition of the AMO Index by Enfield *et al.* (2001) subtracts the linear trend in the North Atlantic mean temperature for the purpose of removing the forced component. There were concerns raised (Trenberth and Shea, 2006; Mann and Emanuel, 2006) that some nonlinear forced trends still remain in the AMO Index. Enfield and Cid-Serrano (2010) showed that removing a nonlinear (quadratic) trend does not affect the multidecadal oscillation. The spatial pattern associated with the secular trend is similar to that of the global warming, while that of the oscillatory component is different (DelSole *et al.*, 2011).

In the Pacific Ocean, the dominant non-ENSO mode of variability is called the Pacific Decadal Oscillation (PDO) (Mantua *et al.*, 1997; Zhang *et al.*, 1997). It has frequencies in the 15–25 year range and the 50–70 year range. The presence of the higher (decadal) frequency makes the PDO Index look somewhat different than the AMO Index. However, the low-frequency portion of the PDO appears to also have approximately the same period as the AMO Index. The cool periods in the PDO in the 20th century were 1890–1924 and 1947–1976, and warm periods were 1925–1946 and 1977–2000. These "regimes shifts" (Minobe, 1997) were approximately in phase with the AMO, and the duration of each "regime" is also about 30 years. It is likely that the origin of the low-frequency part of the PDO may be from the Atlantic (Zhang and Delworth, 2007). Zhang *et al.* (2007) suggested that the atmosphere reacts to the enhanced northward oceanic heat flux in the Atlantic by reducing the poleward eddy atmospheric heat flux in winter over both the North Atlantic and the North Pacific mid-latitudes. The weakening of the winter storm track results in a poleward shift of the westerly wind, and the weakening of the Aleutian Low. The negative wind stress curl anomaly leads to a northward shift of the Kuroshio Current and warm sea-surface temperature in the central and western North Pacific. On the other hand, Xie *et al.* (2008) suggest the influence of the Atlantic on the Pacific can be through the tropics across the narrow Central America isthmus. The above results all suggest connections of AMO and the multidecadal component of the PDO at no or short lags. d'Orgeville and Peltier (2007) found very high correlation of the two at 13-year lag. This is probably due to their unusual methodology of finding the maximum *spatial* correlation of the two phenomena at various assumed lags.

Latif *et al.* (2004) and Knight *et al.* (2005) suggest that the phase shifts of the AMO can be predicted decades in advance. Because of the great inertia of the thermohaline circulation, its variation precedes the sea-surface temperature change in the Atlantic by several years. Using a coupled ocean-atmosphere model Latif *et al.* (2004) found that the strength of the thermohaline circulation at 30°N leads the North Atlantic sea-surface temperature in the extratropics by several years. The correlation is so strong that the authors suggested that the strength of the AMOC can be monitored by observation of the sea-surface temperature index. Knight *et al.* (2005) predicted a cooling phase of the AMO during the next decades due to a weakening AMOC. Keenlyside *et al.* (2008) actually predicted that "global surface temperature may not increase over the next decade, as natural climatic variations in the North Atlantic and tropical Pacific temporarily offset the projected anthropogenic warming." The decade in question is the mean of 2005–2015. The authors claimed no skill beyond 2015. Enfield and Cid-Serrano (2006) and Enfield and Cid-Serrano (2010) projected a 85% probability that a shift to the cooling phase by 2025, like the one we experienced in the middle of the 20th century, by estimating the probability density function of the interval length before phase shifts.

4. AMO as an Explanation for Acceleration of Warming and the Recent Hiatus

Accepting the AMO as part of the factors that affect climate at multi-decadal time scales, we can see that each of its cycles has climate implications, but these could have been misinterpreted in the past. The upswing in the AMO after 1976 led to the conclusion of "accelerating warming trend". The cooling in 1950s and 1960s during the negative phase of the AMO was variously attributed to enhanced aerosols (Rind and Overpeck, 1993; Wigley, 1989; Stott *et al.*, 2006), while the anomalous Early Twentieth Century Warming during the previous positive phase of the AMO was either attributed to solar forcing (Scafetta and West, 2005) or to a singular event of internal variability (Wood *et al.*, 2010), or unexplained. The community is currently facing the task of having to explain the "hiatus" of the global warming rate since 2000. Suggested explanations so far involved coal burning in the emergent economy of China (Kaufmann *et al.*, 2011), increases in stratospheric water (Solomon *et al.*, 2010), or increases in "background" stratospheric aerosol (Solomon *et al.*, 2011). The possibility that various explanations for the observed multi-decadal warming and cooling episodes in history can be consolidated into one that incorporates a recurrent oscillatory version of the AMOC is appealing, though not conclusive. If correct a coherent picture could emerge that accounts for each episode of warming and cooling in the instrumental record, without resorting to a different singular event for each episode.

5. Ocean Heat Uptake

From an energy balance standpoint, the top of the atmosphere radiative imbalance driven by the anthropogenic forcing should be accounted for mostly by the heat uptake in the oceans, as land and ice have much lower heat capacity, and accounts for no more than 10% of the heat storage. So during a hiatus period in surface warming while the anthropogenic forcing is still increasing, the energy must be going into the deeper layers in the ocean; other things being equal a hiatus in surface warming does not generally lead to enhanced outgoing radiation. Meehl *et al.* (2011) used CCSM4 model run with a future scenario (RCP4.5), which does not have oscillatory variations in the forcing or dips such as those by volcano eruptions. With smooth forcing, the 5-member ensemble mean model global mean temperature is also smoothly increasing. However the individual members show variability about the mean of around 0.5 K from peak to trough. They found that during the hiatus period, the composite mean shows that the upper ocean takes significantly less heat whereas the ocean below 300 m takes up significantly more, as compared with the non-hiatus period. In a second paper Meehl *et al.* (2013) compares hiatus periods with accelerated warming periods, and finds the opposite behavior: the upper ocean takes up more heat and the deeper ocean much less.

The question, which has not yet been asked, is, what causes some periods to have more heat going into the deep ocean while some other periods the heat staying more in the upper ocean? We suggest that this is caused by the internal variability. We adopt the view that in models the ensemble mean is mostly the forced solution and the deviation from the ensemble mean by each member is internal variability. There are no hiatus periods in the forced solution under the CCSM4 RCP4.5 experimental setup. In fact this solution is smoothly increasing in its global mean temperature. It is the internal variability, possibly associated with variations in the deep overturning circulations in the oceans, that determines when a portion of the heat should go to the deep ocean and shows up as surfacewarming hiatus, and when it should stay near the surface, and show up as accelerated warming. ENSO, which we generally view as internal variability, also has this kind of vertical distribution variability in the oceans: In La Niña, the upper ocean is cool while the heat goes to the deeper ocean, and opposite behavior in El Niño. The CCSM4 is known to have internal variability that is faster than in the observation. Instead of having hiatus periods of about 2–3 decades as in the observation its cooling periods last 10–15 years. The Interdecadal Pacific Oscillation (IPO) that was discussed by Meehl *et al.* (2013) is the low frequency portion of the Pacific Decadal Oscillation (PDO). In the observation the IPO cooling periods coincides with the cooling periods of the AMO, which led us to suspect that the IPO is just the Pacific manifestation of the AMO, which is ultimately caused by the variations of the Atlantic Meridional Overturning Circulation.

Katsman and van Oldenborg (2011) analyzed 17 simulations for the period 1950–2100 using the ECHAM5-MPI/OM1 coupled climate model, and found that 8-year periods without upper ocean warming are not exceptional despite the increasing greenhouse forcing. Of these hiatus periods, 35% can be explained by an increase in ocean heat content below 700 m, partly due to a decrease in the strength of AMOC, and 45% by an increase in the 8-year mean net outgoing top-of-atmosphere radiation due possibly to previous episode's El Niño upper ocean heat loss. These mechanisms are again those of internal variability involving the Atlantic and the Pacific.

6. The Underlying Anthropogenic Trend

With this new perspective in mind, let us revisit some recent results on the determination of the anthropogenic warming trend from data (see Zhou and Tung, 2013). It is important to point out that the Central England Temperature is not used for this quantitative analysis. Only the latest version of global dataset is used. Foster and Rahmstorf (2011) considered the period 1979–2010 and obtained an estimate of anthropogenic warming after removing El Niño-Southern Oscillation (ENSO), individual volcano eruptions and the solar forcing through multiple regression analysis. The multiple regression analysis needs a predictor for anthropogenic response, which is, however, unknown. They used a linear trend and then added the linear response back to the residual to form the so-called adjusted data. Their adjusted data showed no change in the linear warming rate over the 32-year period. However, the oscillatory component of the AMO, if it existed, was treated as a positive trend in their data since 1979. This is demonstrated in Fig. 6a, where Foster and Rahmstorf's analysis is repeated, but for the longer record 1850–2012, and using HadCRUT4 (Morce *et al.*, 2012). This is an update of our previous result with the recent availability of the 2012 data. The residual plus the regressed linear trend is shown. Approximately the same "anthropogenic trend" since 1979, the start of the satellite era, of 0.16°C/decade is obtained. However, one can see clearly that a 70-year oscillation is still present in the residual (see the orange running mean). DelSole *et al.* (2011) showed that that oscillation is internal variability. It is likely related to the variation in the AMOC, which is mostly natural. So next we would like to redo the multiple linear regression by adding an AMOC as a regressor, with a possible lead. Unfortunately, the available data on AMOC observation is too short for our purpose. The RAPID/MOCHA array at 26.5°N (Rayner *et al.*, 2011) provides monitoring of the AMOC only since 2004. Recently Zhang *et al.* (2011) found that the North Brazil Current (NBC) strength, measured off the coast of Brazil, could be a proxy for the AMOC,

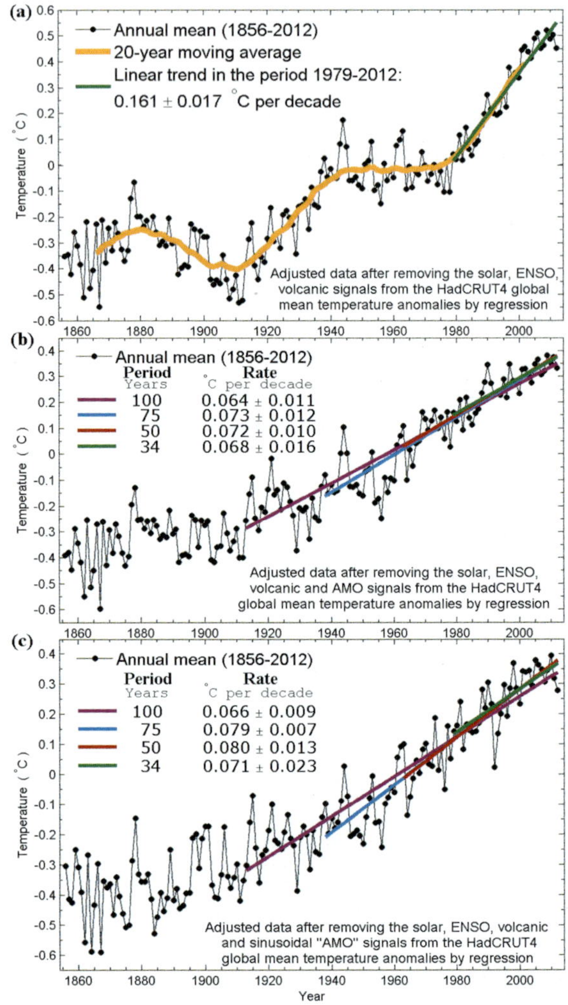

Fig. 6. Multiple-regression analysis of global-mean temperature: (a) adjusted data after the removal of ENSO, volcano and solar influence. A running-time mean (in orange) shows a long-period oscillation still present in the residual. (b) adjusted data after the removal of ENSO, volcano and solar influence, plus the AMO. The AMO index (Enfield *et al.* (2001) is used as a predictor). (c) is the same as (b) except that a sine is fitted to the AMO index and used as the AMO regressor, with not much difference in the results.

and they verified it with a 700-year model run. One could have used NBC as the predictor for the AMO, but that time series is available only for the past 50 years, not long enough for our purpose. They however also found that the NBC variation is coherent with the AMO index. So for our analysis for the past 160 years, the AMO index is used as a compromise. It could be criticized as using part of the response as a predictor, but at least the global mean temperature has more degrees of freedom for its variation than the North Atlantic mean SST.

In Fig. 6b, we repeated the multiple regression analysis by adding the AMO Index (Enfield *et al.*, 2001) as a predictor. The resulting 34-year (since 1979) anthropogenic warming rate, at 0.07°C/decade, is less than half of that found by Foster and Rahmstorf. It is almost a factor of three smaller than that found by Lean and Rind (2008). In fact, the underlying rate is very steady for the past 100 years, at 0.07–0.08°C/decade, through the ups and downs of multi-decadal temperature changes. The results are nearly the same whether the end year is 2010, 2011 or 2012. The same result was obtained using wavelet bandpass in Tung and Zhou (2013). Wu *et al.* (2011) found 0.08°C/decade using Empirical Mode Composition. DelSole *et al.* (2011) found that the multidecadal variability accounts for 0.08°C/decade of the trend in the last 30 years. Subtracting that from the combined trend of 0.16°C/decade in Fig. 6a then also yields the same 0.08°C/decade as the anthropogenic trend. Therefore very different methods and approaches yield approximately the same result for the underlying anthropogenic trend. It appears that the currently accepted picture of anthropogenic warming, with the warming starting in earnest only since the mid-20th century and then accelerated, needs to be revised.

A few clarifications are in order on this somewhat surprising result. One likely question is: How can the anthropogenic warming be linear in time when we know that atmospheric CO_2 has been measured to increase almost exponentially? While the common expectation that the temperature response should follow the greenhouse concentration in the atmosphere is incorrect, because the radiative heating should be proportional to the logarithm of the CO_2, the main reason is that the observed forcing is a net between all greenhouse gases (including methane) and tropospheric sulfate aerosols (anthropogenic pollution), the latter subtracts from the heating, but is highly uncertain. We actually do not know what the net anthropogenic forcing is. This was the reason why we did not simply adopt an anthropogenic *predictor*. Lean and Rind (2008) did use a predictor based on an inventory of individual anthropogenic forcing factors, the same factors as used in the GISS GCM for the simulation of the historical warming. The response shows an accelerated warming after 1978. Zhou and Tung (2013) reexamined that multiple regression model, and found, after taking the AMO into account, that the residual, which should only contain climate noise, still has a trend that is negative after 1978. This is an indication that the warming after 1978 was overestimated by their adopted predictor.

The second question is: Does the small anthropogenic warming that we have found prove that the climate sensitivity is low? The answer is "not necessarily". The result here is a response to the net anthropogenic forcing and does not constrain climate sensitivity, which is a response to CO_2 forcing. To constrain the climate sensitivity we either need to know with more certainty what the aerosol cooling was, or we use other independent data, such as in Tung *et al.* (2008), who used the observed response to the 11-year solar cycle to constrain the Transient Climate Response.

A sensitivity study of the Multiple Linear Regression method to the choice of regressors is discussed in the Appendix.

7. Projection of Global Warming in the 21st Century

Climate projection should not be taken as predictions *per se*, but rather as an exercise to reveal the consequences of the assumptions, which are called scenarios. IPCC AR4 and 5 climate projections are based on scenarios, all of them forced. There was no scenario that incorporates a multidecadal variability. We now introduce one such scenario to complement the set of AR4 and 5 forced scenarios. Our scenario is not any more or less uncertain than the IPCC scenarios.

As we have found in the previous section, the two important predictors, as far as multidecadal episodes of warming are concerned, are the net anthropogenic warming trend, and the AMO index. Volcano aerosols and ENSO indices are not predictable going into the future, but it was shown in Tung and Zhou (2013) that these and the solar forcing have very little secular trend in the 20th century and therefore do not affect the observed global mean temperature when the time series is smoothed. We will assume that they continue not to have a secular trend and include them only in the variance of the predicted warming. We also do not know the AMO index going into the future, because its amplitude and period vary somewhat from cycle to cycle. We instead construct a periodic AMO index by fitting a sinusoidal function to the historical AMO data. We test this index by using it on the historical data in Fig. 6c, with almost the same results as Fig. 6b.

Because the derivative near the top of the AMO cycle is small, it is difficult to pinpoint exactly if 2005 or 2010 was the peak. The prediction of the next down half cycle is probably reasonable, with an error of 5 years, but the timing of the future rising half cycle is probably uncertain by 10 years. That is, we do not have much confidence in the projection past 2040. The small anthropogenic warming that we have deduced in the previous section is probably a result of a cancellation between greenhouse gas warming and tropospheric aerosol cooling. The latter is anthropogenic pollution and it appears to be increasing along with industrial development. There is a limit to how much pollution that citizens in developing countries such as China could tolerate before calls for massive pollution control are heard. When that happens, the net warming will accelerate for a couple decades. In addition, the rate of greenhouse gas emissions may change depending on the degree of implementation of the carbon emission curbs. These social and political effects are unpredictable now. We will simply use the current rates for projection into the future and discuss the sensitivity to changes to that rate.

Figure 7 shows our 40-year projection into the future using a regression model based on the observed global mean temperature from 1930–2012 (the "training" period). In Fig. 7a, the ENSO, solar and volcanic aerosols are treated as an AR(1) noise for the future years and three realizations of the noise model are shown as examples. The light blue shading constitutes the 95% confidence-level error bar. This error bar is smaller in Fig. 7b, where the ENSO, solar and volcanic aerosols are ignored and the variance determined from the historical data where their influences have been removed through multiple linear regression.

In this scenario the current plateauing of warming will likely continue for a few more years. By 2020, the rate of the warming may turn slightly negative if the anthropogenic forcing does not change significantly. The cooling will end sometime in 2040. Because polar ice melt is part of the AMO effects, during the cooling phase there will be some increase of polar ice. Because of the uncertainties listed in the previous paragraph, it is possible that the global warming will simply level off for the next three decades without turning into cooling. After 2040, accelerated warming will ensue with warming rates similar to the episode we have experienced during 1968–2000, but the actual temperature will be starting from a higher level than in 1976. The negative half cycle of the AMO that just started simply delays this accelerated warming by three decades. Hopefully

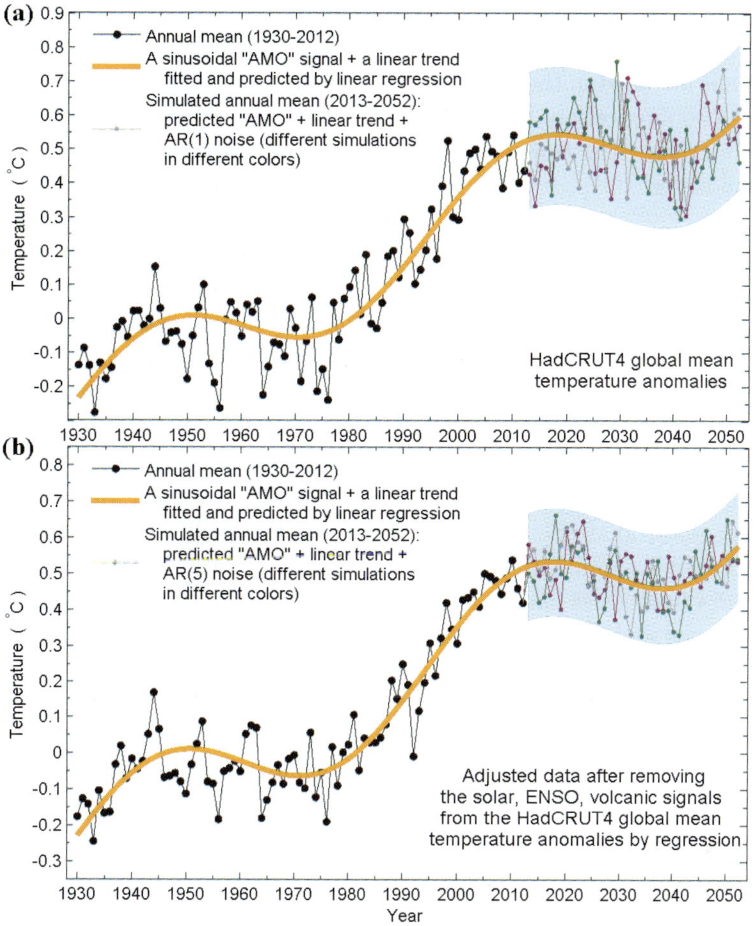

Fig. 7. Multiple linear regression analysis and prediction. The black dots are the observed annual mean global mean surface temperature from HadCRUT4. The orange curve is a fit to the observed data for the period 1930–2012 using two predictors, a linear trend and a sinusoidal "AMO" fit to the observed AMO Index. This statistical model is then extended to future years (2013–2052), using random noise to represent the variances. Three simulations/realizations of the noise model is shown. The light blue band represents two standard deviations of the noise. (a): the noise model for future years is AR(1) and the variance includes ENSO, volcano and solar influences. (b) the noise model is AR(5), and the variance includes only the residual climate noise with ENSO, volcano and solar influences regressed out.

the two to three decades of respite will offer the world time to work out technical and political solutions to the greenhouse warming problem.

Figure 2 earlier shows that the all-model ensemble mean of the GCMs participating in the IPCC AR4 simulated the observed late 20th century warming, meaning that the simulated warming in these models was by forced-anthropogenic response only (assuming that all model ensemble averages out systematic errors in different models). We have argued that about half of the observed warming was caused by the internal multi-decadal variability. If this conclusion is correct then these models probably have an anthropogenic warming rate that is twice as high. When these same models were then used to project into the future, the warming rates would be overestimated. On top of this, the AMO now starts to turn negative, and this effect was not included in the IPCC scenarios. As the AMO cooling offsets the anthropogenic warming it creates the appearance of the "stalling" of the latter.

8. Conclusion

There has been a warming of 0.7–0.8°C for the past 100 years. This fact is not in dispute. If one draws a straight line through the 100 years, the rate of warming is 0.07–0.08°C/decade. That happens to be close to the net rate of anthropogenic warming that we have deduced through multiple linear regression analysis. On top of this anthropogenic trend is a multi-decadal variability. The two and a half cycles can be clearly seen in the observed raw global mean data (see Fig. 1), and this fact is again not in dispute. What is controversial is the interpretation of the cause of the observed multidecadal variability. Here we refer to it as the AMO and have argued that its origin is in the short-term variability of the thermohaline circulation in the Atlantic. A different interpretation is that it is forced in the 20th century by the time-varying anthropogenic aerosols. There are uncertainties unique to each interpretation, and these should be discussed and investigated because the consequence of adopting one or the other interpretation may lead to very different projections of future global warming. Models participating in IPCC AR4 simulated in ensemble mean the observed global warming in the 20th century as almost entirely forced by anthropogenic forcing. Without anthropogenic forcing the simulation would have yielded no warming (see Fig. 2) in the second half of the 20th century. For the past 34 years since the satellite era, the observed warming rate is 0.17°C/decade. If the models continue to attribute this observed warming as almost entirely anthropogenic, and succeed in simulating them as such, then the same models would project in 20 years a warming of 0.34°C, assuming the rate of increase of anthropogenic forcing is unchanged. The uncertainties involved in this approach in projecting future warming lie mainly in the climate sensitivity of the models and the troposphere sulfate aerosols used in the anthropogenic forcing. There is also the uncertain assumption that internal variability played a minor role in the observed global temperature variation in the 20th century. In our data-based approach, we do not need to know the climate sensitivity of the climate system and, since we deduce only the net anthropogenic response, we do not need to know the partition between greenhouse gases that warm and the aerosols that cool, the latter is highly uncertain. We deduced that the net anthropogenic warming rate for the last 34 years, in fact for the last 100 years, is 0.07–0.08°C/decade, about a factor of two smaller than the observed warming rate during the last 34 years. The difference is due to the presence of an internal variability that has been in a warming phase during the past three decades. When projected forward using the same numbers, slight cooling instead of accelerated warming is predicted in 20 years. By 2050, the total warming relative to the pre-industrial era is about 0.8–0.9°C, only about 0.1°C higher than what has already been realized by 2012. The uncertainties with this approach consist of the possibility that the AMO may stop being recurrent, despite its past behavior. If the AMO in the 21st century for some reason fails — does not turn negative in the coming three decades — the predicted warming would then be 0.14°C in 20 years, if the rate of anthropogenic forcing does not change. However, there have been many indications that the Atlantic Meridional Overturning Circulation has already slowed since 2004, with the cooling phase of the AMO following it.

Although the year 2012 was the warmest on record for the contiguous US, according to NOAA's National Climate Data Center, it was only the 9th or 10th warmest globally, depending on which dataset is used. The warmest global-mean year was either 1998 (according to HadCRU) or 2005 (according to GISS). The year 2010 effectively tied with 2005. The warming in 1998 was rather spikey, and could be attributed to a "super" El Niño in 1997/1998, while the peak near 2005 was broader, indicating a top in the multidecadal variability sometime between 2005 and 2010. Proposed explanations for the "stalling of global warming" have been many, but the simplest one appears to be the fact that the AMO has begun its cooling phase. There have been many (at least five in the instrumental record) such cooling periods which have occurred at regular intervals of approximately 70 years apart, and a recurrent AMO may be suggested to account for all of them. There was a previous cycle in the 20th century record (see Fig. 1) when an accelerated warming (during 1910–1940) peaked near 1945 and then was followed by a period of negative rates of change. Seventy years later, we are probably near a similar peak.

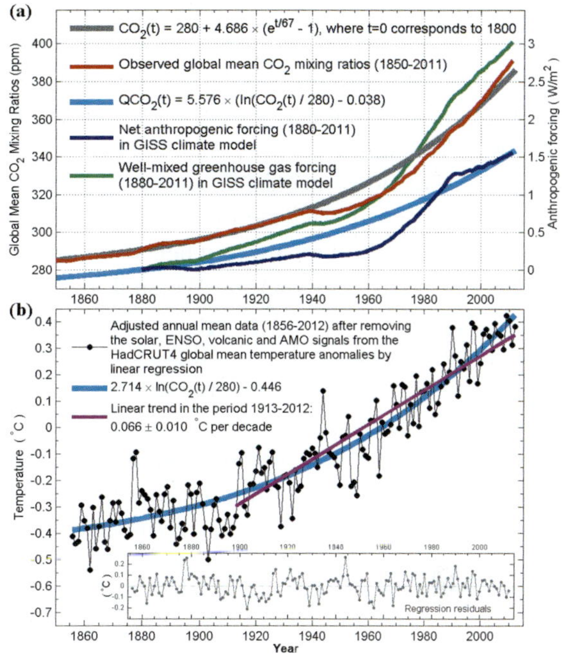

Fig. 8. (a) Various possible greenhouse gas forcing predictors. (b) Adjusted temperature in a regression analysis using QCO_2, solar, ENSO, vocano and AMO indices as regressors. The inset shows that there is still a downward trend in the residual in the recent decade.

Acknowledgments

The research is supported by NASA under NNX11AC75G, and by NSF under grants DMS 0940342 and AGS-1262231.

Appendix: More Discussions on the Multiple Linear Regression

Some may question the result of Zhou and Tung (2013) as being "circular": the result that the net anthropogenic warming rate is steady for the past 100 years might be said to be a direct consequence of the linear form for the net anthropogenic predictor adopted. This criticism is not justified because the result was not based on the part of the response proportional to that regressor. It was based on fitting a linear trend on everything that remains in the adjusted data. The adjusted data is defined as the original data after the removal of ENSO, volcanic aerosols, solar influence and the AMO. The result of the multiple regression analysis actually does not depend much on the exact form of the anthropogenic predictor adopted, as long as it has a secular tend; so we can try other forms of predictors for the anthropogenic forcing.

Figure 8a shows several possible anthropogenic predictors. Since CO_2 is well measured, we will try that first. Using the logarithm of a fitted smooth function of the CO_2 concentration (denoted by QCO_2, which can be interpreted as the index for radiative heating due to CO_2 alone) as the anthropogenic predictor yields a good fit to the adjusted data over the full period of 1850–2012, with the last 100 years still rather linear. Note that the residual in the most recent decade still contains a small negative trend, and when that is added back to the regressed anthropogenic response to yield the adjusted data, the adjusted data becomes more linear. One can use the logarithm of all the well-mixed greenhouse gases as the forcing/predictor. Since the latter in recent decades increases faster than the QCO_2 (Fig. 8a), the negative trend in the residual becomes larger.

Note that no tropospheric aerosol predictor is used in the above examples. The problem of colinearity becomes severe when two secular trends are used as predictors. We shall demonstrate that next by adopting the commonly used anthropogenic forcing indices: one is the radiative forcing from all greenhouse gases, and the other is the aerosols cooling of the reflective tropospheric and aerosol indirect effects. These radiative heating rates are available from GISS at http://data.giss.nasa.gov/modelforce/, and shown in Fig. 9a. After the multiple regression analysis is done, the adjusted data is reconstructed by adding the residual to the sum of the two regressed anthropogenic responses. Due to the colinearity, individual regression coefficients have large error bars. However, when combined to yield the adjusted data, the 100-year, 50-year and 34-year linear trends are still rather steady at 0.07–0.08°C/decade, very close to that found in Fig. 6 using a linear predictor. The net of the two anthropogenic regressors is only slightly collinear with the other (such as the solar) regressors. These examples should be viewed as a sensitivity study to show that our previous result on the anthropogenic response in the past 100 years is not too sensitive to the assumed form of the secular anthropogenic predictor used.

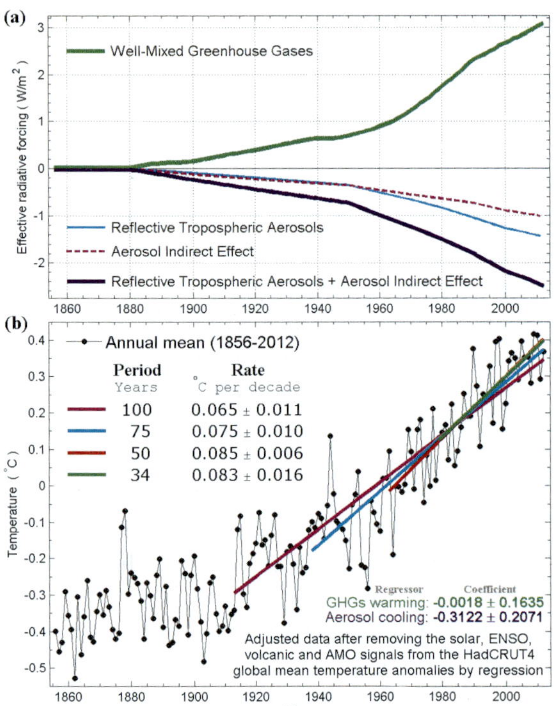

Fig. 9. (a) Radiative forcing of various anthropogenic components. (b) adjusted temperature, which is the same as the combined response to the anthropogenic forcing and the residual.

The question of whether one should use the AMO Index as defined by Enfield *et al.* (2001) or by Trenberth and Shea (2006) was discussed in detail in Enfield and Cid-Serrano (2010), who argued against the latter index as "throwing the baby out with the bath water". To be fair, there is no a priori reason why one index is better than the other. The purpose of the paper by Trenberth and Shea was to argue that if the record active hurricane season in the 2005 in the North Atlantic is attributed to the high SST there, most of the warm SST anomaly relative to 1901–70 is due to the global SST warming, and when the latter is subtracted from the North Atlantic mean SST, only a small contribution remains. Nevertheless, using the Trenberth and Shea index as a regressor changes only slightly the linear trend (slightly larger) in the recent decades in the adjusted temperature response, although we do not recommend that as the AMO index for our purpose.

References

Booth, B. B. B., N. J. Dunstone, P. R. Halloran, T. Andrews, and N. Bellouin, 2012: Aerosols implicated as a prime dirver of twentieth-century North Atlantic climate variability. *Nature*, **484**, 228–232.

Briffa, K. R., F. H. Schweingruber, J. D. Jones, T. J. Osborn, S. G. Shiyatov, and E. A. Vaganov,

1998: Reduced sensitivity of recent tree-growth to temperature at high northern latitudes. *Nature*, **391**, 678–682.

Broecker, W. S., 1975: Climate change: Are we on the brink of a pronounced global warming? *Science*, **189**, 460–463.

Cheng, W., J. C. H. Chiang, and D. Zhang, 2013: Atlantic Meridional Overturning Circulation (AMOC) in CMIP5 models: RCP and historical simulations. *J. Climate*, in press, doi: 10.1175/JCLI-D-12-00496.1.

Chylek, P., C. K. Folland, H. A. Dijkstra, G. Lesins, and M. K. Dubey, 2011: Ice-core data evidence for a prominent near 20 year time-scale of the Atlantic Multidecadal Oscillation. *Geophys. Res. Lett.*, **38**, doi: 10.1029/2011GL047501.

DelSole, T., M. K. Tippett, and J. Shukla, 2011: A significant component of unforced multidecadal variability in the recent acceleration of global warming. *J. Climate*, **24**, 909–026.

Delworth, T. L., and M. E. Mann, 2000: Observed and simulated multidecadal variability in the Northern Hemisphere. *Clim. Dyn.*, **16**, 661–676.

Dima, M., and G. Lohmann, 2007: A hemispheric mechanism for the Atlantic Multidecadal Oscillation. *J. Climate*, **20**, 2706–2719.

d'Orgeville, M., and W. R. Peltier, 2007: On the Pacific Decadal Oscillation and the Atlantic Multidecadal Oscillation: Might they be related? *Geophys. Res. Lett.*, **34**, L23705.

Enfield, D. B., and L. Cid-Serrano, 2006: Projecting the risk of future climate shifts. *Int. J. Climatol.*, **26**, 885–895.

———, and ———, 2010: Secular and multidecadal warmings in the North Atlantic and their relationships with major hurricane activity. *Int. J. Climatol.*, **30**, 174–184.

———, A. M. Mestas-Nunez, and P. J. Trimble, 2001: The Atlantic multidecadal oscillation and its relation to rainfall and river flows in the continental U. S. *Geophys. Res. Lett.*, **28**, 2077–2080.

Folland, C. K., D. E. Parker, and F. E. Kates, 1984: Worldwide marine temperature fluctuations 1856–1981. *Nature*, **310**, 670–673.

Foster, G., and S. Rahmstorf, 2011: Global temperature evolution 1979–2010. *Environ. Res. Lett.*, **6**, 1–8.

Gray, S. T., L. J. Graumlich, J. L. Betancourt, and G. Pederson, 2004: A tree-ring based reconstruction of the Atlantic Multidecadal Oscillation since 1567 A.D.*Geophys. Res. Lett.*, **31**, L12205, doi: 10.1029/2004GL019932.

Huang, N. E., Z. Shen, S. R. Long, M. L. C. Wu, H. H. Shih, Q. N. Zheng, N. C. Yen, C. C. Tung, and H. H. Liu, 1998: The empirical mode decomposition and the Hilbert spectrum for nonlinear and non-stationary time series analysis. *Proc. R. Soc. London Ser. A-Math. Phys. Eng. Sci.*, **454**, 903–995.

Katsman, C. A., and G. J. van Oldenborg, 2011: Tracing the upper ocean's "missing heat". *Geophys. Res. Lett.*, **38**, L14610, doi: 10.1029/2011GL048417.

Kaufmann, R. K., H. Kauppi, M. L. Mann, and J. H. Stock, 2011: Reconciling anthropogenic climate change with observed temperature 1998–2008. *Proc. Natl. Acad. Sci., USA*, **108**, 11790–11793.

Keenlyside, N. S., M. Latif, J. Jungclaus, L. Kornblueh, and E. Roeckner, 2008: Advancing decadal-scale climate prediction in the North Atlantic sector. *Nature*, **453**, 84–88.

Kerr, R. A., 2000: A North Atlantic climate pacemaker for the centuries. *Science*, **288**, 1984–1986.

Knight, J. R., R. J. Allan, C. K. Folland, and M. Vellinga, 2005: A signature of persistent natural thermohaline circulation cycles in observed climate. *Geophys. Res. Lett.*, **32**, doi: 10.1029/2005GL024233.

Latif, M., E. Roeckner, M. Botzet, M. Esch, H. Haak, S. Hagemann, J. Jungclaus, S. Legutke, S. Marsland, and U. Mikolajewicz, 2004: Reconstructing, monitoring and predicting multidecadal-scale changes in the North Atlantic thermohaline circulation with sea surface temperature. *J. Climate*, **17**, 1605–1614.

Lean, J., and D. Rind, 2008: How natural and anthropogenic influences alter global and regional surface temperatures: 1889–2006. *Geophys. Res. Lett.*, **35**, L18701.

Manley, G., 1974: Central England Temperatures: Monthly means 1659 to 1973. *Quart. J. Roy. Meteor. Soc.*, **100**, 389–405.

Mann, M. E., and K. Emanuel, 2006: Atlantic hurricane trends linked to climate change. *EOS*, **87**, 233–244.

Mantua, N. J., S. R. Hare, Y. Zhang, J. M. Wallace, and R. C. Francis, 1997: A Pacific interdecadal climate oscillation with impacts on salmon production. *Bull. Amer. Meteor. Soc.*, **78,** 1069–1079.

Meehl, G. A., J. M. Arblaster, J. T. Fasullo, A. Hu, and K. E. Trenberth, 2011: Model-based evidence of deep-ocean heat uptake during surface-temperature hiatus periods. *Nature Clim. Change*, **1**, 360–364.

———, C. Covey, T. Delworth, M. Latif, B. McAvaney, J. F. B. Mitchell, R. J. Stouffer, and K. E. Taylor, 2007: The WCRP CMIP3 multi-model dataset: A new era in climate change research. *Bull. Amer. Meteor. Soc.*, **88**, 1383–1394.

———, A. Hu, J. M. Arblaster, J. T. Fasullo, and K. E. Trenberth, 2013: Externally forced and internally generated decadal climate variability associated with the Interdecadal Pacific Oscillation. *J. Climate*, in press, doi: 10.1175/JCLI-D-12-00548.1.

Minobe, S., 1997: A 50–70 year climatic oscillation over the North Pacific and North America. *Geophys. Res. Lett.*, **24**, 683–686.

Morce, C. P., J. J. Kennedy, N. A. Rayner, and P. D. Jones, 2012: Quantifying uncertainties in global and regional temperature change using an ensemble of observational estimates: The HadCRUT4 data set. *J. Geophys. Res. Atmos.*, **117**, D08101, doi: 10.1029/2011JD017187.

Parker, D. E., T. P. Legg, and C. K. Folland, 1992: A new daily Central England Temperature series. *Int. J. Climatol.*, **12**, 317–342.

Rayner, D., *et al.*, 2011: Monitoring the Atlantic meridional overturning circulation. *Deep-Sea Res. II*, **58**, 1744–1753.

Rind, D., and J. Overpeck, 1993: Hypothesized causes of decade-to-century-scale climate variability: Climate model results. *Quat. Sci. Rev.*, **12**, 357–374.

Scafetta, N., and B. J. West, 2005: Estimated solar contribution in the global mean surface warming using ACRIM TSI satellite composite. *Geophys. Res. Lett.*, **32**, L18713, doi: 10.1029/2005GL023849.

Schlesinger, M. E., and N. Ramankutty, 1994: An oscillation in the global climate system of period 65–70 years. *Nature*, **367**, 723–726.

Semenov, V. A., M. Latif, D. Dommenget, N. Keenlyside, A. Strehz, T. Martin, and W. Park, 2010: The impact of North Atlantic-Arctic multidecadal variability on Northern Hemisphere surface air temperature. *J. Climate*, **23**, 5668–5677.

Solomon, S., J. S. Daniel, R. R. Neely, J. P. Vernier, E. G. Dutton, and L. W. Thomason, 2011: The persistently variable "background" stratospheric aerosol layer and global climate change. *Science*, **333**, 866–870.

———, D. Qin, M. Manning, Z. Chen, M. Marquis, K. B. Avery, M. Tignor, and H. L. Miller, Eds., 2007: *Climate Change 2007: The Physical Science Basis. Contribution of Working Group I to the Fourth Assessment Report of the IPCC*, Cambridge University Press, Cambridge, 996 pp.

———, K. Rosenlof, R. W. Portmann, J. S. Daniel, S. M. Davis, T. J. Sanflord, and G.-K. Plattner, 2010: Contributions of stratospheric water vapor to decadal changes in the rate of global warming. *Science*, **327**, 1219–1223.

Stott, P. A., *et al.*, 2006: Observational constraints on past attributable warming and predictions of future global warming. *J. Climate*, **19**, 3055–3069.

Taylor, K. E., R. J. Stouffer, and G. A. Meehl, 2012: An overview of CMIP5 and experimental design. *Bull. Amer. Meteor. Soc.*, **93**, 485–498.

Trenberth, K. E., and D. J. Shea, 2006: Atlantic hurricanes and natural variability in 2005. *Geophys. Res. Lett.*, **33**, L12704, doi: 10.1029/2006GL026894.

Tung, K. K., and J. Zhou, 2013: Using data to attribute episodes of warming and cooling in instrumental record. *Proc. Natl. Acad. Sci., USA*, **110**, 2058–2063.

———, ———, and C. D. Camp, 2008: Constraining model transient climate response using independent observations. *Geophys. Res. Lett.*, **35**, L17707.

Wang, C., and L. Zhang, 2013: Multidecadal ocean temperature and salinity variability in the tropical North Atlantic: Linking with the AMO, AMOC, and subtropical cell. *J. Climate*, **26**, 6137–6162.

Wigley, T. M. L., 1989: Possible climate change due to SO2-derived cloud condensation nuclei. *Nature*, **339**, 365–367.

———, and S. C. B. Raper, 1987: Thermal expansion of sea water associated with global warming. *Nature*, **330**, 127–131.

———, and B. Santer, 2012: A probabilistic quantification of the anthropogenic component of twentieth century global warming. *Clim. Dyn.*, **40**, 1087–1102.

Wood, K. R., J. E. Overland, T. Jonsson, and B. V. Smoliak, 2010: Air temperature variations on the Atlantic-Artic boundary since 1802. *Geophys. Res. Lett.*, **37**, L17708, doi: 10.1029/2010GL044176.

Wu, Z., and N. E. Huang, 2009: Ensemble empirical mode decomposition: A noise-assisted data analysis method. *Adv. Adapt. Data Anal.*, **1**, 1–14.

———, ———, J. M. Wallace, B. Smoliak, and X. Chen, 2011: On the time-varying trend in global-mean surface temperature. *Clim. Dyn.*, **37**, 759–773.

Wunsch, C., 1998: The work done by the wind on the oceanic general circulation. *J. Phys. Oceanogr.*, **28**, 2332–2340.

Xie, S.-P., Y. Okumura, T. Miyama, and A. Timmermann, 2008: Influences of Atlantic climate change on the tropical Pacific via the Central America isthmus. *J. Climate*, **21**, 3914–3928.

Zhang, D., R. Msadeck, M. J. McPhaden, and T. Delworth, 2011: Multidecadal variability of the North Brazil Current and its connection to the Atlantic meridional overtuning circulation. *J. Geophys. Res.*, **116**, doi: 10.1029/2010JC006812.

Zhang, R., and T. L. Delworth, 2007: Impact of the Atlantic Multidecadal Oscillation on North Pacific climate variability. *Geophys. Res. Lett.*, **34**, doi: 10.1029/2007GL031601.

———, ———, and I. M. Held, 2007: Can the Atlantic Ocean drive the observed multidecadal variability in Northern Hemisphere mean temperature? *Geophys. Res. Lett.*, **34**, L02709.

———, ———, R. Sutton, D. L. R. Hodson, K. W. Dixon, I. M. H. Held, Y., J. Marshall, Y. Ming, R. Msadeck, J. Robson, A. J. Rosati, M. Ting, and G. A. Vecchi, 2013: Have aerosols caused the observed Atlantic multidecadal variability? *J. Atmos. Sci.*, **70**, 1135–1144, doi: 10.1175/JAS-D-12-0331.1.

Zhang, Y., J. M. Wallace, and D. S. Battisti, 1997: ENSO-like interdecadal variability: 1900–93. *J. Climate*, **10**, 1004–1020.

Zhou, J., and K. K. Tung, 2013: Deducing the multidecadal anthropogenic global warming trend using multiple regression analysis. *J. Atmos Sci*, **70**, 3–8.

Chapter 13

Variability of Sea Ice Extent Over Decadal and Longer Timescales

John E. Walsh[1] and William L. Chapman[2]
[1]*International Arctic Research Center, University of Alaska Fairbanks, Fairbanks, AK 99775, USA*
jwalsh@iarc.uaf.edu
[2]*Department of Atmospheric Sciences, University of Illinois, Urbana, IL 61801, USA*
chapman@atmos.uiuc.edu

Recent syntheses of sea ice and related proxy information have provided an improved picture of Arctic sea ice variability over decadal to century timescales. A spectrum of variability is superimposed on a recent decrease of Arctic sea ice. An outstanding feature is the correspondence with the Atlantic Multidecadal Oscillation, which has timescales of 50–120 years. The linkage appears to arise through the inflow of Atlantic Water to the Arctic Ocean. Less robust, and by all indications non-stationary, associations with atmospheric modes such as the North Atlantic Oscillation have also been documented, primarily in recent decades. One possible reason for the nonstationarity of such associations is that the centers of action of major atmospheric modes may change over the timescale of centuries or even less. While the recent decrease of summer ice in the Arctic appears to be unique in the past 1,450 years, paleo reconstructions also suggest a minimum in Arctic ice coverage during the early Holocene. Unlike the Arctic, Antarctic sea ice shows essentially no trend over the past 30 years. The absence of a trend has been attributed to wind forcing and possibly ocean interactions. Observational information on Antarctic sea ice variability is virtually nonexistent beyond the past 100–150 years, so proxy information provides the only clues to longer-term Antarctic sea ice variability. Such information obtained from ice cores suggests that wintertime ice extent in the East Antarctic sector has decreased by about 20% since 1950, and that multicentury variations also characterize Antarctic ice extent.

1. Introduction

Sea ice in the Arctic has undergone a widely publicized decline in extent in recent decades. The decline of ice extent has been greatest during summer and autumn, during which the rates of decrease of ice extent since 1979 (the advent of routine satellite passive microwave coverage) have been 6.1% and 7.0% per decade respectively (IPCC 2013). The rate of decrease is even greater for September. The corresponding declines in winter and spring have been 2.3% and 1.8% per decade (IPCC 2013). The loss of summer ice has been particularly noteworthy in the past decade, and the summer minima of 2007–2013 have been the lowest of the post-1979 satellite period. In particular, the minimum extent of approximately 3.6 million km^2 reached in September 2012 was more than a half million km^2 below the previous record minimum (reached in September 2007) and more than 50% below the average summer minima of the 1980s and 1990s as shown in Fig. 1. Figure 1 also shows that there was some recovery in the summer of

Climate Change: Multidecadal and Beyond
edited by Chih-Pei Chang et al.

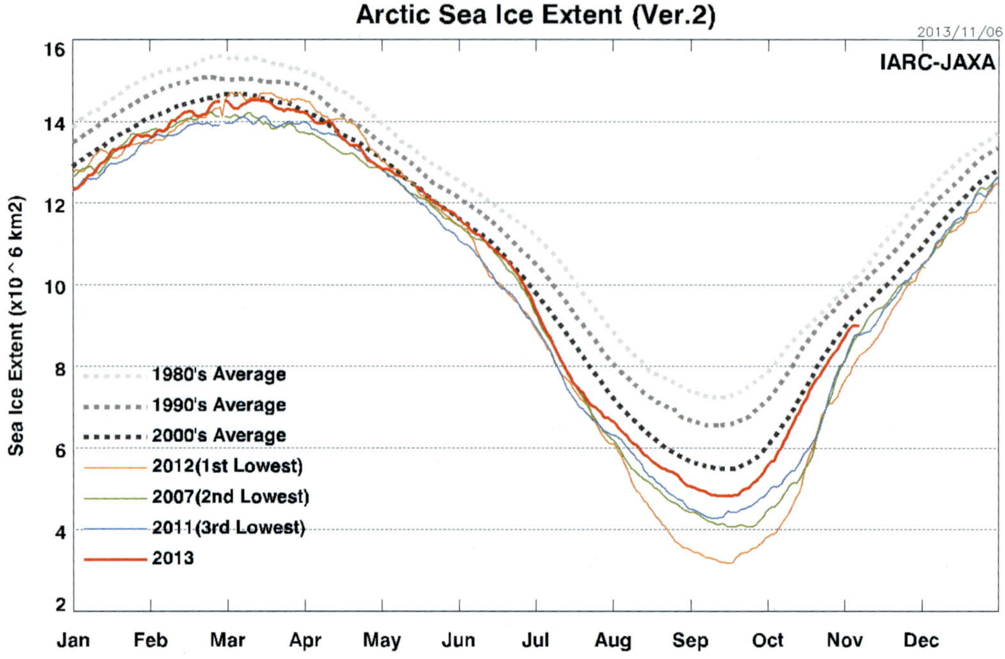

Fig. 1. Seasonal cycle of Arctic ice extent in the four years of least September sea ice extent (2012 in red, 2007 in orange, 2011 in green, 2008 in blue). Decadal means are shown as dashed lines for the 2000s (black), 1990s (dark gray) and 1980s (light gray).

2013, although ice extent in September 2013 was still lower than in any year prior to 2007.

Because ice thickness has also decreased, the loss of sea ice volume has been even greater than the loss of ice extent or area (IPCC 2013). However, prior to the decade of the 2000s, ice thickness measurements are heterogeneous and sparse, as they were limited to submarine measurements of ice draft (primarily in spring) and to point measurements from coastal locations and special field campaigns. For this reason, this present paper focuses on variations in ice extent, which is the most widely documented of all sea ice variables over timescales of decades to centuries.

While the Arctic has warmed in recent decades as global temperatures have increased, there is an emerging consensus that the rapid ice loss of the past decade has been exacerbated by natural variability as well as by warming temperatures. Documentation and diagnosis of the natural variability of sea ice are therefore essential to assessments of the likelihoods of future trajectories of sea ice. The importance of natural variability has also been highlighted in global climate model studies (Kay *et al.*, 2011) which show that rapid ice loss events are triggered by natural variations in the wind and temperature forcing of sea ice. Even the warming that has contributed to the recent ice loss has likely been the result of a combination of greenhouse gas increases and natural variability, which is a major factor in decadal variations of temperature at the regional scale. In the North Atlantic sector, for example, Polyakov *et al.* (2010) have shown that the recent warming is due to roughly comparable contributions from an underlying (greenhouse-related) trend and a multidecadal mode of natural variability.

The variability of sea ice is complicated by its seasonality and by its regional dependencies. Because the loss of Arctic sea ice has been much

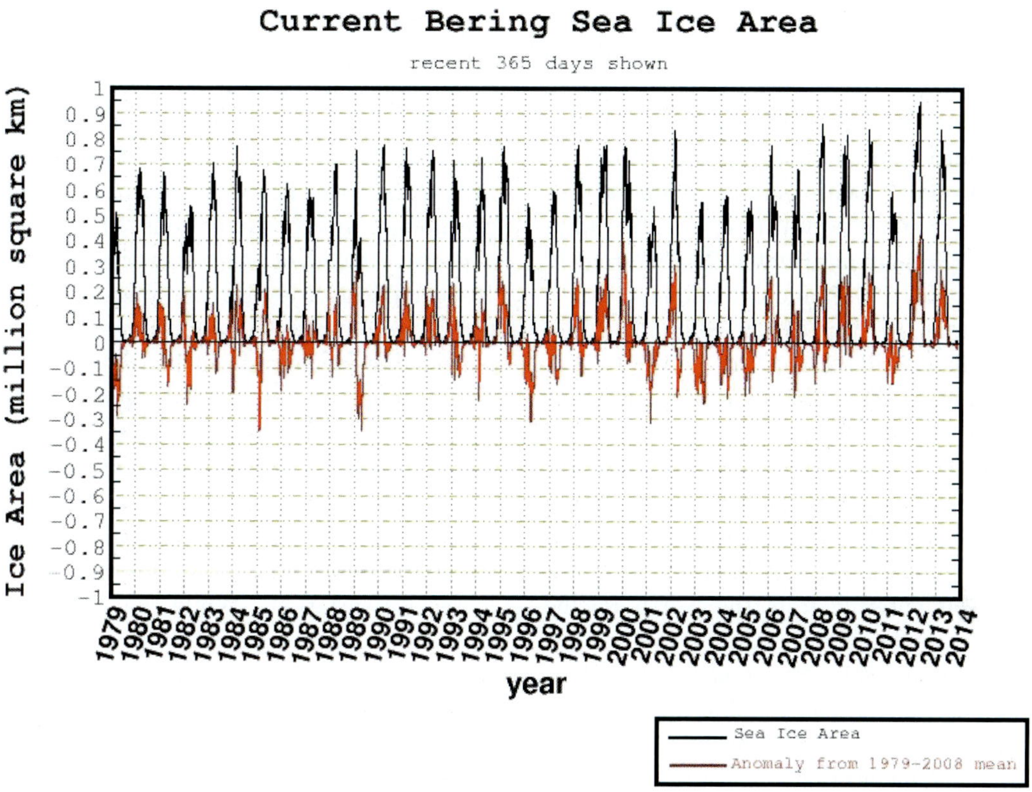

Fig. 2. Time series of Bering Sea ice-covered area, 1979–2013. Black line shows actual area of ice cover, red line shows departure from calendar-date mean for 30-year period 1979–2008. From W. Chapman, Cryosphere Today.

greater in the summer than in the winter, the area of seasonal (first-year) sea ice has increased in winter. Regions such as the Bering Sea show no trend in ice extent over the past 30 years (Fig. 2). In March of 2012, for example, the Bering Sea actually experienced its maximum ice cover of the post-1979 satellite era, only six months before the hemispheric summer ice coverage reached its record minimum. However, the seasonal sea ice in the Bering Sea makes only a small contribution to the total Arctic sea ice volume. Moreover, as will be shown in Section 3, sea ice coverage in the Southern Hemisphere has not decreased since 1979, and has actually shown a slight increase. Natural variability is again a key consideration in the interpretation of the regional and seasonal dependencies of sea ice trends over decadal to multidecadal timescales. Data syntheses in the past several years have advanced our knowledge and understanding of sea ice variability over these timescales.

2. Arctic Sea Ice Extent Variability

Most assessments of interannual variability of sea ice have been for the Arctic and have been based on the post-1979 satellite extent record or, in a few cases, extensions back to the early 1950s. Because this record spans only about three decades, it provides limited information on decadal variability and essentially no information on multidecadal variability. In a recent evaluation of interannual variability of Arctic summer sea ice since 1979 (Ogi *et al.*, 2010), the combined effect of winter and summer wind forcing were shown to have accounted for 50% of the variance of the change in September Arctic sea ice extent from one year to the next, and also for roughly 1/3 of the downward linear trend of ice extent over the past 31 years. The remainder of the variance and trend is presumably due to a combination of variations in air temperatures (including a systematic warming trend), radiative forcing (Francis and Hunter, 2007; Perovich, 2011; Kay *et al.*, 2008), and oceanic heat fluxes that impact the ice from below (Shimada *et al.*, 2006; Polyakov *et al.*, 2012).

In the Pacific sector of the Arctic, for example, the Bering Strait is the entry corridor for warmer water of subarctic origin. After entering the Chukchi Sea, this water generally moves eastward offshore of the northern Alaskan coast. There are indications of a positive feedback whereby heat entering from the North Pacific reduces the thickness and coverage of sea ice north of Alaska (Shimada *et al.*, 2006). This thinner, looser ice is more mobile and susceptible to movement by winds, driving the Beaufort Gyre and facilitating the transport of the warmer Pacific water from the Alaskan shelf to the deeper waters offshore. This transport leads to a reduction of the ice cover farther offshore, and the reduction then may be further enhanced by the albedo effect of reduced reflection of solar radiation. This mechanism may be especially relevant to the ice loss of the past decade because the sector containing the Beaufort, Chukchi and East Siberian Seas has experienced the greatest reduction of summer sea ice extent.

Attribution of the recent loss of sea ice has been addressed with climate models. One such study (Kay *et al.*, 2011) is based on an ensemble of simulations by a global climate model, the Community Climate System Model, version 4 (CCSM4). In this set of experiments, control runs as well as simulations with historical and projected forcing were used to assess the probability that a change as extreme as the recent summer sea ice loss in the Arctic could have occurred by chance. The simulations run with 20th Century forcing indicated that the observed decrease of recent decades cannot be explained by natural variability alone. However, natural variability was found to be prominent on decadal and even multidecadal variations of sea ice. For example, in 21st Century greenhouse-driven simulations, the chances of a positive trend of sea ice over any particular 10-year period (despite the long-term negative trend) were more than 40%. Even for a 20-year period, the chances of an increase of ice cover were about 5% despite the external (RCP8.5 greenhouse scenario) forcing (Fig. 3).

A recent study by Day *et al.* (2012) used output from multicentennial control simulations by an ensemble of different global climate models to address the role of internal variability in sea ice variations. While the Arctic Oscillation showed little relationship with pan-Arctic sea ice extent in the models, both the Atlantic Meridional Oscillation (AMO) and the Atlantic Meridional Overturning Circulation (AMOC) were found to be significantly correlated with sea ice extent in the models. The relationships obtained from the models were used to deduce that 5–30% of the recent decline of Arctic sea ice is due to AMO-driven variability. While this finding is model-based, it is consistent with the observational results summarized later in this section. Polyakov *et al.* (2010), for example, argue that the portion of Arctic variability associated with the AMO may be even larger than 5–30%.

Internal variability notwithstanding, attribution studies of the recent trend in sea ice area have pointed to the importance of external forcing from increasing concentrations of greenhouse gases. In a study using a record of March and September Arctic ice extent back to 1953 (Notz and Marotzke, 2012), the observational record was found to be "sufficient to virtually exclude internal variability and self-acceleration as an explanation for the observed long-term trend, clustering and magnitude of recent sea ice minima" — and this study was

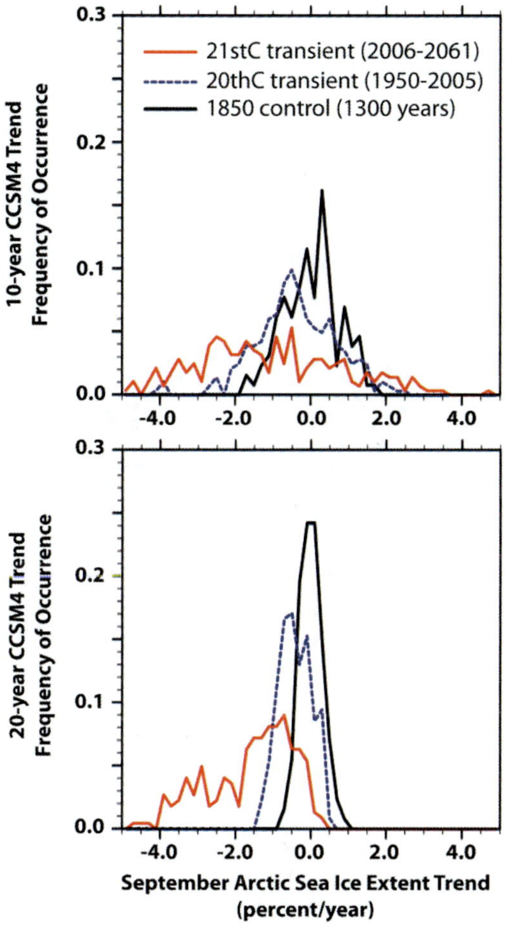

Fig. 3. Distribution of 10-year (upper) and 20-year (lower) trends of September Arctic sea ice coverage in CCSM4 model simulations with 21st Century transient (RCP 8.5) forcing (red lines), 20th Century transient forcing (blue lines) and 1850 control forcing (black lines). From Kay *et al.*, 2011.

completed prior to the record minimum of 2012 noted above.

Annual indices of atmospheric modes of variability such as the Arctic Oscillation and the Pacific Decadal Oscillation indices were only weakly associated with the recent pan-Arctic ice retreat, while the slight downward trend of solar irradiance in recent decades would have favored a cooling and hence an increase of ice extent (Fig. 4). The only external driver that correlated highly with pan-Arctic ice extent of recent decades was the concentration of atmospheric CO_2. This correlation, together with the plausibility of a causal relationship between atmospheric CO_2 and sea ice extent, makes greenhouse gases the most likely explanation for the recent trend of sea ice (Notz and Marotzke, 2012).

Time series of sea ice over longer timescales are necessary in order to assess multiyear to decadal variability in the observational record. Figure 5 shows February and August ice variations in pan-Arctic ice coverage based on a recent augmentation of a digital sea ice dataset derived from historical charts, ship and aircraft reports, merged with the post-1979 satellite record. Because of the absence of pan-Arctic coverage in the pre-satellite period, uncertainties in the yearly values are greater prior to 1979. In its Fifth Assessment Report, the Intergovernmental Panel on Climate Change (IPCC 2013) has estimated the uncertainties to be one standard deviation of the interannual variability in the post-1979 satellite period, 1.5 times the standard deviation for the period 1953–1979, and twice the standard deviation for the period prior to 1953. While the accelerating downward trend of the recent decades dominates the time series, multiyear and decadal variations are apparent, *e.g.*, in the 1910s and 1920s of the August time series. Models capture this multiyear and even multidecadal variability although, because such variations are largely internal variability, their phasing does not correspond with observed variations (Kay *et al.*, 2008; Wang *et al.*, 2007).

On a regional basis, multidecadal variability is even more apparent. A retreat of sea ice in the Atlantic sector has been documented as part of an early 20th Century warming trend (Ifft, 1922; Johannessen *et al.*, 2004). An analysis of historical ice charts compiled by the Arctic and Antarctic Research Institute has shown that summer sea ice in the Russian seas is characterized by three periods since the 1930s: (A) declining ice extent until the mid-1950s, (B) generally increasing or stable ice extent from the

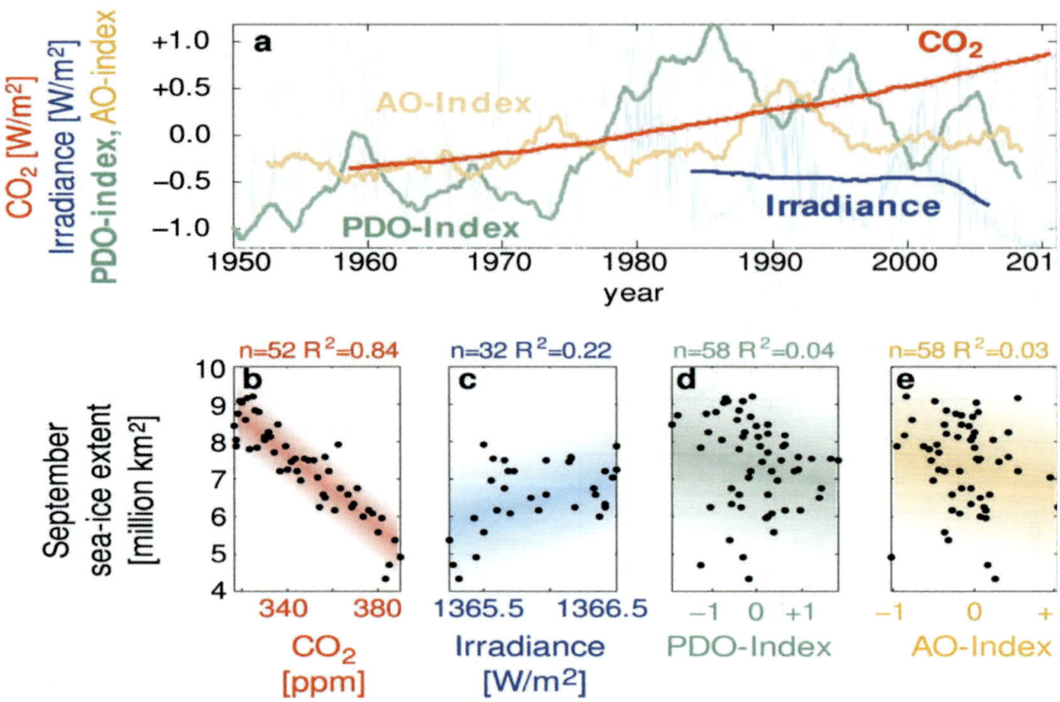

Fig. 4. (a) Time series of candidate drivers to explain the evolution of Arctic sea ice, 1953–2010: CO_2 concentrations (red), solar irradiance (black), Pacific Decadal Oscillation (PDO) index (green), Arctic Oscillation Index (yellow). (b–e) Values of September ice extent plotted against annual mean values of various forcings; linear regressions have indicated R^2 values, with $\pm 2\sigma$ ranges indicated by colored shading. From Notz and Marotzke, 2012.

mid-1950s to the mid-1980s, and (C) decreasing extent from the mid-1980s onward (Mahoney *et al.*, 2008).

A recent synthesis of ship reports and other information (Divine and Dick, 2006) has led to a reconstruction of North Atlantic sea ice variations back to 1750, with a focus on the region between Greenland and Novaya Zemlya. While ice variations in the subpolar North Atlantic are among the best documented over the past several hundred years, much of the ice in the North Atlantic marginal ice zone is seasonal and hence is not a major contributor to the total volume of sea ice in the Northern Hemisphere. Nevertheless, the differences between the ice positions of 1989–2002 and 1870–1920 in the Greenland Sea (spring) and the Barents Sea (summer) ranged from 250 to 375 km. These differences are twice as large as the standard deviation of the interannual variation of the ice edge in the respective regions and seasons. As noted below, the uncertainty in interannual variations in the estimated pan-Arctic ice extent during the 1870–1920 timeframe was also about twice the interannual standard deviation. However, the relatively good coverage of the sea ice observations in the eastern North Atlantic reduces the uncertainty in this region.

Over the full 260-year timeframe in the study of Divine and Dick, multidecadal variations become more prominent, as shown in Fig. 6 for the Greenland Sea region. The April and June time series of Greenland Sea ice extent show prominent decreases in the early 20th Century, while several decades of low ice coverage in the late 1700s are apparent in June and August. Perhaps more noteworthy in this study is the detection of oscillations of North Atlantic ice

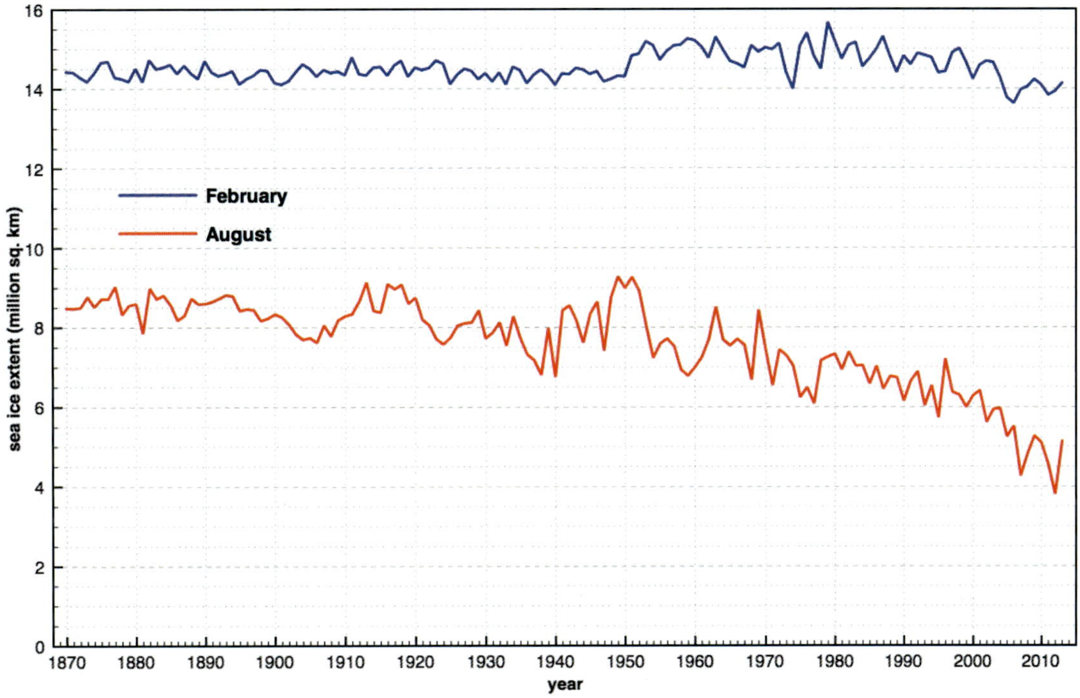

Fig. 5. Time series of Northern Hemisphere sea ice extent for the calendar months of March through August. Updated from Chapman and Walsh, 1993.

cover with periods of 60–80 years and 20–30 years, superimposed on the negative trend of the past century or two. The higher-frequency (20–30 year) oscillations are dominant in the Barents Sea (not shown here). The 60–80 year oscillations are more prominent in the Greenland Sea, as shown by the results of a wavelet analysis in Fig. 6.

The results suggest that the acceleration of the ice loss in recent decades may be partly attributable to the phase of the 60–80 year oscillation, in association with a warming of the subpolar North Atlantic and the Arctic. The timescale of the 60–80 year oscillation is consistent with that of the Atlantic Multidecadal Oscillation (Schlesinger and Ramankutty, 1994; Delworth and Mann, 2000), for which a signature has been found in the subpolar North Atlantic and Arctic Ocean water temperatures (Polyakov *et al.*, 2010; Polyakov *et al.*, 2004). Higher-frequency (decadal and shorter) variability that dominates the Barents Sea sector of the North Atlantic appears to be related to the North Atlantic Oscillation (Lassen and Thejll, 2005), although such associations may not be stationary over millennial timescales (see below).

An even longer record for a specific portion of the North Atlantic is based on reports from the coast of Iceland (Deser *et al.*, 2000). The so-called "Koch Index," is based on the number of weeks during which ice was observed along the Icelandic coast weighted by the length of the coastline affected by ice. Figure 7 shows the annual values of this index extending back to 1200. Large variations are apparent over interannual, decadal and centennial timescales. In the 20th Century, the late 1960s–early 1970s, as well as the 1910s, stand out as decades with heavy ice, while the 1920s through the 1950s were generally free of ice. On centennial scales, the 1600s through the 1800s had generally heavy

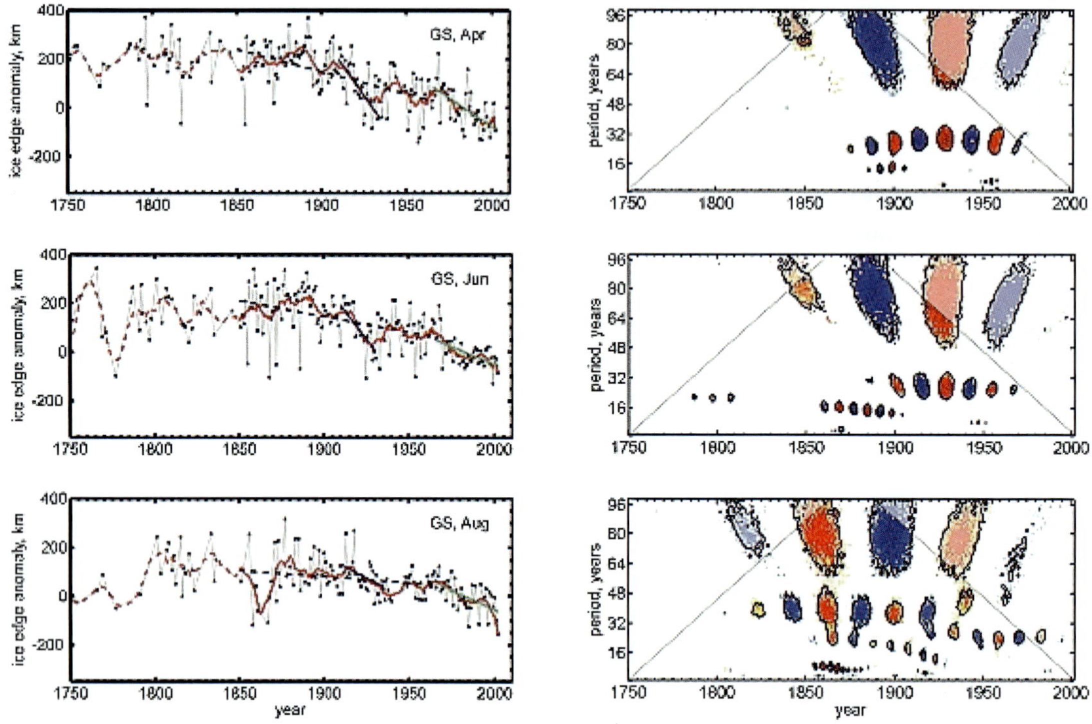

Fig. 6. Left: Time series of April, June and August ice extent anomalies for 1750–2002, with linear trends for historical (1850–1966, black line) and contemporary (1967–2002, green lines) subperiods. Right: Normalized wavelet power spectra for the time series of the ice extent anomalies shown in the left panels. Wavelet power is statistically significant in colored areas. From Divine and Dick, 2006.

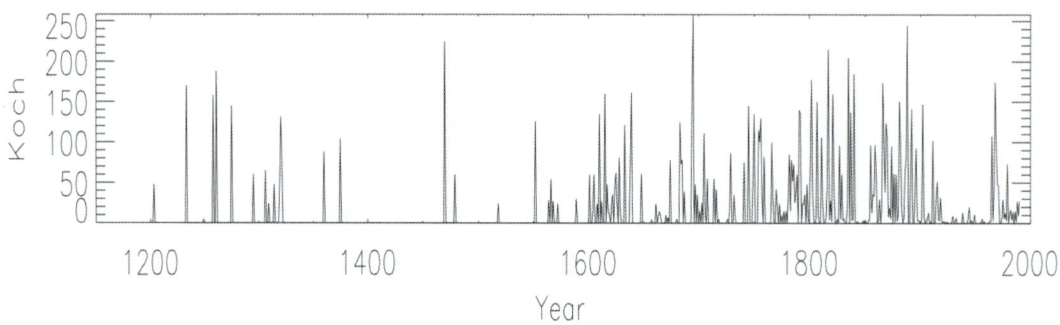

Fig. 7. Koch Index of sea ice severity on the coast of Iceland. From Deser *et al.*, 2000.

ice conditions, consistent with the timing of the Little Ice Age in the North Atlantic and Europe. By contrast, most winters of the 1200s through the 1500s were generally ice-free along the Iceland coast. The Icelandic time series highlights the embedding of multiyear to decadal variations within longer-term (multi-century) climate variations.

Paleoclimatic proxy information from tree rings and ice cores has enabled Macias Fauria *et al.* (2010) to extend the record of North Atlantic sea ice variability back to 1200 AD.

Similar to Divine and Dick (2006), the focus of this study was the Nordic Seas region from Greenland to the Barents Sea. The squared correlation between the reconstructed and observed time series of North Atlantic ice extent for 1864–1997 was 0.59, indicating that more than half the observed variance was captured by the proxy reconstruction. A wavelet analysis and a multi-taper method (MTM) spectral analysis showed multidecadal low-frequency (50–120 year) oscillations, consistent with the Atlantic Multidecadal Oscillation and with the results of Divine and Dick (2006). These oscillations account for most of the multidecadal variability seen in the latter portion of the record: large extent in the second half of the 19th century, low values in the 1920s and 1930s, higher values in the 1960s, and low values in the present period for which the extent is the lowest in the entire 800-year record. This oscillation appears throughout the eight-century record. In addition, decadal-scale (7–32 year) oscillations were detectable throughout the reconstructed record. While these decadal-scale variations correlated significantly (negatively) with the North Atlantic Oscillation (NAO) in the decades since the 1920s and especially during the second half of the 20th century, no relationship was found prior to the 1920s. The non-stationarity of the correlation between Nordic Seas ice extent and the NAO is shown in Fig. 8.

A more recent synthesis of paleo information has expanded the reconstruction of ice extent to the pan-Arctic domain and to the past 1450 years, extending back to 560 AD (Kinnard *et al.*, 2011). This reconstruction is based on high-resolution terrestrial proxies from the circum-Arctic domain: ice cores, tree rings and lake sediments. Because these proxies provide information primarily about the warm season, the reconstructions depict summer sea ice

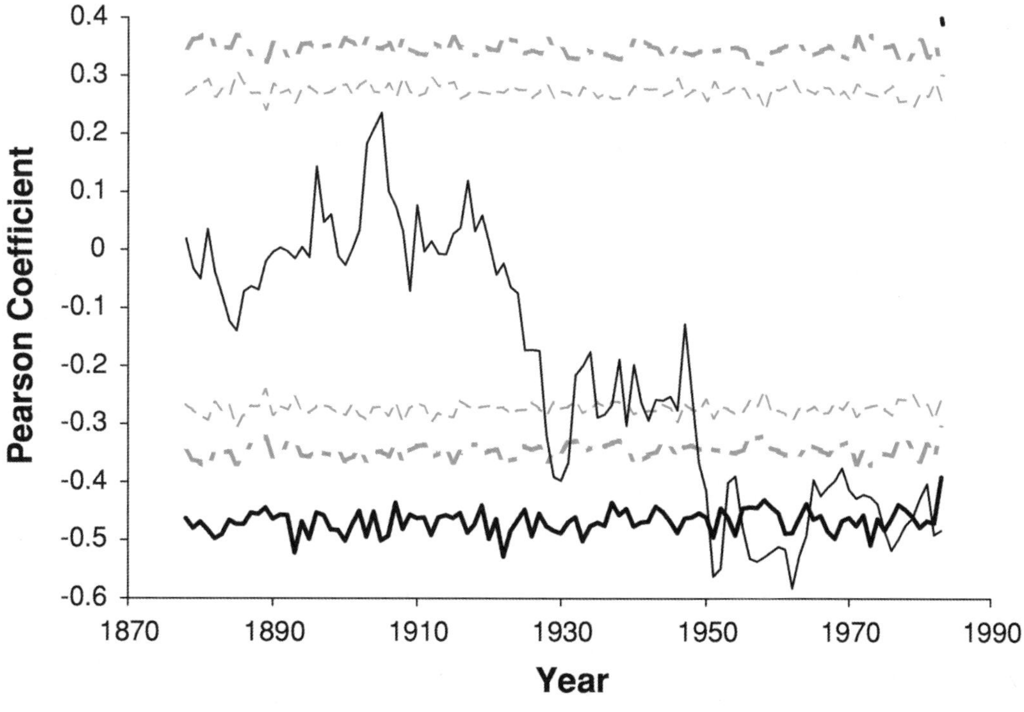

Fig. 8. Thin black line is 30-year running correlation between the winter NAO and sea ice extent in the western Nordic Seas, with thresholds (horizontal lines) for 10%, 5% and 1% significance. From Hurrell (1995).

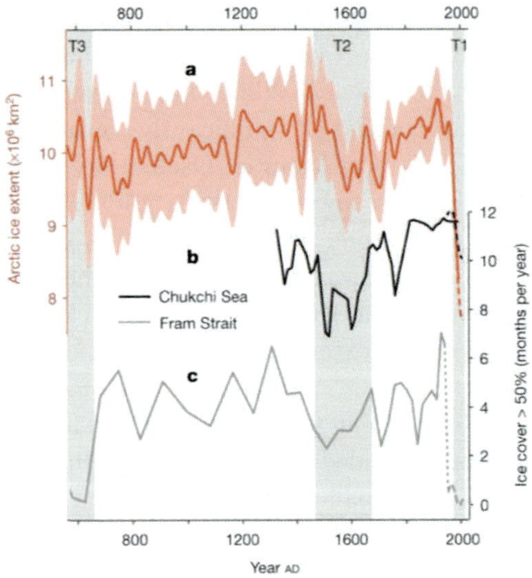

Fig. 9. Reconstructed late-summer ice extent for the Arctic Ocean (red), the Chukchi Sea (black) and Fram Strait (gray). From Kinnard *et al.*, 2011.

variations. The pan-Arctic time series in Fig. 9 shows that the recent decline of sea ice is unprecedented in the 1450-year reconstruction. The abrupt decline in the Fram Strait region during the past several decades is also unique in the 1450-year time series for that region. However, the reconstructions for the Chukchi and Barents Seas indicate that sea ice coverage in these regions was comparable to the present during the late 1500s and early 1600s. All the reconstructed time series show multidecadal variations, although the uncertainties in the reconstructions are large, especially prior to the 16th century.

Kinnard *et al.* (2011) suggest that the main factor driving the decline of sea ice extent on the multidecadal timescales is the advection of warm North Atlantic water into the Arctic. This suggestion is supported by the paleo reconstruction of ocean temperatures based on marine sediments (Spielhagen *et al.*, 2011). The reconstructed ocean temperatures west of Svalbard are warmer in the early 21st century than at any time in the past 2000 years. The planktic foraminiferal data on which the reconstruction is based show multicentennial changes underlying a higher-frequency multidecadal variability.

In summary, recent syntheses of sea ice and related proxy information have provided an improved picture of Arctic sea ice variability over decadal to century timescales. While there is a spectrum of variability, an outstanding feature is the correspondence with the Atlantic Multidecadal Oscillation, which has timescales of 50–120 years. The linkage appears to arise through the inflow of Atlantic Water to the Arctic Ocean. Less robust, and by all indications non-stationary, associations with atmospheric modes such as the North Atlantic Oscillation have also been documented, primarily in recent decades. One possible reason for the nonstationarity of such associations is that the centers of action of major atmospheric modes may change over the timescale of centuries or even less (Jung and Hilmer, 2001).

On even longer timescales, there are indications of millennial variations of ice cover in the Arctic. Funder *et al.* (2011) used the abundance and origins of driftwood along the Greenland and Canadian Arctic coasts to deduce that the limit of year-round sea ice along the Greenland coast 6000–8500 years ago was about 1000 km farther north than at present. Corresponding model simulations by Funder's group suggest that summer minima at this time were even lower than the extremes of recent years (2007, 2012). Other modeling studies have suggested that the minimum ice cover around 8000 B.P. can be used to constrain the choice of parameterizations and models, thereby reducing uncertainty in model projections of future sea ice evolution (Goosse *et al.*, 2007). This minimum in multiyear ice north of Greenland was followed by an increase to maximum values in the past 2500 years, particularly during the Little Ice Age. The latter finding is consistent with the time series of the Koch Index discussed earlier. The 8500–6000 B.P. minimum north of Greenland was accompanied by an increase of ice around Ellesmere Island, implying a role of atmospheric and oceanic forcing patterns associated with ice dynamics.

The history of Arctic sea ice back through the pre-Quaternary has been synthesized by Polyak *et al.* (2010), who drew upon a variety of studies based on sediment records, ice cores, driftwood, terrestrial and marine proxies and historical records. Key conclusions from Polyak's synthesis are that sea ice has been a major feature of the Arctic as far back as 47 Ma B.P., and that sea ice has covered at least part of the Arctic Ocean for at least 13–14 million years. While sea ice appears to have been widespread for the past 2–3 Myr, "episodes of considerably reduced sea ice extent or even seasonally ice-free conditions occurred during warmer periods linked to orbital variations" (Polyak *et al.*, 2010, p. 1757). The most recent such example is the early Holocene (8500–6000 B.P.) minimum summarized above.

3. Antarctic Sea Ice Extent Variability

The summary provided here has focused on the Arctic. The Southern Hemisphere also has a sea ice cover, although it is highly seasonal, ranging in extent from about 2 million km^2 in March to 15–16 million km^2 in September. While this ice cover is found in the Southern Ocean and the subpolar seas adjacent to the Antarctic continent, we choose for simplicity in this discussion to refer to Southern Hemisphere sea ice as "Antarctic" sea ice. The homogeneous record of Antarctic sea ice extent variations is limited to the passive microwave satellite record that began in 1979. While too short to depict multidecadal and longer variations, there is a spectrum of shorter fluctuations over timescales of months to several years (Fig. 10). Two conclusions have emerged from analyses of the satellite record. First, there is a small positive trend in total Antarctic sea ice extent over the past 33 years, although the pan-Antarctic time series and trend obscure regional and seasonal differences. As in the Arctic, trends are generally

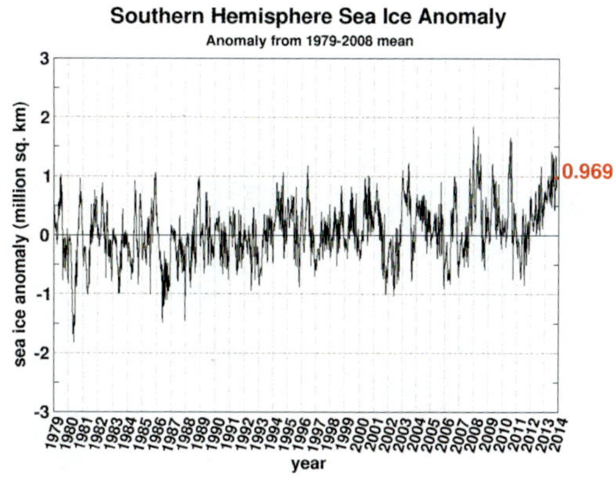

Fig. 10. Time series of anomalies (departures from calendar-date means, 1979–2008) of Southern Hemisphere ice-covered area.

Source: W. Chapman, Cryosphere Today.

larger in summer than in winter. As shown in Fig. 11, the positive trends are most apparent in the Ross Sea, Weddell Sea, the Indian Ocean sector and the western Pacific Ocean sector; the trends are negative in the Bellingshausen and Amundsen Seas as well as near the Antarctic Peninsula (Shu *et al.*, 2012). Second, there are associations between Antarctic sea ice variations and (a) the Southern Annular Mode (Simpkins *et al.*, 2012) and the El Niño/Southern Oscillation teleconnection (Simpkins *et al.*, 2012; Yuan, 2004). However, trends in Antarctic sea ice coverage do not appear to be significantly related to trends in either of these modes in any season (Simpkins *et al.*, 2012). A recent broader assessment of atmosphere-ice-ocean interactions in a framework of recent trends concludes that changes in wind-driven ice advection and temperature advection explain the Antarctic ice variations of the past several decades, as well as the absence of a decrease of overall ice coverage in the Southern Hemisphere (Holland and Kwok, 2012). The large and statistically significant trends in Antarctic sea ice drift, driven largely by local winds, point to the importance of including atmospheric variability in evaluations of decadal-scale and longer Antarctic sea ice variations. The role of the ocean in

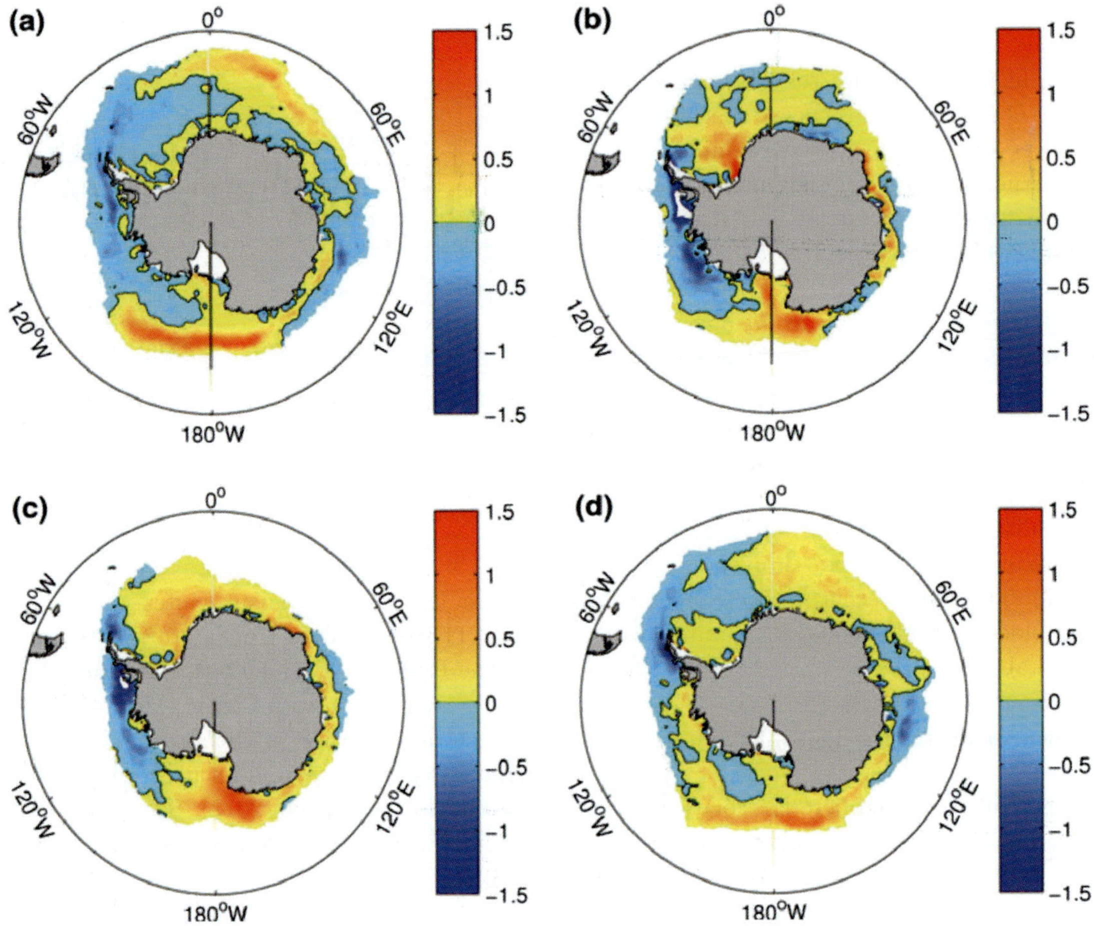

Fig. 11. Linear trends (% per year) of Antarctic sea ice concentration over the period 1979–2009 for (a) spring, (b) summer, (c) autumn and (d) winter. Yellows and orange shades indicate increases, blue shades denote decreases. From Simkins *et al.*, 2012.

these variations remains to be fully elucidated. Prior to the 1970s, there is little information other than records from whaling ships. While attempts have been made to use the whaling ship logs to infer a decrease of Antarctic sea ice since the 1800s (de la Mare, 1997), concerns have been raised about the adequacy of the derived sea ice information (Ackley *et al.*, 2003).

Observational information on Antarctic sea ice variability is virtually nonexistent beyond the past 100–150 years, so proxy information provides the only clues to longer-term Antarctic sea ice variability. A correlation between methanesulphonic acid (MSA) concentrations in a Law Dome ice core and satellite-derived ice extent has been used to produce a proxy record of wintertime ice extent for 1841–1995 in the East Antarctic sector (80–140°E) (Curran *et al.*, 2003). The results suggest that Antarctic sea ice coverage has decreased by about 20% since 1950, in general agreement with de la Mare's (1997) conclusion based on whaling ship logs. Some indication of an 11-year cycle was also reported by Curran *et al.* (2003). The use of MSA concentrations as proxies for ice extent is complicated by the fact that correlations between MSA and sea ice vary in magnitude and sign among regions around Antarctica (Foster *et al.*, 2006). Sea salt sodium (ssNA+) has also been used as a proxy for Antarctic sea ice changes, although the derived changes are averaged over longer timescales, i.e., several centuries to millennia (Rothlisberger and Abram, 2009).

4. Conclusion

This review has emphasized sea ice variability over multiyear to multidecadal timescales, in keeping with the theme of this volume. In the case of the Arctic, an assessment of sea ice variability based on recent data is challenged by the rapidity of the recent decrease of sea ice. Arctic sea ice is indeed at the vanguard of climate change, and the consensus of the recent research summarized is that greenhouse forcing is a major driver of the recent changes of sea ice. In this respect, the past few decades are not representative of the longer record of sea ice. While a declining (or even vanishing, in summer) sea ice cover will still exhibit variability, the temporal variance and other characteristics of sea ice coverage will change. Consequently, nonstationarity confounds assessments of Arctic sea ice variability.

In the Antarctic, there is little trend of hemispheric total coverage over the relatively short satellite period. The absence of a trend has been attributed to the relative importance of winds in controlling the extent of Antarctic sea ice, and ocean interactions may also be contributing factors. Nevertheless, there are regional trends of Antarctic ice extent over the past several decades, particularly in the winter season. There may also have been longer-term variations that are simply not rigorously demonstrable in the limited observational information available for sea ice in the Southern Hemisphere.

The contrasting trends in Arctic and Antarctic sea ice must also be viewed in the context of the extent of multiyear sea ice, which has historically been much greater in the Arctic than in the Antarctic. Because of its greater thickness, multiyear sea ice has reduced the seasonal amplitude of Arctic sea ice extent, at least until the most recent decade. The loss of multiyear sea ice is perhaps the most notable aspect of the recent Arctic sea ice decline. As multiyear sea ice disappears in the Arctic, one may expect a new sea ice regime in the Arctic similar to the Antarctic with a large seasonal cycle and with variability strongly influenced by wind and oceanic forcing.

In both hemispheres, it is apparent that sea ice variability is already intertwined with variability in the atmosphere (winds and temperature) and ocean (currents, temperature and salinity). It follows that variations of sea ice over decadal and longer timescales are part of corresponding broader climate variations that are the subject of this volume. A system-scale perspective on decadal and longer climate variations remains an ultimate objective of climate science, and the inclusion of sea ice represents an important step towards a system-scale perspective.

Acknowledgments

This work was supported by the National Oceanic and Atmospheric Administration through Grant NA11OAR4310172 and by the National Science Foundation through Grant ARC-1023131.

References

Ackley, S. F., P. Wadhams, J. C. Comiso, and A. P. Worby, 2003: Decadal decrease of Antarctic sea ice extent inferred from whaling records revisited on the basis of historical and modern sea ice records. *Polar Res.*, **22**, 19–25.

Chapman, W., L., and J. E. Walsh, 1993: Recent variations of sea ice and air temperature in high latitudes. *Bull. Amer. Meteor. Soc.*, **74**, 33–47.

Curran, M. A. J., T. D. Van Ommen, V. I. Morgan, K. L. Phillips, and A. S. Palmer, 2003: Ice core evidence for sea ice decline since the 1950s. *Science*, **303**, 1203–1206.

Day, J. J., J. C. Hargreaves, J. D. Annan, and A. Abe-Ouchi, 2012: Sources of multi-decadal variability in Arctic sea ice extent. *Environ. Res. Lett.*, 7, 034011, doi: 10.1088/1748-9326/7/3/034011.

de la Mare, W. K., 1997: Abrupt mid-twentieth-century decline in Antarctic sea ice extent from whaling records. *Nature*, **389**, 57.

Delworth, T., and M. Mann, 2000: Observed and simulated multi-decadal variability in the Northern Hemisphere. *Clim. Dyn.*, **16**, 661–676.

Deser, C., J. E. Walsh, and M. S. Timlin, 2000: Arctic sea ice variability in the context of recent atmospheric circulation trends. *J. Climate*, **13**, 617–633.

Divine, D. V., and C. Dick, 2006: Historical variability of the sea ice edge position in the Nordic Seas. *J. Geophys. Res.*, **111**, C01001, doi: 10.1029/2004JC002851.

Foster, A., M. A. J. Curran, T. D. Van Ommen, and V. Morgan, 2006: Covariation of sea ice and methane sulphonic acid in Wilhelm II Land, East Antarctica. *Ann. Glaciol.*, **44**, 429–432.

Francis, J. A., and E. Hunter, 2007: Drivers of declining sea ice in the Arctic winter: A tale of two seas. *Geophys. Res. Lett.*, **34**, L17503, doi: 10.1029/2007GL030995.

Funder, S., H. Goosse, H. Jepsen, E. Kaas, K. H. Kjær, N. J. Korsgaard, N. K. Larsen, H. Linderson, A. Lyså, P. Möller, J. Olsen, and E. Willerslev, 2011: A 10,000-year record of Arctic Ocean sea-ice variability — View from the beach. *Science*, **333**(6043), 757–750.

Goosse, H., E. Driesschaert, T. Fichefet, and M.-F. Loutre, 2007: Information on the early Holocene climate constrains the summer sea ice projections for the 21st century. *Clim. Past*, **3**, 683–692.

Holland, P. R., and R. Kwok, 2012: Wind-driven trends in Antarctic sea ice drift. *Nat. Geosci*, **5**, 872–875, doi: 10.1038/ngeo1627.

Hurrell, J. W., 1995: Decadal trends in the North Atlantic Oscillation: Regional temperatures and precipitation. *Science*, **269**, 676–679.

Ifft, G. N., 1922: The changing Arctic. *Mon. Wea. Rev.*, **50**, 589.

Intergovernmental Panel on Climate Change. *Climate Change, 2013: The Physical Science Basis. Contribution of Working Group I to the Fifth Assessment Report of the IPCC*, Cambridge University Press, Cambridge, U.K., in press.

Johannessen, O. M., L. Bengtsson, M. W. Miles, S. L. Kuzmina, V. E. Semenov, G. V. Alekseev, A. P. Nagurny, V. K. Zakharov, L. P. Bobylev, L. H. Pettersson, K. Hasselmann, and H. P. Cattle, 2004: Arctic climate change: Observed and modeled temperature and sea-ice variability. *Tellus*, **56A**, 328–341.

Jung, T., and M. Hilmer, 2001: The link between the North Atlantic Oscillation and Arctic sea ice export through Fram Strait, *J. Climate*, **14**, 3932–3943.

Kay, J. E., T. L'Ecuyer, A. Gettelman, G. Stephens, and C. O'Dell, 2008: The contribution of cloud and radiation anomalies to the 2007 Arctic sea ice extent minimum. *Geophys. Res. Lett.*, **35**, L08503, doi: 10.1029/2008GL033451.

———, M. M. Holland, and A. Jahn, 2011: Inter-annual to multi-decadal Arctic sea ice extent trends in a warming world. *Geophys. Res. Lett.*, **38**, L15708, doi: 10.1029/2011GL048008.

Kinnard. C., C. M. Zdanowicz, D. A. Fisher, E. Isaksson, A. de Vernal, and L. G. Thompson, 2011: Reconstructed changes in Arctic sea ice over the past 1,450 years. *Nature*, **479**, 509–512.

Lassen, K., and P. Thejll, 2005: *Multi-decadal Variation of East Greenland Sea-ice Extent: AD 1500–2000*. Danish Meteorological Institute, Sci. Rep. 05–02, 26 pp.

Macias Fauria, M., A. Grinsted, S. Helama, J. Moore, M. Timonen, T. Martma, E. Isaksson, and M. Eronen, 2010: Unprecedented low twentieth century winter sea ice extent in the western Nordic Seas. *Clim. Dyn.*, **34**, 781–795.

Mahoney, A. R., R. G. Barry, V. Smolyanitsky, and F. Fetterer, 2008: Observed ice extent in the Russian Arctic. *J. Geophys. Res.*, **113**, C11005, doi: 10.1029/2008JC004830.

Notz D., and J. Marotzke, 2012: Observations reveal external driver for Arctic sea-ice retreat. *Geophys. Res. Lett.*, **39**, L08502, doi: 10.1029/2012GL051094.

Ogi, M., K. Yamazaki, and J. M. Wallace, 2010: Influence of winter and summer surface wind anomalies on summer Arctic sea ice extent. *Geophys. Res. Lett.*, **37**, L07701, doi: 10.1029/2009GL042356.

Perovich, D. K., 2011: The changing Arctic sea ice cover. *Oceanography*, **24**(3), 162–173.

Polyak, L., R. B. Alley, J. T. Andrews, J. Brigham-Grette, T. M. Cronin, D. A. Darby, A. S. Dyke, J. J. Fitzpatrick, S. Funder, M. Holland, A. E. Jennings, G. H. Miller, M. O'Regan, J. Savelle, M. Serreze, K. St. John, J. W. C. White, and E. Wolff, 2010: History of sea ice in the Arctic. *Quaternary Sci. Rev.*, **29**, 1757–1778.

Polyakov, I. V., G. V. Alekseev, L. A. Timokhov, U. S. Bhatt. R. L. Colony, H. L. Simmons, D. Walsh, J. E. Walsh, and V. F. Zakharov, 2004: Variability of the intermediate Atlantic water of the Arctic Ocean over the last 100 years. *J. Climate*, **17**, 4485–4497.

———, V. A. Alexeev, U. S. Bhatt, E. I. Polyakova, and X. Zhang, 2010: North Atlantic warming: Fingerprints of climate change and multidecadal variability. *Clim. Dyn.*, **34**, 439–457.

———, J. E. Walsh and R. Kwok, 2012: Recent changes of Arctic multiyear sea ice cover and the likely causes. *Bull. Amer. Meteor. Soc.*, **93**, 145–151.

Rothlisberger, R., and N. Abram, 2009: Sea-ice proxies in Antarctic ice cores. *PAGES News*, **17**, 24–26.

Schlesinger, M. E., and N. Ramankutty, 1994: An oscillation in the global climate system of period 65–70 years. *Nature*, **367**, 723–726.

Shimada, K., T. Kamoshida, M. Itoh, S. Hishino, E. Carmack, F. McLaughlon, S. Zimmerman, and A. Proshutinsky, 2006: Pacific Ocean inflow: Influence on catastrophic reduction of sea ice cover in the Arctic Ocean. *Geophys. Res. Lett.*, **33**, L086–5, doi: 10.1029/2005GL025624.

Shu, Q., F. Qiao, and Z. Song, 2012: Sea ice trends in the Antarctic and their relationship to surface air temperature during 1979–2009. *Clim. Dyn.*, **28**, 2355–2363, doi: 10.1007/s00382-011-1143-9.

Simpkins, G. R., L. M. Ciasto, D. W. J. Thompson, and M. H. England, 2012: Seasonal relationships between large-scale climate variability and Antarctic sea ice concentration. *J. Climate*, **25**, 5451–5469.

Spielhagen, R. F., K. Werner, S. A. Serensen, K. Zamelczyk, E. Kandiano, G. Budeus, K. Husum, T. M. Marchitto, and M. Hald, 2011: Enhanced modern heat transfer to the Arctic by warm Atlantic water. *Science*, **331**, 450–453.

Wang, M., J. E. Overland, V. M. Kattsov, J. E. Walsh, X. Zhang, and T. Pavlova, 2007: Intrinsic versus forced variations in coupled climate model simulations over the Arctic during the 20th century. *J. Climate*, **20**, 1093–1107.

Yuan, X., 2004: ENSO-related impacts on Antarctic sea ice: A synthesis of phenomena and mechanisms. *Antarct. Sci.*, **15**, 415–425.

Chapter 14

Multi-year Prediction and Predictability

Timothy DelSole,* Michael K. Tippett,† and Liwei Jia‡

*George Mason University, Fairfax, VA
and Center for Ocean-Land-Atmosphere Studies, Calverton, MD, USA
tdelsole@gmu.com

†Department of Applied Physics and Applied Mathematics,
Columbia University, New York, NY, USA and
Center of Excellence for Climate Change Research, Department of Meteorology,
King Abdulaziz University, Jeddah, Saudi Arabia

‡Atmospheric and Oceanic Sciences Program, Princeton University, Princeton, USA

This chapter reviews climate prediction and predictability on multi-year time scales. There is clear evidence from coupled atmosphere-ocean general circulation models that the climate system can be predicted on multi-year time scales based on changes in greenhouse gas concentration, aerosol concentration, and solar insolation, and on initial condition information primarily from the ocean subsurface. New statistical optimization techniques have substantially clarified the space-time structure of predictability on multi-year time scales. These techniques reveal not only predictability of ocean quantities, but also predictability of temperature and precipitation over land on multi-year time scales. A forecast system based on these predictable components is verified to have skill on multi-year time scales over the entire twentieth century observational record.

1. Introduction

This chapter presents an overview of climate prediction on multi-year time scales. More precisely, this chapter presents rigorous arguments that certain spatial patterns in the climate system are predictable on multi-year time scales. The physics of these predictable components are not well understood and hence are not discussed in detail, but some discussion of mechanisms can be found in other chapters of this book (*e.g.*, chapter 9). Multi-year climate predictions are in high demand by policy makers concerned with resource management and infrastructure investment. In addition, climate predictions are of inherent scientific value because they provide a test of our understanding of the climate system.

It is well established that atmospheric predictability is limited to less than a month in the absence of changes in radiative forcing or boundary conditions (Shukla, 1981). Specifically, numerical calculations suggest that small perturbations in the atmosphere amplify with a doubling time of about 1.8 days (Simmons and Hollingsworth, 2002), suggesting an upper limit of predictability due to atmospheric dynamics of less than a month (Shukla, 1981). Beyond a month, slowly varying components of the climate system, such as sea surface temperature, soil moisture, snow cover, or sea ice thickness, are predictable owing to their slower time scale. These slower components can influence the atmosphere and hence give rise to predictability of atmospheric variables beyond a month. One of

Climate Change: Multidecadal and Beyond
edited by Chih-Pei Chang et al.

the great accomplishments in climate research is the demonstration that the coupled atmosphere-ocean system supports a mode of variability in the tropical Pacific, known as the El Niño Southern Oscillation (ENSO), characterized by autocorrelation time scales (and hence linear predictability) exceeding several months (Philander, 1990). Furthermore, the change in tropical precipitation and latent heating patterns associated with ENSO generates "teleconnection patterns" that can affect weather in midlatitudes (Lau, 1997). Although state-of-the-art climate models do not fully reproduce the observed ENSO teleconnection patterns, especially in monsoonal regions (Yang and DelSole, 2012), dynamical models and empirical statistical models can simulate this variability with such fidelity that they can be used to construct skillful forecasts of seasonal mean temperature and precipitation a few months in advance (Barnston *et al.*, 2012).

The feasibility of multi-year predictions has been established along two independent lines of argument. First, climate models reveal that changes in greenhouse gas concentration, aerosol concentration, and solar insolation during the twentieth century force multidecadal changes in the globally averaged surface temperature. The anticipated response to these forcing changes, which we call the "forced component," is sufficiently large as to be distinguishable from natural variability and sufficiently consistent with observations as to form a plausible explanation of many (but not all) observed changes on multidecadal time scales (Stott *et al.*, 2000; Meehl *et al.*, 2003; Hegerl *et al.*, 2007). Second, coupled atmosphere-ocean general circulation models reveal that, even in the absence of interannual variations of natural or anthropogenic forcing, certain components of subsurface oceanic variability are predictable on decadal scales (Griffies and Bryan, 1997; Delworth and Mann, 2000; Latif *et al.*, 2006; Boer and Lambert, 2008; Branstator *et al.*, 2012). However, the dynamical mechanisms responsible for "unforced" multi-year predictability in the ocean are not completely understood. Inferences from the distribution of water properties suggest that the deep ocean circulation varies on time scales longer than millennial time scales (Broecker and Peng, 1982). Wind forcing from the atmosphere and differences in temperature and salinity act to generate an overturning circulation in the deep ocean (Wunsch and Ferrari, 2004), but the character of this variability on multidecadal time scales is sensitive to modeling assumptions (Tulloch and Marshall, 2012). Thus, while there is substantial evidence that the deep ocean varies on multidecadal time scales, the precise space-time structure of this variability is unclear, and how these variations in the deep ocean are communicated to the ocean surface and subsequently to the atmosphere is even less clear.

Other climate components with the potential to generate atmospheric predictability beyond a month include soil moisture, snow cover, sea ice, glaciers, and ice sheets. While a wide range of modeling studies suggest that soil moisture anomalies can give rise to predictability in the first three months (Guo *et al.*, 2012), there is little evidence that such anomalies can give rise to predictability beyond a year. Similar comments can be made regarding snow cover (Cohen and Entekhabi, 2012). Arctic sea ice extent can persist for several months at certain times of the year and influence the general circulation by reducing the amount of sunlight reflected back into space and changing the sensible and latent heat fluxes at the ocean surface, thereby changing the energy and water budget over the Arctic (Deser *et al.*, 2010). Changes in Arctic sea ice also alter the surface temperature gradient, which in turn influences the strength of the westerlies through thermal wind balance (Budikova, 2009; Blanchard-Wrigglesworth *et al.*, 2011). However, predictions of interannual and decadal variations of Arctic sea ice extent are notoriously difficult with dynamical models but have advanced rapidly in recent year (Tietsche *et al.*,

2014). Glaciers and ice sheets appear to vary significantly on multidecadal-to-centennial time scales or longer, but our understanding of the dynamics of this variability has not advanced to the point of being able to produce reliable forecasts of this variability. Volcanic eruptions give rise to multi-year predictability in subsequent years, usually associated with a cooling due to enhanced aerosol loading in the stratosphere, but volcanic eruptions themselves are unpredictable. Thus, while multi-year predictability conceivably could arise from components "outside" the atmosphere-ocean system, the dynamics of these components are so poorly understood or unpredictable that their role in multi-year predictability cannot be established definitively.

In this chapter, we review a statistical framework for quantifying predictability and explicitly identify patterns that are predictable on multi-year time scales. There are different methods for isolating predictable patterns depending on whether the predictability arises from forced or unforced variability. We will apply these methods to show unforced components dominate predictability in the first couple years of a forecast while forced components dominate after five or ten years, consistent with Branstator and Teng (2010).

2. Measuring Predictability

To diagnose predictability, it is necessary to define the concept precisely. Consider the univariate case in which the variable being predicted is y, and the observations on which this prediction is based are denoted $\mathbf{x}$. In predictability studies, y and $\mathbf{x}$ are lagged in time; let τ denote the lag. Then the most complete prediction of y given $\mathbf{x}$ is the conditional distribution $p(y|\mathbf{x})$. *The variable y is said to be* ***unpredictable*** *if it is independent of* $\mathbf{x}$, *for all possible values* $\mathbf{x}$. That is,

$$p(y|\mathbf{x}) = p(y) \text{ for unpredictability.} \quad (1)$$

It follows that a necessary condition for predictability is that $p(y|\mathbf{x}) \neq p(y)$ for some $\mathbf{x}$. In statistics, the distributions $p(y|\mathbf{x})$ and $p(y)$ are called the conditional and unconditional distributions; in predictability, the distributions are called the *forecast* and *climatological* distributions. It is natural to measure predictability by some measure of the difference between these two distributions. A common measure of predictability is one minus the ratio of forecast to climatological variances:

$$P_\tau = 1 - \frac{\text{var}[y|\mathbf{x}]}{\text{var}[y]}. \quad (2)$$

This measure is between 0 and 1 (inclusive) and often is referred to as the "fraction of predictable variance." For Gaussian distributions, this measure is independent of the realization of $\mathbf{x}$.

An essential aspect of measuring predictability is significance testing. By significance testing, we mean assessing whether the value of P_τ estimated from finite samples is much larger than the range of values expected under the hypothesis of no predictability (1). For joint Gaussian distributions, one assumes, either implicitly or explicitly, that y is linearly related to $\mathbf{x}$ in the form

$$y = \beta_1 x_1 + \beta_2 x_2 + \cdots + \beta_K x_K + \mu + \epsilon, \quad (3)$$

where ϵ is a Gaussian random variable. In this case, the null hypothesis of no predictability is equivalent to the hypothesis

$$H_0 : \beta_1 = \beta_2 = \cdots = \beta_K = 0. \quad (4)$$

There exist essentially two testing procedures, depending on data availability. If the data is available in the form of $(K+1)$-tuples $[y, x_1, \ldots, x_K]$, then the estimated predictability measure P_τ corresponds to the coefficient of determination R_τ^2, which can be estimated by regression methods and whose significance can be tested by standard methods. Alternatively, if the data is available in the form of $(K+E)$-tuples $[y_1, y_2, \ldots, y_E, x_1, \ldots, x_K]$, where there are E multiple realizations of y for each fixed

$\mathbf{x}$, often called *ensemble members*, then the conditional and unconditional variances can be estimated directly, without setting up the regression model (3), and the null hypothesis can be tested by a standard procedure known as Analysis of Variance (ANOVA).

Just as linear regression and ANOVA provide two methods for testing and measuring predictability in the univariate case, the two analogous methods in the multivariate case are Canonical Correlation Analysis (CCA) and Predictable Component Analysis (PrCA), respectively (Schneider and Griffies, 1999; DelSole and Tippett, 2007). Each method can be interpreted as finding a linear combination of y-variables that maximizes predictability. To clarify this, let the weights of the linear combination be collected in the vector $\mathbf{q}$. Then, the predictability measure (2) for the linear combination $\mathbf{q}^T\mathbf{y}$, where superscript T denotes the matrix transpose, becomes

$$P_\tau = 1 - \frac{\mathbf{q}^T \boldsymbol{\Sigma}_{Y|X}\mathbf{q}}{\mathbf{q}^T \boldsymbol{\Sigma}_Y \mathbf{q}}, \tag{5}$$

where $\boldsymbol{\Sigma}_{Y|X}$ and $\boldsymbol{\Sigma}_Y$ are the conditional and unconditional covariance matrices of $\mathbf{y}$. Maximizing P_τ over all possible $\mathbf{q}$ leads to the generalized eigenvalue problem

$$\left(\boldsymbol{\Sigma}_Y - \boldsymbol{\Sigma}_{Y|X}\right)\mathbf{q} = \lambda \boldsymbol{\Sigma}_Y \mathbf{q}, \tag{6}$$

where λ is an eigenvalue equal to the value of P_τ for the corresponding eigenvector. If the data is available in the form $(\mathbf{y}, \mathbf{x})$ (in particular, only one realization of $\mathbf{y}$ exists for each $\mathbf{x}$), then we fit the variables to a model, usually of the linear form

$$\mathbf{y} = \mathbf{L}_\tau \mathbf{x} + \boldsymbol{\epsilon}, \tag{7}$$

where $\mathbf{L}_\tau$ is a least squares operator that depends on time lag and $\boldsymbol{\epsilon}$ is a multivariate white noise process. It can be shown that the conditional covariance matrix of the linear model is

$$\boldsymbol{\Sigma}_{Y|X} = \boldsymbol{\Sigma}_Y - \mathbf{L}_\tau \boldsymbol{\Sigma}_X \mathbf{L}_\tau^T, \tag{8}$$

where $\boldsymbol{\Sigma}_X$ is the covariance matrix for $\mathbf{x}$. Substituting (8) into (6) and solving leads to CCA (DelSole and Chang, 2003). If, instead, the data is available in the form of multiple "repetitions" or ensemble members under the same $\mathbf{x}$, i.e., in the form $(\mathbf{y}_1, \mathbf{y}_2, \ldots, \mathbf{y}_E, \mathbf{x})$, then the conditional covariance matrix is

$$\boldsymbol{\Sigma}_{Y|X} = \boldsymbol{\Sigma}_Y - \boldsymbol{\Sigma}_S \tag{9}$$

where $\boldsymbol{\Sigma}_S$ is the covariance matrix of conditional means $\boldsymbol{\mu}_{Y|X}$, which can be estimated from the ensemble. Substituting the conditional covariance matrix (9) into (6) and solving leads to Predictable Component Analysis.

These methods find not just the weights that maximize P_τ, but actually a complete, linearly independent set of weight vectors $\mathbf{q}$. These vectors provide the basis for systematically decomposing predictability into an uncorrelated set of components ordered such that first component maximizes predictability, the second maximizes predictability subject to being uncorrelated with the first, and so on (Schneider and Griffies, 1999; DelSole and Tippett, 2007). This decomposition is such that the random vector $\mathbf{y}(t)$ can be expressed equivalently as a sum of predictable components $\mathbf{p}_1, \mathbf{p}_2, \ldots, \mathbf{p}_M$ as

$$\mathbf{y}(t) = \sum_{m=1}^{M} \mathbf{p}_m r_m(t), \tag{10}$$

where the time series are $r_m(t) = \mathbf{q}_m^T \mathbf{y}(t)$ and the loading patterns are $\mathbf{p}_m = \boldsymbol{\Sigma}_Y \mathbf{q}_m$. This decomposition is especially useful when only a few components explain most of the predictability.

In practice, the above covariance matrices are singular because the number of grid points or variables being analyzed exceeds the number of samples. The resulting maximization problem is therefore ill-posed and requires regularization methods for solution. A standard approach is to project the data onto a truncated set of principal components and then maximize predictability in the reduced space.

One drawback of decomposing predictability by maximizing P_τ is that the resulting decomposition depends on lead time τ. That is, decomposing predictability at two distinct lead times

τ_1 and τ_2 leads to two distinct decompositions. Such lead-time dependent decompositions are awkward for diagnosing predictability in an overall sense. To diagnose predictability in a manner that does not depend on lead time, one can maximize an *integral* of the average predictability over lead time (DelSole and Tippett, 2009a):

$$APT = 2\int_0^\infty P_\tau d\tau. \qquad (11)$$

Since P_τ is dimensionless, the resulting integral has units of time. If the process is a first-order Markov process, then the autocorrelation decays exponentially and (11) defined above equals an e-folding time scale. In various other statistical analyses (Davis, 1976; DelSole, 2001), the integral of a correlation emerges naturally as a time scale of a process. For these reasons, (11) is called the Average Predictability Time (APT). The decomposition of APT and tests for significance have been worked out (DelSole and Tippett, 2009b; Jia and DelSole, 2011). This technique has been demonstrated to extract weather, intraseasonal, interseasonal, decadal, and climate predictability in a single step (DelSole and Tippett, 2009b). Thus, this technique seamlessly diagnoses predictability on all time scales, from days to decades.

3. Unforced Variability

In this section, we explicitly identify patterns of annual mean near-surface temperature that are predictable in multiple climate models on multi-year time scales. To do this, we find patterns that maximize APT in a set of pre-industrial control simulations, which simulate unforced climate variability in the absence of interannual variations of anthropogenic or natural forcing. The particular simulations are drawn from the fifth phase of the Coupled Model Intercomparison Project (CMIP5) (Taylor *et al.*, 2012). We include a control run only if it simulates 500 or more years, and use the penultimate 250 years for training and the last 250 years for assessment. In later sections, we want to derive the forced response from the same models, so we consider a model only if, in addition to a 500-year control run, it has at least four simulations with anthropogenic and natural forcing. These constraints leave eight models. The control simulations from the eight models were pooled to generate a multi-model data set, separately for training and assessment. The total sample size for training or assessment is 2000 years, with 250 years from each model. See Jia and DelSole (2012a) for further analysis details.

Only one control simulation from each model is available, so a linear model of the form (7) is estimated using 15 leading principal components of the control simulations, and then the resulting regression model is used to maximize APT. The loading vector of the most predictable pattern, shown in Fig. 1a, suggests that the North Atlantic and North Pacific are predictable on multi-year time scales, consistent with previous studies (Boer, 2004; DelSole *et al.*, 2011; Branstator *et al.*, 2012; Yang *et al.*, 2013). The APT of this component is well separated from those of the other components and statistically significant in each model individually.

To clarify the predictability, we project the component onto independent control simulations and evaluate the squared multiple correlation for each model and lead time separately. The multiple correlation is the correlation between the component time series r, and the least squares prediction of r based on 15 time-lagged principal components. The result, shown in Fig. 1b, reveals that one model has significant predictability for only one year, while other models have significant predictability for as long as ten years. This large range in predictability limits across models is consistent with previous studies (DelSole *et al.*, 2011; Branstator *et al.*, 2012; Jia and DelSole, 2012a).

The time series produced by projecting the component onto observational estimates of sea surface temperature from the Extended

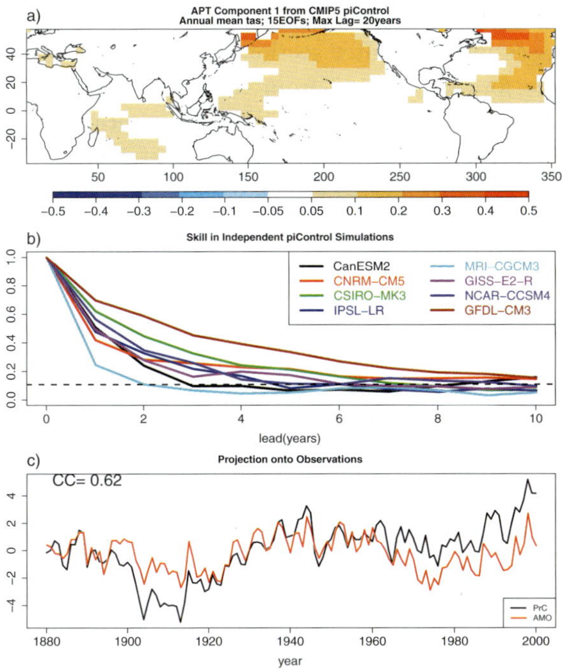

Fig. 1. (a) The most predictable component of annual mean near-surface air temperature over the ocean between 40°S and 60°N derived from eight CMIP5 control simulations; (b) the squared multiple correlation of the most predictable component in each model; (c) the (scaled) AMO index (red), and the time series produced by projecting the pattern onto observations (black). The dashed line in (b) indicates the 5% significance level. The spatial structure shows the change in temperature (in degree kelvin per unit variate) relative to the climatological mean. The correlation coefficient between the two time series in (c) is indicated in the top left of the panel.

Reconstruction data set (ERSSTv3) (Smith *et al.*, 2008) is shown in Fig. 1c. The figure shows that the time series fluctuates strongly on multidecadal time scales. This result is noteworthy because there is no guarantee that the most predictable component in climate models is predictable (or slowly varying) in observations. Fig. 1c also shows the observed Atlantic Multidecadal Oscillation (AMO) index. The strong correlation between the two time series confirms that the most predictable component in the CMIP5 models projects on a well documented component of multi-decadal variability in observations.

It should be recognized that observations contain both forced and unforced variability, so projecting any component on unfiltered observations will produce time series with a mix of forced and unforced variability. Indeed, the projection of the unforced component on observations shown in Fig. 1c has a clear trend, presumably because the forced response projects on this particular component. Recall also that the AMO index is detrended, which aggravates the discrepancy between the two time series. Detrending the predictable component time series improves the match with the AMO index (not shown). Ideally, the forced response ought to be removed from observations before performing the projection. Such a method is discussed in the next section. For simplicity of interpretation, we do not remove the forced response from observations. Accordingly, the reader should keep in mind that projecting predictable components on observations may produce time series with forced variability, especially if forced variability resembles the predictable component.

The second most predictable component, shown in Fig. 2, has loadings in the Pacific basin and exhibits multidecadal variations in observations which are correlated with the PDO index. Again, it is noteworthy that a component that is predictable in control simulations has a counterpart that varies on multi-decadal time scales in observations. DelSole *et al.* (2013) show that the predictability of the leading component comes primarily from persistence, while predictability of the second component comes from interactions with other components. To clarify this point, the lead-lag correlations between the second predictable component and the others were calculated. For the model derived in this chapter, the second predictable component has its strongest correlation with the fourth predictable component. The lead-lag correlation between the second and fourth components, shown in Fig. 3, suggests that the two components are involved in a coupled propagation: component 2

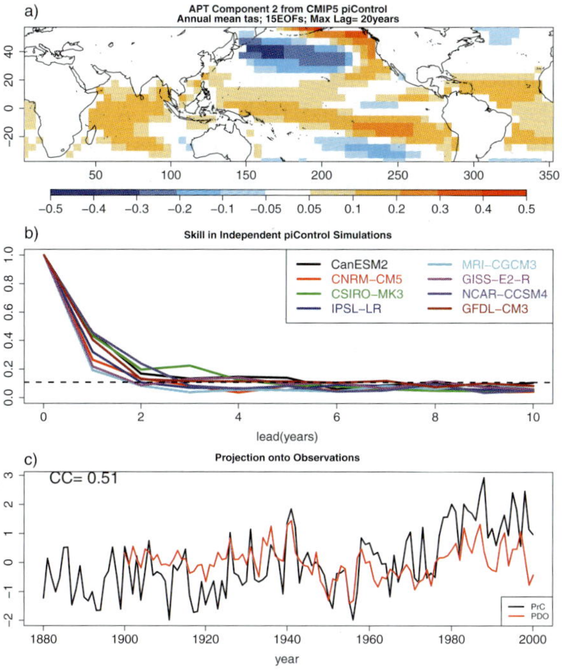

Fig. 2. Same as Fig. 1, except for the *second* most predictable component.

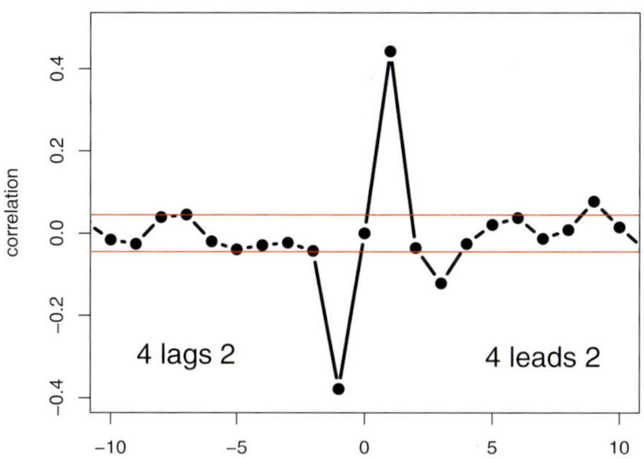

Fig. 3. Lead-lag correlation between the 2nd and 4th most predictable components of annual mean surface air temperature derived from eight CMIP5 control simulations. Horizontal lines indicate the 5% significance level.

tends to evolve into component 4, which in turn evolves into component 2 of opposite sign. A similar relation, though noisier, is seen in observations (DelSole *et al.*, 2013). As a consequence, the predictability of Pacific components cannot be explained as mere persistence

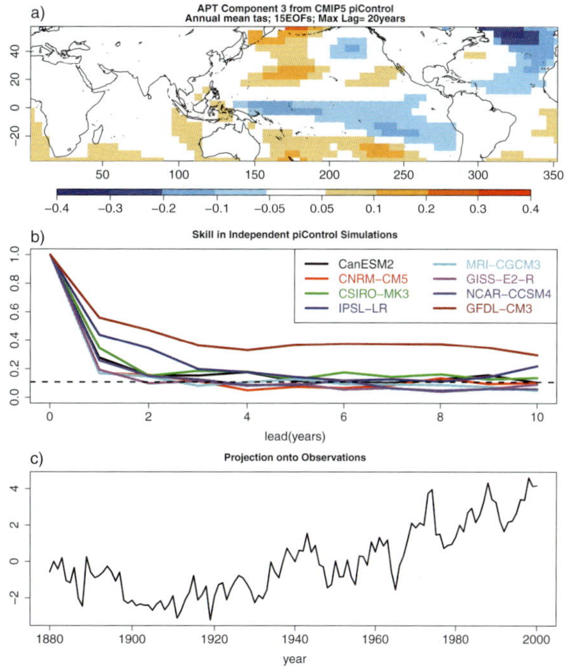

Fig. 4. Same as Fig. 1, except for the *third* most predictable component.

of a single component, but instead requires the interaction of at least two components, consistent with the suggestion of Teng and Branstator (2011). The third component is shown in Fig. 4 and dominated by a strong trend, presumably because the forced response in observations projects on this pattern. The fourth predictable component is shown in Fig. 5 and has large loadings in the equatorial Pacific, suggesting a connection to ENSO. In fact, the correlation of the projected time series and NINO3.4 is 0.62. Thus, the above results suggest that ENSO interacts with a PDO-like structure on multi-year time scales.

The above analysis was repeated using data only over land, and for precipitation (Jia and DelSole, 2012a). This analysis demonstrates that both temperature and precipitation are predictable *over land* for 3–5 years. *These conclusions contradict the widely held belief that temperature and precipitation over land are unpredictable beyond seasonal time scales.* One

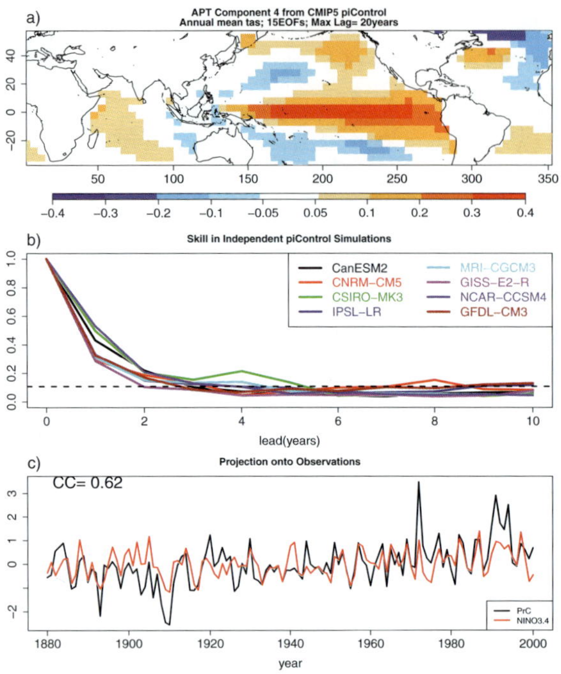

Fig. 5. Same as Fig. 1, except for the *fourth* most predictable component.

possible reason that this predictability has not been detected previously is that these components explain relatively little variance on short time scales and hence are difficult to detect by methods that do not explicitly maximize predictability.

4. Forced Variability

Forced variability refers to the response to interannual variations of natural and anthropogenic forcing, including volcanic and solar forcing. If y is the variable to be predicted, and $\mathbf{f}$ specifies the time history of climate forcing with natural and anthropogenic forcing, then the variable y is said to be unforced if

$$p(y|\mathbf{f}) = p(y). \tag{12}$$

It follows that a necessary condition for y to have a component of forced predictability is for $p(y|\mathbf{f}) \neq p(y)$. Unfortunately, y generally comprises a mixture of forced and unforced variability, thereby obscuring the structure of forced predictability. To separate forced and unforced variability, some assumptions about their interaction need to be made. A common starting point is to assume that forced variability is an independent and additive perturbation to unforced variability, in which case observed variability can be modeled as

$$\mathbf{o} = \mathbf{F}\mathbf{a} + \mathbf{u}, \tag{13}$$

where $\mathbf{o}$ is the observation vector, the columns of $\mathbf{F}$ contain the response to particular climate forcings, called *climate change signals*, $\mathbf{a}$ is a vector of amplitude coefficients, and $\mathbf{u}$ is a random vector representing unforced variability. This model provides the basis of optimal fingerprinting techniques, which are standard techniques for detecting climate change in observations. However, these techniques assume that the climate change signal one wishes to detect, and the statistics of the background unforced variability, are *pre-specified*. Unfortunately, estimation of time-dependent climate change signals is generally a nontrivial problem, owing to the fact that climate models from which they are estimated simulate these signals embedded in a background of unforced variability. Thus, identification of time-varying climate change signals already constitutes a signal detection problem that must be solved before optimal fingerprinting techniques can be applied to observations (Hasselmann, 1993).

Jia and DelSole (2012b) develop an optimal approach to estimating time-dependent climate change signals by finding the linear combination of variables that maximizes the mean square detection statistic

$$D = \frac{\mathbf{q}^T\left(\boldsymbol{\Sigma}_{Y|F} + \Delta\right)\mathbf{q}}{\mathbf{q}^T\boldsymbol{\Sigma}_Y\mathbf{q}}, \tag{14}$$

where $\boldsymbol{\Sigma}_{Y|F}$ and $\boldsymbol{\Sigma}_Y$ are the covariance matrices of the forced and unforced system, respectively,

$$\Delta = \left(\boldsymbol{\mu}_{Y|F} - \boldsymbol{\mu}_Y\right)\left(\boldsymbol{\mu}_{Y|F} - \boldsymbol{\mu}_Y\right)^T, \tag{15}$$

and $\boldsymbol{\mu}_{Y|F}$ and $\boldsymbol{\mu}_Y$ are the means of the forced and unforced systems, respectively. The weights for the most detectable vector is found by solving the generalized eigenvalue problem

$$\left(\boldsymbol{\Sigma}_{Y|F} + \Delta\right)\mathbf{q} = \lambda \boldsymbol{\Sigma}_Y \mathbf{q}. \tag{16}$$

The eigenvalue problem actually yields a complete set of weight vectors $\mathbf{q}_1, \mathbf{q}_2, \ldots, \mathbf{q}_M$, and corresponding loading vectors $\mathbf{p}_i = \boldsymbol{\Sigma}_Y \mathbf{q}_i$. If the vectors are ordered by decreasing eigenvalue, then the first maximizes the mean square detection statistic, the second maximizes mean square detection statistic subject to being uncorrelated with the first, and so on. If there exists only a single forced response vector $\mathbf{f}$, then only one eigenvalue will differ from unity and the corresponding eigenvector will be $\mathbf{p} = \mathbf{f}$. If two forced response vectors exist, say $\mathbf{f}_1$ and $\mathbf{f}_2$, then only two eigenvalues will differ from unity, and the corresponding eigenvectors $\mathbf{p}_1$ and $\mathbf{p}_2$ *span the same space as the climate change signals* $\mathbf{f}_1$ and $\mathbf{f}_2$. That is, the optimally detectable patterns generally are not in one-to-one correspondence with specific climate forcings. This "mixing" of forced responses is unavoidable in optimization problems in which the forced responses are neither independent nor orthogonal in space.

To illustrate the technique, we apply it to determine whether annual mean anomalies are detectable. This analysis should not be construed as suggesting that such signals are ideal for detecting climate change signals. They are not. Stronger signal-to-noise ratios would be obtained by taking longer term averages and including temporal information. The purpose of this exercise is to obtain an estimate of the forced response on time scales that are short enough to assess decadal prediction skill (considered in the next section). The result of maximizing the mean square detection statistic of annual mean near-surface air temperature over the oceans in eight CMIP5 models are shown in Fig. 6. Full details of this analysis can be found in Jia and DelSole (2012b).

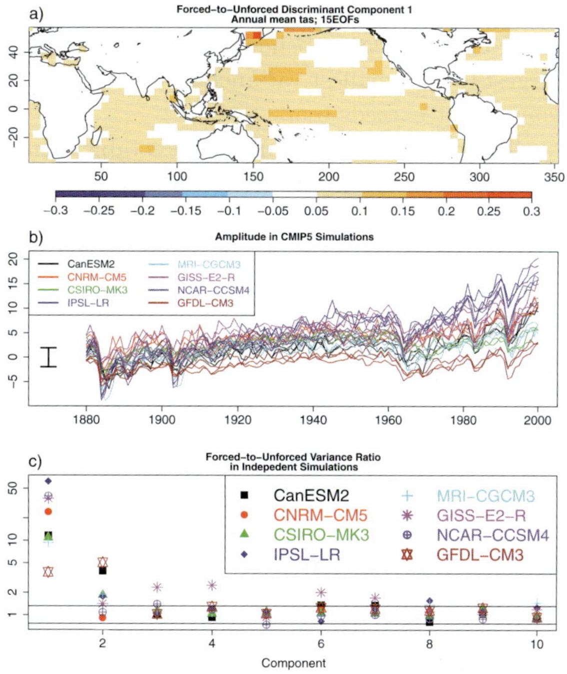

Fig. 6. (a) The most detectable component, on average, of annual mean near-surface air temperature over the ocean between 40°S and 60°N derived from eight CMIP5 control simulations; (b) the projection of the most detectable pattern on independent twentieth century simulations; (c) the forced-to-unforced variance ratios of each discriminant component in each model. The error bar in (b) indicates the 2.5 and 97.5 percentiles in the pre-industrial control simulations. For plotting (b), each model time series has been normalized by the standard deviation of the pre-industrial control runs of the respective model, and ensemble members from the same model have the same color. The two lines in (c) show the 2% confidence interval of the variance ratio under the null hypothesis of equal variance. The spatial pattern has units of degree Kelvin per unit variate relative to the pre-industrial mean.

The mean square detection statistic D is essentially the ratio of the variances in the forced and unforced simulations. Accordingly, we call the resulting component the *forced-to-unforced discriminant*. The forced-to-unforced variance ratio is shown in Fig. 6c. The leading component is detectable in all models, as indicated by the fact that all values for the first component lie above the upper significance threshold. In contrast, the second is not detectable in some models. Also, the degree of detectability is strongly model dependent: the ratio of the largest to the smallest variance ratio for the first component exceeds ten.

Time series of the most detectable component in independent twentieth century runs are shown in Fig. 6b. Each time series shows a clear trend, interrupted by sudden coolings coincident with major volcanic eruptions. Moreover, the variability due to the trend greatly exceeds the variability of the pre-industrial control simulations (indicated by the error bar), demonstrating that this component is significantly detectable in the models. Note that the model ensembles become separated at the end of the twentieth century, indicating that a large fraction of the uncertainty in forced response comes from model differences.

The spatial pattern of the most detectable component is shown in Fig. 6a. The pattern corresponds to temperature changes in the same direction nearly everywhere. Combining this fact with the increasing trend in each time series implies that the bulk of the detectability is due to twentieth century warming. Note that the loadings are relatively uniform in the tropics, but are very weak in the North Atlantic, in contrast to the most predictable component shown in Fig. 1. These differences in spatial structure provide the basis for separating forced and unforced variability in observations (DelSole *et al.*, 2011).

5. Multi-year Predictions

The previous sections showed that certain components in climate models are predictable on multi-year time scales. However, these results, derived from climate models, do not necessarily imply that such predictability exists in the real climate system, or that the predictability in nature has the same spatio-temporal structure as in models. The most compelling way to demonstrate the reality of this predictability is to show that climate models can make skillful predictions on multi-year time scales. Unfortunately, true predictions of future, independent data on multi-year time scales are scarce. Recently, coupled atmosphere-ocean models and empirical statistical models have produced skillful, multi-year *hindcasts* using observation-based initial conditions (Keenlyside *et al.*, 2008; Smith *et al.*, 2008; Pohlmann *et al.*, 2009; van Oldenborgh *et al.*, 2012; Zanna, 2012; Ho *et al.*, 2013; Newman, 2013). The purpose of this section is to further clarify this multi-year prediction skill by showing that a forecast model containing the above predictable components has skill, and that this skill arises from these predictable components.

In this new forecast system, the forced component is predicted using the most detectable component shown in Fig. 6, with the amplitude determined by a least squares fit of this pattern for each year to the ensemble mean, multi-model mean of eight CMIP5 twentieth century simulations. These simulations were initialized at some random point from a pre-industrial simulation and then forced with anthropogenic and natural forcing characteristic of the twentieth century. As a result, all prediction skill from these simulations come from the forced response, not from observational initial condition information. The use of dynamical models to predict forced variability is attractive because dynamical models account for a wide variety of complex physical processes and avoid making simplistic (*e.g.*, constant trend) assumptions regarding the response. Of course, dynamical models are imperfect and may fail to capture key physical processes. The skill of the model-predicted forced component therefore provides a measure of how well the forced component is predicted by dynamical models.

To predict the unforced component, we use the multivariate regression model (7), which predicts the mean sea surface temperature in a given year based on the state of this field at an antecedent year. Ideally, observations should be used to train and verify the regression model, but unfortunately the available observational record for global scale temperature variations is about 100 years (Smith *et al.*, 2008), which severely constraints the estimation and validation of decadal prediction models. One might use cross validation to demonstrate skill (Zanna, 2012; Newman, 2013), but cross validation implies that different statistical models are used to predict different decades, and it is difficult to ensure that all characteristics of the regression model (*e.g.*, domain, number of PCs) are specified independently of the cross validation procedure. We avoid these issues by constructing a single multivariate regression model trained on pre-industrial control simulations. The advantage of deriving a regression model from model simulations is that a much larger sample size is available for estimation, the entire observational record is independent of the training data and hence provides genuinely independent verification data, and all questions about the validity of cross validation in decadal prediction are avoided. On the other hand, a regression model trained on numerical simulations (instead of observations) may be unable to capture realistic SST variability. In a sense, the regression model represents a simplified version of the dynamical models from which it was trained, and hence provides a way to validate dynamical models without performing forecasts with the dynamical models themselves. To mitigate sensitivity to the choice of dynamical model, a single regression model is trained on control runs from eight different climate models. The resulting prediction model is exactly the same model used in Sec. 3, and so it contains the same predictable components as discussed in that section.

To validate the above forecast, observational estimates of sea surface temperature from the Extended Reconstruction data set (ERSSTv3) (Smith *et al.*, 2008) are used. Since the hindcasted forced component comes from a simulation, the same forced component estimate is used at every lead time, hence its skill

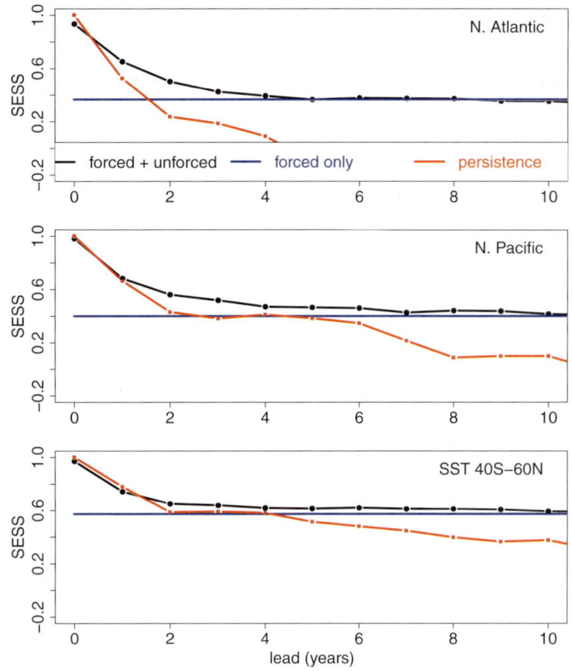

Fig. 7. The skill of predicting observed annual mean near-surface temperatures over the ocean in the North Atlantic (top), North Pacific (middle), and between 40°S and 60°N (bottom) during 1910–2000, using an estimate of the forced component derived from twentieth-century simulations from eight CMIP5 models (blue line), and using this estimate of the forced response plus a multivariate regression model to predict unforced variability (black curve). The skill of a persistence forecast also is shown (red curve). Skill is measured by the squared error skill score (SESS), defined as one minus mean square error divided by the observed climatological variance.

is lead time independent. To quantify prediction skill, we select three indices, namely the area-weighted average over the North Pacific, North Atlantic, and globe, and calculate one minus the ratio of the mean square prediction error to the climatological variance of these indices. The skill of the forced response estimate in predicting 20C observations is shown in Fig. 7 as the blue line. The sum of forced and unforced temperature anomalies is predicted by adding the estimated forced response to the regression model prediction. The skill of the resulting predictions is shown as the black curve in Fig. 7. The skill is positive at all lead times.

However, the skill of the forced plus unforced prediction approaches the skill of the forced component alone. On the other hand, the skill generally exceeds the skill of a persistence forecast, demonstrating that the regression model "adds value" beyond mere persistence. The skill is largest for the global domain, reflecting the dominance of forced predictability on global space scales. The fact that the skill of the combined forced-unforced prediction is larger than the skill of the forced response prediction, especially in the first few lead years, demonstrates that *initial condition information contributes to skill during this time scale.*

The above results demonstrate that our forecast system can skillfully predict SSTs on multi-year time scales. However, our forecast system combines different patterns together so it is unclear whether its skill comes from the predictable patterns discussed in previous sections. To clarify this point, we show in Fig. 8 the skill of our forecast system in predicting the first four predictable components discussed in Sec. 3. The skill is calculated by first forecasting the entire SST field, then projecting each predictable component onto the forecast field and observations, then calculating the skill of the resulting indices.

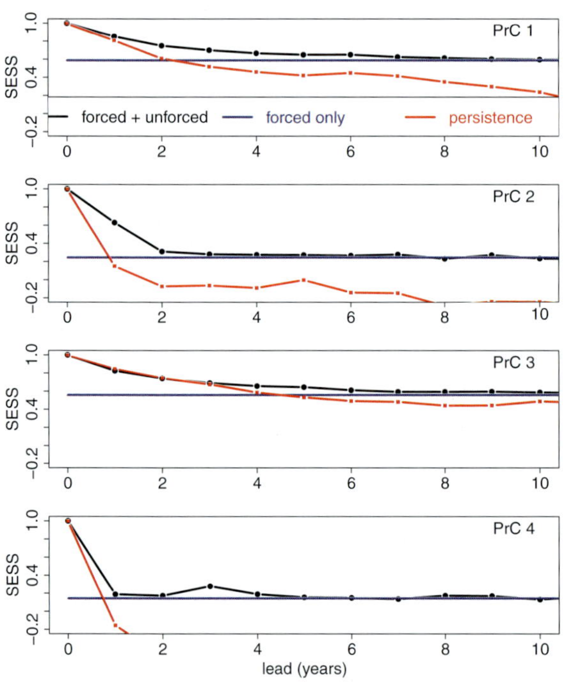

Fig. 8. Same as Fig. 7, except for predicting the first four predictable components obtained by maximizing APT.

The figure shows that the skill of each predictable component exceeds the skill of the forced component alone, and exceeds the skill of a persistence forecast. In the case of the third predictable component, the skill of a persistence forecast is nearly the same as that of our model, implying that the skill of this component comes primarily from persistence. In contrast, our forecast model predicts the other components with more skill than a persistence forecast. In fact, for the second and fourth predictable components, the skill of a persistence forecast is *negative*. This negative skill is consistent with our hypothesis that the predictability in the Pacific comes from the *interaction* of components rather than persistence.

6. Challenges in Multi-year Prediction

The above results are quite promising for multi-year prediction, but significant challenges remain. Note that the skill of predictions based on twentieth century simulations is artificially high because it includes observed volcanic influences, which generally are not predictable on multi-year time scales. Thus, the actual skill of forced predictability will be lower than that estimated here. Also, predicting the response to climate forcing requires accurate predictions of anthropogenic and natural forcing and accurate predictions of the response to such forcing. Neither of these requirements are trivial: the future concentration of greenhouse gases and aerosols depends on mankind's socio-economic development, and the response of the climate system depends on, among other things, how clouds respond to changes in aerosols and to a warmer climate. Predicting unforced components in the climate system requires solving an initial-value problem. However, the most important information for initializing decadal predictions is believed to be the subsurface ocean, which is poorly observed, especially prior to deployment of the Argo array in 2007 (Roemmich and Owens, 2000). Moreover, even if the sub-surface ocean were well observed, constructing initial conditions from these observations without also generating "initialization shock" or "climate drift" due to dynamical imbalances in the initial state has proven difficult. Finally, even if the sub-surface ocean were well observed and could be used to initialize dynamical models, different models imply widely different levels of predictability on decadal time scales (DelSole *et al.*, 2011; Branstator *et al.*, 2012).

Forced and unforced variability can be separated relatively cleanly using suitable climate simulations, but validating these components in the real climate system is challenging. For instance, if one's forced response is incorrect, then removing that (incorrect) response from observations produces a residual that will lead to incorrect conclusions about forced and unforced variability. Also, state-of-the-art climate models do not fully reproduce observed variations on multi-decadal time scales, and demonstrating that models simulate unforced variability with the correct magnitude, spatial structure, and temporal variation to explain all the discrepancies has proven difficult. Another problem is that forced and unforced variability are difficult to separate on short space and time scales due to low signal-to-noise ratios and similarity in spatio-temporal structure. For instance, Zhang *et al.* (2007) and Semenov *et al.* (2010) show that multi-decadal variations in the North Atlantic could force Northern Hemisphere multi-decadal changes in surface air temperature, which dominate the global changes and project strongly on forced changes.

Finally, the dynamics of the predictable components identified here still remain to be clarified. For instance, the precise reason for why predictability is larger in the North Atlantic rather than the North Pacific remains obscure to us. Despite these challenges, the results presented here provide compelling evidence that the actual climate system can be predicted on multi-year time scales.

Acknowledgments

TD and LJ were supported from grants from the NSF (ATM0830068, ATM0830062), the National Oceanic and Atmospheric Administration (NA09OAR4310058), the National Aeronautics and Space Administration (NNX09AN 50G), and the Department of Energy (ER64633, ER65095). MKT was supported by grants from the National Oceanic and Atmospheric Administration (NA05OAR4311004 and NA08OAR 4320912) and the Office of Navel Research (N00014-12-1-0911).

References

Barnston, A., M. K. Tippett, M. L. L'Heureux, S. Li, and D. G. DeWitt, 2012: Skill of real-time seasonal ENSO model predictions during 2002–11: Is our capability increasing? *Bull. Am. Meteor. Soc.*, **93**, ES48–ES50.

Blanchard-Wrigglesworth, E., K. C. Armour, C. M. Bitz, and E. DeWeaver, 2011: Persistence and inherent predictability of Arctic sea ice in a GCM ensemble and observations. *J. Climate*, **24**, 231–250.

Boer, G. J., 2004: Long time-scale potential predictability in an ensemble of coupled climate models. *Clim. Dyn.*, **23**, 29–44.

Boer, G. J., and S. J. Lambert, 2008: Multi-model decadal potential predictability of precipitation and temperature. *Geophys. Res. Lett.*, **35**, L05706, doi:10.1029/2008GL033 234.

Branstator, G., and H. Teng, 2010: Two limits of initial-value decadal predictability in a CGCM. *J. Climate*, **23**, 6292–6311.

Branstator, G., H. Teng, G. A. Meehl, M. Kimoto, J. R. Knight, M. Latif, and A. Rosati, 2012:

Systematic estimates of initial value decadal predictability for six AOGCMs. *J. Climate*, **25**, 1827–1846.

Broecker, W. S., and T.-H. Peng, 1982: *Tracers in the Sea*. Eldigio Press Lamont Doherty Geological Observatory, 690 pp.

Budikova, D., 2009: Role of Arctic sea ice in global atmospheric circulation: A review. *Global and Planetary Change*, **68**, 149–163.

Cohen, J., and D. Entekhabi, 2012: Eurasian snow cover variability and northern hemisphere climate predictability. *Geophys. Res. Lett.*, doi: 10.1029/1998GL900321.

Davis, R. E., 1976: Predictability of sea surface temperature and sea level pressure anomalies over the North Pacific Ocean. *J. Phys. Oceanogr.*, **6**, 249–266.

DelSole, T., 2001: Optimally persistent patterns in time-varying fields. *J. Atmos. Sci*, **58**, 1341–1356.

DelSole, T., and P. Chang, 2003: Predictable component analysis, canonical correlation analysis, and autoregressive models. *J. Atmos. Sci.*, **60**, 409–416.

DelSole, T., L. Jia, and M. K. Tippett, 2013: Decadal prediction of observed and simulated sea surface temperatures. *Geophys. Res. Lett.*, **40**, 2773–2778.

DelSole, T., and M. K. Tippett, 2007: Predictability: Recent insights from information theory. *Rev. Geophys.*, **45**, RG4002, doi:10.1029/2006RG000202.

DelSole, T., and M. K. Tippett, 2009a: Average predictability time: Part I. Theory. *J. Atmos. Sci*, **66**, 1172–1187.

DelSole, T., and M. K. Tippett, 2009b: Average predictability time: Part II: Seamless diagnosis of predictability on multiple time scales. *J. Atmos. Sci.*, **66**, 1188–1204.

DelSole, T., M. K. Tippett, and J. Shukla, 2011: A significant component of unforced multidecadal variability in the recent acceleration of global warming. *J. Climate*, **24**, 909–926.

Delworth, T. L., and M. E. Mann, 2000: Observed and simulated multidecadal variability in the Northern Hemisphere. *Clim. Dyn.*, **16**, 661–676.

Deser, C., R. Tomas, M. Alexander, and D. Lawrence, 2010: The seasonal atmospheric response to projected Arctic sea ice loss in the late twenty-first century. *J. Climate*, **23**, 333–351.

Griffies, S. M., and K. Bryan, 1997: A predictability study of simulated North Atlantic multidecal variability. *Clim. Dyn.*, **13**, 459–487.

Guo, Z., P. A. Dirmeyer, and T. DelSole, 2012: Land surface impacts on subseasonal and seasonal predictability. *Geophys. Res. Lett.*, **38**, doi:10.1029/2011GL049 945.

Hasselmann, K., 1993: Optimal fingerprints for the detection of time-dependent climate change. *J. Climate*, **6**, 1957–1971.

Hegerl, G. C., *et al.*, 2007: Understanding and attributing climate change. *Climate Change 2007: The Physical Science Basis. Contribution of Working Group I to the Fourth Assessment Report of the Intergovernmental Panel on Climate Change*, S. Solomon, D. Qin, M. Manning, Z. Chen, M. Marquis, K. Averyt, M. Tignor, and H. Miller, Eds., Cambridge University Press, 663–745.

Ho, C. K., E. Hawkins, L. Shaffrey, and F. M. Underwood, 2013: Statistical decadal predictions for sea surface temperatures: a benchmark for dynamical GCM predictions. *Clim. Dyn.*, **41**, 917–935.

Jia, L., and T. DelSole, 2011: Diagnosis of multiyear predictability on continental scales. *J. Climate*, **24**, 5108–5124.

Jia, L., and T. DelSole, 2012a: Multi-year predictability of temperature and precipitation identified in climate models. *Geophys. Res. Lett.*, **39**, doi:10.1029/2012GL052 778.

Jia, L., and T. DelSole, 2012b: Optimal determination of time-varying climate change signals. *J. Climate*, **25**, 7122–7137.

Keenlyside, N. S., M. Latif, J. Jungclaus, L. Kornblueh, and E. Roeckner, 2008: Advancing decadal-scale climate prediction in the North Atlantic sector. *Nature*, **453**, 84–88.

Latif, M., M. Collins, H. Pohlmann, and N. Keenlyside, 2006: A review of predictability studies of Atlantic sector climate on decadal time scales. *J. Climate*, **19**, 5971–5987.

Lau, N.-C., 1997: Interactions between global SST anomalies and the midlatitude atmospheric circulation. *Bull. Am. Meteor. Soc.*, **78**, 21–33.

Meehl, G. A., W. M. Washington, T. M. L. Wigley, J. M. Arblaster, and A. Dai, 2003: Solar and greenhouse gas forcing and climate response in the twentieth century. *J. Climate*, **16**, 426–444.

Newman, M., 2013: An empirical benchmark for decadal forecasts of global surface temperature anomalies. *J. Climate*, **26**, 5260–5269.

Philander, S. G. H., 1990: *El Niño, La Niña, and the Southern Oscillation*. Academic Press.

Pohlmann, H., J. H. Jungclaus, A. Kohl, D. Stammer, and J. Marotzke, 2009: Initializing decadal climate predictions with the GECCO oceanic synthesis: Effects on the North Atlantic. *J. Climate*, **22**, 3926–3938.

Roemmich, D., and W. B. Owens, 2000: The Argo project: Global ocean observations for understanding and prediction of climate variability. *Oceanography*, **13**, 45–50.

Schneider, T., and S. Griffies, 1999: A conceptual framework for predictability studies. *J. Climate*, **12**, 3133–3155.

Semenov, V. A., M. Latif, D. Dommenget, N. S. Keenlyside, A. Strehz, T. Martin, and W. Park, 2010: The impact of North Atlantic-Arctic multidecadal variability on northern hemisphere surface air temperature. *J. Climate*, **23**, 5668–5677.

Shukla, J., 1981: Dynamical predictability of monthly means. *Mon. Wea. Rev.*, **38**, 2547–2572.

Simmons, A. J., and A. Hollingsworth, 2002: Some aspects of the improvement in skill of numerical weather prediction. *Q. J. R. Meteorol. Soc.*, **128**, 647–677.

Smith, D. M., S. Cusack, A. W. Colman, C. K. Folland, G. R. Harris, and J. M. Murphy, 2008: Improved surface temperature prediction for the coming decade from a global climate model. *Science*, **317**, 796–799.

Stott, P. A., S. F. Tett, G. S. Jones, M. R. Allen, J. F. B. Mitchell, and G. J. Jenkins, 2000: External control of 20th century temperature by natural and anthropogenic forcings. *Science*, **290**, 2133–2137.

Taylor, K. E., R. J. Stouffer, and G. A. Meehl, 2012: An overview of CMIP5 and the experimental design. *Bull. Am. Meteor. Soc.*, **93**, 485–498.

Teng, H., and G. Branstator, 2011: Initial-value predictability of prominent modes of North Pacific subsurface temperature in a CGCM. *Clim. Dyn.*, **36**, 1813–1834.

Tietsche, S., J. J. Day, V. Guemas, W. J. Hurlin, S. P. E. Keeley, D. Matei, R. Msadek, M. Collins, and E. Hawkins, 2014: Seasonal to interannual Arctic sea ice predictability in current global climate models. *Geophys. Res. Lett.*, **41**, 1035–1043.

Tulloch, R., and J. Marshall, 2012: Exploring mechanisms of variability and predictability of atlantic meridional overturning circulation in two coupled climate models. *J. Climate*, **25**, 4067–4080.

van Oldenborgh, G. J., F. J. Doblas-Reyes, B. Wouters, and W. Hazeleger, 2012: Decadal prediction skill in a multi-model ensemble. *Clim. Dyn.*, **38**, 1263–1280.

Wunsch, C., and R. Ferrari, 2004: Vertical mixing, energy, and the general circulation of the oceans. *Annu. Rev. Fluid Mech.*, Annual Reviews, 281–314.

Yang, X., and T. DelSole, 2012: Systematic comparison of ENSO teleconnection patterns between models and observations. *J. Climate*, **25**, 425–446.

Yang, X., *et al.*, 2013: A predictable AMO-like pattern in GFDL's fully-coupled ensemble initialization and decadal forecasting system. *J. Climate*, **26**, 650–661.

Zanna, L., 2012: Forecast skill and predictability of observed North Atlantic sea surface temperatures. *J. Climate*, **25**, 5047–5056.

Zhang, R., T. L. Delworth, and I. M. Held, 2007: Can the Atlantic Ocean drive the observed multidecadal variability in the Northern Hemisphere mean temperature? *Geophys. Res. Lett.*, **34**, doi:10.1029/2006GL028 683.

Chapter 15

Decadal Hydroclimate Variability Across the Americas

Richard Seager*
Lamont Doherty Earth Observatory of Columbia University, Palisades, NY 10964, USA
seager@ldeo.columbia.edu

Decadal hydroclimate variability in North America and tropical and extratropical South America is analyzed and possible mechanisms for its origin discussed. Focus is on southwestern North America (including Mexico) and the Great Plains, the northeast United States, northeast Brazil and southeastern South America. The varying roles of ocean forcing, internal atmospheric variability and radiatively-forced hydroclimate change are analyzed. In some regions such as southwest North America and the Plains, and northeast Brazil, decadal variations of hydroclimate are quite well understood and can be attributed to variations in tropical Pacific and tropical North Atlantic sea surface temperatures. The mechanisms of tropical ocean influence are reviewed and a case is made that the precipitation anomalies across the Americas associated with the so-called Pacific Decadal Oscillation (PDO) are essentially the same as those associated with the El Niño-Southern Oscillation and also derive from the tropical component of the PDO SST anomalies. In other regions, such as the north east United States, strong decadal timescale variations are present but cannot be explained in terms of ocean forcing. Both there and in southeast South America decadal variations cannot easily be distinguished from secular wetting trends given the length of the observational record. Finally, it is shown that across much of the Americas near term future radiatively-forced precipitation change will be of the amplitude of historical decadal precipitation variability indicating an important and predictable change.

1. Introduction

Decadal variations of hydroclimate across the Americas have left a substantial mark on the social, economic and agricultural character of the continent. For example, in perhaps the most famous example, the Colorado River Pact of 1922 apportioned river flow between the upper basin and lower basin states based on recorded flows in the immediately preceding decades. However these decades happened to coincide with the early twentieth century North American pluvial (Fye *et al.*, 2003; Cook *et al.*, 2011) and the flows of 22 billion cubic meters (BCM) per year were, as we now know, about the highest to occur in the past millennium, way in excess of both the long term 20th Century average of 18.6 BCM and the tree ring reconstructed average flow over the last 500 years of only 16.7 BCM (Christensen and Lettenmaier, 2007; Meko *et al.*, 2007). Seven U.S. states and Mexico are still trying to grapple with this coincidence of decision making and the quirks of climate variability. Later, the tremendous population growth of the southwest U.S. occurred within the decades of the 1980s and 1990s and was no doubt made easier by the fact that this was another wetter than normal (pluvial) period

Climate Change: Multidecadal and Beyond
edited by Chih-Pei Chang et al.

(Seager *et al.*, 2005b; Swetnam and Betancourt, 1998). On the dry side, the 1930s Dust Bowl drought forced millions to leave their homes, an exodus from the Plains states and a permanent shift in the agricultural, social and economic structure of the Plains and the wider U.S. (Worster, 1979). Further back, the 1890s drought was a defining event in the decision to create the Bureau of Reclamation and Federalize the development of the West (Reisner, 1986). At the other end of the Americas a shift in the last few decades to a wetter climate has enabled a vast expansion of agriculture in the subhumid to semi-arid Pampas of southeast South America (Seager *et al.*, 2010a) with tremendous benefit to local agricultural interests and the global food supply.

While it is remarkably easy to draw convincing connections between decadal hydroclimate variability and matters of social import and interest that effect millions, it remains almost impossible to predict how hydroclimate will evolve in coming decades due to natural causes. That may be because it is truly unpredictable. However, while the mechanisms of atmospheric response to SST variations are reasonably well understood, the ocean mechanisms of decadal variability remain poorly understood, and efforts in decadal prediction, while no longer embryonic, are still in their infancy. Hence it might be rash at this point to conclude that decadal variability is unpredictable. Regardless, at this point our ability to predict hydroclimate change in coming decades relies almost entirely on the response to increasing radiative forcing which in some regions of the Americas is likely to quickly become a serious matter (*e.g.* Seager *et al.* (2007, 2012b); Vano *et al.* (2013)). On the other hand the amplitude of projected human-induced hydroclimate change in the coming decades is not expected to be outside the range of natural variability (Seager *et al.*, 2012b). Of course, unlike natural decadal variability, human-induced change is uni-directional as long as the responsible changes in forcing (*e.g.* greenhouse gases) do not reverse in sign. Nonetheless, this does imply that for the coming decades probabilistic projections of future hydroclimate need to account for both forced change and evolving natural variability. Which brings us back to the character and cause of natural decadal variability.

In the absence of any proven ability to predict the evolution of naturally occurring modes of hydroclimate variability, the very least we can do is to better understand the nature of this phenomena and reliably attribute past change between natural variability and forced change as one means, together with model projections, of providing guidance, in a probabilistic manner, as to future hydroclimate in key regions of concern. This is what we will attempt to do here. The regions we will focus on are southwestern North America and the Plains, the northeast U.S., northeast Brazil and southeast South America which span a range of tropical to midlatitude regions, semi-arid to humid climates and states of understanding. While recognizing that droughts and pluvials are composed of seasonally varying anomalies (Seager *et al.*, 2005b; Mo and Berberry, 2011) that can have different forcing mechanisms here, except for Northeast Brazil, in the interest of simplicity and brevity, we only analyze annual mean anomalies. In all cases we aim to illustrate the character and causes of decadal hydroclimate variability across the Americas using nothing more than a simple set of time series and maps. This deliberately simple approach is based on the idea that, to be worth worrying about, decadal variability of hydroclimate must be visible in the raw climate record and evident in the personal experiences of anyone living through it.

2. Southwestern North America; A Reasonably Well Understood Example of Strong Oceanic Control

We begin with probably the most studied region, that of southwestern North America which for our purposes includes the Plains and regions to the West from central Mexico to the latitude of the California-Oregon border (95–125°W, 25–40°N). Considerable work to date has shown that the hydroclimate history of these regions on the interannual to decadal timescales can be quite well reproduced by atmosphere models forced by observed SSTs (Schubert *et al.*, 2004b; Huang *et al.*, 2005; Seager *et al.*, 2005b; Herweijer *et al.*, 2006). Here we make this same point again but take the opportunity to show a comparison of soil moisture as simulated by the atmosphere model against that simulated by a land hydrology model forced by the observed history of air temperature and precipitation.

The land hydrology model is the Variable Infiltration Capacity (VIC) model of Liang *et al.* (1994) and covers 1915 to 2003 with forcing meteorological data as described in Hamlet and Lettenmaier (2005). The atmosphere model is a 16 member ensemble of simulations with the National Center for Atmospheric Research (NCAR) Community Climate Model 3 (CCM3) (Kiehl *et al.*, 1998). In the first configuration (Global Ocean Global Atmosphere — GOGA) the model is forced by Kaplan *et al.* (1998) SSTs in the tropical Pacific and Hadley Centre SSTs elsewhere (Rayner *et al.*, 2003). In the second configuration (tropical Pacific Ocean Global Atmosphere — POGA-ML) the model is forced only by tropical Pacific Ocean observed SSTs and is coupled to a mixed layer ocean which computes the SST elsewhere. The reader is referred to Seager *et al.* (2005b) for more details and explanation of these choices. Both simulations cover 1856 to 2011.

Figure 1 shows that, for the common period, the GOGA modeled soil moisture tracks the VIC observationally-based estimates remarkably well including the early to mid 1950s drought, the overall wet conditions in the 1980s and 1990s and the shift back to drier conditions following the 1997/98 El Niño. That this is so indicates that hydroclimate in southwest North America is strongly influenced by variations of SSTs.

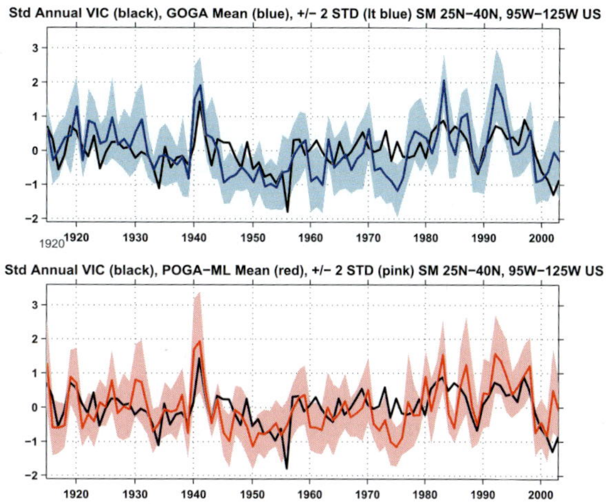

Fig. 1. Top, the soil moisture anomaly in southwestern North America as computed by the VIC model forced by observed meteorology (black) and that computed by the global climate model forced by observed global SSTs (GOGA configuration) shown as the 16 member ensemble mean (blue) and the plus and minus two standard deviation spread of the ensemble (blue shading). Bottom, same as top but with the global climate model forced by only tropical Pacific SSTs (POGA-ML configuration).

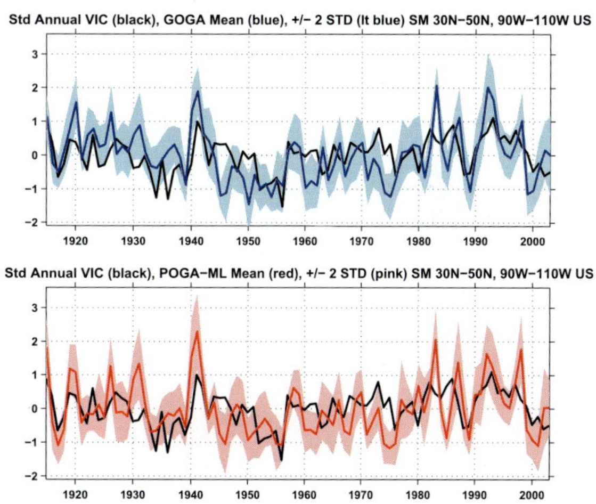

Fig. 2. Same as Figure 1 but for the Great Plains region.

The correlation coefficient between the observed and modeled time series is 0.52 indicating that a quarter of the variance in the former is controlled by ocean variability. Figure 2 shows the same comparison for the Great Plains (30–50°N,

110–90°W) where the GOGA model tracks VIC soil moisture with only slightly less skill than in the southwest and with both regions sharing many of the same decadal scale droughts and pluvials. For both regions the lower panels show the POGA-ML simulations which also have high skill and indicate that the part of the global SST forcing that matters most in the GOGA simulations is the tropical Pacific Ocean. Note that forcing by SST variability can create trends in precipitation of multidecadal length and that, therefore, assuming that late twentieth century trends, or even trends over the whole twentieth century, will necessarily represent radiatively-forced hydroclimate change are almost certainly misguided.

The driving role of the tropical Pacific is further illustrated in Fig. 3 where the regression

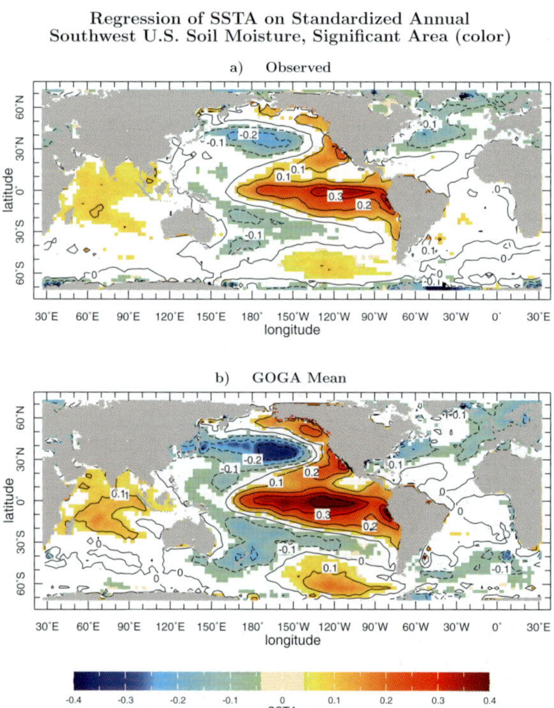

Fig. 3. The regression of annual mean global SST anomaly on the VIC (top) and GOGA modeled (bottom) annual mean soil moisture anomaly in southwest North America. Areas of significance at the 5% level are colored.

of global SST anomalies on annual mean southwest North America soil moisture is shown for observations (with VIC soil moisture) and the GOGA model. Colors are only shown where the regression is significant at the 5% level. The association of wet (dry) conditions with warm (cold) tropical Pacific SSTs is clearly seen. The relationship is stronger in the model due to the isolation in the ensemble mean of the SST forced component. These relations between soil moisture and SSTs are weaker than those between precipitation and SSTs (see Seager *et al.* (2005b)) but still easily retain significance indicating a clear association between the tropical Pacific Ocean and land surface hydrological conditions. While model simulations are particularly useful for identifying cause and effect, observational studies also indicate the key driving role of the Pacific on Plains and southwest hydroclimate variability (Nigam *et al.*, 1999; Barlow *et al.*, 2001; Ting and Wang, 1997).

Although not evident from Figs. 1 through 3, dry conditions in the southwest and Plains are also associated with warm tropical Atlantic SST anomalies and model studies have shown these to play a generally secondary role relative to the tropical Pacific anomalies but are important for generating the 1930s and 1950s droughts (Schubert *et al.*, 2004; Seager, 2007; Seager *et al.*, 2008). Observational analyses also support the idea that warm phases of Atlantic multidecadal variability favor drought over the Plains (Nigam *et al.*, 2011). The mechanisms of the tropical Atlantic-North America precipitation link are detailed in Kushnir *et al.* (2010) and it appears that it is the tropical SST anomalies in the Atlantic that are most important for driving the response (see as well Schubert *et al.* (2004) and Sutton and Hodson (2005)).

To make the links between tropical SSTs and decadal hydroclimate variations over North America more clear we show in Fig. 4 the observed SST anomaly and the observed and

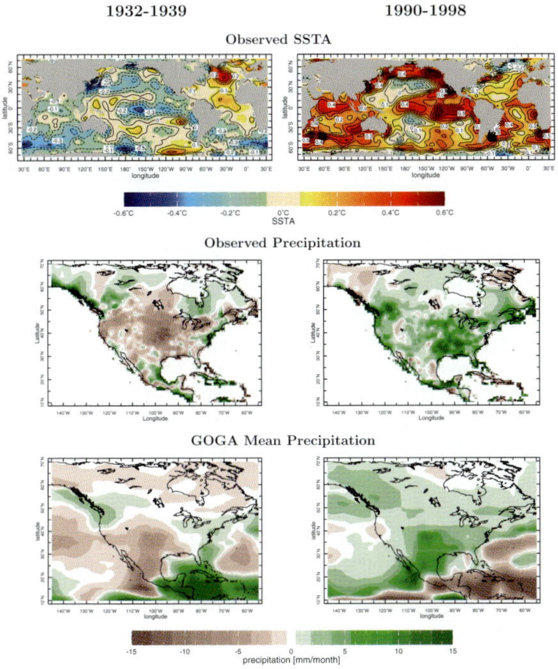

Fig. 4. The observed SST (top) and precipitation (middle) and modeled precipitation (bottom) anomalies for the 1932-39 Dust Bowl period (left column) and 1990-98 pluvial period (right column) in North America.

GOGA modeled precipitation over North America for the dry decade of 1932 to 1939 and the wet decade of 1990 to 1998. The observed precipitation is the gridded data from the University of East Anglia Climatic Research Unit (CRU, http://www.cru.uea.ac.uk/data/). Much like the regression derived pattern, which accounted for both interannual and longer timescales, these decadal timescale maps emphasize the role of the tropical Pacific: the cold SSTs in the 1930s and the warm SSTs in the 1990s. The warm tropical North Atlantic, partly responsible for the Dust Bowl drought, is also seen in the 1930s. The degree to which the 1930s and 1990s were the opposite of each other in SST and precipitation anomalies (apart from a general warming of the oceans) is quite striking. The model precipitation anomalies also reasonably reproduce the observed patterns, emphasizing the extent of oceanic control on precipitation anomalies at these timescales.

However it must be noted that the modeled Dust Bowl drought is centered in the southern Plains and northern Mexico whereas the observed drought was centered in the Central to northern Plains. The model drought pattern is typical of droughts forced by cool tropical Pacific and warm tropical North Atlantic SST anomalies and is, for example, very similar to the modeled (and observed) pattern of the 1950s drought. The more northern-centered Dust Bowl drought was quite unusual (Seager *et al.*, 2008) but this drought was also unusual in that it was the only one that was associated with widespread soil erosion and dust storms which have been attributed to the post World War I expansion of agriculture into the region and the farming practices of the time (Worster, 1979; Hansen and Libecap, 2004). Recently, Cook *et al.* (2008, 2009, 2010) have performed simulations of the Dust Bowl period with imposed SSTs, loss of crop cover and exposure of bare earth (in areas as indicated by contemporary Soil Conservation Service surveys). The model had an interactive dust module that simulates the emission of dust, its transport, interaction with radiation and ultimate deposition. That work found that the dust storms reflect solar radiation and induce an anomalous radiative sink which the atmosphere balances by subsidence and low level divergence, suppressing precipitation. This intensifies the drought in the regions of dust aerosol loading which also means the model drought shifts its center from the southwest into the central Plains better matching the observed pattern. Comparison of modeled circulation anomalies to those in the 20th Century Reanalysis (Compo *et al.*, 2010) shows that it is only when active dust storms are included that the model can catch the observed continental scale subsidence that occurred during the Dust Bowl.

If the dust storms were important to the character and intensity of the Dust Bowl drought then the drought was not just a natural disaster but one for which human activity was in part, however unwittingly, responsible. In contrast, Hoerling *et al.* (2009) have argued that the SST forcing of the Dust Bowl drought was weak and that the strength arose from internal atmospheric variability unconnected to oceanic conditions. Of course, it is likely that some combination of SST forcing, internal atmospheric variability and dust aerosol feedbacks on both of these was ultimately responsible. Future research may be able to unravel the relative roles of each which will probably require a more fine grained approach than has been attempted to date, looking at individual seasons and years in the Dirty Thirties.

3. North Eastern North America: A Poorly Understood Example of Strong Multidecadal Variability

Northeastern North America is a wet region of the continent and does not experience the frequent multiyear droughts that the sub-humid to semi-arid regions of the Plains and southwest do. However hydroclimate has undergone serious decadal to mulitidecadal fluctuations over the past century. In a recent paper Seager *et al.* (2012b) examined the precipitation history of the Catskill Mountains region (where 87% of New York City's water comes from) which is representative of a wider northeast U.S. region. Here the major features of the last century are a severe multiyear drought between 1962 and 1966 followed by an early 1970s abrupt transition to a wetter climate that has persisted to the present day.

These are clearly seen in the observed precipitation data as well as in the VIC soil moisture data (Fig. 5). The amplitude of the multidecadal variability in this region is truly impressive: in fall the precipitation has increased by about

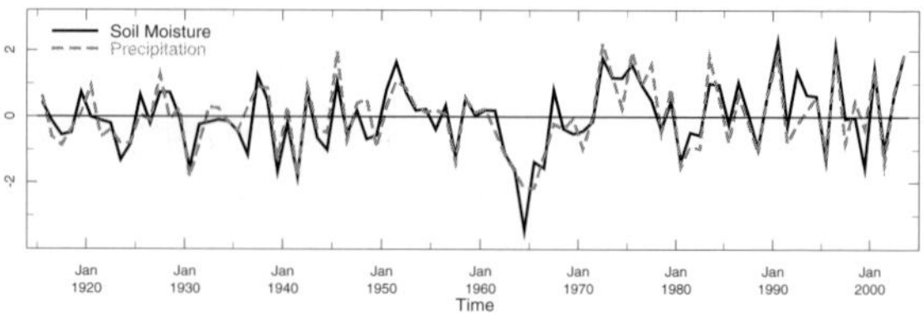

Fig. 5. The observed precipitation and soil moisture (as modeled by the VIC land surface hydrology model forced by observed precipitation and temperature) in the Catskill Mountains region of the northeast United States which provides 87% of the New York City water supply. The unique 1960s drought and the subsequent shift in the early 1970s to a wetter climate are clear with the latter being an unexplained mystery.

25% over the last century. None of this observed hydroclimate history (precipitation and soil moisture) is reproduced in atmosphere models forced by observed SSTs indicating that, unlike for western North America, the observed history has not been strongly influenced by the past variations of the oceans (Seager *et al.*, 2012b). Further, while northeast North America is expected to get wetter in response to rising greenhouse gases (Hayhoe *et al.*, 2007), the wetting trend in the region up to 2007 is an order of magnitude greater than that which the models participating in the Coupled Model Intercomparison Project 3 (CMIP3), and assessed by Intergovernmental Panel on Climate Change Assessment Report 4 (IPCC AR4), indicate should have occurred in response to changes in radiative forcing. The CMIP5 model simulations do not alter this disagreement.

The inability to simulate the past century of northeast precipitation variability in terms of either ocean or radiative forcing suggests that it instead arose from internal and natural atmospheric variability. The 1960s drought was diagnosed by Seager *et al.* (2012b) to have arisen from circulation anomalies that persistently placed northerly descending air over the region with a role for the highly negative North Atlantic Oscillation (NAO) during these years.

In regard to the subsequent shift to a wetter climate — pluvial — Seager *et al.* (2012b) provide some tantalizing evidence that this was associated with a strengthening of the mid-latitude storm track from the east Pacific across North America to the Atlantic that occurred quite abruptly in the early 1970s. This storm track strengthening was identified earlier by Chang and Fu (2002) and Harnik and Chang (2003) in reanalyses and radiosonde observations and Seager *et al.* (2012b) identified it again in the 20th Century reanalysis (Compo *et al.*, 2010) which only assimilates surface pressure data while being SST forced. After the early 1970s storm track strengthening it remained stronger until the present. This storm track behavior appears real in that it is seen in reanalyses, radiosonde observations and also surface pressure data but does not align with any well known climate transitions such as the 1976/77 tropical Pacific climate shift (Zhang *et al.*, 1997), trends in the NAO (Hurrell, 1995) or the Atlantic Multidecadal Oscillation (Kushnir, 1994; Ting *et al.*, 2009). It has also received very little attention and appears almost like an embarrassing uncle in the climate family. However its reality, impacts and causes deserves more attention. It might be that the mid-latitude atmosphere on its own, absent the influence of the ocean and changes in composition, can generate variability on decadal and longer timescales that, viewed from our short period of instrumental observations, appear as 'climate change'. Another clue is that, according to a recent tree ring reconstruction of Palmer Drought Severity Index in the nearby Hudson River Valley region, the 1960s drought was a brief interruption in a multi century trend towards a wetter climate (Pederson *et al.*, 2013) that is also found across the eastern U.S. (Seager *et al.*, 2009; Pederson *et al.*, 2012). However it is not known if these changes are caused by centennial or millennial variability, or if the wetting trend has gone on for millennia and is potentially of orbital origin.

4. Northeast Brazil or 'Nordeste': A Very Well Understood Case of Strong Ocean Control

Northeast Brazil (or the 'Nordeste' as it is known in Brazil) is a climate oddity: a semi-arid equatorial region placed to the west of a warm tropical ocean and under the influence of easterly trade winds. The Nordeste sits between the Atlantic Intertropical Convergence Zone (ITCZ) to its north and the South American Monsoon and South Atlantic Convergence Zone to its southwest. As such it has a climatic affinity to dry regions of the equatorial West Pacific Ocean but why this oceanic character extends over land into the Nordeste is not clear. (Another region of aridity to the west of a tropical ocean exists in East Africa but appears to have different causes (Yang *et al.*, 2015).) The aridity of the Nordeste, together with an endless cycle of disrupting climate anomalies, has long ensured that the Nordeste is one of the poorest regions in Brazil and the source of migrants to wealthier regions. A harrowing description of the late 1870s drought and famine in the Nordeste is given by Davis (2001). It is estimated that about 500,000 people died then and that between the 1930s and 1950s 1.5 million inhabitants fled the region with drought as a prime, but not sole, cause (Siegel, 1971).

The essential dynamics of Nordeste rainfall variability were deduced from observations by Hastenrath and Heller (1977) and, during drought years, involve an anomalous cross equatorial SST gradient with the warm anomaly to the north, and/or cold anomaly to the south, and a northward displaced ITCZ that prevents rains from reaching their southernmost latitude over the Nordeste in southern hemisphere summer to fall. This relation was then modeled successfully with a GCM by Moura and Shukla (1981). El Niño events also create a tendency to dry conditions in the Nordeste, as noted early on by Covey and Hastenrath

(1978), which occurs both by warming the tropical troposphere creating widespread stability, by altering the Walker Circulation and, also, by remotely warming the tropical North Atlantic Ocean and displacing the ITCZ northwards (Chiang *et al.*, 2002; Giannini *et al.*, 2004; Hastenrath, 2006).

Giannini *et al.* (2004) showed that the oceanic control on Nordeste precipitation is sufficiently strong that an atmosphere model forced by observed SSTs was able to accurately reproduce the variability of March–April–May (MAM, the Nordeste wet season) precipitation in the 1950 to 1994 period. We expand upon this by showing, in the top panel of Fig. 6, the modeled (with global SST forcing) and CRU observed MAM precipitation history for the much longer period of 1901 to 2008. In a departure from our usual plotting practice, we show both the ensemble mean and the individual ensemble members, which is made visually tractable here because of the strong correlation between ensemble members. The tight relationship between model and observed precipitation from 1950 to 1994 noted by Giannini *et al.* (2004) extends up through the last two decades and also through the first half of the 20th Century. To our knowledge this is the first presentation of an SST-forced model simulation of Nordeste precipitation during the first half of the last century. Many observed strong precipitation departures (*e.g.* 1932) are reproduced in almost all ensemble members indicating a quite high SST-forcing-signal-to-internal-atmospheric-variability-noise ratio. The lower panel in Fig. 6 shows the same comparison but for the tropical Pacific SST-only forced ensemble and makes clear that much of the good agreement in the case with global SST-forcing derives from SST anomalies outside the tropical Pacific Ocean.

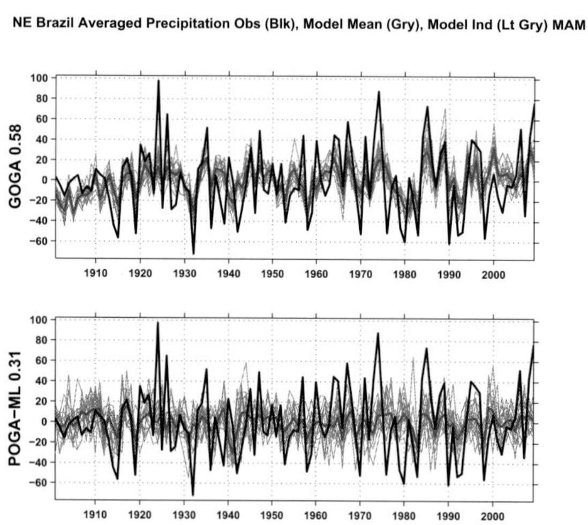

Fig. 6. The observed (solid black line) precipitation anomaly for the March–April–May season in Northeast Brazil and that modeled with global SST forcing (GOGA, top) and tropical Pacific only SST forcing (POGA-ML, bottom) shown as both the ensemble mean (thick gray line) and the 16 individual ensemble members (thin gray lines). Correlation coefficients between the observed and ensemble mean time series are shown at left.

Figure 7 shows the regression of global SST onto MAM Nordeste precipitation for the 1900 to 2008 period with colors applied where the relation is statistically significant for both the observations (top) and the GOGA model (bottom). There actually is a significant relationship between dry conditions and El Niños in the Pacific which is reliably simulated by the model. In addition, the association between dry conditions and warm SSTs in the tropical north Atlantic and/or cold SSTs south of the mean ITCZ is also very clear and this is again well represented by the model.

The Nordeste is obviously a region of strong interannual precipitation variability but much longer timescale variations exists too. For example the early to mid 1920s appear to have been a wet period which was immediately followed by a dry period culminating in the 1932 drought while the mid 1970s to mid 1980s were an extended dry period. Since the early 20th Century has not received the attention of the later period here we show in Fig. 8 the SST anomalies and observed and modeled precipitation anomalies for the MAM seasons of the 1920 to 1926 wet

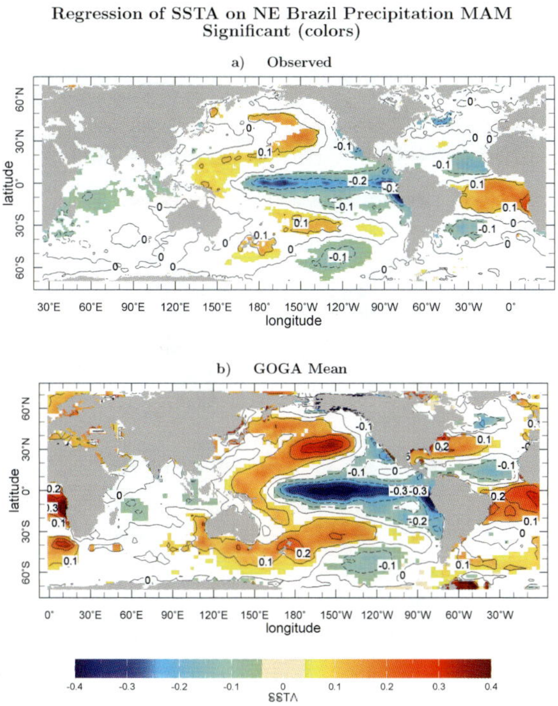

Fig. 7. The regression of March–April–May global SST anomaly on the observed (top) and GOGA modeled (bottom) March–April–May precipitation anomaly in Northeast Brazil. Areas of significance at the 5% level are colored.

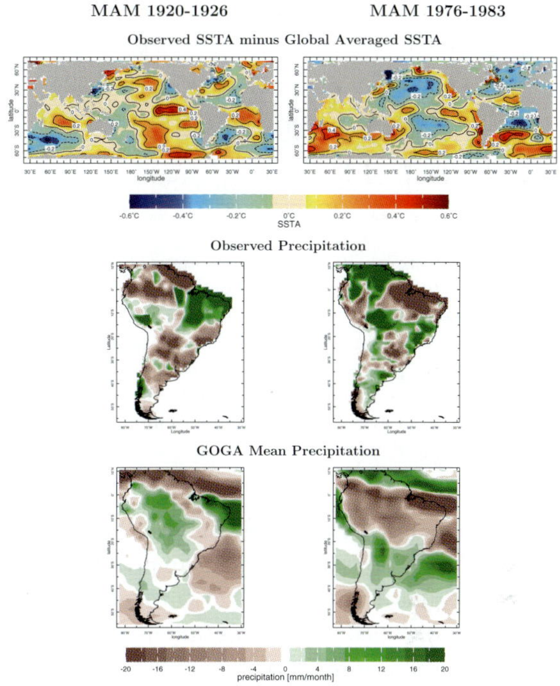

Fig. 8. The observed SST (top) and precipitation (middle) and modeled precipitation (bottom) anomalies for the 1920 to 1926 wet period (left column) and 1976 to 1983 dry period (right column) in Northeast Brazil.

period and contrast this with the 1976 to 1983 dry period. The observed decadal patterns of precipitation anomalies across South America, with wet in the Nordeste in the earlier period and dry in the later period, are reproduced with some degree of success in the global SST forced model. The 1920-26 period was perhaps weakly El Niño-like, which would not be expected to cause wet conditions in the Nordeste, while the tropical Pacific anomalies were indistinct in the later period. Instead it appears that the tropical Atlantic was the dominate control on these decadal Nordeste precipitation anomalies with opposite signed north-south cross equatorial SST anomalies in the two periods consistent with precipitation-SST relationships expected on the basis of the regression analysis.

5. Southeast South America: A Mixed Message of Strong Oceanic Forcing and a Mysterious Long Term Wetting Trend

Our final example is the southeast of South America, a region that is influenced by both mid-latitude dynamics, the southern part of the South American monsoon and the South Atlantic Convergence Zone. This region has become tremendously important to the global food production system in recent years as the growing of grains and legumes has expanded into the sub-humid to semi-arid Pampas in recent decades to take advantage of a long term wetting trend (Seager *et al.*, 2010a). This is also a region where the tropical oceans exert a considerable influence on interannual and decadal timescale

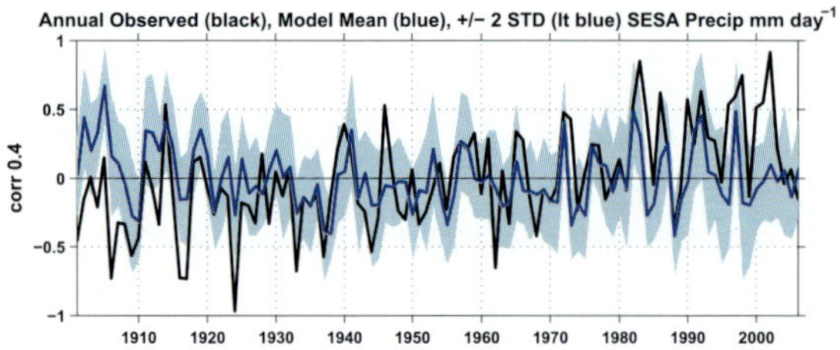

Fig. 9. The observed precipitation anomaly for southeastern South America and that computed by the global climate model forced by observed global SSTs (GOGA configuration) shown as the 16 member ensemble mean (blue) and the plus and minus two standard deviation spread of the ensemble (blue shading). The correlation coefficient between the observed and ensemble mean time series is shown at left.

fluctuations in precipitation. Figure 9 shows the observed variations of annual mean precipitation for the region from 1901 to 2010 together with the ensemble mean and plus/minus two standard deviation spread of the model forced by global SSTs. The agreement between the observations and ensemble mean is impressive and, as shown in Seager *et al.* (2010a), does not derive from the tropical Pacific forcing alone, even though El Niño events are typically wet in the region and La Niña events dry (see Seager *et al.* (2010a) and Mo and Berberry (2011) and references therein). Seager *et al.* (2010a) in fact argue that the multidecadal precipitation variability in southeast South America precipitation is influenced by multidecadal tropical North Atlantic SST variability. Their argument is that warm tropical North Atlantic SSTs cause an intensified Atlantic ITCZ with anomalous divergent winds aloft which interact with the mean flow vorticity field to force a remote response that includes subsidence, and precipitation suppression, over southeast South America. This mechanism draws some support from the modeling experiments examined by Mo and Berberry (2011). Thus much of the decadal precipitation variability in this region can be explained in terms of a combination of tropical Pacific and tropical north Atlantic SST variability, rather like for the Nordeste.

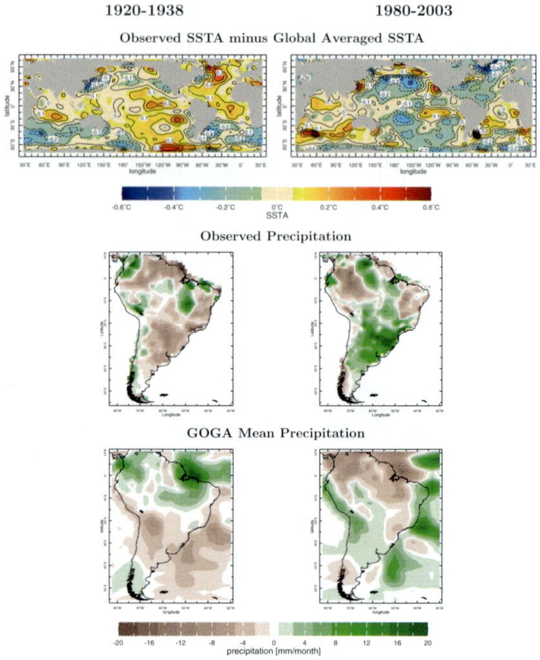

Fig. 10. The observed SST with global mean removed (top) and precipitation (middle) and modeled precipitation (bottom) anomalies for the 1920 to 1938 dry period (left column) and 1980 to 2003 wet period (right column) in southeast South America.

However we cannot explain the long term wetting trend easily in terms of tropical SSTs. Figure 10 shows the SST anomalies and observed and modeled precipitation anomalies for the 1920 to 1938 period of a drier climate and 1980 to 2003 period of a wetter climate in southeast South America. The model does a good job of reproducing the early 20th century drought in southeast South America but can only produce the sign, not the amplitude, of the late 20th century pluvial. In the earlier period the tropical Pacific was actually weakly warm, which is not expected to go along with dry in southeast South America, while the later period has only weak tropical Pacific SST anomalies. However the tropical North Atlantic was warm in the earlier period and cold in the later period which would be expected to go along with dry and then wet in southeast South America according to Seager *et al.* (2010a) and Mo and Berberry (2011). That is, on the decadal timescale, the tropical North Atlantic appears to exert a strong influence on southeast South America precipitation.

Despite this evidence of ocean influence, the late 20th century was much wetter than can be explained in terms of SST forcing alone. There is no model-based evidence that such a strong long term wetting trend in the region is a response to rising greenhouse gases (Seager *et al.*, 2010a; Gonzalez *et al.*, 2014). On the other hand model evidence does suggest that ozone depletion caused a wetting in the region over the late 20th Century that can explain perhaps a quarter of the observed trend (Gonzalez *et al.*, 2014). Even so, that would leave the majority of the wetting trend unexplained and, as for the northeast U.S., perhaps it is evidence of very long timescale natural variability that arises from processes, perhaps atmospheric, unrelated to SST variability.

6. Mechanisms of Decadal Hydroclimate Variability Across the Americas

The dynamics that link ENSO SST anomalies to North American hydroclimate anomalies have become increasingly well understood in recent years. Tropical Pacific SST anomalies force convective heating anomalies above with upper level divergence anomalies that force Rossby waves that propagate north and east towards North America. The waves alter the mean flow which then alters the propagation of transient eddies whose altered momentum fluxes then feedback onto the mean flow. The final transient and mean flow response to ENSO SST anomalies is, therefore, a result of the mean flow anomalies forced directly by tropical heating anomalies and the subsequent eddy-mean flow interaction it brings into play (Seager *et al.*, 2010; Harnik *et al.*, 2010). How these circulation anomalies relate to the moisture divergence anomalies that force precipitation anomalies has also been studied and it appears that the mean flow divergence anomalies are the most important and themselves dominated by the change due to the circulation change (as opposed to the humidity change). Transient eddy moisture convergence anomalies appear to play a secondary role (Seager *et al.*, 2005a; Seager and Naik, 2012). (This work could, however, be done with the most up to date atmospheric reanalyses and in more detail in order to clearly determine the causes of ENSO-forced hydroclimate anomalies over the Americas.) Also, the dominance of mean flow moisture divergence anomalies over transient eddy moisture flux divergence anomalies does not suggest that transient eddies play a secondary role overall. This is because transient momentum flux anomalies are themselves a main driver of the mean flow circulation anomalies. A breakdown of ENSO hydroclimate anomalies into components due to mean and transient circulations that takes account of this coupling has not been attempted (perhaps for good reason!).

It has been possible to use the increased understanding of the ENSO teleconnection to hydroclimate in the Americas to develop improved understand of tropical Pacific forcing of decadal hydroclimate. For the case of persistent North American droughts, such as the 1930s Dust Bowl, the 1950s southwest drought, the 1998-2002 turn-of-the-century drought, the 1856-65 'Civil War' drought and the 1870s and 1890s western droughts, the causes were essentially the same as those during a La Niña season. That is because in all these six cases the tropical Pacific adopted an extended (though weak) La Niña like state with no El Niño events. Similarly, the wet North American decade of the 1990s went along with persistent El Niño-like conditions between 1990 and 1998 except for a weak cold interval in winter 1996/97. Huang *et al.* (2005) extended this reasoning to show that the overall wetter climate in North America and extratropical South America between 1977 and 1998, relative to the 15 years before 1976, could be reproduced within an atmosphere model forced by the tropical Pacific SST difference between the two periods. This was so whether the time average SST difference was imposed or whether the atmosphere model saw the actual time history of SST indicating the essential linearity of the hydroclimate response. That is, the wet northern and southern mid-latitudes in the post 1976 two decades can be understood as a response to the change in the long term decadal mean tropical Pacific SSTs and is not a rectified effect of the difference in El Niño and La Niña events between the decades before and after 1976 or influenced by any non-linearity between SST anomalies and the precipitation response. This was a significant work in advancing and simplifying our understanding of decadal variations of hydroclimate across the Americas.

The longer timescales of Atlantic Ocean SST variability make interpretation of Atlantic forcing of hydroclimate more straightforward. Going back to Schubert *et al.* (2004), it has been recognized that it is SST anomalies in the tropical North Atlantic that exert an influence on North American hydroclimate. More recently Kushnir *et al.* (2010) have shown that, in the summer season, the response is essentially that given by the Gill-model solution to a subtropical heating anomaly: warm SSTs induce deep atmospheric heating and poleward and ascending flow above with equatorward and descending air on the western flank of the forced cyclonic anomaly. The scale is set by Rossby wave dispersion and is such that the equatorward, descending air is over the southern Plains and southwest where precipitation is suppressed. In contrast winter season tropical North Atlantic SST anomalies appear to first impact tropospheric temperatures and moist static stability and then tropical Pacific precipitation which then forces a wave response, akin to that during ENSO, that reaches North America. This winter response has a certain Rube Goldberg character but does appear robust across models, has some observational support and may indeed be one means of inter-basin coupling of relevance to hydroclimate variations in the Americas (Kushnir *et al.*, 2010).

However ENSO is possibly not the only thing that happens in the Pacific Ocean. A decade and a half ago Zhang *et al.* (1997) showed that, in addition to the ENSO cycle with its characteristic irregular, interannual period, there is a longer timescale 'ENSO-like' pattern of variability. This became known as the Pacific Decadal Oscillation (PDO, *e.g.* Mantua and Hare (2002)) and is different from interannual ENSO in that its North Pacific SST anomalies are larger relative to its tropical SST anomalies, the tropical SST anomalies are more meridionally broad and it has a timescale of a few decades. Famously the Pacific Ocean shifted from a phase of the PDO with cold tropical Pacific SST anomalies to the phase with warm anomalies (and opposite signed anomalies in the central North Pacific Ocean) during 1976/77. This is the shift whose associated precipitation anomalies were modeled by Huang *et al.* (2005). Shifts to the cold phase occurred in the early to mid 1940s (Deser *et al.*, 2004) and, seemingly, in 1997/98. The problem with the PDO is that it is so highly correlated to ENSO, and shares the same hemispheric symmetry (Garreaud and Battisti, 1999), that it is not clear yet whether it is anything more than a basin-wide low frequency response to tropical Pacific conditions.

Here we put aside the matter of the origin of the PDO and restrict attention to the precipitation anomalies across the Americas associated with its characteristic SST pattern. The PDO is, as usual, defined as the first empirical orthogonal function of monthly Pacific Ocean SST anomalies north of 20°N after the global mean SST anomaly has been subtracted from each point to remove longer term trends due to climate change. The associated Principal Component (PC, or the time series describing the temporal evolution of the EOF pattern) was formed into annual means and standardized. Global SSTs, observed and modeled precipitation, for years with values greater than one standard deviation, were composited into a warm phase PDO composite and, for years with values less than one standard deviation, into a cold phase PDO composite. For comparison the observed and modeled SST and precipitation anomalies for years with greater and less than one standard deviation of the NINO3.4 SST index were averaged to derive El Niño and La Niña composites of SST and precipitation anomalies.

The results are shown in Fig. 11. The PDO SST anomalies are only a fraction of a Kelvin and quite meridionally broad, in agreement with the earlier studies. Also, the warm phase and cold phase PDO-associated precipitation anomalies across the Americas are essentially the opposite of each other indicating a first order linearity. The sign is such that during the warm phase the tropics are dry and the extra tropics of North and South America are wet, all within patterns that are large scale and strikingly coherent. In contrast to the PDO, the SST anomalies for the El Niño and La Niña composites have much stronger equatorial Pacific features. The ENSO precipitation response is again, to first order, essentially linear and, despite the differences in SST forcing patterns, is extremely similar to that of the PDO. Of course it might be thought that the PDO-ENSO precipitation pattern similarity arises from the response to the large ENSO events contained within the PDO composites, a result of the strong temporal correlation between the PDO and ENSO indices. (The annual mean PDO and

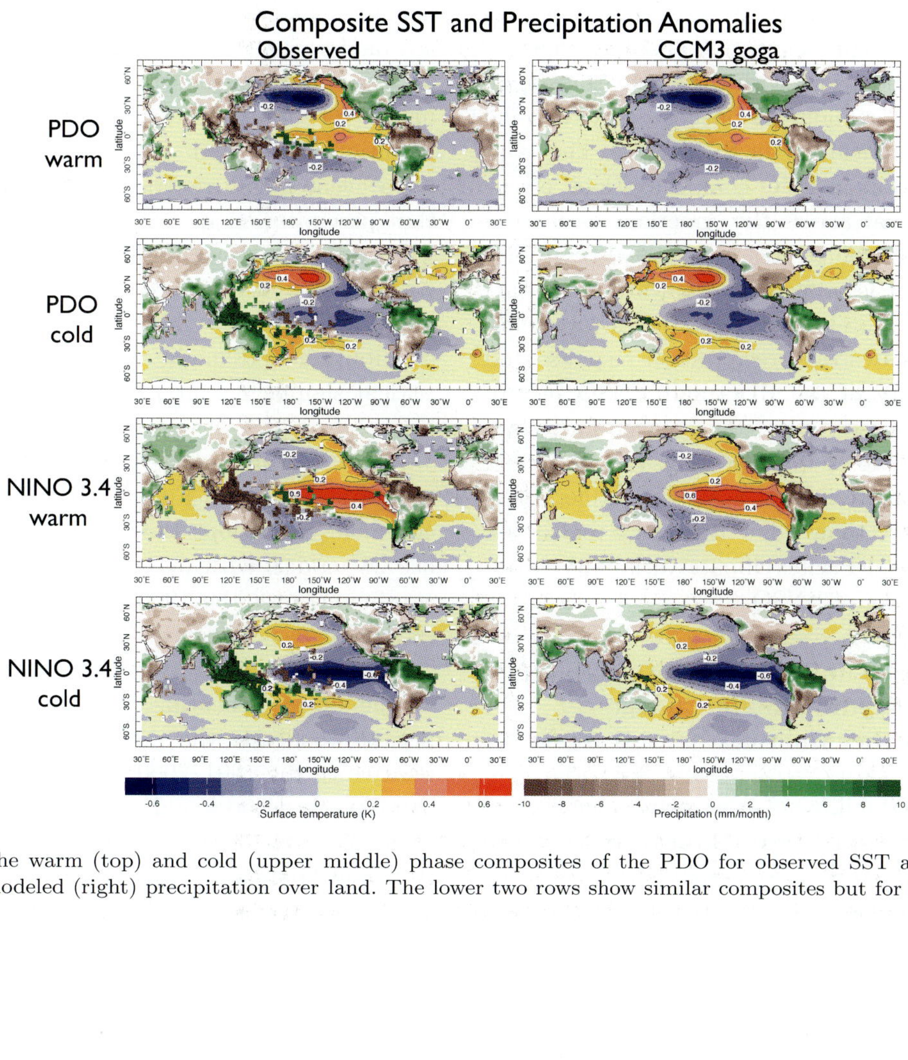

Fig. 11. The warm (top) and cold (upper middle) phase composites of the PDO for observed SST and observed (left) and modeled (right) precipitation over land. The lower two rows show similar composites but for El Niño and La Niña.

NINO3.4 time series have a correlation coefficient of 0.64.) However the work of Huang *et al.* (2005) shows that the decadal timescale precipitation response would be the same if the details of the ENSO events in any one period were replaced with the decadal timescale SST anomalies. That is, the PDO precipitation response shown in Fig. 11 is an actual physical response to the weak PDO SST anomalies.

Of course one might wonder which part of the PDO SST anomalies is responsible for the precipitation response across the Americas. Huang *et al.* (2005), in looking at the decadal precipitation change across the 1976/77 climate shift, addressed this by forcing an atmosphere model first with the global SST change and then with the tropical Pacific change alone and found that the latter explained essentially all the precipitation change in the former. We have also made PDO precipitation anomaly composites from simulations with the same atmosphere model used for Figure 11 but forced by the time history of tropical Pacific SSTs alone. The tropical Pacific-forced PDO composites look essentially identical across the Americas as for the case of PDO composites with global SST forcing. Hence it is in fact the tropical Pacific part of the PDO SST anomalies that overwhelmingly dominates the precipitation response and the similarity of the PDO and ENSO precipitation responses is no coincidence. This comparison reinforces the idea that the PDO and its hydroclimate impact across the Americas is nothing more than a low frequency realization of the tropical Pacific climate variability familiar in the form of ENSO.

7. Decadal Timescale Radiatively-driven Hydroclimate Change Across the Americas

Natural variability of the atmospheres and oceans is a potent source of hydroclimate change on decadal timescales across the Americas. Another source is radiatively-forced climate change and climate models robustly predict that hydroclimate will change in systematic ways in the coming decades with the wet regions of the tropics and mid to high latitudes getting wetter and the dry regions of the subtropics getting drier and expanding poleward (Held and Soden, 2006; Seager *et al.*, 2007, 2010). Any reliable, probabilistic, projection of hydroclimate in coming decades must take account of the possible evolution of natural decadal variability and the human-induced, radiatively-forced, change over the same time period. Therefore, given the presence of the natural variability, it is reasonable to wonder when the radiatively-forced change will make its presence known. In this regard it is common to assess whether the forced changes are statistically significant or not (*e.g.* Deser *et al.* (2013)) but this might be too demanding a criteria. In contrast, in operational seasonal-to-interannual prediction the probability of a season's precipitation being in the lower, middle or upper tercile of the historical distribution is reported. For example a strong El Niño event might generate, say, a 60% chance that precipitation in southwest North America will be in the upper tercile and this would be considered a useful forecast even though the associated anomaly falls within the historical distribution and might well fall short of statistical significance at the usual 5% or higher levels.

Here we turn to the Coupled Model Intercomparison Project Five (CMIP5) multimodel ensemble and adopt the tercile approach. We use 28 models with one run from each that is continuous over the 1900 to 2050 period using estimated past radiative forcings and the Representative Concentration Pathway 8.5 (rcp85) as the future scenario. For the 1901 to 2000 period we divide the precipitation anomalies, relative to the 1901–2000 period, into ten non-overlapping decades and compute the means for each. The ten decadal means are ranked and the lower boundary of the upper tercile is taken to be the seventh highest value and the upper boundary of the lower tercile as the fourth highest value. The near-term climate change for each model is taken to be the precipitation in the 2031–2040 decade minus the climatological average of the second half of the twentieth century (1951–2000). We then determine for each model whether the precipitation change is in the upper, middle or lower tercile of that model's distribution of decadal means. We also determine the multimodel ensemble mean climate change precipitation signal as the average across the models of the 2031–2040 precipitation minus that for 1951–2000. We then count up the number of models that have a precipitation change in the upper or lower terciles of the distributions and the same sign change as the multimodel ensemble mean climate change precipitation signal.

In the absence of any forced climate trends it would be expected that for each model there would be equal probabilities of the 2031–2040 decadal mean anomaly falling in the three terciles while the multimodel mean change would have equal probability of being positive or negative. Hence more than 17% of the models having a precipitation change in the upper or lower tercile, and the same sign change as the multimodel ensemble mean, is the very minimum required to identify a climate change signal with an amplitude equivalent to this measure of natural variability.

The tercile results, along with the multimodel mean precipitation shift, are shown in Fig. 12. Quite remarkably, across vast regions of the Americas, more than 60% of the models have precipitation shifts in the outer terciles of the historical variability. In some parts of the northeast U.S., Canada and Alaska the number of models reaches up to 100%. This result makes clear that for most of the Americas radiatively forced changes in precipitation in the near term future will be of equivalent magnitude to the decadal variations that have occurred in the past, and which caused significant social impacts. The difference is that the past decadal variations were temporary and were followed by decades of more normal precipitation totals or opposite sign anomalies, while the radiatively-forced changes are of the same sign decade to decade and, indeed, intensify as the century advances.

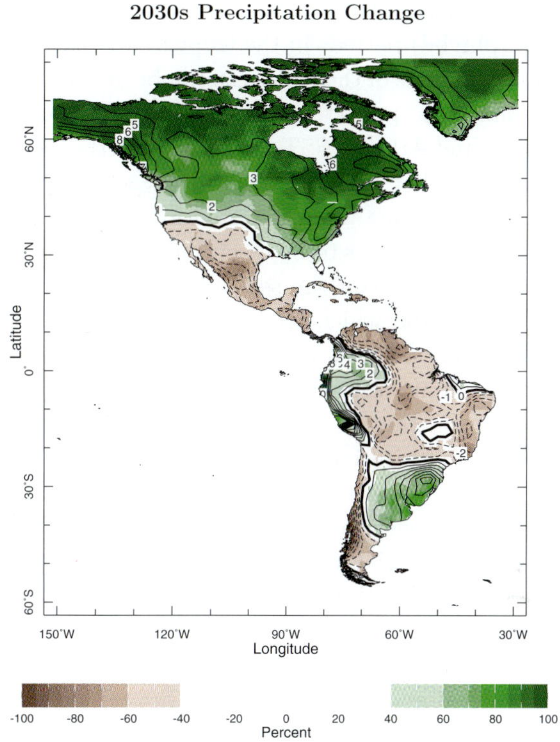

Fig. 12. The percentage of models that have a 2031–2040 less 1951–2000 precipitation shift in the upper (green colors) or lower (multiplied by minus one, brown colors) of that model's distribution of decade mean precipitation anomalies over the 1901 to 2000 period *and* with the shift having the same sign as the multimodel ensemble mean. Single continuous runs of 28 CMIP5 models were used. The multimodel ensemble mean precipitation shift is shown as contours.

8. Conclusions

Thanks to much research in recent years decadal variability of hydroclimate across the Americas is not as much of a mystery as it was a decade or so ago. To a very large extent decadal periods of drier than normal or wetter than normal (pluvial) periods in the tropical and extratropical regions of the Americas can be attributed to similarly persistent anomalies of tropical Pacific and Atlantic SST anomalies. Further, the dynamics that link the tropical oceans to precipitation anomalies in the Americas on decadal timescales appears to be essentially the same as those operating on the seasonal to interannual timescales and which are becoming increasingly well understood. In addition, as for the case of seasonal to interannual variability, it is the tropical oceans that play the dominant role influencing precipitation on decadal timescales. Of course this is expected given the lack of evidence of an atmospheric response, at any timescale, to extratropical SST anomalies that can compete in amplitude with the response to tropical SST anomalies. It may be that on decadal timescales extratropical SST anomalies are larger, relative to tropical anomalies, than they are on shorter timescales but the physics of atmosphere-ocean coupling still biases the atmosphere to being most sensitive to the tropical components of the anomalies. This understanding of the causes of decadal hydroclimate variability works best for, of the regions studied, southwestern North America and the Plains, northeast Brazil and southeast South America. Indeed a quite astonishing amount of longer than interannual variability in these regions can be explained in terms of tropical ocean forcing.

However decadal hydroclimate variability across the America still contains some mysteries. Although southeast South America precipitation is strongly influenced by the tropical Pacific and Atlantic Oceans it also has experienced a century long wetting trend, most marked since the 1960s, that cannot be so explained. Similarly the northeast U.S. — a region where the oceans exert negligible influence on precipitation variability — shifted to a much wetter climate in the early 1970s. Atmosphere models forced by observed global SSTs cannot reproduce these trends. It does seem as if ozone depletion caused a small portion of the wetting trend in southeast South America (Gonzalez *et al.*, 2014). However, in both extratropical regions, models forced by changes in trace gases, solar irradiance, aerosol concentrations etc. do not produce wetting trends anywhere near as strong as those that happened. Consequently, it is probably worth examining climate model simulations to determine if atmosphere variability alone can generate such long term changes in precipitation. This unexplained variability also highlights the need for longer term climate records that can be gained from, for example, tree ring records. In regard to the north east U.S., for example, Pederson *et al.* (2013) show that the late 20th Century shift to a wetter climate is only the end of a wetting trend that has been ongoing since at least the early 16th Century. Knowing that does not make it any easier to explain but does help us know what we need to explain. In this case of the northeast U.S., for example, it appears we need to examine decadal variability in the light of hydroclimate changes over the last millennium and, probably, the Holocene and examine possible contributions from changes in solar irradiance, volcanism, orbital configuration and unforced variability at longer than decadal timescales.

We have not concerned ourselves here with the predictability of the ocean states that drive decadal variability across the Americas. It remains to be demonstrated whether tropical Pacific Ocean SST variations beyond the interannual timescale are predictable. The one suggestion that they might be is that of Karspeck *et al.* (2003) and Seager *et al.* (2004). They used an ensemble prediction with the Zebiak-Cane ENSO model initialized in 2003 to claim that the equatorial east Pacific in the period from 1998 to 2013 would be cooler than during the 15 years before 1993, which it has been. Our own efforts to locate multiyear to decadal predictability of tropical Pacific SSTs in initialized predictions with state-of-the-art global models have been unsuccessful (although Meehl *et al.*

(2013) suggest some shorter term predictability of the mid 1970s shift in the tropical Pacific). In contrast there does seem some success in developing predictive capacity for decadal Atlantic Ocean variations (Meehl *et al.*, 2013). However, whether this translates into prediction of the SST variations of most importance for forcing Americas hydroclimate is not clear. Hence, certainly for now, naturally occurring decadal variations in hydroclimate across the Americas will remain unpredicted other than in a probabilistic way (*e.g.* Hoerling *et al.* (2011)). The steadily evolving changes in precipitation across the Americas forced by rising greenhouse gases appear to be more predictable in the coming decades and reach the level of natural decadal variability in the near term future. In terms of hydroclimate, the radiatively-forced changes in precipitation shown here will combine with the even more predictable effects of rising temperatures to cause notable decreases in precipitation less evaporation in many subtropical regions of the Americas. Assessments of the range of natural decadal variations can be merged with the range of forced changes to prepare probabilistic assessments of potential future hydroclimates in the Americas that might be of use in scenario developments for applications purposes.

Acknowledgments

I would like to thank Jennifer Nakamura for preparing the figures and helping with the analysis, Naomi Henderson for help with the CMIP5 archive and the terciles analysis, Haibo Liu for help with the CMIP5 data, and Yochanan Kushnir for many stimulating discussions. I also thank C.P. Chang for encouraging this work and his patience in waiting for the submission. John Walsh and one anonymous reviewer provided very useful comments and criticisms that I hope have led to an improved chapter. This work was supported by NOAA awards NA08OAR432092 and NA10OAR4310137 (Global Decadal Hydroclimate Variability and Change).

References

Barlow, M., S. Nigam, and E. H. Berberry, 2001: ENSO, Pacific decadal variability, and U.S. summertime precipitation, drought, and stream flow. *J. Climate*, **14**, 2105–2128.

Chang, E. K. M., and Y. Fu, 2002:. Interdecadal variations in northern hemisphere winter storm track intensity. *J. Climate*, **15**, 642–658.

Chiang, J. C. H., Y. Kushnir, and A. Giannini, 2002: Deconstructing Atlantic ITCZ variability: Influence of the local cross-equatorial SST gradient, and remote forcing from the eastern equatorial Pacific. *J. Geophys. Res.*, **107**, 10.1029/ 2000JD000307.

Christensen, N., and Lettenmaier, D. P., 2007:. A multimodel ensemble approach to assessment of climate change impacts on the hydrology and water resources of the Colorado River basin. *Hydrol. Earth Syst. Sci.*, **3**, 1 44.

Compo, G. *et al.*, 2010: The Twentieth Century reanalysis project. *Quart. J. Roy. Meteor. Soc.*, **137**, 1–28.

Cook, B. I. H., R. L. Miller, and R. Seager, 2008: Dust and sea surface temperature forcing of the 1930s Dust Bowl drought. *Geophys. Res. Lett.*, **35**, doi:10.1029/2008GL033486.

Cook, B. I. H., R. L. Miller, and R. Seager, 2008: Amplification of the North American Dust Bowl drought through human-induced land degradation. *Proc. Nat. Acad. Sci.*, **106**, 4997–5001.

Cook, B. I. H., R. L. Miller, and R. Seager, 2010: Atmospheric circulation anomalies during two persistent North American droughts: 1932–1939 and 1948–1957. *Clim. Dyn.*, **36**, 2339–2355.

Cook, B. I. H., R. Seager, and R. L. Miller, 2011: On the causes and dynamics of the early Twentieth Century North American pluvial. *J. Climate*, **24**, 5043–5060.

Covey, D. L., and S. Hastenrath, 1978:.The Pacific El Nim̃o phenomenon and the Atlantic circulation. *Mon. Wea. Rev.*, **106**, 1280–1287.

Davis, M., 2001: *Late Victorian Holocausts: El Niño, Famines and the Making of the Third World*, Verso, London, 464pp.

Deser, C., A. S. Phillips, and J. W. Hurrell, 2004: Pacific interdecadal climate variability:

Linkages between the tropics and the North Pacific during boreal winter since 1900. *J. Climate*, **17**, 3109–3124.

Deser, C., A. S. Phillips, V. Bourdette, and H. Teng, 2013: Uncertainty in climate change projections: the role of internal variability. *Clim. Dyn.*, **38**, 527–546.

Fye, F. K., D. W. Stahle, and E. R. Cook, 2003: Paleoclimatic analogs to twentieth-century moisture regimes across the United States. *Bull. Amer. Meteor. Soc.*, **84**, 901–909.

Garreaud, R. D., and D. S. Battisti, 1999: Interannual (ENSO) and interdecadal (ENSO-like) variability in the southern hemisphere tropospheric circulation. *J. Climate*, **12**, 2113–2123.

Giannini, A., R. Saravanan, and P. Chang, 2004: The preconditioning role of tropical Atlantic variability in the development of the ENSO teleconnection: implications for the prediction of Nordeste rainfall. *Clim. Dyn.*, **22**, 839–855.

Gonzalez, P., L. M. Polvani, R. Seager, and G. Correa, 2014: Stratospheric ozone depletion: a key driver of recent precipitation trends in South Eastern South America. *Clim. Dyn.*, **42**, 1775–1792, doi: 10.1007/s00382-013-1777-x.

Hamlet, A. F., and D. P. Lettenmaier, 2005: Production of temporally consistent gridded precipitation and temperature fields for the continental U.S.. *J. Hydromet.*, **6**, 330–336.

Hansen, Z. K., and G. D. Libecap, 2004: Small farms, externalities, and the Dust Bowl of the 1930s. *J. Pol. Econ.*, **112**, 665–694.

Harnik, N., and E. K. M. Chang, 2003: Storm track variations in radiosonde observations and reanalysis data. *J. Climate*, **16**, 480–495.

Harnik, N., R. Seager, N. Naik, M. Cane, and M. Ting, 2010: The role of linear wave refraction in the transient eddy mean flow response to tropical Pacific SST anomalies. *Quart. J. Roy. Meteor. Soc.*, **136**, 2132–2146.

Hastenrath, S., 2006: Circulation and teleconnection mechanisms of Northeast Brazil droughts. *Prog. Oceanogr.*, **70**, 407–415.

Hastenrath, S., and L. Heller, 1977: Dynamics of climate hazards in northeast Brazil. *Quart. J. Roy. Meteor. Soc.*,**103**, 77–92.

Hayhoe, K., C. P. Wake, T. G. Huntington, L. Luo, M. D. Schwartz, J. Sheffield, E. Wood, B. Anderson, J. Bradbury, A. DeGaetano, T. J. Troy, and D. Wolfe, 2007: Past and future changes in climate and hydrological indicators in the US Northeast. *Clim. Dyn.*, **28**, 381–407.

Held, I. M., and B. J. Soden, 2006: Robust responses of the hydrological cycle to global warming. *J. Climate*, **19**, 5686–5699.

Herweijer, C., R. Seager, and E. R. Cook, 2006: North American droughts of the mid to late Nineteenth Century: History, simulation and implications for Medieval drought. *J. Climate*, **11**, 1131–1149.

Hoerling, M. P., J. Hurrell, A. Kumar, L. Terray. J. Eischeid, P. Pegion, T. Zhang, X. Quan, and T. Xu, 2011: On North American decadal climate for 2011–20. *J. Climate*, **24**, 4519–4528.

Hoerling, M. P., X. W. Quan, and J. Eischeid, 2010: Distinct causes for two principal U.S. droughts of the 20th century. *Geophys. Res. Lett.*, **36**, doi:10.1029/2009GL039860.

Huang, H. P., R. Seager, and Y. Kushnir, 2005: The 1976/77 transition in precipitation over the Americas and the influence of tropical sea surface temperature. *Clim. Dyn.*, **24**, 721–740.

Hurrell, J. W., 1995: Decadal trends in the North Atlantic Oscillation: Regional temperatures and precipitation. *Science*, **269**, 676–679.

Kaplan, A., M. A. Cane, Y. Kushnir, A. C. Clement, M. B. Blumenthal, and B. Rajagopalan, 1998: Analyses of global sea surface temperature: 1856–1991. *J. Geophys. Res.*, **103**, 18567–18589.

Karspeck, A., R., R. Seager, and M. A. Cane, 2003: Predictability of tropical Pacific decadal variability in an intermediate model. *J. Climate*, **17**, 2842–2850.

Kiehl, J. T., J. J. Hack, G. B. Bonan, B. A. Bovile, D. L. Williamson, and P. J. Rasch, 1998: The National Center for Atmospheric Research Community Climate Model: CCM3. *J. Climate*, **11**, 1131–1149.

Kushnir, Y., 1994: Interdecadal variations in North Atlantic sea surface temperature and associated atmospheric conditions. *J. Climate*, **7**, 141–157.

Kushnir, Y., R. Seager, M. Ting, N. H. Naik, and J. Nakamura, 2010: Mechanisms of tropical Atlantic SST influence on North American hydroclimate variability. *J. Climate*, **23**, 5610–5628.

Liang, X., D. P. Lettenmaier, E. F. Wood, and S. J. Burges, 1994: A simple hydrologically based model of land surface water and energy fluxes for GCMs. *J. Geophys. Res.*, **99**, 14415–14428.

Mantua, N. J., and S. R. Hare, 2002: The Pacific Decadal Oscillation. *J. Oceanogr.*, **38**, 35–44.

Meehl, G. A. *et al.*, 2013: Decadal climate prediction: an update from the trenches. *Bull. Amer. Meteor. Soc.*, **95**, 243–267.

Meko, D. M., C. A. Woodhouse, C. A. Baisan, T. Knight, J. J. Lukas, M. K. Hughes, and M. W. Salzer, 2007: A 1,200 year perspective on 21st Century drought in southwestern North America. *Proc. Nat. Acad. Sci.*, **58**, doi:10.1029/2007GL029988.

Mo, K. C., and E. H. Berberry, 2011: Drought and persistent wet spells over South America based on observations and the U.S. CLIVAR drought experiments. *J. Climate*, **24**, 1801–1820.

Moura, A. D., and J. Shukla, 1981: On the dynamics of droughts in northeast Brazil: Observations, theory and numerical experiments with a general circulation model. *J. Atmos. Sci.*, **38**, 2653–2675.

Nigam, S., M. Barlow, and E. H. Berberry, 1999: Analysis links Pacific decadal variability to drought and streamflow in United States. *EOS Trans. Amer. Geophys. Union*, **80**, 621–625.

Nigam, S., B. Guan, and A. Ruiz-Barradas, 2011: Key role of the Atlantic Multidecadal Oscillation in 20th century drought and wet periods over the Great Plains. *Geophys. Res. Lett*, **38**, L16713, doi:10.1029/2011GL048650.

Pederson, N., A. Bell, E. R. Cook, U. Lall, N. Devenini, R. Seager, K. Eggelston, and K. Vranes (2013). Is an epic pluvial masking the water security of the greater New York City region?. *J. Climate*, **26**, 1339–1354.

Pederson, N., A. R. Bell, T. A. Knight, C. Leland, N. Malcomb, K. J. Anchukaitis, K. Tackett, J. Scheff, A. Brice, B. Catron, W. Blozan, and J. Riddle, 2012: A long-term perspective on a modern drought in the American Southeast. *Env. Res. Lett.*, **7**, 014034, doi:10.1088/1748-9326/7/1/014034.

Rayner, N. A., D. E. Parker, E. B. Horton, C. K. Holland, L. V. Alexander, D. P. Rowell, E. C. Kent, and A. Kaplan, 2003: Global analyses of sea surface temperature, sea ice, and night marine air temperature since the late nineteenth century. *J. Geophys. Res.*, **108**, dpi:10.1029/2002JD002670.

Reisner, R., 1986: *Cadillac Desert*, Penguin Books, New York, 582 pp.

Schubert, S. D., M. J. Suarez, P. J. Pegion, R. D. Koster, and J. T. Bacmeister, 2004: On the cause of the 1930s Dust Bowl. *Science*, **303**, 1855–59

Schubert, S. D., M. J. Suarez, P. J. Pegion, R. D. Koster, and J. T. Bacmeister, 2004b: Causes of long-term drought in the United States Great Plains. *J. Climate*, **17**, 485–503.

Seager, R., 2007: The turn-of-the-century North American drought: Dynamics, global context and prior analogs. *J. Climate*, **20**, 5527–5552.

Seager, R., N. Harnik, W. A. Robinson, Y. Kushnir, M. Ting, H. P. Huang, and J. Velez, 2005a: Mechanisms of ENSO-forcing of hemispherically symmetric precipitation variability. *Quart. J. Roy. Meteor. Soc.*, **131**, 1501–1527.

Seager, R., A. R. Karspeck, M. A. Cane, Y. Kushnir, A. Giannini, A. Kaplan, B. Kerman, and J. Velez, 2004: Predicting Pacific decadal variability, In *Earth Climate: The ocean-atmosphere interaction*, C. Wang, S.-P. Xie and J. A. Carton, Eds., 105–120.

Seager, R., Y. Kushnir, C. Herweijer, N. Naik, and J. Velez, 2005b: Modeling of tropical forcing of persistent droughts and pluvials over western North America: 1856–2000. *J. Climate*, **18**, 4068–4091.

Seager, R., Y. Kushnir, M. Ting, M. A. Cane, N. Naik, and J. Velez, 2008: Would advance knowledge of 1930s SSTs have allowed prediction of the Dust Bowl drought? *J. Climate*, **21**, 3261–3281.

Seager, R., and N. Naik, 2012: A mechanisms-based approach to detecting recent anthropogenic hydroclimate chnage. *J. Climate*, **25**, 236–261.

Seager, R., N. Naik, W. Baethgen, A. Robertson, Y. Kushnir, J. Nakamura, and S. Jurburg, 2010a: Tropical oceanic causes of interannual to multidecadal precipitation variability in southeast South America. *J. Climate*, **23**, 5517–5539.

Seager, R., N. Naik, M. A. Cane, N. Harnik, M. Ting, and Y. Kushnir, 2010: Adjustment of the atmospheric circulation to tropical Pacific SST anomalies: Variability of transient eddy propagation in the Pacific-North America sector. *Quart. J. Roy. Meteor. Soc.*, **136**, 277–296.

Seager, R., N. Naik, and G. A. Vecchi, 2010: Thermodynamic and dynamic mechanisms for large-scale changes in the hydrological cycle in response to global warming. *J. Climate*, **23**, 4651–4668.

Seager, R., N. Pederson, Y. Kushnir, J. Nakamura, and S. Jurburg, 2012: The 1960s drought and subsequent shift to a wetter climate in the Catskill Mountains region of the New York City watershed. *J. Climate*, **25**, 6721–6742.

Seager, R., M. Ting, I. M. Held, Y. Kushnir, J. Lu, G. Vecchi, H.-P. Huang, N. Harnik, A. Leetmaa, N.-C. Lau, C. Li, J. Velez, and N. Naik, 2007: Model projections of an imminent transition to a more arid climate in southwestern North America. *Science*, **316**, 1181–1184.

Seager, R., M. Ting, C. Li, B. Cook, and J. Nakamura, 2012b: Projections of declining surface water availability in the southwest U.S. *Nature Clim. Change*, **3**, 482–486.

Seager, R., A. Tzanova, and J. Nakamura, 2009: Drought in the southeastern United States: Causes, variability over the last millennium and the potential for future hydroclimate change. *J. Climate*, **22**, 5021–5045.

Siegel, B. J., 1971: Migration dynamics in the interior of Ceara, Brazil. *Southwest J. Anthropol.*, **27**, 234–258.

Sutton, R. T., and D. L. R. Hodson, 2005: Atlantic Ocean forcing of North American and European summer climate. *Science*, **309**, 115–118.

Swetnam, T. W., and J. L. Betancourt, 1998: Mesoscale disturbance and ecological response to decadal climatic variability in the American southwest. *J. Climate*, **11**, 3128–3147.

Ting, M., Y. Kushnir, R. Seager, and C. Li, 2009: Forced and internal Twentieth Century SST trends in the North Atlantic. *J. Climate*, **22**, 1469–1481.

Ting, M., and H. Wang, 1997: Summertime U.S. precipitation variability and its relation to Pacific sea surface temperature. *J. Climate*, **10**, 1853–1873.

Vano, J. A., B. Udall, D. R. Cayan, J. T. Overpeck, L. D. Brekke, T. Das, H. C. Hartmann, H. G. Hidalgo, M. Hoerling, G. J. McCabe, K. Morino, R. S. Webb, K. Werner, and D. P. Lettenmaier, 2013: Understanding uncertainties in future Colorado River streamflow. *Bull. Amer. Meteor. Soc.*, in press.

Worster, D., 1979: *Dust Bowl: The Southern Plains in the 1930s*, Oxford Univ. Press, New York, 416 pp.

Yang, W., R. Seager, M. A. Cane and B. Lyon, The annual cycle of East African precipitation. *J. Climate*, in press.

Zhang, Y., J. M. Wallace, and D. S. Battisti, 1997: ENSO-like interdecadal variability. *J. Climate*, **10**, 1004–1020.

Chapter 16

The Interhemispheric Pattern and Long-Term Variations in the Tropical Climate over the 20th and 21st Centuries

John C. H. Chiang

Department of Geography and Berkeley Atmospheric Sciences Center, University of California, Berkeley, CA 94720-4740, USA
jch_chiang@berkeley.edu
http://climate.geog.berkeley.edu/~jchiang/

Several lines of evidence — paleoclimate observations, model simulations, and theory — suggest that thermal forcing originating from the extratropics have a profound effect on tropical rainfall. Cold thermal forcing located in the Northern extratropics cools air and surface temperatures over the entire hemisphere, generating an interhemispheric temperature asymmetry that shifts the latitudinal position of the Intertropical Convergence Zone southwards and weaken rainfall over Northern Hemisphere summer monsoon regions. This change is associated with a northward anomalous cross-equatorial atmospheric energy transport induced by differences in the energy flux entering each hemisphere. I summarize how this 'interhemispheric pattern' could be usefully applied to understanding multidecadal and longer-timescale forced changes in tropical rainfall over the 20th and 21st centuries, focusing on three specific causes: anthropogenic aerosols, the late 1960s climate shift, and hemispheric asymmetric feedbacks to greenhouse warming.

1. Introduction

Paleoclimate evidence examining the worldwide occurrence of millennial events during the last deglacial reveals a systematic global reorganization of climate. These millennial events — the Dansgaard/Oeschger (D/O) and Heinrich events — punctuated climate throughout the last glacial and into the deglacial period. They are most strongly expressed in proxy temperature records from Greenland ice core measurements (Fig. 1a). Figure 1b shows a schematic of a typical D/O cycle as recorded by Greenland cores: starting from a cold ('stadial') phase, Greenland temperatures undergo a rapid warming to a warm ('interstadial') phase, where they stay (with a small drift to cooler temperatures) for several hundred years before shifting back down to a colder stadial phase (Ganopolski and Rahmstorf, 2001).

These events are remarkable for two reasons: (i) the rapidity with which the temperature changes from cold to warm phases, typically on the order of a few decades; and (ii) the magnitude of the temperature change is on the order of 10°C, comparable to the difference in annual mean temperatures between Moscow and Madrid (Alley, 2004). Heinrich events in particular are associated with some of the coldest millennial events in the Greenland ice core record (Fig. 1a, red dots).

These events are associated with global climate changes, in particular impacting Northern Hemisphere temperatures and tropical rainfall. Northern Hemisphere temperatures are cooler during cold D/O stadials (Voelker, 2002),

Climate Change: Multidecadal and Beyond
edited by Chih-Pei Chang et al.

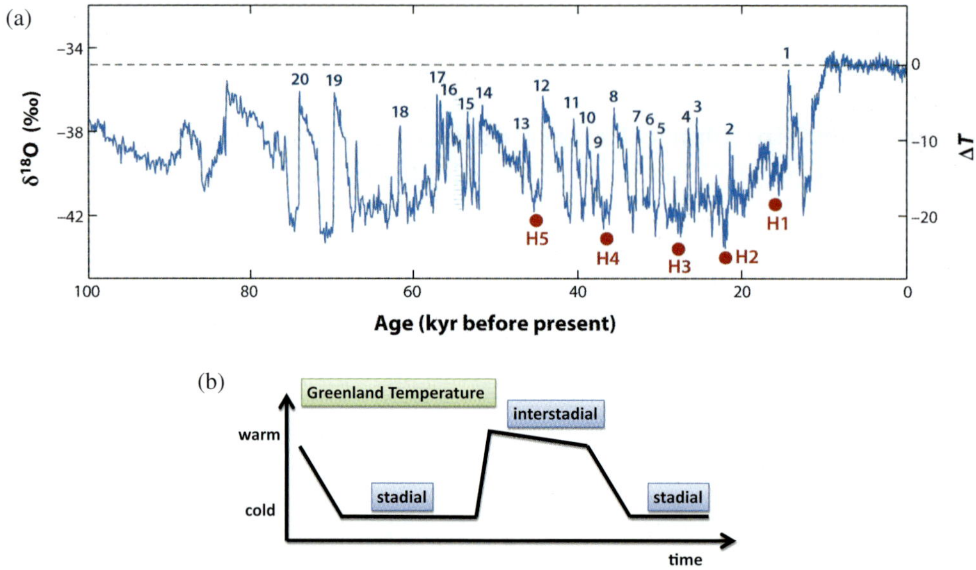

Fig. 1. (a) Greenland temperatures during the last glacial period, as recorded by stable oxygen isotopes in the Greenland Ice Core Project (GRIP) core (Grootes *et al.* 1993). The Dansgaard-Oeschger (D/O) events are numbered, and the timing of the Heinrich events (H1-5) are marked by dots. Figure originally from Ganopolski and Rahmstorf (2001), and as reproduced in Chiang (2009). (b) Schematic of a typical D/O cycle. Starting from a cold phase (stadial), Greenland temperatures abruptly warm typically over a few decades to a warm (interstadial) phase. It will stay in the interstadial for a few hundred years (and with a slight drift to cooler conditions), before returning to stadial conditions. A typical temperature difference between a stadial and interstadial is ~10°C.

the tropical Atlantic Intertropical Convergence Zone (ITCZ) shifted southwards (Peterson *et al.*, 2000), and the northern hemisphere monsoons were weaker (Wang *et al.*, 2001). There is also proxy evidence for changes to the Southern Hemisphere climate, in particular warming (cooling) over Antarctica during stadials (interstadials) (Barbante *et al.*, 2006); and a shift in the Southern Hemisphere westerlies during Heinrich events (Anderson *et al.*, 2009). Taken together, the proxies show strong evidence for a global rearrangement of climate during these abrupt climate changes.

The dominant hypothesis for the driver of these millennial events is a weakening of the Atlantic Meridional Overturning Circulation (AMOC), as first proposed by Broecker *et al.* (1985). Since then, a wealth of proxy and modeling evidence has accumulated in support of this hypothesis (Alley, 2007). Most significantly, coupled model simulations whereby the AMOC is artificially slowed down (commonly through input of freshwater in the North Atlantic) show a global rearrangement of climate with a colder Northern Hemisphere, southward-shifted ITCZ, and weakens northern hemisphere monsoon rainfall, in agreement with paleoclimate proxies (Fig. 2). Thus, an abrupt slowdown of the AMOC appears to drive a global reorganization of climate (Cheng *et al.*, 2007; Vellinga and Wood, 2002; Zhang and Delworth, 2005).

The purpose of this paper is to summarize an emerging hypothesis for extratropical thermal influence on the tropical rainfall climate via a global teleconnection pattern — the so-called ***interhermispheric pattern*** — mediated by the atmosphere. We focus here on the specific applicability of this hypothesis to multidecadal and longer-term variations in tropical rainfall changes over the 20th and 21st centuries.

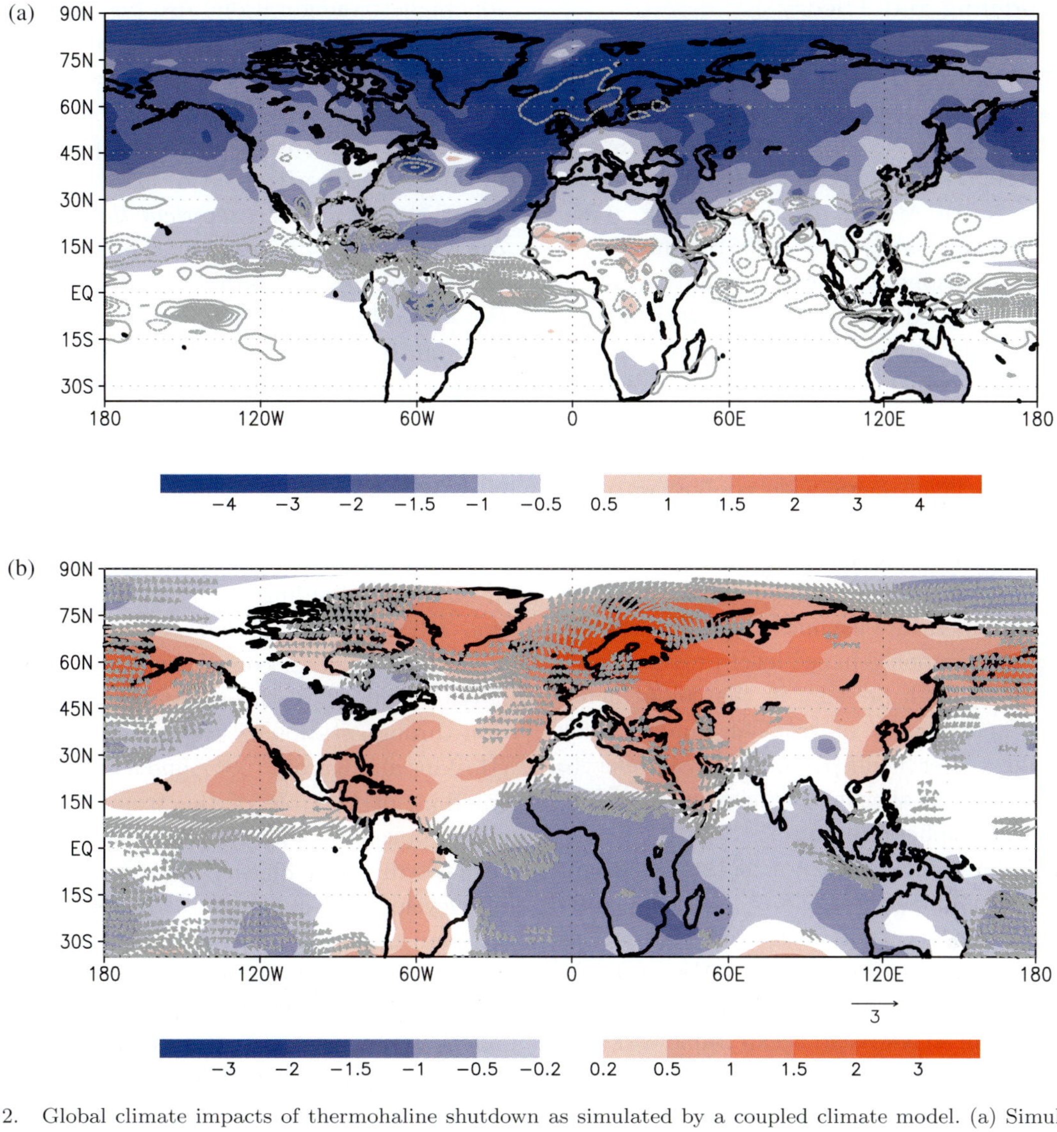

Fig. 2. Global climate impacts of thermohaline shutdown as simulated by a coupled climate model. (a) Simulated July–September surface temperature (shaded; units are K) and rainfall (contour interval 0.5 mm day−1; dashed contours are negative) changes during years 10–14 after the abrupt introduction of freshwater to the high-latitude North Atlantic ocean and resultant slowdown of the Atlantic meridional overturning circulation. (b) Same as (a), but for sea level pressure (shaded; units are Pa) and surface winds (reference vector is 3m/s, and anomalies less than 1m/s in magnitude are not plotted). The global climate response exhibits the "interhemispheric pattern": a northern hemisphere-wide cooling, a interhemispheric seesaw in atmospheric mass, and a southward shift of the tropical rainbands. Specifically, there is a southward displacement of the Atlantic Intertropical Convergence Zone (ITCZ), a weakening of the northern branch of the Pacific ITCZ, and drying over the Asian and West African monsoon regions. The model results are as described in Cheng *et al.* (2007) using the Community Climate System Model, version 3; in this figure, the average of three ensemble members are used.

2. Modeling and Theory

So, what is the essential dynamics of this teleconnection? The pattern of global changes seen in the AMOC shutdown appear to largely result from atmospheric teleconnections of high-latitude North Atlantic cooling (Cheng *et al.*, 2007; Chiang and Bitz, 2005): simulations with an atmospheric general circulation model (AGCM) coupled to a slab ocean with a cooling artificially applied in the extratropical North Atlantic show climate anomalies that are similar to the AMOC shutdown (cf. Fig. 2). We point specifically to these gross features of the climate response: a hemispheric-wide cooling/warming in the same hemisphere as the extratropical forcing and little temperature response in the opposite hemisphere, producing an interhemispheric temperature asymmetry; a hemispheric asymmetric response in sea level pressure, with anomalously high pressure in the colder hemisphere; and a shift of tropical rainfall away from the cooler hemisphere. We will refer to this response as the 'interhemispheric pattern' or 'interhemispheric response'.

This gross global pattern of changes is not just a consequence of North Atlantic cooling, but also occurs in many other situations of extratropical thermal forcing. For example, Chiang and Bitz (2005) applied extratropical cooling in an AGCM-slab ocean framework separately to the North Atlantic and North Pacific (using an alteration to the q-flux), and also to Northern Hemisphere land regions (by applying a last glacial maximum land ice-sheet boundary condition, but with zero thickness) — all showed an interhemispheric pattern of climate response. Rotstayn and Lohmann (2002) examined the global climate response to the direct and indirect effects of 20th century anthropogenic aerosols — largely occurring over the extratropical northern hemisphere — in an AGCM-slab ocean framework, also finding an interhemispheric response. Kang (2009) applied a localized extratropical thermal forcing in an AGCM-slab ocean in an aquaplanet configuration, also finding that the climate response to be approximately zonally symmetric, and exhibiting an interhemispheric temperature asymmetry and ITCZ shifted away from the anomalously colder hemisphere.

These observations lead us to propose a hypothesis: ***that extratropical thermal forcing (localized or otherwise) can drive a global climate response characterized by the interhemispheric pattern, mediated through the atmosphere and aided by thermodynamic interactions with the surface ocean.*** This hypothesis is conceptually analogous to atmospheric teleconnections resulting from the other great global teleconnection of Earth's climate, those occurring during El Nino-Southern Oscillation (ENSO) events. Pursuit in our understanding of the ENSO teleconnection advanced our understanding of tropical-to-extratropical teleconnections, including the development of a theory for the teleconnection through stationary Rossby wave dynamics (Hoskins and Karoly 1981), and a greater understanding of the role of the tropics in driving extratropical circulation and climate changes (Trenberth *et al.*, 1998).

Development of this extratropical-to-tropical teleconnection from extratropical thermal forcing is still in its infancy, with most of our knowledge coming from modeling studies — primarily those using an AGCM-slab ocean configuration, where the effect on the tropical circulation and rainfall is quite apparent and pronounced. A plausible mechanistic theory, akin to Hoskins and Karoly (1981) for the ENSO teleconnection, remains to be developed. However, a recent trend has been to interpret the teleconnection from an energy flux framework (Yoshimori and Broccoli, 2008; Kang *et al.*, 2009). The idea is that there is an interhemispheric contrast in energy flux between the hemispheres that lead to a compensation by the cross-equatorial heat transport in moving energy from one hemisphere to the other. This, in turn, is tied to the Hadley circulation response, as that circulation is responsible for meridional energy transports in the tropics. Specifically, meridional atmospheric energy transports are in the same direction as the upper branch of the Hadley circulation, with a magnitude approximately proportional to the mass flux; so in the northern summer (for example), the winter Hadley cell straddles the equator, and the energy flux is directed to the Southern Hemisphere (Fig. 3). This cross-equatorial energy flux is reduced if the winter cell weakens; this is achieved readily by an equatorial shift in the latitude of the uplift (Lindzen and Hou, 1988). Given the change in cross-equatorial atmospheric heat transport, and climatological

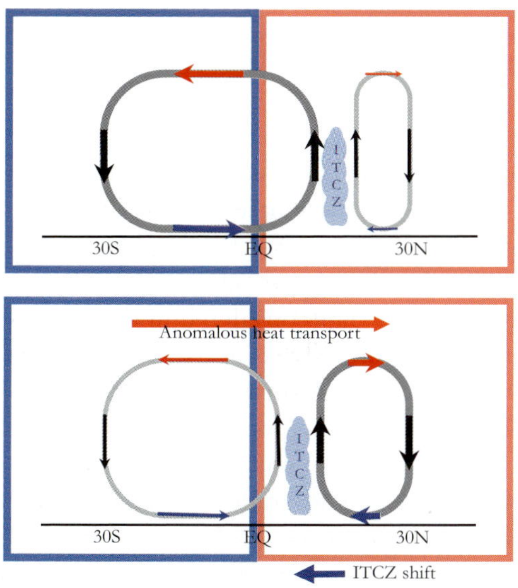

Fig. 3. Schematic of energy transports by the Hadley circulation. **Top panel:** Hadley circulation as depicted for northern hemisphere summer. Energy transport by the Hadley circulation is in the direction of the upper branch of the Hadley cells (red arrows). In this case, the southern (winter) Hadley cell is stronger, and transports energy from the northern (summer) hemisphere to the southern (winter) hemisphere. **Bottom panel:** with a southward displacement of the ITCZ, the southern cell becomes weaker and the northern cell becomes stronger, with commensurate changes to the energy transported by each. The net change in the energy transport is northward, so that a southward shift in the ITCZ leads to an anomalous northward cross-equatorial heat transport (orange arrow). Figure courtesy of Yen-Ting Hwang (University of Washington).

values of the gross moist stability and gross moisture stratification, the change in zonal mean tropical rainfall can be quantified (Kang *et al.*, 2009).

While hemispheric asymmetric forcings drive the cross-equatorial flux changes, there are radiative feedbacks (positive and negative) that modify the cross-equatorial flux requirements and thus the magnitude of the tropical response. We illustrate this with a study by Cvijanovic and Chiang (2013) who examined the energy flux changes for the specific case of a surface extratropical North Atlantic cooling, in an AGCM-slab ocean simulation. In this case, cooling was applied to the slab ocean over the North Atlantic north of 50°N as a modification to the Q-flux (and with no compensating warming). The forcing drives an interhemispheric pattern of response, and a new energy balance configuration with altered top-of-atmosphere and horizontal energy fluxes are achieved by the model climate in equilibrium (Fig. 4). If the surface flux changes over the northern extratropics are normalized to 1, then around 1/3 of the cooling is locally compensated for by TOA radiative changes in the northern extratropics, whereas the other 2/3 is obtained from increased northward energy transports from the northern tropics. The local compensation is driven largely by negative temperature- outgoing longwave radiation (OLR) feedbacks.

Interestingly, the energy derived to supply the increased northward atmospheric heat transport to the northern extratropics comes not from TOA changes in the northern tropics (there, cloud longwave feedbacks are almost perfectly compensated for the cloud shortwave feedbacks), but from the *southern* tropics (this was also noted in Chiang and Bitz, 2005). The TOA flux changes in the southern tropics comes largely from cloud longwave feedbacks — the southward ITCZ shift leads to more deep convective clouds there the replace shallow cumulus, resulting in a decrease to the outgoing longwave radiation (OLR). This anomalous

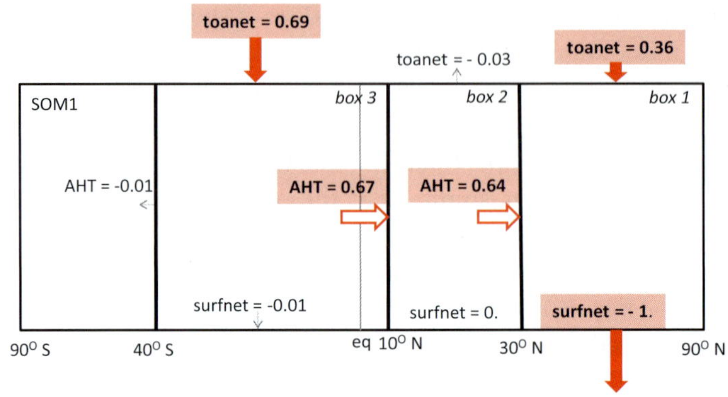

Fig. 4. Box diagram of (normalized) energy flux changes into the atmosphere in an AGCM-SOM simulation given extratropical North Atlantic cooling. The cooling is applied as a q-flux anomaly in the North Atlantic north of 50N (see Cvijanovic and Chiang 2013 for details). Shown is the summary of the equilibrated energy flux changes divided into latitudinal boxes, chosen to reflect the gross changes in the response. TOA and surface net energy budget anomalies are given at the top and bottom of each box respectively, and anomalous atmospheric heat transports are shown at the borders between the boxes. The arrows indicate the directions of the anomalous fluxes, and flux values are normalized relative to the surface flux anomaly in box 1 (which comes mostly from the applied North Atlantic cooling). The major flux changes are emphasized in boldface and in red. Adapted from Cvijanovic and Chiang (2013).

increase in the southern tropics TOA flux gets fed into the northern tropics via cross-equatorial atmosphere energy transports; this cross-equatorial transport is tied to the southward ITCZ shift. Note that the southern extratropics do not play a role in the energy flux changes, though this does not preclude circulation changes there (*e.g.*, Lee *et al.*, (2011) found significant changes to the Southern Hemisphere westerlies in a model simulation with applied North Atlantic cooling).

The radiative feedbacks associated with this extratropical cooling and ITCZ shift is generally quite varied and complex, and no simple story emerges from it. However, it does indicate that radiative feedbacks play an important role in determining the character of the teleconnection, specifically the sensitivity of the tropical response to extratropical forcing. For example, a number of relatively small but positive radiative feedbacks occur in the northern extratropics; if the strength of (say) the cloud shortwave feedback increased, then it could lead to a larger poleward energy flux requirement and hence larger tropical response. (Kang *et al.*, 2009) showed that cloud radiative feedbacks modulate the strength of the tropical response to extratropical forcing.

On the other hand, changes to the ocean heat transport (not considered in the AGM-slab ocean framework) could act to reduce the sensitivity of the tropical response to extratropical cooling. For example, were the ocean allowed to also move heat, then part of the cross-equatorial energy flux demand could be taken up by the ocean heat transport. For example, Cai *et al.* (2006) found that the northward ocean heat transport increased in a coupled model simulation in response to anthropogenic aerosol forcing, largely from changes to the AMOC. The actual sensitivity of the tropical ITCZ climate to extratropical thermal forcing is an outstanding question that needs to be addressed.

3. Drivers of 20th Century and Future Interhemispheric Gradients

Our purpose is to explore whether the hypothesis presented — extratropical thermal forcing leading to tropical rainfall changes via the interhemispheric pattern — can be usefully applied to interpreting and understand decadal and longer-term changes in 20th century and future tropical climate. Two questions are addressed here:

(i) What are the extratropical thermal forcings in historical and future climate that can drive an interhemispheric response?
(ii) Is there evidence that such forcings impact tropical climate through an interhemispheric pattern of response?

We address the latter question in Sections 4 and 5. For the first question, we highlight three forcings — anthropogenic aerosols, North Atlantic SST, and feedbacks to anthropogenic greenhouse warming — that appear to be relevant to our hypothesis for analyzing 20th century and future climate changes.

Anthropogenic Aerosols. Anthropogenic aerosol emissions, in particular sulfates, increased dramatically since the 1960s (Boucher and Pham, 2002; Lamarque *et al.*, 2010) (Fig. 5). The emissions occur largely over the northern extratropics where the industrialized nations are located, and because of the relatively short lifetime the aerosols remain largely in the northern hemisphere. Aerosol emissions are dominated by sulfates that have a direct cooling influence, but black carbon (a warming influence) has become relatively more important towards the end of the 20th century. A plateauing of sulfate emissions occurred in the 1970s with introduction of legislation in North America and Europe to combat air pollution; as we will show later, this plateau apparently has a marked effect on the tropical changes to anthropogenic aerosols. Unlike sulfates however, black carbon emissions have steadily increased throughout the 20th century.

Anthropogenic aerosols generally provide a cooling influence over the Northern Hemisphere. Several idealized studies imposing direct and indirect effects of late 20th century aerosol emissions in AGCMs coupled to slab ocean models reveal a global climate response similar to the interhemispheric pattern (*e.g.*, Rotstayn and Lohmann, 2002; Williams *et al.*, 2001). The potential range in magnitude of these effects are however large because of uncertainty in the representation of indirect effects.

Abrupt North Atlantic SST changes in the late 1960s. Thompson *et al.*(2010) showed that the interhemispheric SST difference (the difference in SST averaged over the each hemisphere) underwent a significant and abrupt shift in the late 1960s with the north cooling relative to the south by around 0.5 K (Fig. 6). The interhemispheric SST difference stayed this way until the late 1990s before partially recovering. While the source of the shift remains unknown, the spatial pattern reveals high loadings over the northern North Atlantic, reminiscent of changes associated with a weakening of the AMOC.

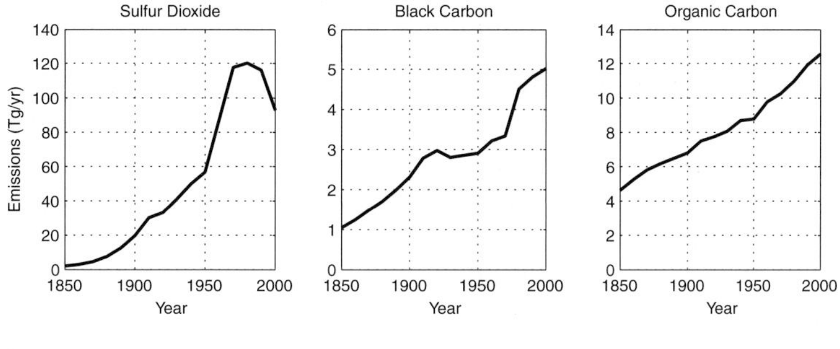

Fig. 5. Timeseries of global anthropogenic aerosol emissions, from L to R: Sulfur Dioxide, Black Carbon, and Organic Carbon. Data from Lamarque *et al.* (2010).

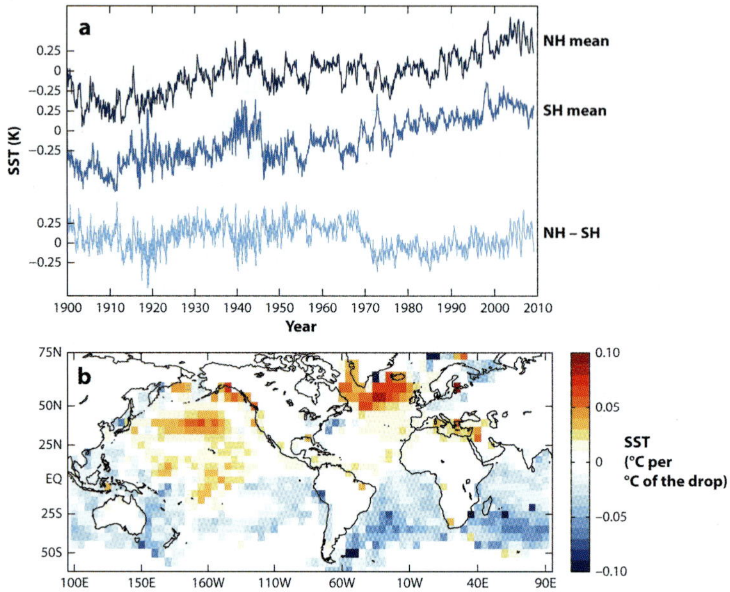

Fig. 6. Figure from Thompson *et al.* (2010). (a) Northern Hemisphere (NH) mean sea-surface temperatures (SSTs) (*top curve*), Southern Hemisphere (SH) mean SSTs (*middle curve*), and NH mean SST minus SH mean SST (*bottom curve*). (b) Spatial regression formed by regressing SST data that have been weighted by a triangular function centered about 1970, with end points in 1940 and 2000, onto a time series of the drop. The time series of the drop was generated by fitting a third-order polynomial fit to the NH-minus-SH time series from panel (a) for the period 1968–1972 and then by extending the resulting fitted time series backward and forward over various time intervals centered about 1970 using fixed pre- and postdrop values for years before 1968 and after 1972. Units for the regressed SST are °C per degree of the drop.

This late 1960s shift is also associated with multidecadal SST variations over the entire North Atlantic, as measured by the Atlantic Multidecadal Oscillation (AMO — a detrended and temporally-smoothed measure of SST averaged over the North Atlantic basin). The AMO also exhibits a decline in the late-1960s to colder conditions. Detection and attribution studies have generally indicated that while the 20th century warming trend over the North Atlantic is due to external forcing, no definitive consensus has emerged on the origins of the multidecadal swings. Several studies have argued that the AMO is largely due to internal variations (Terray, 2012; Ting *et al.*, 2009), while others have argued that it is largely anthropogenically forced (*e.g.*, Mann amd Emanuel, 2006). On the other hand, coupled modeling studies of natural variations in the AMOC have shown that the surface temperature anomalies associated with such changes resemble those of the AMO, suggesting a link between the two (*e.g.*, Knight *et al.*, 2005; Latif *et al.*, 2004).

Hemispheric asymmetric responses to greenhouse warming. Radiative forcing by long-lived greenhouse gases is generally symmetric about the equator. However, the hemispheric asymmetry of earth leads to an uneven response to anthropogenic greenhouse warming. Unlike oceans, land has negligible heat capacity, so ocean heat uptake from greenhouse warming is larger in the South. There is an interhemispheric contrast in surface fluxes resulting from this; and the North is expected to warm faster than the south. Second, summertime Arctic sea ice extent started decreasing since around the 1960 (Stroeve *et al.*, 2007), leading to more solar absorption in the Northern Hemisphere; on the other hand, Antarctic sea ice extent has increased slightly over 1979–2006 (Cavalieri and Parkinson, 2008). Finally, wind-driven upwelling and mixing over the southern ocean means that the ocean there is thermally connected to greater depths, thus warms up slower under transient warming. In all these cases, they lead to greater warming in the North over the South. However, a slowdown of the AMOC in response to greenhouse warming could lead to the opposite effect of greater warming in the South than North (*e.g.*, Feulner *et al.*, 2013).

4. Regional Tropical Climate Changes over the 20th Century

I now address the second question posed in Section 3, whether interhemispheric forcings affect tropical climate. I focus on 20th century changes in two regions with particularly strong connections to Greenland during abrupt climate changes — the tropical Atlantic, and the Asian monsoon region.

4.1. *Tropical Atlantic*

Chang *et al.* (2011) investigated long-term trends of the Atlantic ITCZ over the 20th century, using a measure of the difference in SST between the north and south tropical Atlantic (south minus north; the Atlantic Interhemispheric SST gradient, or AITG) as a proxy. The AITG (Fig. 7) shows multidecadal behavior associated with the AMO, on top of an upward trend indicating that the south tropical Atlantic has been warming up relatively faster than the North. Given the strong coupling between the Atlantic ITCZ and the AITG (*e.g.*, Chiang *et al.*, 2002), this trend indicates a progressive southward displacement in the Atlantic ITCZ.

Chang *et al.* (2011) analyzed coupled model simulations of the AITG in the Coupled Model Intercomparison Project (CMIP) phase 3 multimodel simulations of the 20th century. They extracted the common behavior of the model AITG indices through an empirical orthogonal function analysis (Barnett, 1977), finding that a majority of the models exhibit a common upward AITG trend until around 1980, followed by a reversal. The ensemble-mean AITG upward trend from 1900–1982 was found to be statistically significant, but significantly smaller in magnitude than the observed. It meant that either there is a strong unforced component to the actual trend (which was entirely possible, as indicated by the AITG behavior in CMIP3 preindustrial simulations), but also that

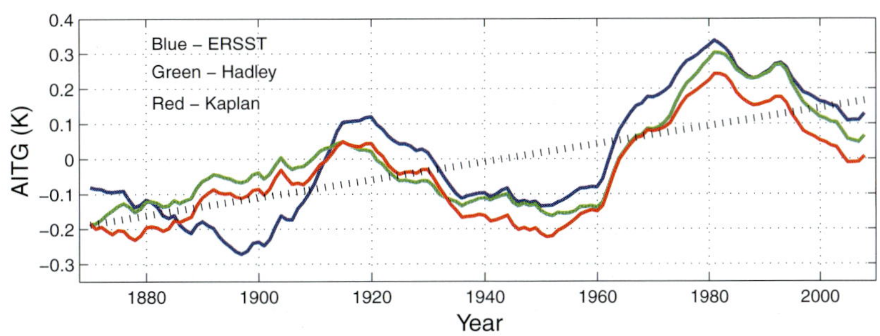

Fig. 7. The observed Atlantic Interhemispheric Gradient Index, defined as the SST difference between the North Tropical Atlantic (5°N–35°N, 0°W–80°W, northern box in top panel) and the South Tropical Atlantic (5°S–35°S, 60°W–20°E, southern box in top panel), South minus North. Each line is from an observational SST dataset as denoted: ERSST v3b (Smith *et al.*, 2008), Hadley Centre SST (Rayner *et al.*, 2003), and Kaplan Extended SST v2 (Kaplan *et al.*, 1998). The black dashed line is the linear least-square best-fit line to the average of the three curves. Figure from Chiang *et al.* (2013).

the models could be systematically underestimating the forced trend. Finally, an attribution analysis using a small set of single forcing runs indicated that the forced trend was predominantly due to anthropogenic sulfate aerosols.

Recently, Chiang *et al.* (2013) repeated the same Chang *et al.* (2011) analysis on the AITG with the CMIP5 multimodel ensemble, essentially confirming the broad conclusions of the CMIP3 study — namely, that there is a forced component to the trend, and primarily by anthropogenic aerosols. The attribution analysis in Chiang *et al.* (2013) was significantly more robust than Chang *et al.* (2011) because of the inclusion of many more single forcing runs in the analysis. Like the CMIP3, the CMIP5 also simulated a reversal of the trend around 1980, confirming that this reversal is also forced. The attributed causes of the reversal arises from two sources: (i) from the leveling of anthropogenic aerosol emissions in the 1970s with the introduction of legislation limiting atmospheric pollution; and (ii) to the increasing influence of anthropogenic greenhouse gases that warmed the northern hemisphere faster than the southern hemisphere.

The influence of anthropogenic aerosols may also have weakened the Atlantic cold tongue, via its influence on the AITG. Tokinaga and Xie (2011) examined observations of SST and surface winds over the past six decades in the equatorial Atlantic, showing a warming of the equatorial cold tongue relative to the SST around it, combined with a weakening of the easterly trades. Their analysis of CMIP3 historical simulations over the same time period showed a similar change to the equatorial cold tongue and trades, and correlated to changes in the interhemispheric SST gradient (Fig. 8). The link between the interhemispheric and equatorial zonal SST gradients are dynamically linked: a colder North/warmer South pattern leads to a weakening of the southeasterly trades, the latter which have a zonal component at the equator. The resulting weakening of surface equatorial easterlies reduces the east-west equatorial thermocline gradient and thus the equatorial zonal SST contrast.

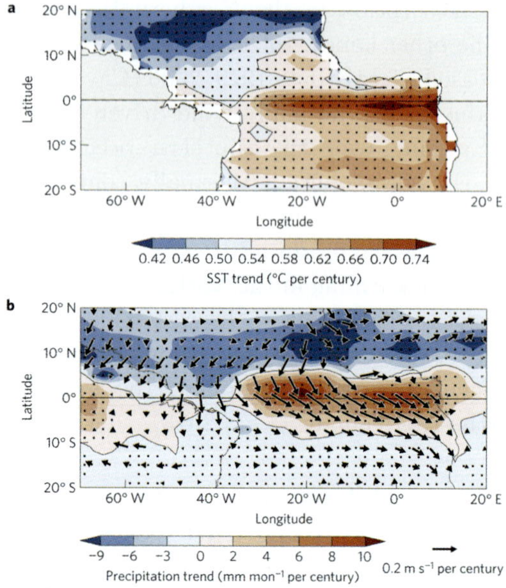

Fig. 8. From Tokinaga and Xie (2011), Figure 4. As quoted from their figure caption: "Patterns of simulated climate for 1900–1999. **a**, **b**, SST (colour) (a), surface wind (vector) and precipitation (colour) (b) for June–August, based on the ensemble mean of the climate of the twentieth-century experiments from 19 CMIP3 models. For SST and precipitation trends, grid points marked with filled circles exceed the 95% confidence level based on the Mann–Kendall test. For surface wind trend, only vectors exceeding the 95% confidence level are plotted."

4.2. *Monsoons*

Several recent papers have argued for the role of the interhemispheric thermal gradient in driving late 20th century weakening of to the North African and South Asian summer monsoons, in particular associated with aerosol forcing. The earliest suggestions came from AGCM-slab ocean simulations with imposed aerosol concentrations (Rotstayn *et al.*, 2000; Williams *et al.*, 2001). In particular, Rotstayn and Lohmann (2002) explored the effect of anthropogenic aerosols on the global climate in an AGCM-slab ocean model simulation with prescribed late 20th century aerosol concentrations and a parameterization for indirect effects. They noted a complex pattern of rainfall changes in the tropics, but grossly reflecting a southward shift in the tropical rainbands. They showed that the observed tropical rainfall trends resembled the modeled changes, albeit with smaller amplitudes. In particular, they focused on the Sahel drying which was prominent both in the model simulations and observational records. They concluded that the spatially-varying aerosol forcing can substantially alter tropical rainfall and low latitude circulations, and specifically proposed it as a mechanism for the observed Sahel drought in the latter half of the 20th century.

More recently, the Sahel hypothesis proposed by Rotstayn and Lohmann (2002) gained support from more comprehensive modeling study. Kawase *et al.* (2010) analyzed a fully coupled model simulation of the 20th century that showed a drying of the summer rainfall over North Africa. Their attribution analysis using single-forcing runs showed that the drying trend was primarly due to anthropogenic aerosols. Through a moisture-budget analysis, they inferred that aerosols reduced West African rainfall both through a thermodynamic effect of tropospheric cooling brought about by the aerosol forcing; and a dynamic effect through altering the local atmospheric circulation from changing the meridional SST gradient in the tropical Atlantic. The analysis of Kawase *et al.* (2010) thus showed the possible influence of anthropogenic aerosols on Sahel drought in a comprehensive coupled climate model, and furthermore pointed to specific mechanisms of how remote anthropogenic aerosol forcing alters tropical rainfall.

Anthropogenic aerosols have also been implicated in the weakening of the South Asian monsoon rainfall over the latter half of the 20th century. Using a coupled modeling approach similar to Kawase *et al.* (2010), Bollasina *et al.* (2011) showed that the all-forcing simulations was able to reproduce the downward trend in the South Asian monsoon rainfall over the latter half of the 20th century, and with comparable magnitude. Their attribution using single-forcing runs showed anthropogenic aerosols to be the cause of the simulated drying. Bollasina *et al.* (2011) explained the drying effect by aerosols to be a consequence of a slowdown in the tropical meridional overturning circulation, compensating for the aerosol-induced energy imbalance between the hemispheres; by this interpretation, they highlight the potential contribution of remote aerosols to the drying (although their analysis did not explicitly show this). In this sense, their interpretation bear resemblance to energy flux interpretation of Kang *et al.* (2009) for extratropical thermal forcing on tropical rainfall. Interestingly, a modeling study by Cowan and Cai (2011) found a similar Asian monsoon response to anthropogenic aerosols in the CSIRO coupled model, but they also tested the relative roles of Asian and non-Asian roles in this drying. They show that Asian (i.e., local) aerosols only drive a weak drying of rainfall, with a larger contribution by remote aerosols.

While anthropogenic aerosols have been shown to drive a weakening of the monsoons over the latter half of the 20th century, its influence is gradual (a trend), reflecting the temporal history of anthropogenic aerosol emissions. This is in direct contrast to changes in the Sahel rainfall that abruptly weakened in the late 1960s, suggesting that the cause of this abrupt shift was not due to aerosols. Folland *et al.* (1986) showed in 1986 that the Sahel drought was linked to global SST temperature changes resembling an interhemispheric pattern. Given that the interhemispheric SST gradient underwent an abrupt shift in the late 1960s towards anomalously cooler conditions in the North (Thompson *et al.*, 2010), the suggestion is that the SST changes drove the Sahel drought.

Liu and Chiang (2012) explored this idea from observations and model simulations. They attempted to extract the signatures of a North Atlantic influence on the Sahel and Asian monsoon by focusing on surface temperature and pressure anomalies over Eurasia and North Africa — AGCM-slab ocean model simulations of extratropical North Atlantic cooling showed that Eurasian and North African temperatures cooled, and sea-level pressure increased (similar to the pattern seen in the AMOC shutdown simulation of Fig. 2). Using a combined EOF method on surface temperature, pressure, and rainfall over Eurasia and North Africa, they extracted a leading mode that spatially resembled the impact of North Atlantic cooling, and with a temporal signature of a late 1960s shift. They were able to achieve a similar result with analyzing an ensemble of AMIP-type runs where the observed SST was prescribed; this indicated that the information of the 1960s shift resided in the surface ocean temperature. Liu and Chiang (2012) concluded that the observed climate changes associated with Sahel drought were consistent with the influence of North Atlantic cooling, though they did not establish a causal relationship.

There are observational suggestions of a global climate shift in the late 1960s, not just with Sahel rainfall. Baines and Folland (2007) found evidence for a rapid shift in atmospheric circulation around that time associated with changes to the interhemispheric SST difference, and largely in tropical regions like rainfall Amazon basin and Northeast Brazil; and sea level pressure and SST changes over the tropical North Atlantic and west and central Pacific. They also find changes to the Southern Hemisphere midlatitude circulation — subtropical jet and storm track — and possible changes to the Antarctic sea ice boundary. Interestingly, the global climate changes described in Baines and Folland (2007) are reminiscent of those seen during the abrupt events of the last glacial; it suggests a common origin with global teleconnections resulting from North Atlantic cooling.

5. Global Interhemispheric Changes in the 20th Century and Future

Future forced changes to the interhemispheric pattern will dominated by anthropogenic greenhouse gas and aerosol emissions, as highlighted in Section 3. However, unlike the 20th century forced changes to the interhemispheric pattern where the two influences acted in opposition, in the future they will likely both act in the same direction as anthropogenic aerosol emissions are projected to decrease in the 21st century (Lamarque *et al.*, 2011).

Friedman *et al.* (2013) examined simulations of an index in the global interhemispheric surface temperature (land + ocean) difference — the Northern Hemisphere minus Southern Hemisphere mean surface air temperature, which Friedman *et al.* (2013) coined the 'Interhemispheric Temperature Asymmetry' (ITA) index — in the Coupled Model Intercomparison Project (CMIP) phase 3 and phase 5 simulations of historical and future climates. Over the 20th century, the observed ITA stayed relatively flat with no significant trend; the most pronounced feature was an abrupt shift in the late 1960s, similar to what Thompson *et al.* (2010) found in the interhemispheric SST gradient. However, around 1980 a significant upward trend towards more positive ITA started to emerge (Fig. 8, green line), with the North warming significantly faster than the South.

The CMIP3 and 5 multimodel simulations of the ITA was able to reproduce the observed relative lack of trend in the 20th century, and the emergence of a significant positive trend starting around 1980; this trend continues upward in all future simulations examined (Fig. 9), reaching values that well exceed the 20th century range by the end of the 21st century. Interestingly, the late 1960s shift in the ITA is not simulated, suggesting that the shift was not a response to climate forcing.

Attribution studies using single-forcing runs showed that prior to 1980, a positive trend

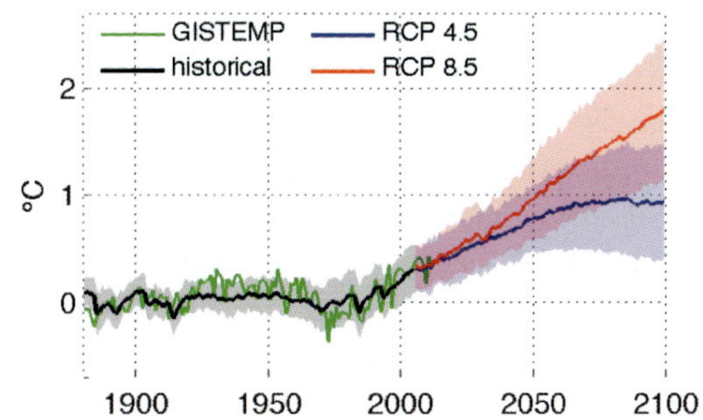

Fig. 9. Interhemispheric Temperature Asymmetry (ITA), defined as the surface air temperature averaged over the northern hemisphere, minus that for the southern hemisphere. The observed ITA is shown in green (derived from GISTEMP), and CMIP5 multimodel ensemble mean in black. The multimodel ensemble mean for the future scenarios RCP 4.5 and 8.5 are in blue and red, respectively. Shading represents one ensemble standard deviation. The simulated ITA remains relatively flat throughout most of the 20th century, then trends upward from around 1980, extending into the future scenarios. From Friedman *et al.* (2013).

induced by increasing greenhouse gases was essentially negated by anthropogenic aerosol emissions that drove the ITA trend in the opposite direction; it was only after the leveling-off of sulfate aerosol emissions in the mid-1970s and its subsequent decline, that the greenhouse-gas induced ITA trend could finally manifest itself (Chiang *et al.* (2013) came to a similar interpretation for the reversal in the AITG trend after the late 1970s). While Friedman *et al.* (2013) did not extend their attribution study to future projections, anthropogenic greenhouse gas and aerosols are both likely to contribute significantly to the projected ITA trend, and in the same direction, as anthropogenic aerosol emissions are projected to decrease in the 21st century. As an indication of the relative importance of aerosols, Levy *et al.* (2013) show that the reduction to anthropogenic aerosol emission in the RCP 4.5 scenario contributes around 0.9K global warming in the new GFDL coupled chemistry-climate model CM3 by the end of the 21st century. Since this warming is concentrated largely in the Northern Hemisphere (see their Fig. 6), it indicates a ~1K contribution to the interhemispheric difference.

What does the positive future trend in the ITA mean for tropical rainfall? Naively, one would expect a more northern position of the mean tropical rainfall position; and indeed Friedman *et al.* (2013) shows that the CMIP5 simulated rainfall indeed shift progressively more northwards into the 21st century. However, from energy flux arguments the tropical rainfall should more strictly follow the interhemispheric contrast in energy fluxes and not temperature; predicting the latter, however, is nontrivial. A study by Frierson and Hwang (2012) is instructive in this regard: they examined nine AGCM-SOM simulations of doubled CO_2 simulations from several modeling groups, finding a range of latitudinal ITCZ shifts in these simulations, and in both directions. They attributed the range in the simulated ITCZ responses to modeled differences in the cloud and ice responses in the extratropics of both hemispheres, which in turn altered the nature of the interhemispheric energy flux contrast.

The variety in the AGCM-slab ocean ITCZ responses to global warming as found by Frierson and Hwang (2012) are likely to carry over into fully coupled model simulations of future warming, since similar radiative feedbacks operate in both. Another factor to consider in future projections is the anthropogenic aerosol influence, both in the nature of the future decrease, as well as the strength of the influence especially with regards to the indirect effect (Levy *et al.*, 2013). Moreover, changes to the cross-equatorial ocean heat transports, in particular the response of the AMOC to global warming, is likely to factor into the future projections: most CMIP5 coupled model simulations of future scenarios show weakening of the AMOC (Cheng *et al.*, 2013), and which would drive a *southward* shift in the ITCZ. Finally, other non-atmospheric effects such as land use change may also be able to alter the interhemispheric contrast in forcing due to changes in albedo (Swann *et al.*, 2012), though this effect still has to be quantified.

6. Discussion

This paper outlines an emerging hypothesis for the influence of extratropical thermal forcing on the tropical climate through a global atmospheric teleconnection, one characterized by an interhemispheric pattern in temperature and surface pressure and a shift in the tropical rainband to the anomalously warmer hemisphere. I specifically focus on the potential applicability of this hypothesis to interpreting multidecadal changes to 20th century and future projected changes to tropical rainfall, in particular over the Atlantic ITCZ and North African and Asian monsoons. The plausibility for this hypothesis rests of several lines of evidence:

1. Empirical evidence from paleoproxies that such a global rearrangement of climate indeed occurred during abrupt climate changes during the last glacial, and tied specifically to pronounced cooling events over Greenland.
2. Consistent modeling evidence from various models (especially in the AGCM-slab ocean configuration) and various forcing types for an interhemispheric pattern of response to extratropical thermal forcing.
3. A quantitative and consistent interpretation for the influence of extratropical thermal forcing on the tropical climate from energy flux constraints, specifically tying tropical circulation changes to cross-equatorial heat transport changes to compensate for hemispheric differences the TOA and surface energy flux.

Further advancement of this hypothesis rests on two separate lines of development. First, there is a need to put this teleconnection on a more secure mechanistic footing. The energy flux constraint, whilst able to provide quantitative interpretation for the tropical rainfall change, is still essentially a diagnostic analysis, and applicable only to zonal means. The development of a mechanistic theory — akin to Hoskins and Karoly (1981) for the ENSO teleconnection — will be important in expanding the acceptance and applicability of this hypothesis. In particular, the theory will have to reconcile the energy flux interpretation of tropical rainfall changes on the one hand, with the more traditional viewpoint of tropical SST control of the structure of tropical rainfall (*e.g.*, Xie *et al.*, 2010).

Advancement will also rest on how well the hypothesis is able to lend insight into the dynamics of the real tropical rainfall — both the mean state structure and its change with time. There has been some success already in interpreting 20th century tropical rainfall changes — as highlighted in this review — but further advances in this regard may be limited by the lack of reliable observational rainfall data over the long-term. A more likely avenue for advancement would be from applying the hypothesis to climate model simulations of the tropical climate and its changes. For example, a recent and interesting result by Hwang and Frierson (2013) showed that coupled model simulations with more energy flux into the southern hemisphere atmosphere have stronger double-ITCZ bias. They specifically find that cloud biases over the Southern Ocean explain the most model-to-model differences in the excess precipitation in the southern Tropics. The hypothesis presented here offers a potential explanation for this in that the ITCZ is drawn towards the southern hemisphere extratropical heating. Thus in this view, the tropical ITCZ bias problem rests not with tropical ocean-atmosphere processes, but with the Southern hemisphere midlatitude processes! Should this prove to be correct, it would provide an important breakthrough to a longstanding problem in climate simulation.

Acknowledgments

JC acknowledges the National Science Foundation (AGS-1143329) and the Office of Science Biological and Environmental Research, Department of Energy (DE-FG02-08ER64588) for funding. He thanks Leon Rotstayn and a reviewer for helpful comments, Yen-Ting Hwang for providing Fig. 3, and Ivana Cvijanovic for providing Fig. 4.

References

Alley, R. B., 2004: Abrupt climate change. *Sci. Am.*, **291**, 62–69.

———, 2007: Wally was right: Predictive ability of the North Atlantic "Conveyor belt" hypothesis for abrupt climate change. *Annu. Rev. Earth Pl. Sc.*, **35**, 241–272.

Anderson, R. F., S. Ali, L. I. Bradtmiller, S. H. H. Nielsen, M. Q. Fleisher, B. E. Anderson, and L. H. Burckle, 2009: Wind-driven upwelling in the Southern Ocean and the deglacial rise in atmospheric CO_2. *Science*, **323**, 1443–1448.

Baines, P. G., and C. K. Folland, 2007: Evidence for a rapid global climate shift across the late 1960s. *J. Climate*, **20**, 2721–2744.

Barbante, C., *et al.*, 2006: One-to-one coupling of glacial climate variability in Greenland and Antarctica. *Nature*, **444**, 195–198.

Barnett, T. P., 1977: Principal time and space scales of Pacific trade wind fields. *J. Atmos. Sci.*, **34**, 221–236.

Bollasina, M. A., Y. Ming, and V. Ramaswamy, 2011: Anthropogenic aerosols and the weakening of the South Asian summer monsoon. *Science*, **334**, 502–505.

Boucher, O., and M. Pham, 2002: History of sulfate aerosol radiative forcings. *Geophys. Res. Lett.*, **29**, 22-1–22-4.

Broecker, W. S., D. M. Peteet, and D. Rind, 1985: Does the ocean-atmosphere system have more than one stable mode of operation? *Nature*, **315**, 21–26.

Cai, W., D. Bi, J. Church, T. Cowan, M. Dix, and L. Rotstayn, 2006: Pan-oceanic response to increasing anthropogenic aerosols: Impacts on the Southern Hemisphere oceanic circulation. *Geophys. Res. Lett.*, **33**(21).

Cavalieri, D. J., and C. L. Parkinson, 2008: Antarctic sea ice variability and trends, 1979–2006. *J. Geophys. Res.*, **113**, C07004.

Chang, C. Y., J. C. H. Chiang, M. F. Wehner, A. R. Friedman, and R. Ruedy, 2011: Sulfate aerosol control of tropical Atlantic climate over the twentieth century. *J. Climate*, **24**, 2540–2555.

Cheng, W., C. M. Bitz, and J. C. H. Chiang, 2007: Adjustment of the global climate to an abrupt slowdown of the Atlantic meridional overturning circulation. *Past and Future Changes of the Ocean's Meridional Overturning Circulation: Mechanisms and Impacts*, A. Schmittner, J. C. H. Chiang, and S. Hemming, Eds., American Geophysical Union, 295–313.

———, J. C. H. Chiang, and D. Zhang, 2013: Atlantic Meridional Overturning Circulation (AMOC) in CMIP5 models: RCP and historical simulations. *J. Climate*, **26**, 7187–7197.

Chiang, J. C. H., 2009: The Tropics in paleoclimate. *Annu. Rev. Earth. Pl. Sc.*, **37**, 263–297,

———, and C. M. Bitz, 2005: Influence of high latitude ice cover on the marine Intertropical Convergence Zone. *Clim. Dyn.*, **25**, 477–496.

———, C.-Y. Chang, and M. F. Wehner, 2013: Long-term behavior of the Atlantic interhemispheric SST gradient in the CMIP5 historical simulations. *J. Climate*, in press.

———, Y. Kushnir, and A. Giannini, 2002: Deconstructing Atlantic Intertropical Convergence Zone variability: Influence of the local cross-equatorial sea surface temperature gradient and remote forcing from the eastern equatorial Pacific. *J. Geophys. Res. Atmos.*, **107**.

Cowan, T., and W. Cai, 2011: The impact of Asian and non-Asian anthropogenic aerosols on 20th century Asian summer monsoon. *Geophys. Res. Lett.*, **38**.

Cvijanovic, I., and J. C. H. Chiang, 2013: Global energy budget changes to high latitude North Atlantic cooling and the tropical ITCZ response. *Clim. Dyn.*, **40**, 1435–1452.

Feulner, G., S. Rahmstorf, A. Levermann, and S. Volkwardt, 2013: On the origin of the surface air temperature difference between the hemispheres in Earth's present-day climate. *J. Climate*, **26**, 7136–7150, doi: 10.1175/JCLI-D-12-00636.1.

Folland, C. K., T. N. Palmer, and D. E. Parker, 1986: Sahel rainfall and worldwide sea temperatures, 1901–85. *Nature*, **320**, 602–607.

Friedman, A. R., Y.-T. Hwang, J. C. H. Chiang, and D. M. W. Frierson, 2013: Interhemispheric temperature asymmetry over the 20th century and in future projections. *J. Climate*, **26**, 5419–5433.

Frierson, D. M. W., and Y. T. Hwang, 2012: Extratropical influence on ITCZ shifts in slab ocean simulations of global warming. *J. Climate*, **25**, 720–733.

Ganopolski, A., and S. Rahmstorf, 2001: Rapid changes of glacial climate simulated in a coupled climate model. *Nature*, **409**, 153–158.

Grootes, P. M., M. Stuiver, J. W. C. White, S. Johnsen, J. Jouzel, 1993: Comparison of oxygen-isotope records from the Gisp2 and Grip Greenland ice cores. *Nature*, **366**, 552–554.

Hoskins, B. J., and D. J. Karoly, 1981: The steady linear response of a spherical atmosphere to thermal and orographic forcing. *J. Atmos. Sci.*, **38**, 1179–1196.

Hwang, Y.-T., and D. M. W. Frierson, 2013: A link between the double intertropical convergence zone problem and cloud biases over the Southern Ocean. *Proc. Nat. Acad. Sci.*, **110**(13), 4935–4940.

Kang, S. M., 2009: *The Response of Tropical Precipitation to Extratropical Thermal Forcing.* Program in Atmospheric and Oceanic Sciences, Princeton University, 156 pp.

Kang, S. M., D. M. W. Frierson, and I. M. Held, 2009: The tropical response to extratropical thermal forcing in an idealized GCM: The importance of radiative feedbacks and convective parameterization. *J. Atmos. Sci.*, **66**, 2812–2827.

Kawase, H., M. Abe, Y. Yamada, T. Takemura, T. Yokohata, and T. Nozawa, 2010: Physical mechanism of long-term drying trend over tropical North Africa. *Geophys. Res. Lett.*, **37**(9).

Knight, J. R., R. J. Allan, C. K. Folland, M. Vellinga, and M. E. Mann, 2005: A signature of persistent natural thermohaline circulation cycles in observed climate. *Geophys. Res. Lett.*, **32**(20).

Lamarque, J. F., G. P. Kyle, M. Meinshausen, K. Riahi, S. J. Smith, D. P. van Vuuren, A. J. Conley, and F. Vitt, 2011: Global and regional evolution of short-lived radiatively-active gases and aerosols in the Representative Concentration Pathways. *Climatic Change*, **109**(1–2), 191–212.

———, *et al.*, 2010: Historical (1850–2000) gridded anthropogenic and biomass burning emissions of reactive gases and aerosols: Methodology and application. *Atmos. Chem. Phys.*, **10**, 7017–7039.

Latif, M., *et al.*, 2004: Reconstructing, monitoring, and predicting multidecadal-scale changes in the North Atlantic thermohaline circulation with sea surface temperature. *J. Climate*, **17**, 1605–1614.

Lee, S. Y., J. C. H. Chiang, K. Matsumoto, and K. S. Tokos, 2011: Southern Ocean wind response to North Atlantic cooling and the rise in atmospheric CO_2: Modeling perspective and paleoceanographic implications, *Paleoceanography*, **26**, PA1214, doi: 10.1029/2010PA002004.

Levy II, H., L. W. Horowitz, M. D. Schwarzkopf, Y. Ming, J.-C. Golaz, V. Naik, and V. Ramaswamy, 2013: The roles of aerosol direct and indirect effects in past and future climate change, *J. Geophys. Res. Atmos.*, **118**, doi: 10.1002/jgrd.50192.

Lindzen, R. S., and A. Y. Hou, 1988: Hadley circulations for zonally averaged heating centered off the equator. *J. Atmos. Sci.*, **45**, 2416–2427.

Liu, Y., and J. C. H. Chiang, 2012: Coordinated abrupt weakening of the Eurasian and North African monsoons in the 1960s and links to Eextratropical North Atlantic cooling. *J. Climate*, **25**, 3532–3548.

Mann, M. E., and K. A. Emanuel, 2006: Atlantic hurricane trends linked to climate change. *EOS, Transactions American Geophysical Union*, **87**, 233–241.

Peterson, L. C., G. H. Haug, K. A. Hughen, and U. Rohl, 2000: Rapid changes in the hydrologic cycle of the tropical Atlantic during the last glacial. *Science*, **290**, 1947–1951.

Rotstayn, L. D., and U. Lohmann, 2002: Tropical rainfall trends and the indirect aerosol effect. *J. Climate*, **15**, 2103–2116.

Rotstayn, L. D., B. F. Ryan, and J. E. Penner, 2000: Precipitation changes in a GCM resulting from the indirect effects of anthropogenic aerosols. *Geophys. Res. Lett.*, **27**, 3045–3048.

Stroeve, J., M. M. Holland, W. Meier, T. Scambos, and M. Serreze, 2007: Arctic sea ice decline: Faster than forecast. *Geophys. Res. Lett.*, **34**(9).

Swann, A. L. S., I. Y. Fung, and J. C. H. Chiang, 2012: Mid-latitude afforestation shifts general circulation and tropical precipitation. *P. Natl. Acad. Sci. USA*, **109**, 712–716.

Terray, L., 2012: Evidence for multiple drivers of North Atlantic multi-decadal climate variability. *Geophys. Res. Lett.*, **39**(19).

Thompson, D. W. J., J. M. Wallace, J. J. Kennedy, and P. D. Jones, 2010: An abrupt drop in Northern Hemisphere sea surface temperature around 1970. *Nature*, **467**, 444–447.

Ting, M. F., Y. Kushnir, R. Seager, and C. H. Li, 2009: Forced and internal twentieth-century SST trends in the North Atlantic. *J. Climate*, **22**, 1469–1481.

Tokinaga, H., and S. P. Xie, 2011: Weakening of the equatorial Atlantic cold tongue over the past six decades. *Nat. Geosci.*, **4**, 222–226.

Trenberth, K. E., G. W. Branstator, D. Karoly, A. Kumar, N. C. Lau, and C. Ropelewski, 1998: Progress during TOGA in understanding and modeling global teleconnections associated with tropical sea surface temperatures. *J. Geophys. Res. Oceans*, **103**, 14291–14324.

Vellinga, M., and R. A. Wood, 2002: Global climatic impacts of a collapse of the Atlantic thermohaline circulation. *Climate Change*, **54**, 251–267.

Voelker, A. H. L., 2002: Global distribution of centennial-scale records for Marine Isotope Stage (MIS) 3: A database. *Quat. Sci. Rev.*, **21**, 1185–1212.

Wang, Y. J., H. Cheng, R. L. Edwards, Z. S. An, J. Y. Wu, C. C. Shen, and J. A. Dorale, 2001: A high-resolution absolute-dated Late Pleistocene monsoon record from Hulu Cave, China. *Science*, **294**, 2345–2348.

Williams, K. D., A. Jones, D. L. Roberts, C. A. Senior, and M. J. Woodage, 2001: The response of the climate system to the indirect effects of anthropogenic sulfate aerosol. *Clim. Dyn.*, **17**, 845–856.

Xie, S. P., C. Deser, G. A. Vecchi, J. Ma, H. Y. Teng, and A. T. Wittenberg, 2010: Global warming pattern formation: Sea surface temperature and rainfall. *J. Climate*, **23**, 966–986.

Yoshimori, M., and A. J. Broccoli, 2008: Equilibrium response of an atmosphere-mixed layer ocean model to different radiative forcing agents: Global and zonal mean response. *J. Climate*, **21**, 4399–4423.

Zhang, R., and T. L. Delworth, 2005: Simulated tropical response to a substantial weakening of the Atlantic thermohaline circulation. *J. Climate*, **18**, 1853–1860.

Chapter 17

Climate of China in the Holocene

Wang Shaowu[1], Wen Xinyu[1], and Huang Jianbin[2]

[1] *Laboratory for Climate and Ocean-Atmosphere Studies, Department of Atmospheric and Oceanic Sciences, School of Physics, Peking University, Beijing 100871, China*

[2] *The Center for Earth System Science and the Institute for Global Change Studies, Tsinghua University, Beijing 100084, China*

swwang@pku.edu.cn

Studies of climatic change and variability in China during the Holocene are reviewed in this paper. The following key issues are discussed; (1) long-term climate trend in the Holocene, (2) East Asian summer and winter monsoons (EASM and EAWM) in the Holocene, (3) abrupt change of monsoons in the Holocene, (4) climate and monsoons in the last 2k-yr, (5) climate changes since AD 1880.

1. Introduction

Progress in the study of paleo-climate largely depends on the availability of proxy data. There are plentiful documentations in China, in which many anomalous climatic events were recorded. The study of the documentary data in China has been extensively developed since the 1970s and 1980s, following the pioneering paper from Zhu (1972), a milestone in the studies of climate change in China. For example, an Atlas of Drought and Flood in China for the last 500 years was issued in 1981 (CMI, 1981). Furthermore, the variations of seasonal mean temperature and precipitation in China for the last 500 yr were reconstructed by Wang and Zhao (1981) and Wang (1991). In addition, studies on the climate change in China were extended to two millennia by Zhang (1991, 2005), Zhang *et al.* (1997), Yang *et al.* (2002), and Ge *et al.* (2003). An Atlas of seasonal mean temperature and precipitation anomalies in China for the period 1880–2007 was compiled by combining the documentary data with observations (Wang *et al.*, 2009b).

However, there are also important difficulties in using documentary data. First, documentary data are confined to eastern China and therefore cannot present an effective climate signal in the west. Second, the temporal gaps and the spatially uneven geographical distribution of documentations prevent us from reconstructing the climate in the past, especially before AD 1470, near the beginning of Ming Dynasty. Finally, digitization of the historical data and its interpretation is still a big challenge to the paleo-climate community.

Despite the historical data, ice cores also present unique long proxy records in west China with annual resolution (Xiao *et al.*, 2008; Wang *et al.*, 2006). It differs from the tree-ring data, which gives little information about low frequency climate variability, and therefore is usually applied to examine the high frequency variability of climate such as inter-annual to decadal time scales (Shao *et al.*, 2005; Xu and

Climate Change: Multidecadal and Beyond
edited by Chih-Pei Chang et al.

Shao, 2007). Understanding of climate change during the Holocene has been greatly improved in recent decades. For example, pollen data has been analyzed in detail by Shi and Kong (1992) in the reconstruction of the Holocene climate of China (Wang *et al.*, 2001). The use of stalagmite data has increased significantly in recent years. The records of δ^{18}O provide clear information on the intensity of Asian summer monsoon (Wang *et al.*, 2005; Zhang *et al.*, 2008). A comprehensive review of the studies of climate change by using stalagmite data has been made by Man Wenmin (personal communication). Sediment data in the coastal areas of the Asian continent (Xiao *et al.*, 2005a,b; Xiang *et al.*, 2006) also provide useful information about East Asian monsoons.

China is located on the east coast of the Eurasian continent, where monsoons dominate the annual cycle of local climate. Temperatures and precipitations vary significantly depending on the development of monsoons, especially in eastern China. Therefore, climate change in China should be closely related to the activities of the monsoons over East Asia.

Climate Change during the Holocene is analyzed in the present paper along different time scales. First, an introduction is given in Section 1. The climatic trend during the Holocene is examined in Section 2. Next, studies of the variations and variability of the East Asian monsoon are synthesized in Sections 3 and 4. The climate and monsoons in China during the last 2k-yr, and since AD 1880, are discussed in Sections 5 and 6. Finally, a discussion is made in Section 7.

2. Climate Trend During the Holocene

Figure 1 gives temperature changes during the last 10k years (C^{14} year) based on pollen records (Wang *et al.*, 2001). Original pollen data are taken from "The Climates and Environments of Holocene Megathermal in China" edited by Shi and Kong (1992). It shows the Megathermal occurred from 8.5 to 3.0k years BP. Maxima of temperature were found in between 7.5~ 5.5k years BP (C^{14} year), which is about 8.0~ 6.0k years in calendar years. Unfortunately, there is no comprehensive temperature series during the Holocene that covers all of China and synthesizes all the different types of proxy data presently available, such as δ^{18}O of ice core, δ^{18}O of stalagmite and so on. Therefore, Fig. 1 provides only an estimate of the climate in China during the Holocene with the following caveats to be noted. First, the pollen data primarily reflects to a great extent the temperatures in the warm season, rather than the whole year. Temperature variability is usually smaller in summers than in winters, so Fig. 1 may underestimate the amplitude of temperature changes of the Holocene. Furthermore, winter temperature may or may not be positively correlate with that in summers, and other seasons. So, representativeness of pollen data to the annual mean temperatures is still controversial.

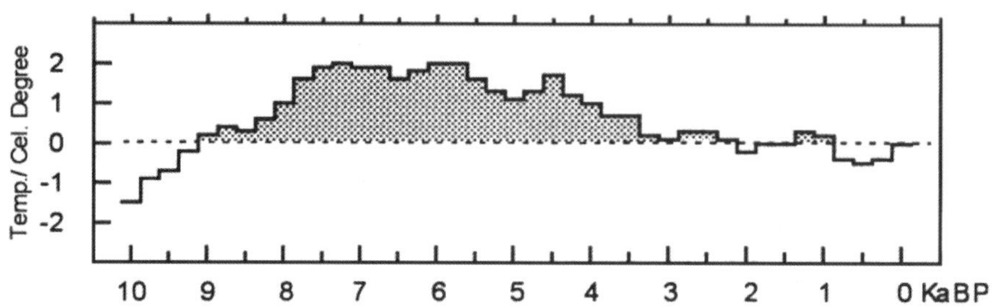

Fig. 1. Temperature anomalies in China in the last 10k years BP (C^{14} year) (Wang *et al.*, 2001).

Second, the temperature series of China shown in Fig. 1 is synthesized from ten regional series, in which the borders between the regions are defined by a strictly statistical analysis (Wang *et al.*, 1998). So, this result differs from a number of other series, which cover only a part of the territory or the coverage changes significantly with time. Finally, it should be noted that the phase and the amplitude of temperature maximum in the Holocene varies from region to region. For example, temperature maximum in Xinjiang is found in early Holocene rather than in early-middle Holocene as shown in Fig. 1.

Precipitation in China shows a greater variability than does temperature. So, it is more difficult to construct a series representative of China as a whole. Precipitation formation varies as monsoon changes from summer, when winds come from oceans to the southwest or southeast of China, to winter when winds blow from northwest and north. Therefore, we have to examine precipitation changes by seasons. Since the summer monsoon accompanies far more moisture than the winter monsoon, we shall concentrate on the analysis of the summer precipitation in China. Zhou (2011) has provided a stacked monsoon precipitation index based on stalagmite $\delta^{18}O$ signals (Fig. 2). The black thick line in Fig. 2 shows the changes of moisture in monsoonal China. The maxima were found between 10 to 7k-yr BP. The other reconstructions (dashed line, dashed and solid lines in grey in Fig. 2a) support to some extent the results of the stalagmite $\delta^{18}O$ analysis.

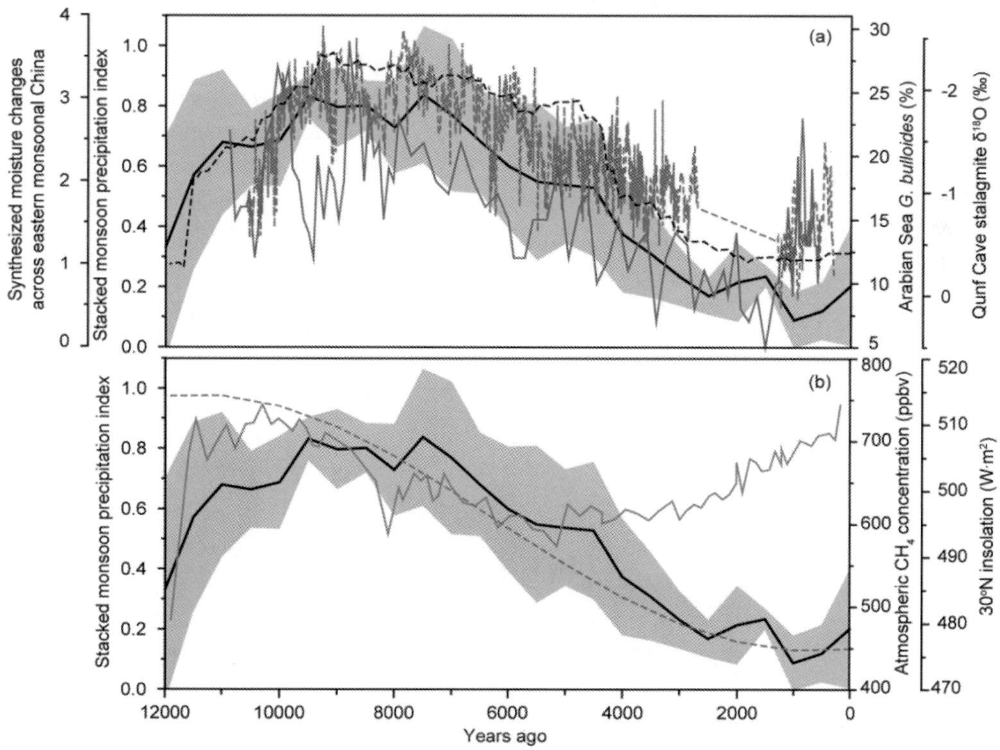

Fig. 2. (a) The stacked monsoon precipitation index of monsoonal China (black thick line) with shading indicating one standard deviation ($\pm\sigma$). (b) Dashed and solid lines in grey show atmospheric CH_4 and summer insolation at 30°N, respectively (Zhou, 2011) along with the precipitation index from a.

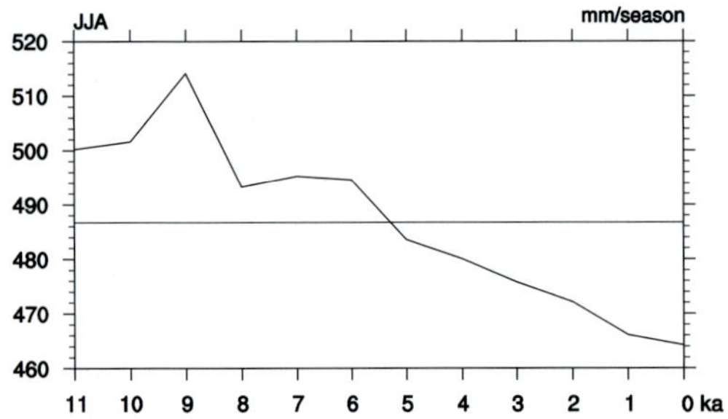

Fig. 3. Simulated summer precipitation (mm) over China (Wang *et al.*, 2009a).

While the representativeness of stalagmite $\delta^{18}O$ to the monsoonal precipitation has been questioned (Tan, 2009, 2011), the decreasing trend of the precipitation index during the Holocene clearly follows the insolation changes at 30°N. Both temperature and precipitation in China during the Holocene vary simultaneously with the changes of insolation, which can be considered as the predominant role taken by the orbital factors. Numerical simulation using the CAM2 model, forced by insolation only associated with the orbital factors (Wang *et al.*, 2009a), gives the similar precipitation changes in China (Fig. 3). It is indicated that decreasing insolation in the Holocene may cause cooling over land, reduce the land-sea temperature difference, and weaken summer monsoon rainfall in China.

The maximum of summer rainfall was found before 8 k-yr (Figs. 2 and 3). However, the maximum of temperature appeared since 8 k-yr (Fig. 1). It must note that the timing in Fig. 1 was shown in ^{14}C years. Therefore, the real time (timing in calendar years) of maximum may occur earlier than that shown in Fig. 1.

3. East Asian Monsoons in the Holocene

It is suggested that stalagmite $\delta^{18}O$ characterizes the East Asian summer monsoon (Wang *et al.*, 2005). Figure 4 indicates the decreasing trend of summer monsoon during the Holocene. Increasing of $\delta^{18}O$ corresponds to decreasing of precipitation, and associated monsoon intensity. Further, decreasing of summer monsoon intensity over East Asia is in accordance with monsoon changes in other regions around the world, such as in the Arabian Sea (Gupta *et al.*, 2003), South Asia (Fleitmann *et al.*, 2003), North Africa (deMenocal *et al.*, 2000), and North America (Haug *et al.*, 2001). The opposite change (increasing monsoon precipitation) observed in South America (Maslin and Burns, 2000) supports the conclusion that the summer monsoon has changed during the Holocene under the control of orbital factors.

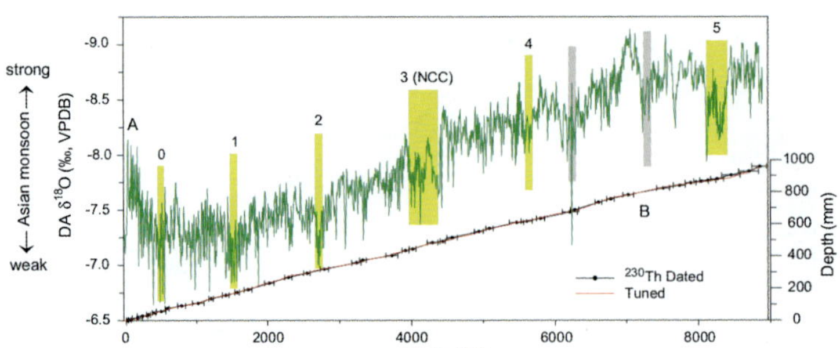

Fig. 4. A. The time series of Dongge cave stalagmite $\delta^{18}O$. The vertical yellow bars denote the timing of Bond events 0 to 5, two grey bars show the weak summer monsoon events that were not numbered by Bond *et al.* (1997), and NCC means the collapse of the Neolithic Culture of China. B. Age-depth relations shown by ^{230}Th dates with 2σ errors (Wang *et al.*, 2005).

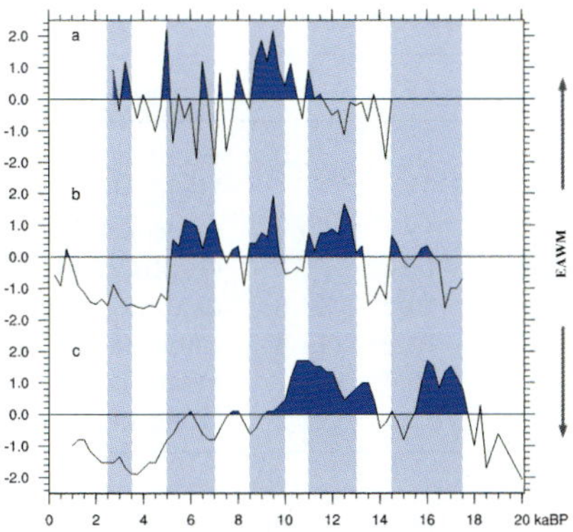

Fig. 5. Normalized anomalies of East Asian winter monsoons since the last deglaciation. (a) Grain size >63 μm% in Lake Tuolekale, Xinjiang, China (An *et al.*, 2011), (b) A · g in Huguang Maar Lake, South China (Wang *et al.*, 2012), and (c) Δ SST in the South China Sea (Huang *et al.*, 2011). Periods with strong winter monsoons are shaded.

It is usually postulated that the monsoon varies in winter opposite to that in summer. This implies that the winter monsoon will intensify when the summer monsoon weakens. However, this may or may not be true depending on the time scales. During the Holocene, the winter monsoon over East Asia seems to be weakened, rather than intensified, with time. Figure 5 provides three time series that characterize the winter monsoon intensity since the last deglaciation. In Fig. 5, the grain size of lacustrine sediments (An *et al.*, 2011), sedimentary diatom assemblages (Wang *et al.*, 2012), and alkenone-based sea surface temperature (Huang *et al.*, 2011) are used to form the winter monsoon intensity series. The original data are normalized. The periods with strong winter monsoon are shaded. The weakening trend of the winter monsoon during the Holocene is more clearly illustrated in Figs. 5b and 5c, though uncertainties remain in the series. Increasing of the winter insolation in the Northern Hemisphere during the Holocene (Kutzbach and Guetter, 1986) coincide with the weakening of winter monsoon over East Asia. Therefore, both the summer monsoon and the winter monsoon were weakened throughout the Holocene following the decline of the solar insolation. No opposite trend was observed between the winter and summer monsoons over East Asia during the Holocene.

4. Abrupt Changes of the Monsoons

The long-term trend of the monsoons over East Asia has been punctuated by a number of abrupt climate change events, each characterized by abrupt increasing of ice-rafted debris (IRD). Hematite-stained grains and Icelandic glass were used to identify the ice-rafting events in the North Atlantic, which were widely accepted as abrupt climate change events in the Holocene and numbered by Bond *et al.* (1997). Examination of ~50 globally distributed paleoclimate records reveals a comprehensive response of global climate to the cold events around the North Atlantic (Mayewski *et al.*, 2004).

Signals of abrupt changes in the monsoons over East Asia are presented in Tables 1 and 2. It is found that the winter monsoon was stronger than normal and the summer monsoon was weaker than normal during the periods of cold events in the North Atlantic. However, the number of events with weaker summer monsoons or stronger winter monsoons is greater than that listed in Tables 1 and 2. For example, weakening summer monsoons was also found in 6.2–6.3k-yr BP and 7.2–7.3k-yr BP as shown in Fig. 4 (gray bars), but these two time intervals were not included in the cold event list of Bond *et al.* (1997), although some relatively weak maxima of IRD were also seen in the curves of Bond *et al.* (1997). The predominant influence of cold events in the North Atlantic on the weakening of summer monsoons and intensifying of winter monsoon has been proven by a series of modeling studies by the GCM model (Lu *et al.*, 2006; Li and Bates, 2007; Wang *et al.*, 2009c; Yu *et al.*, 2009; Cheng *et al.*, 2010; Sun *et al.*, 2012). It shows that weaker summer monsoons as events of abrupt change usually accompany stronger winter monsoons. It differs from the

Table 1. Weak summer monsoon in response to the North Atlantic cold events (k-yr BP).

No.	Number of cold events	0	1	2	3	4	5	6	7	8	Authors
1	Cold events	0.4	1.4	2.8	4.1	5.5	8.2	9.4	10.4	11.0	Bond *et al.*, 1997
2	SCS monsoon	0.3	1.2	3.1	4.3	6.0	8.3	9.5	/	/	Wang *et al.*, 1999
3	SW monsoon	/	1.8	3.0	4.0	5.8	8.2	9.4	/	/	Overpeck *et al.*, 1996
4	SW monsoon	0.3	1.4	3.4	4.4	5.8	8.2	9.6	10.4	/	Gupta *et al.*, 2003
5	African monsoon	0.5	1.9	3.0	4.6	6.0	8.0	x	10.2	11.8	deMenocal *et al.*, 2000
6	Plateau monsoon	0.3	1.5	2.8	4.4	5.9	8.2	9.5	10.2	11.3	Zhou *et al.*, 2002
7	EA monsoon	0.4	1.6	2.8	4.2	5.6	8.2	x	10.4	/	Wang *et al.*, 2005
8	SA monsoon	0.3	1.5	2.8	4.1	5.9	8.3	9.5	x	11.2	Hong *et al.*, 2003

Table 2. Strong winter monsoon in response to the North Atlantic cold events (k-yr BP).

No.	Number of cold events	0	1	2	3	4	5	6	7	8	Authors
1	Cold events	0.4	1.4	2.8	4.1	5.5	8.2	9.4	10.4	11.0	Bond *et al.*, 1997
2	Eastern Tibetan Plateau	0.5	1.5	3.0	4.3	5.7	8.1	9.2	10.5	/	Yu *et al.*, 2011
3	Central Asia	/	/	3.1	4.0	5.1	8.0	9.6	10.2	11.1	An *et al.*, 2011
4	Okinawa	0.6	1.7	3.3	4.6	5.9	8.1	9.4	/	/	Jian *et al.*, 2000
5	East China Sea	0.4	1.9	2.7	4.6	5.8	/	/	/	/	Xiao *et al.*, 2005a

longer timescale trend in the Holocene, where positive correlation was observed between summer and winter monsoons.

5. Climate and Monsoons in the Last 2k-yr

The last 2000 years is a critical time interval, since it includes both the Medieval Warm Period (MWP) and the Little Ice Age (LIA). It is believed that these two key climate periods were caused by natural factors, such as solar and volcanic activities. So, examination of the climate of the MWP and the LIA will improve the understanding of the formation mechanisms of natural climate variability.

Man Wenmin (personal communication) in her Doctor's thesis synthesized the stalagmite $\delta^{18}O$ data in South China, and reconstructed the East Asian Summer Monsoon index series for the last 2k years based on the definition of Guo *et al.* (2003). The summer monsoon index was calculated by Guo *et al.* (2003) according to the change in sea level pressure (ΔSLP) between 160°E and 110°E from 10°N to 50°N averaged for June, July and August (JJA). Stalagmite

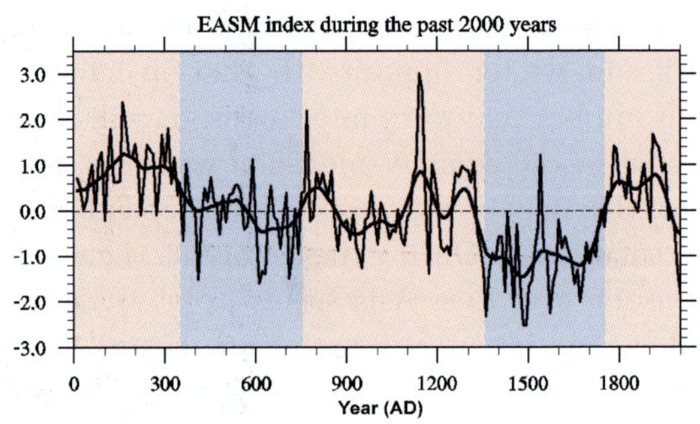

Fig. 6. East Asian Summer Monsoon index series reconstructed by using stalagmite $\delta^{18}O$ in South China, smooth curve gives 50-yr running mean (Man Wenmin, 2012, Personal communication).

$\delta^{18}O$ data in south China show a significant negative correlation with the monsoon index, allowing the index series of monsoons for the last 2k-yr to be reconstructed by using $\delta^{18}O$ data. In Fig. 6, a positive index shows lower $\delta^{18}O$ and greater summer precipitation from a stronger than normal summer monsoon. A highly smoothed approximation of the periods with stronger or weaker than normal monsoon index is shaded in red or blue, respectively. The Roman Warm Period (RWP), The Dark

Ages Cold Period (DACP), MWP, LIA, and the Present Warm Period (PWP) are demonstrated clearly in Fig. 6, although uncertainties remain in the selection of timing of the periods.

Unfortunately, no index similar to the summer monsoon series shown in Fig. 6 is available for the winter monsoon, especially with a high time resolution. Therefore, only a few single winter monsoon series are shown in Fig. 7(a, b, c). Two temperature series of China are also given in Figs. 7(d and e) for comparison. It is found that the periods with stronger winter monsoon (shaded blue) coincide with colder period in China, while the temperatures are high in the periods with weaker winter monsoon (shaded red).

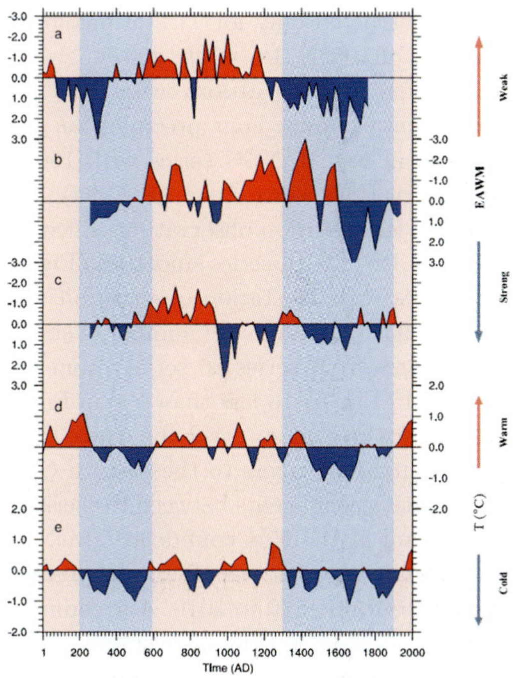

Fig. 7. East Asian Winter Monsoon and temperatures of China in the last 2k-yrs. The monsoons were reconstructed by using: (a) Normalized grain-size in southwest of Cheju Island (Xiang *et al.*, 2006), (b) Normalized grain-size in PC6 of East China Sea (Xiao *et al.*, 2005a), and (c) Normalized grain-size in DD2 of East China Sea (Xiao *et al.*, 2005b). Temperatures shown are: (d) Temperature anomalies of China (Yang *et al.*, 2002) and (e) Winter temperature anomalies of East China (Ge *et al.*, 2003).

Table 3. Timing of climate periods in the last 2k-yr (AD).

Climate period	RWP	DACP	MWP	LIA	PWP
Global	~400	400–900	900–1300	1300–1900	1900~
Summer monsoon	~350	350–750	750–1350	1350–1750	1750~
Winter monsoon	~200	200–600	600–1300	1300–1900	1900~

The five climate periods identified in Figs. 6 and 7 are outlined in Table 3, in which the timing of the periods found from the viewpoint of global scale climate change (Jones and Mann, 2004; Wang, 2010, 2011) are also given for comparison. It is found that the time interval with weak summer monsoons was shorter than that with strong winter monsoons in the LIA, while the period with strong summer monsoons was longer than those with weak winter monsoons in the MWP. Therefore, the LIA (MWP) is characterized by strong (weak) winter monsoons and weak (strong) summer monsoons respectively.

6. Climate Changes Since AD 1880

Instrumental observations of temperatures in China started in the late 19th century. The number of stations with more or less consistent observations increased to a few dozens in the early 20th century. However, almost all of the stations were limited to eastern China, mainly to the east of 105°E. The approximately full coverage of stations over China was not available until 1951. Therefore, the new temperature series (Fig. 8a) constructed by the National Climate Center (NCC) (Editorial Committee, 2011), which consists of about 2200 stations, only extends back to 1951. The other two temperature series in Fig. 8 were reconstructed by Wang *et al.* (1998) (b) and Tang and Ren (2005) (c). Wang *et al.* (1998) reconstructed temperature series for ten regions, which covers the entire land territory of China, including

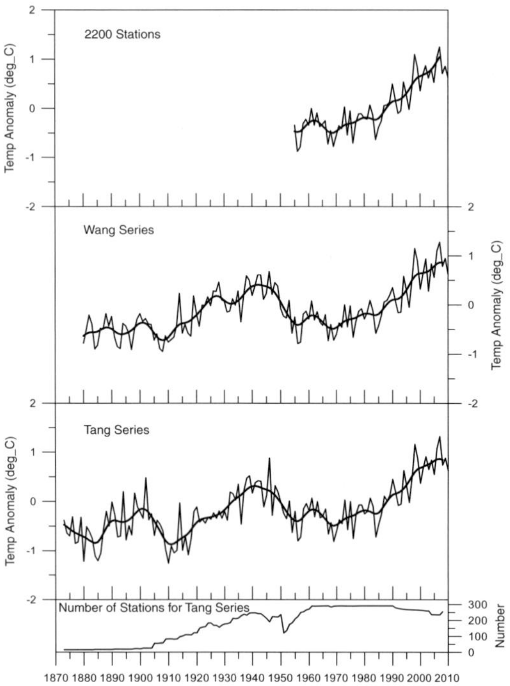

Fig. 8. Temperature anomalies of China relative to the mean of 1971-2000: (a) NCC (Editorial Committee, 2011), (b) Wang *et al.* (1998), (c) Tang and Ren (2005), (d) Number of stations used in (c).

Xinjiang, Tibet, and Taiwan. The regional series were averaged to form a national series with weightings proportional to the area of the regions. Temperatures were averaged for five key stations in each region. However, observations were often missing, especially before 1951. At those times, Wang *et al.* (1998) filled the gaps using documentary data. Tang and Ren (2005) on the other hand, constructed the average temperature series by using only observational data, thus avoiding the uncertainty in the interpretation of documentary data, while suffering an unevenness of data coverage. In each part of the record, for example in the late 19th century, observations are only available at a few stations, which significantly reduces the representativeness of the series. Therefore, somewhat greater discrepancy was found between the W and T series in the 19th century. Nevertheless, the correlation coefficient (c.c) between these two series is 0.91 proving their general reliability. By the way, the NCC series has a c.c. higher than 0.95 with W series based on the data since 1951, thus supporting the technique used in the reconstruction of the W series. Both series of W and T show a general warming trend since 1880. However, the warming rate was 0.5°C/100-yr of W, and 0.8°C/100-yr of T. Therefore, one can conclude that temperature in China increased 0.5°C/100-yr to 0.8°C/100-yr during the 20th century.

The availability of precipitation observations is better than that of temperatures, especially in the first half of the 20th century over eastern China. In western China, precipitation observations are available only since 1951. Therefore, precipitation series for the last hundred years were reconstructed only for 71 stations located to the east of 105°E (Pu *et al.*, 2007).

The gaps in observations were also filled by documentary data. Four precipitation series are shown in Fig. 9, NCC series with 160 stations observations since 1951 (Fig. 9a), NCC series with 2200 station observations since 1951 (Fig. 9b), CRU-TS 2.1 series since1901 (Fig. 9c), and a series with 71 station observations since 1880 (Fig. 9d). Although the number of stations varies greatly from series to series, from more than 2000 in Fig. 9b to less than 100 in Fig. 9d, and the first three series cover the whole China, but the fourth one refers to the eastern China, the correlation coefficients between the series are all significant at the 95% confidence level.

No significant trend was found in precipitation changes from 1880 to 2010. A predominant 20-year periodicity is evident over East China, while the precipitations increased during the last 30 years in western China.

Monsoons clearly play an important role in controlling climatic change over China. However, the definition of the monsoon varies from author to author. In this paper, an original one was accepted according to Guo *et al.* (2003), who defined the monsoon index as the normalized sea level pressure (SLP) difference between land

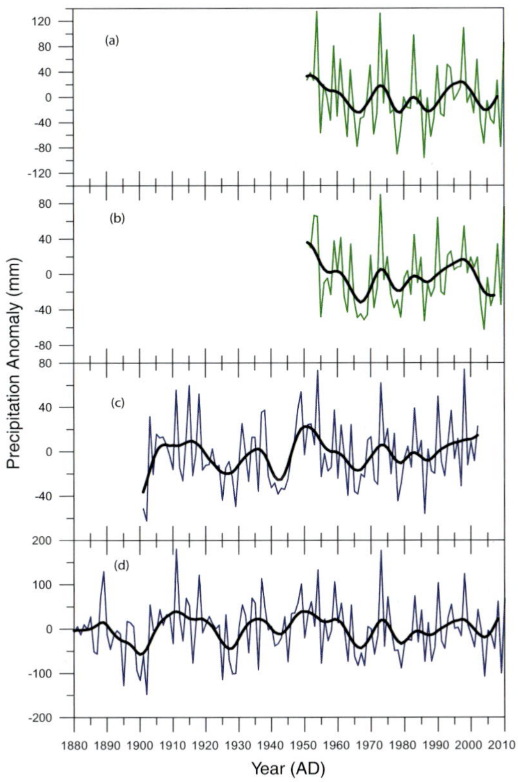

Fig. 9. Precipitation series of China: (a) NCC series with 160 stations, (b) NCC series with 2200 stations, (c) CRU series, (d) Series with 71 stations (Editorial Committee, 2011).

(110°E) and sea (160°E), averaged from 20°N to 50°N for summer (I_{SM}) and winter (I_{WM}) (Fig. 10).

It is found that the winter monsoon was relatively weak before 1925, at which time it increased rapidly and remained strong until 1985. It then weakened significantly near the end of the 20th century. The summer monsoon was stronger than normal from 1900 to 1965, after which it weakened rapidly and then remained at a very low level till the end of the 20th century. Both winter and summer monsoons show change from weak to strong and again from strong to weak since 1880. However, the timing of changes varied from winter to summer. It infers that the summer monsoon may or may not vary in phase with winter monsoon during the last 130 years.

7. Discussion

Changes of summer and winter monsoons show a similar decreasing trend during the Holocene. However, the monsoons vary out of phase from winter to summer on shorter timescales. For example, summer monsoons were strong and winter monsoons were weak in the MWP, but

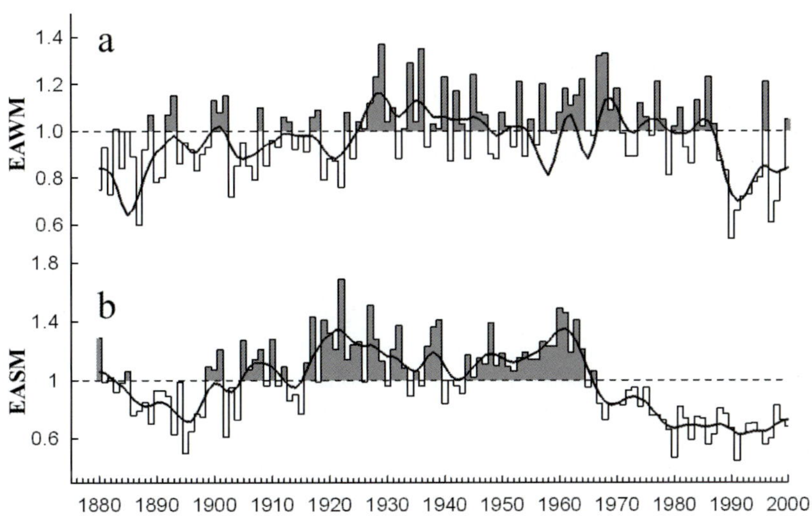

Fig. 10. East Asian winter (a) and summer (b) monsoon index 1880–2000 (Relative to the average of the 20th century).

the opposite is true in the LIA. The in-phase or out of phase relationship between summer monsoons and winter monsoons depend on the difference of the mechanism which control them. The trend of monsoon changes in the Holocene may relate to the orbital forcing, but the abrupt changes of monsoon may be attributed to the cold events over the North Atlantic Ocean.

Monsoons in the 20th century do not show any linear correlation to the global warming. Therefore, both anthropogenic (global) and natural (regional) forcing are working together to control the monsoon variability in the 20th century, although it is difficult to separate the role of one from that of the other.

Synthesis of studies on climate change in China and monsoon variability indicated that mechanism of climate change varies with time scale. Monsoons play an important role in control the climate in China, but variations of monsoon are under the influence of orbital, global and regional forcing. All of these distinct forcings forms the complexity of climate change in China.

Acknowledgments

This work is supported by the National Natural Science Foundation of China under Grant No. 41130105, the Program of Chinese Academy of Science No. 9-XDA05090104, and the National Program on Key Basic Research Project (973 Program) under Grant No. 2010CB950104.

References

An, C.-B., J. Zhao, S. Tao, *et al.*, 2011: Dust variation recorded by lacustrine sediments from arid Central Asia since-15 cal ka BP and its implication for atmospheric circulation. *Quat. Res.*, **75**, 566–573.

Bond, G., W. Showers, M. Cheseby, *et al.*, 1997: A pervasive millennial-scale in North Atlantic Holocene and glacial climates. *Science*, **278**, 1257–1266.

Central Meteorological Institute (CMI), China Meteorological Administration, *et al.*, 1981: *The Drought and Flood Charts in China for the last 500 years.* Beijing: China Atlas Press, 332 pp.

Cheng, J., Z. Y. Liu, F. He, *et al.*, 2010: Model evidence for climatic impact of thermohaline circulation on China at the century scale. *Chinese Sci. Bull.*, **55**, 3215–3221, doi: 10.1007/s 11434-010-4089-0.

deMenocal, P., J. Ortiz, T. Guilderson, *et al.*, 2000: Abrupt onset and termination of the African Humid Period: Rapid climate responses to gradual insolation forcing. *Quat. Sci. Rev.*, **19**, 347–361.

Editorial Committee, 2011: *National Second Assessment Report of Climate Change.* Beijing: China Science Press, pp. 710.

Fleitmann, D., S. J. Burns, M. Mudelsee, *et al.*, 2003: Holocene forcing of the Indian monsoon recorded in a stalagmite from southern Oman. *Science*, **300**, 1737–1739.

Ge, Q., J. Zheng, X. Fang, *et al.*, 2003: Winter half-year temperature reconstruction for the middle and lower reaches of the Yellow River and Yangtze River, China, during the past 2000 years. *The Holocene*, **13**, 933–940.

Guo, Q., J. Cai, X. Shao, *et al.*, 2003: Interdecadal variability of East Asian summer monsoon and its impact on the climate of China. *Acta Geographica Sinica*, **58**, 569–576.

Gupta, A. K., D. M. Andersen, and J. T. Overpeck, 2003: Abrupt changes in the Asian southwest monsoon during the Holocene and their links to the North Atlantic Ocean. *Nature*, **421**, 354–357.

Haug, G. H., K. A. Hughen, D. M. Sigman, *et al.*, 2001: South ward migration of the Intertropical Convergence Zone through the Holocene. *Science*, **293**, 1304–1308.

Hong, Y. T., B. Hong, Q. H. Lin, *et al.*, 2003: Correlation between Indian Ocean summer monsoon and North Atlantic climate during the Holocene. *Earth Planet. Sc. Lett.*, **211**, 371–380.

Huang, E., J. Tian, and S. Steinke, 2011: Millennial-scale dynamics of the winter cold tongue in the southern South China Sea over the past 26ks and the East Asian winter-monsoon. *Quat. Res.*, **75**, 196–204.

Jian, Z., P. Wang, Y. Saito, *et al.*, 2000: Holocene variability of the Kuroshio Current in the Okinawa Trough, northwestern Pacific Ocean. *Earth Planet. Sc. Lett.*, **184**, 305–319.

Jones, P. D., and M. E. Mann, 2004: Climate over past millennium. *Rev. Geophys.*, **42**, 1–42.

Kutzbach, J. E., and P. J. Guetter, 1986: The influence of changing orbital parameters and surface boundary conditions on climate simulations for the past 18,000 years. *J. Atmos. Sci.*, **43**(16), 1726–1759.

Li, S., and G. T. Bates, 2007: Influence of the Atlantic Multidecadal Oscillation on the winter climate of East China. *Adv. Atmos. Sci.*, **24**, 126–135.

Lu, R., B. Dong, and H. Ding, 2006: Impact of the Atlantic Multidecadal Oscillation on the Asian summer monsoon. *Geophys. Res. Lett.*, **33**, L24701.

Maslin, M. A., and S. J. Burns, 2000: Reconstruction of the Amazon Basin effective moisture availability over the past 14,000 years. *Science*, **290**, 2285–2287.

Mayewski, P. A., E. E. Rohling, J. C. Stager, *et al.*, 2004: Holocene climate variability. *Quat. Res.*, **62**, 243–255.

Overpeck, J., D. Anderson, S. Trumbore, *et al.*, 1996: The southwest Indian Monsoon over the last 18000 years. *Clim. Dyn.*, **12**, 213–225.

Pu, B., S. Wang, and J. Zhu, 2007: Spatial pattern of seasonal precipitation over the eastern part of China. *Acta Sci. Nat. Univ. Pekin*, **43**(5), 620–629.

Shao, X., L. Huang, H. Liu, *et al.*, 2005: Reconstruction of precipitation variation from tree rings in recent 1000 yr in Delingha, Qinghai. *Science China Earth Sciences*, **48**(7), 939–949.

Shi, Y., and Z. Kong, 1992: *The Climates and Environments of Holocene Megathermal in China.* Beijing: China Ocean Press, 212 pp.

Sun, Y., S. C. Clemens, C. Morrill, *et al.*, 2012: Influence of Atlantic meridional overturning circulation on the East Asian winter Monsoon. *Nat. Geosci.*, **5**, 46–49.

Tan, M., 2009: Circulation effect: Climatic significance of the short term variability of the oxygen isotopes in stalagmites from monsoonal China-Dialogue between paleoclimate records and modern climate research. *Quat. Sci.*, **29**(5), 851–862. (Chinese).

———, 2011: Trade-wind driven in verse coupling between stalagmite δ18O from monsoon region of China and large scale temperature-circulation effect on decadal to processional time scales. *Quat. Sci.*, **31**(6), 1086–1097. (Chinese).

Tang, G., and G. Ren, 2005: Reanalysis of surface temperature variation trend of China in the last 100 years. *Studies of Climate and Environment*, **10**, 791–798.

Wang, L., M. Sarnthein, H. Erlenkeuser, *et al.*, 1999: Holocene variations in Asian monsoon moisture: A bidecadal sediment record from the South China Sea. *Geophys. Res. Lett.*, **26**, 2889–2892.

Wang, L., J. Li, H. Lu, *et al.*, 2012: The East Asian winter monsoon over the last 15,000 years: Its links to high-latitudes and tropical climate systems and complex correlation to the summer monsoon. *Quat. Sci. Res.*, **32**, 131–142.

Wang, N., T. Yao, J. Pu, *et al.*, 2006: Climatic and environmental changes over the last millennium recorded in Malan ice core from the northern Tibetan Plateau. *Science China Earth Sciences*, **49**(10), 1079–1089.

Wang, S., 1991: Reconstruction of palaeotemperature series in China from the 1380s to the 1980s. *Würzburger Geographische Arbeiten*, **80**, 1–19.

———, 2010: Medieval Warm Period and Little Ice Age. *Adv. Climate Change Res.*, **6**(5), 388–390.

———, 2011: Dark Ages Cold Period. *Adv. Climate Change Res.*, **7**(3), 233–234.

———, D. Gong, and J. Zhu, 2001: Twentieth-century climatic warming in China in the context of the Holocene. *The Holocene*, **11**(3), 313–321.

———, J. Huang, X. Wen, *et al.*, 2009a: Two modes of summer precipitation variation of Holocene in China. *Quat. Sci.*, **29**(6), 1086–1094. (Chinese).

———, J. Ye, D. Gong, *et al.*, 1998: Construction of mean annual temperature series for the last hundred years in China. *Quart. J. Appl. Meteor.*, **9**, 392–401.

———, and Z. Zhao, 1981: Droughts and floods in China, 1470–1979. *Climate and History: Studies in Past Climates and Their Impact on Man*, T. N. L. Wigley, M. J. Ingram, and G. Farmer, Eds., Cambridge: Cambridge University Press, 271–288.

———, Z. G. Zhao, and W. J. Li, 2009b: *Atlas of Seasonal Temperature and Precipitation Anomalies over China.* Beijing: China Meteorological Press, 257 pp.

Wang, Y., H. Cheng, R. L. Edwards, *et al.*, 2005: The Holocene monsoon: Links to solar changes and North Atlantic climate. *Science*, **308**, 854–857.

Wang, Y., S. Li, and D. Luo, 2009c: Seasonal response of Asian monsoonal climate to the Atlantic Multidecadal Oscillation. *J. Geophys. Res.*, **114**, D02112.

Xiang, R., Z. Yang, Y. Saito, *et al.*, 2006: East Asian Winter Monsoon changes inferred from environmentally sensitive grain-size component records during the last 2300 years in mud area south west of Cheju Island. *Science China Earth Sciences*, **49**, 604–614.

Xiao, C., D. Qin, T. Yao, *et al.*, 2008: Progress on observation of cryospheric components and climate-related studies in China. *Adv. Atmos. Sci.*, **25**(2), 164–180.

Xiao, S., A. Li, M. Chen, *et al.*, 2005a: Recent 8ka mud records of the East Asian Winter Monsoon from the Inner shelf of the East China Sea. *Earth Sci. J. China Un.: Geosci.*, **30**(5), 573–581.

———, ———, F. Jiang, *et al.*, 2005b: Recent 2000-year geological records of mud in the inner shelf of the East China Sea and their climatic implications. *Chinese Sci. Bull.*, **50**, 466–471.

Xu, Y., and X. Shao, 2007: Standardiazation of Qilian Juniper ring-width series in the eastern margin of the Qaidam Basin. *J. Geograph. Sci.*, **17**, 175–186.

Yang, B., A. Braeuning, K. R. Johnson, *et al.*, 2002: General characteristics of temperature variation in China during the last two millennia. *Geophys. Res. Lett.*, **29**(9), 381–384.

Yu, L., Y.-Q. Gao, H.-J. Wang, *et al.*, 2009: The responses of East Asian Summer monsoon to the North Atlantic Meridional Overturning Circulation in an enhanced freshwater input simulation. *Chinese Sci. Bull.*, **54**, 4724–4732.

Yu, X., W. Zhou, Z. Liu, *et al.*, 2011: Different patterns of changes in the Asian summer and winter monsoon on the eastern Tibetan Plateau during the Holocene. *The Holocene*, **21**(7), 1031–1036.

Zhang, D., 1991: The Little Ice Age in China and its correlations with global change. *Quat. Sci.*, **2**, 104–112.

———, 2005: Severe drought events as revealed in the climate records of China and their temperature situations over the last 1000 years. *Acta Meteor. Sinica*, **19**(4), 485–491.

———, C. Liu, J. Jiang, 1997: Reconstruction of six regional dry/wet series and their abrupt changes during the last 1000 years in East China. *Quat. Sci.*, **1**, 1–11.

Zhang, P., H. Cheng, R. L. Edwards, *et al.*, 2008: A test of climate, sun, and culture relationship from an 1810-year Chinese cave record. *Science*, **322**, 940–942.

Zhou, W., X. Lu, Z. Wu, *et al.*, 2002: Peat record reflecting Holocene climatic change in the Zoigê Plateau and AMS radiocarbon dating. *Chinese Sci. Bull.*, **47**, 66–70.

Zhou, X., 2011: Asian monsoon precipitation changes and the Holocene methane anomaly. *The Holocene*, **22**(7), 731–738.

Zhu, K.-Z., 1972: Preliminary study on the climate change in China during the last 5000 years. *Acta Archeological Sinica*, **2**(1), 15–38. (Chinese).

Chapter 18

North Atlantic Hurricane Activity: Past, Present and Future

Rym Msadek*, Gabriel A. Vecchi and Thomas R. Knutson
NOAA/GFDL Princeton University Forrestal Campus
Princeton, NJ 08542, USA
**Rym.Msadek@noaa.gov*

We review past, present and future North Atlantic hurricane activity based on analysis of observational records and models projections. When adjusted for likely missed tropical cyclones, the observational record does not show any significant increase or decrease of North Atlantic hurricane frequency. Downscaling results for most available CMIP5 models show a decrease or little change in overall frequency of tropical storms and hurricanes, although in the Atlantic basin, previous studies by other investigators report a wider range of change (+/−60%). Some model projections of late 21st century hurricane activity indicate an increase in frequency of the strongest storms (category 4–5 hurricanes). The projected increase is substantial (+100% per century) in the CMIP3 ensemble model downscaling, but much smaller (+40%) and only marginally significant in the CMIP5 ensemble model downscaling. Rainfall rates in the inner core of the hurricanes are projected to increase with potentially a substantial damage impact. The largest source of uncertainty in predicting changes in Atlantic tropical storms activity over the first half of the 21st century arises from the internal variability of the climate system. Nonetheless, some of these natural fluctuations appear to be predictable beyond seasonal time scale. We review recent predictability assessment results based on two CMIP5 models. Initializing these models with observational estimates leads to encouraging results in predicting multi-year variations in North Atlantic hurricane frequency. However the short record and the persistent character of the time series limits the ability to confidently predict North Atlantic hurricane activity for now. Remaining model biases, despite the tremendous improvement over the recent decades, and the changing observational system make it an ongoing challenge to simulate past hurricane activity and project or predict its future behavior.

1. Introduction

A growing scientific effort has been devoted recently for detecting and attributing changes in hurricane activity. Indeed, a potential increase of North Atlantic tropical cyclones (TCs), which includes tropical storms and hurricanes, in response to changing climate conditions could have an important impact on society, particularly if the most intense hurricanes become more frequent, coupled with ongoing sea level rise. A number of studies have discussed the character of and causes behind past changes in North Atlantic TC frequency (*e.g.*, Solow and Moore, 2002; Mann and Emanuel, 2006; Chang and Guo, 2007; Landsea, 2007; Mann *et al.*, 2007; Holland and Webster, 2007; Vecchi and Knutson, 2008; Landsea *et al.*, 2010; Knutson *et al.*, 2010; Villarini *et al.*, 2011b) and in its future evolution (Knutson *et al.*, 2010 and references therein). Having a reliable data record of past hurricane

Climate Change: Multidecadal and Beyond
edited by Chih-Pei Chang et al.

activity, along with a quantitative assessment of its uncertainty is essential for assessing any long-term change in hurricane activity.

Besides the projections of hurricane activity over the coming century, there is a large interest in hurricane climate prediction at shorter time scales, from seasonal to decadal. Reliable seasonal forecasts have become feasible thanks to the tremendous technological improvement over the last decades along with improved understanding of the global climate system and in particular the climate factors that influence hurricane activity. Seasonal prediction efforts are being undertaken at a number of institutions, with a variety of techniques being applied, each exhibiting retrospective skill (Saunders and Lea, 2005; Vitart and coauthors, 2007; Klotzbach, 2007; Wang *et al.*, 2009; Camargo and Barnston, 2009; LaRow *et al.*, 2010). It would be a useful exercise, beyond the scope of the present study, to quantitatively compare the retrospective skills of these various systems across common periods, and to assess the relationship between retrospective skill and real-time skill for seasonal hurricane prediction, and the extent to which a multi-system approach could outperform individual systems. While a number of methods have been developed over the years for seasonal predictions of Atlantic TCs using high-resolution hurricane models (Camargo *et al.*, 2007; Zhao *et al.*, 2010), the use of statistical models combined with downscaled dynamical models proved to be very successful. Despite being less computationally demanding, the methods seem to generally reproduce the results of high-resolution dynamical models (*e.g.* Villarini *et al.*, 2011b). The predictors used for the statistical model can also provide useful information to explore hurricane activity uncertainty. The extent to which such forecasts can be extended beyond the seasonal time scale has been recently investigated and is discussed in this paper, with the assumption that initializing dynamical models with observations could lead to skill in predicting natural variability.

While warm local sea surface temperatures (SSTs) favor the formation and intensification of Atlantic TCs, it is not entirely clear how increasing global SSTs will influence TC activity. Emanuel (2007) showed that the power dissipation of Atlantic storms increased by about 60% since the 1970s and that the average duration and maximum speed also increased. Landfalling storms are only a fraction of the total number of storms. Category 4 and 5 storms account for only about 6% of US landfalls but caused 48% of all hurricane damage between 1900 and 2005 (Blake *et al.*, 2007; Pielke *et al.*, 2008). It is therefore essential to determine how these potentially very destructive hurricanes are projected to evolve as climate warms and sea level rise continues. This historical apparent increase in Atlantic hurricane activity has been linked to observed changes in SST averaged over the Atlantic main development region (MDR) (*e.g.*, Mann and Emanuel, 2006; Emanuel, 2007; Holland and Webster, 2007; Mann *et al.*, 2007). The potential for hurricane formation is enhanced by low values of wind shear in the troposphere. In the tropical Atlantic, warmer SSTs tend to increase wind shear, leading to fewer storms overall. It has been suggested that the observed increase in TCs could also be the result not of global warming *per se* but of the warming of the Atlantic relative to the other tropical ocean basins, or "relative SSTs" (Vecchi and Soden, 2007; Vecchi *et al.*, 2008; Swanson, 2008). This distinction between absolute and relative SST has important implications for projections of future hurricane activity since model projections indicate a substantial warming of the tropical SSTs globally in response to increased greenhouse gases concentrations, but relatively small changes in Atlantic relative SSTs (Vecchi *et al.*, 2008).

The continuous advance of climate models and the public release of a large amount of climate data through the Coupled Model Intercomparion Projects that recently produced its fifth (CMIP5) assessment has allowed us to revisit the Atlantic TC projections that were produced with previous (CMIP3) models (*e.g.*, Bender *et al.*, 2010). We present in this chapter recent advances in assessing the change of North Atlantic hurricane frequency that was observed over the 20th century (Sec. 2) and distinguish between fluctuations that appear to be natural and those resulting from changes in external forcing (Sec. 3). In Sec. 4, we review the projected changes in Atlantic hurricane activity given by the most recent climate model studies and we present in Sec. 5 the uncertainties associated with these projections. Recent results about multi-year predictions of hurricane variability are described in Sec. 6 before the concluding remarks in Sec. 7.

2. Observed Hurricane Activity and Record Uncertainty

The historical record of hurricane activity in the Atlantic dates back to the late 19th century and has been more widespread than in any other basin owing to the aircraft reconnaissance that started in the late 1940s. A widely used reconstructed record, the Hurricane Database (HURDAT) has been since provided by the National Oceanic and Atmospheric Administration (NOAA; Jarvinen *et al.*, 1984; MacAdie *et al.*, 2009). It provides a historical record of Atlantic-wide tropical storm and hurricane counts, tracks, and intensities from 1851 to the present. As shown in Fig. 1a, this original record shows a significant upward trend in the annual numbers of hurricanes since the late 1800s, suggesting a substantial increase of hurricane activity over the last century, consistent with the analysis of the 20th historical record by (Holland and Webster, 2007).

However, the methodology and distribution of tropical storm counts observations have considerably changed since the late 19th century

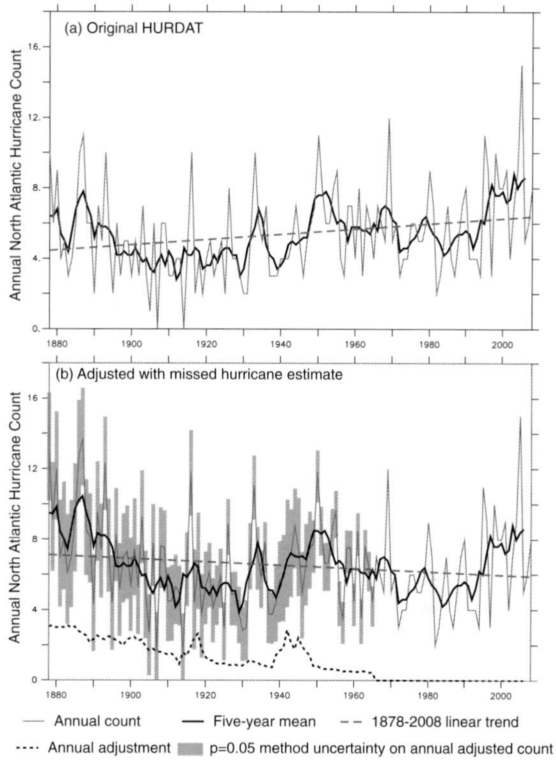

Fig. 1. From (Vecchi and Knutson, 2011) "© Copyright [15 March 2011] AMS": Time series of Atlantic basin hurricane counts over the period 1878–2008 from (a) the unadjusted HURDAT and (b) after the adjustment for estimated missed hurricanes and the magnitude of the hurricane count adjustments. Plots show the annual (light lines) and 10-yr running mean (dark lines) counts; (b) gray shading indicates the 95% method uncertainty on the adjustment. Dashed lines depict the linear least squares trends computed over the period 1878–2008.

along with the technology and recording practices that have been continuously evolving with time (Landsea *et al.*, 2010). Therefore, the secular increase in the observed record could partly reflect our increased observational capabilities. Recent analyses have attempted to adjust the HURDAT database to account for missed tropical storms or hurricanes (Vecchi and Knutson, 2008, 2011; Villarini *et al.*, 2011a), particularly during the pre-satellite era (Fig. 1b). For example, the above studies showed that after adjustment, the updated estimate of Atlantic hurricane counts does not show any significant trend (Fig. 1b). Therefore, although global mean temperatures show a significant warming trend since the late 19th century, after accounting for known changes to the observing network there is no significant increasing century-scale trend in Atlantic tropical storm or hurricane frequency. It is hence premature to conclude from present observations that Atlantic hurricane frequency has changed (increase or decrease) due to increased atmospheric CO_2 concentrations. Detecting such a change, were it to happen in the future, will require a reliable record of past and future hurricane activity. This is complicated by the fact that long-term changes in hurricane activity are also spatially heterogeneous, partly reflecting a real non-homogeneity and partly because of changes in ship tracks density and location over time. Closer examination (Vecchi and Knutson, 2008, 2011) indicates that areas with increased storm frequency in the raw/unadjusted data are far from US landfall regions (i.e., in the open Atlantic), with decreases present in the western Atlantic. The historical record also includes many storms of relatively brief duration (<2 days) in more recent years (Landsea *et al.*, 2010). Neither the longer-lived storms nor hurricanes that hit the US coast show any significant increase in frequency since 1878.

3. Internal and Forced Components of Hurricane Variability

In addition to the trend that was discussed in the previous section, Fig. 1a shows clear multi-decadal variability in the Atlantic hurricane frequency that remains pronounced in the adjusted time series (Fig. 1b). Multiple interpretations of these multi-decadal changes are possible. Mann and Emanuel (2006) attributed the decadal variability in TCs to the impact of anthropogenic forcing on Atlantic SSTs, and anthropogenic aerosol forcing was suggested as the origin of the cooling observed between 1950 and 1970. Decadal fluctuations in hurricane activity have also been related to the basin-wide Atlantic Multidecadal Variability (AMV or AMO, Goldenberg *et al.*, 2001; Zhang and Delworth, 2006), which is thought to be primarily driven by the natural fluctuations of the Atlantic Meridional Overturning Circulation (MOC, Knight *et al.*, 2005; Zhang *et al.*, 2012). The AMV, defined as a basin-averaged SST index, is also influenced by external forcing (*e.g.* greenhouse gases and aerosols), adding difficulty to the problem of attributing the AMV-related climate impacts to internal or forced variability. Few studies have attempted to distinguish in the AMV between the component attributable to external forcing and that due to natural variability (*e.g.* Ting *et al.*, 2009; DelSole *et al.*, 2011; Terray, 2012).

A recent analysis by Camargo *et al.* (2013) explored the relative importance of externally forced and internal SST variability on tropical Atlantic cyclone potential intensity. They showed that the late twentieth century increase in North Atlantic potential intensity (*e.g.* Kossin and Camargo, 2009), while closely related to the tropical Atlantic SST increase, included a considerable influence of internal variability and hence could be attributed purely to anthropogenic warming (Fig. 2). Their study highlighted the importance of remote SST on hurricane activity, which had been previously identified by Vecchi and Soden (2007); Swanson (2008). Using sensitivity experiments to isolate the role of the Atlantic MDR and the other oceans, they showed that when non-local SST effects are taken into account, the influence of global mean SST changes on potential intensity are smaller than those forced by Atlantic SST only (Fig. 2). The remote influence of Indian and tropical Pacific Ocean and to a smaller extent of extratropical North Atlantic reduces the local SST forcing in potential intensity for both forced changes and natural fluctuations. The remote SST influence tends to reduce the

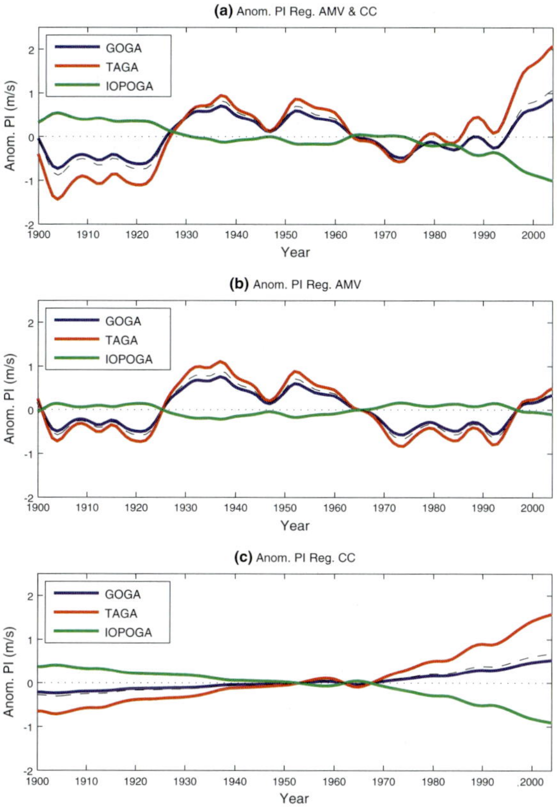

Fig. 2. From Camargo *et al.*, 2013 "© Copyright [March 2013] Springer": Regression of August–September–October (ASO) potential intensity (PI) MDR anomalies onto the sum of AMV and Climate Change (CC) indices (a), onto the AMV index (b), and onto the CC index (C). This decomposition aims at separating the effect of forced variability from that due to internal variability associated with AMV following the method described in Ting *et al.*, 2009. The three colored lines correspond to three different sensitivity experiments in which the global atmospheric model is forced using SST anomalies prescribed over the global ocean (GOGA), over tropical Atlantic ocean only (TAGA) and over the Indian and tropical Pacific Ocean (IOPOGA). The thin black dashed line is the sum of TAGA and IOPOGA to check the linearity of the experiments.

potential intensity relative to the impact of tropical Atlantic SST alone, through changes in the upper level atmospheric temperatures (Vecchi *et al.*, 2008; Camargo *et al.*, 2013). Indeed, the non-local SSTs tend to warm the tropical Atlantic upper troposphere, which increases the atmospheric stability in the tropics, reducing the potential intensity. This is consistent with the idea of relative SST and its relationship with hurricane activity (Vecchi and Soden, 2007; Ramsey and Sobel, 2011). From these studies, it appears that the recent increase in Atlantic potential intensity between 1990 and 2005 is due to both internal and forced variability (Fig. 2).

4. Projected Changes of Hurricane Activity

While warm local SSTs favor the formation and intensification of Atlantic tropical storms, it is not entirely clear how increasing global SSTs will influence TC activity because of the large number of other key features that impact hurricane formation and development. For instance, a lack of vertical wind shear between the surface and upper-level winds will tend to favor hurricane formation. Depending on the nature of the circulation changes, the more favorable conditions associated with warmer SSTs can be offset by a higher magnitude vertical wind shear. As mentioned in Sec. 1, Atlantic hurricane activity has been linked to changes in SST averaged over the Atlantic MDR relative to changes that occur over the whole tropical ocean. In particular, it has been suggested that the increase in tropical storm counts since about 1980 could be the result of the warming of the Atlantic relative to the other tropical ocean basins, i.e. an increase of relative SST (Vecchi and Soden, 2007; Vecchi *et al.*, 2008; Swanson, 2008).

This distinction between local absolute SST and relative SST has implications for projections of future hurricane activity since model projections indicate a warming of the tropical SSTs globally (and hence in the tropical Atlantic) in response to increased greenhouse gases concentrations, but relatively small changes in relative SSTs. As shown by Vecchi *et al.*, 2008, using relative SST to project hurricane changes over the 21st century indicates a much smaller

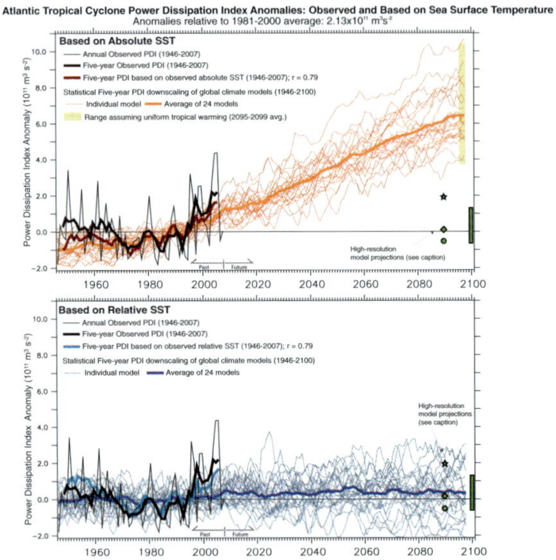

Fig. 3. From (Vecchi *et al.*, 2008) "© Copyright [31 October 2008] Science': Observed (1946–2007) and simulated (1946–2100) changes in PDI anomalies. The observed PDI anomalies (relative to the 1981–2000 average) are regressed onto the absolute (upper panel) and relative (lower panel) SST. These regression models are used to build estimates of PDI from output of global climate models for historical and future conditions.

projected change in Atlantic PDI over the coming century than suggested by using a statistical model based on MDR SST alone (Fig. 3). A recent statistical analysis by Villarini *et al.* (2011b) indicates that both local MDR SSTs and tropical-mean SSTs modulate TCs activity with similar weights but opposite signs, supporting the notion that relative SST is a robust indicator of Atlantic hurricane activity changes in response to climate change (*e.g.* Latif *et al.*, 2007; Vecchi and Soden, 2007; Swanson, 2008; Knutson *et al.*, 2008; Bender *et al.*, 2010; Zhao *et al.*, 2010; Ramsey and Sobel, 2011; Villarini and Vecchi, 2012) .

A number of current model projections indicate a potential for decrease in Atlantic hurricane frequency over the 21st century (Bender *et al.*, 2010; Sugi and Yoshimura, 2012; Knutson *et al.*, 2013). Besides assessing the changes in overall hurricane frequency, it is important to understand, and if possible make accurate projections of the behavior of the strongest landfalling hurricanes because historically they have caused the largest damage and thus have the greatest potential impact on society (Emanuel, 2011). Based on the Coupled Model Intercomparison Project 3 (CMIP3) multi-models ensemble, (Bender *et al.*, 2010) projected a significant (100%) increase of basin-wide category 4–5 Atlantic hurricane frequency by the end of the 21st century. These results were recently revisited by (Knutson *et al.*, 2013) using the Coupled Model Intercomparison Project 5 (CMIP5) models and the associated RCP4.5 climate change emission scenarios. For the CMIP5 projections, a distinction between the early and late part of the 21st century was made to account for the possible non-linearity of the changes in hurricane frequency (Villarini and Vecchi, 2012). In both CMIP3 and CMIP5 ensembles, [Knutson *et al.* (2013)] found a reduction in the frequency of TCs in a warmer climate when all storm categories are combined and an increase in the frequency of the strongest storms (Fig. 4). A robust projected increase in hurricane rainfall rates was also found as the climate gets warmer and the atmosphere holds more water vapor. The projected changes in precipitation appear to scale better with the increase in Atlantic temperature rather than with the decrease in hurricane frequency (Fig. 5).

The increase of the strongest storms was however smaller (+40% by the end of the 21st century) for the CMIP5 multi-model ensemble than for CMIP3 and only marginally significant. For the early 21st century, the projected increase is larger (+45% in 30 yr) but again is only marginally significant. The differences between CMIP3 and CMIP5 ensembles could also be the result of different forcing (IPCC A1B scenario vs. RCP4.5) or a different model response to the same forcing, as will be discussed in the next section. To explore the spread of the results among models, Knutson *et al.* (2013) additionally downscaled ten of the individual

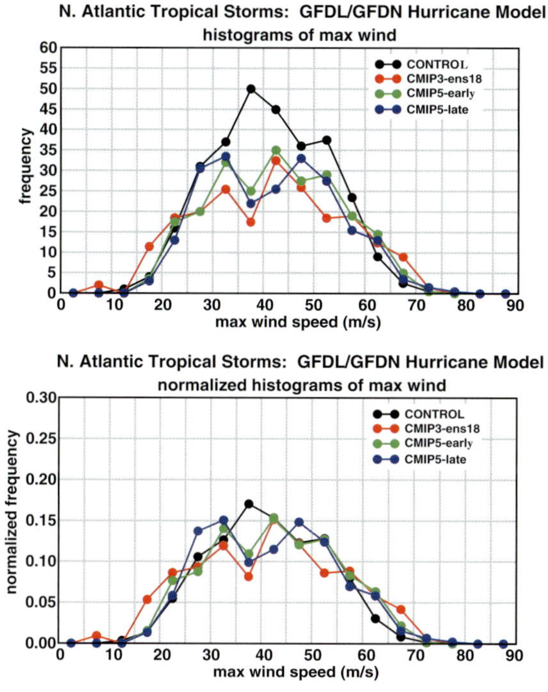

Fig. 4. From (Knutson *et al.*, 2013) "© Copyright [2013] AMS": Upper panel: Frequency histograms for lifetime maximum surface wind speeds projected by the GFDL hurricane model downscaling experiments based on the CMIP3/A1B scenarios and the more recent CMIP5/representative concentration pathway (RCP) 4.5 ensemble mean climate change experiments relative to a control run. A distinction between early and late 21st century projections is made for the CMIP5 experiments as detailed in the text. Lower panel: same as upper panel for the normalized frequency.

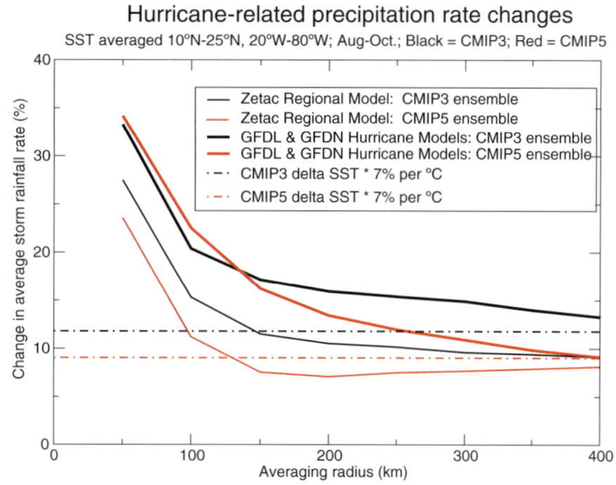

Fig. 5. From (Knutson *et al.*, 2013) "© Copyright [2013] AMS": Change (in %) between the control to warm climate in average hurricane rainfall rate for various averaging radii about the storm center (in km) for the CMIP3/A1B (black) and CMIP5/RCP4.5 (red) late 21st century multi-model ensemble climate changes, based on the Zetac regional model (thin solid lines) or the GFDL/GFDN hurricane model ensemble (thick solid lines). The dashed lines illustrate idealized water vapor content scalings, obtained by multiplying the average SST change in the region 10–25° N, 20–80° W by 7% per degree Celsius.

CMIP3 models using the GFDL hurricane model and found a significant increase of the strongest storms (category 4–5) in three of the ten downscaled models, based on 13-yr samples.

A decrease in the number of US landfalling storms was also projected for the late 21st century but these changes were not statistically significant (Knutson *et al.*, 2013). Villarini and Vecchi (2013) projected, using a statistical model, an increase in Atlantic PDI in the CMIP5 climate models, which they attributed to an intensification of Atlantic TCs in response to both greenhouse gases and aerosols (Fig. 6). An increase in the mean lifetime maximum intensity of hurricanes was also found in the CMIP3 and CMIP5 experiments analyzed by Knutson *et al.* (2013). Due to their experimental design (five-day integrations of storm cases), Knutson *et al.* (2013) were however not able to clearly address the issue of PDI changes in their higher resolution model.

5. Uncertainty in Decadal-to-Centennial Scale Projections of Atlantic Hurricane Frequency

Climate model projections provide the best estimate of likely upcoming changes in hurricane activity, but as in any forecasts these projections are associated with large uncertainty. While the analysis by Knutson *et al.* (2013) shows relatively robust projections of decreasing Atlantic tropical storm frequency over the 21st century in the CMIP3 and CMIP5 ensemble mean projections, their results differ from

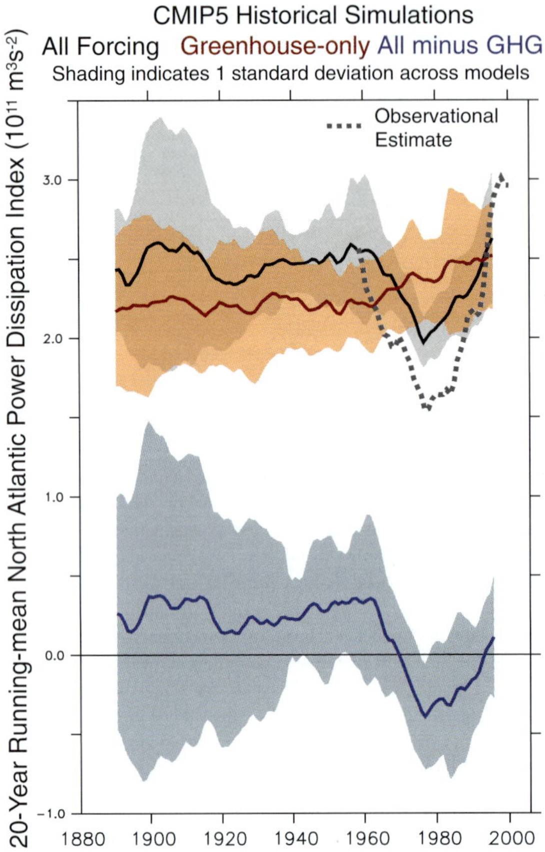

Fig. 6. From (Villarini and Vecchi, 2013) "© Copyright [May 2013] AMS": PDI evolution in response to external forcing (black) in the CMIP5 ensemble mean of 13 climate models. The impact on PDI of greenhouse forcing (in red) and non-greenhouse forcing (in blue) is distinguished. Shading indicates one inter-model standard deviation. Non-greenhouse forcing includes natural (*e.g.* solar forcing and volcanoes) and anthropogenic (ozone, aerosols) forcing.

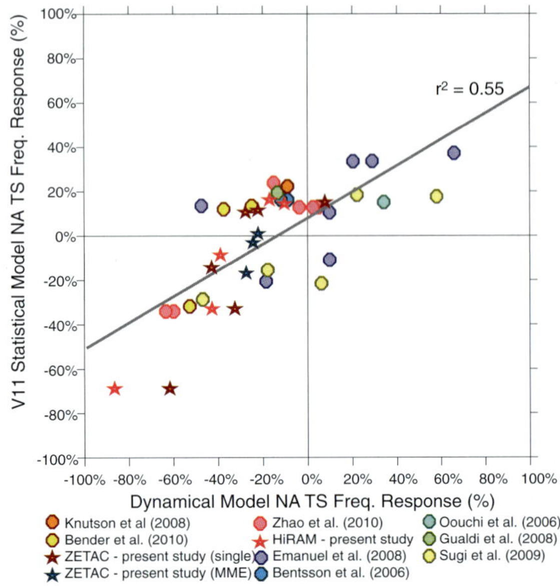

Fig. 7. From (Knutson *et al.*, 2013) "© Copyright [2013] AMS": Comparison of published dynamical model projections of Atlantic hurricane frequency (x-axis) with projections based on relative SST using the statistical downscaling model defined in Villarini *et al.* (2010, 2011b). Blue stars correspond to the CMIP3, CMIP5-Early and CMIP5-Late multi-model ensemble results.

previous studies by other investigators. Comparing the results of Knutson *et al.*, 2013 with previously published results (Fig. 7) indicates a lack of robustness in the sign of the projected change in Atlantic tropical storm frequency given that a number of other model studies indicate an increased frequency in a warmer climate (Sugi *et al.*, 2002; Oouchi *et al.*, 2006; Chauvin *et al.*, 2006; Emanuel *et al.*, 2008; Sugi *et al.*, 2009; Murakami *et al.*, 2012). Despite these discrepancies in the projected change in tropical storm frequency, models are more consistent if their tropical storm projections are viewed in terms of the projected change in relative SST. Indeed models that project an increase in TC frequency in a warmer climate are generally forced with (or simulate) a SST in the Atlantic that exceeds the tropical-mean warming (Fig. 7, Villarini and Vecchi, 2012). Relating the changes in tropical storm frequency to changes in relative SST thus helps us understand and interpret the differences in model projections. It also suggests that a better constraint on Atlantic relative SSTs could improve the confidence in Atlantic tropical storm and hurricane frequency projections.

Further, even though climate models have improved greatly in recent years, they still contain large biases that limit their reliability. Emanuel *et al.* (2013) stressed that historical simulations with current climate models fail to reproduce the observed cooling trend near the tropopause over the Atlantic, and they suggested that this could lead to an underestimation of future TC increase. One major source of model errors comes from the representation of tropical convection, which is itself linked to the simulation of clouds that rely on complex processes (Randall *et al.*, 2007). The simulation of hurricane activity is also tightly linked to that of El Niño Southern Oscillation (ENSO), adding it as a source of uncertainty for the projections of future hurricane activity. Indeed, ENSO is unevenly represented among climate models and because of the models spread, there is no consensus on its future evolution (Guilyardi *et al.*, 2009; Vecchi and Wittenberg, 2010; Guilyardi *et al.*, 2012) and therefore on the subsequent future hurricane changes due to ENSO.

Currently the resolution of climate models is usually not sufficient to simulate the key features of atmospheric circulation that impact hurricane formation and development, projections as in (Knutson *et al.*, 2013) presented in Sec. 4 rely on downscaling techniques. Knutson *et al.* (2013) explored the dependence of the hurricane projections on the downscaling model, using two different downscaling models: an 18-km grid regional model (Zetac) and a 50-km grid global model (HiRAM C180) with similar large-scale climate projections from global climate models as boundary conditions. While the projection of fewer Atlantic tropical storms was overall robust between the two downscaling experiments, little consistency was found when looking at individual CMIP3 models.

Vecchi *et al.* (2011) showed that the forcing could also have a strong impact on the downscaled circulation with biased trends in the reanalysis (*e.g.* NCEP) imprinting on the simulation of historical trends in storm counts in the Zetac model. This indicates that further work is needed to assess the role of reanalysis uncertainty in historical simulations and assessments of TC activity.

The downscaling experimental design can limit the applicability of a study for landfalling storm statistics. For example, in Knutson *et al.* (2013) the downscaling framework used (i.e., an operational hurricane prediction system) was limited to five-day runs since that is how the model is used operationally. For the study, the time of maximum intensity of each storm was identified in the parent regional atmospheric model (*e.g.* the Zetac model) and then the starting point for the five day experiment was determined by backing up three days from that time of maximum intensity. Since the hurricane model was integrated forward for a maximum of five days, the simulations were generally too short for most of the case-study storms to reach the US coast. In future studies, a focus on landfalling systems for climate impacts applications, as well as increased computational resources, should help address this limitation.

Given the numerous sources of uncertainty, it is important to determine which one might dominate on the time scale of interest, to put more effort into reducing it. Villarini and Vecchi (2012), following the methodology developed by Hawkins and Sutton (2009), have partitioned the sources of uncertainty in projections of North Atlantic hurricane frequency into three main components: the forcing uncertainty that arises from imperfect knowledge of future radiative forcing, the response uncertainty arising from imperfect knowledge of how the climate system will respond to changes in radiative forcing, and the internal variability uncertainty arising from chaotic variations in the climate system that are not driven by radiative forcing. Interestingly, they showed that even though tropical Atlantic and tropical mean SSTs are both predictors of tropical storm activity in the statistical model, the sources of uncertainties for projected tropical Atlantic SST and tropical mean SST are very different from the uncertainty of projecting changes in relative SST and TCs (Fig. 8).

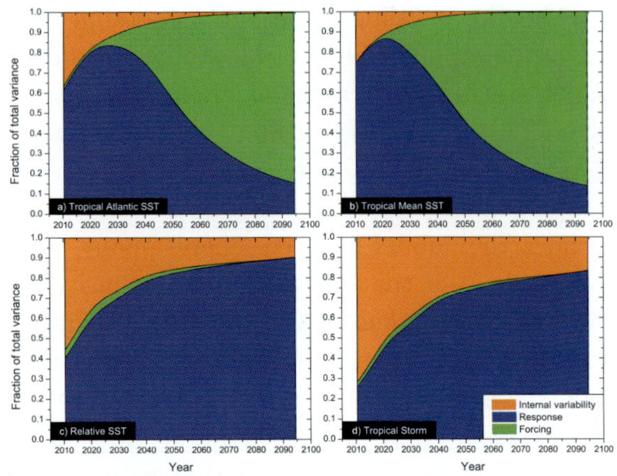

Fig. 8. From (Villarini and Vecchi, 2012) "© Copyright [2012] Nature": Fractional contribution to uncertainties in CMIP5 projections of Atlantic tropical storm frequency and its predictors as defined in the statistical-dynamical model of (Villarini *et al.*, 2010). The decomposition follows the method defined by (Hawkins and Sutton, 2009).

For tropical Atlantic and tropical mean SSTs, the response uncertainty dominates until the mid-21st century, after which the climate forcing becomes the dominant source of uncertainty. In contrast, internal variability is the largest source of uncertainty in projections of hurricane frequency during the first half of the 21st century with an important contribution of the response uncertainty as well, which becomes dominant at the end of the century. The uncertainty of forcing (i.e. future greenhouse gas emissions and other radiative forcing) does not play a large role for hurricane frequency whereas it does for projections of tropical Atlantic SST. This again shows that a better constraint on future changes in Atlantic SST, and in particular relative SST, could help reduce the uncertainty in future projection of Atlantic hurricane frequency.

6. Multi-year Predictions of Hurricane Activity

The large internal variability of hurricane activity is a large source of uncertainty in future projections, as described in the previous section, and it complicates the assessment of whether an observed change within a relatively short record is due to anthropogenic forcing or natural variability. In an attempt to quantify or predict the part that is caused by internal climate variability, recent coordinated experiments have been performed as part of CMIP5. In these experiments, besides imposing historical and future projected time-varying forcings as boundary conditions, the models used for the predictions were initialized with estimates of the observed state of the climate system (Taylor *et al.*, 2012). The purpose of these experiments was to determine where initialization could increase skill by either predicting some of the internal variability or improving the forced response or both. While for SST, most of the skill on multi-year timescales arises from predicting the warming trend associated with radiative forcing changes (van Oldenborgh *et al.*, 2011), Smith *et al.* (2010) suggested that initialization can increase the skill in multi-year hurricane forecasts. This is consistent with and motivated by a number of studies, which suggested that internal variations of the climate system could play a large role in changes of hurricane frequency on decadal timescales (*e.g.* Goldenberg and Shapiro, 1996; Zhang and Delworth, 2006, 2009; Knight *et al.*, 2006; Latif *et al.*, 2007; Dunstone *et al.*, 2011; Villarini *et al.*, 2010; Villarini and Vecchi, 2012).

Vecchi *et al.* (2013) investigated the predictability of North Atlantic hurricane frequency in two climate prediction systems based on the GFDL and UKMetOffice coupled models using a statistical-dynamical model based on tropical Atlantic SST and relative SST as predictors (Vecchi *et al.*, 2011). In this way, statistical-dynamical predictions could be made without a direct count of the exact number of storms produced by the model. The retrospective 5-yr mean and 9-yr mean predictions from these two models are encouraging with the best results

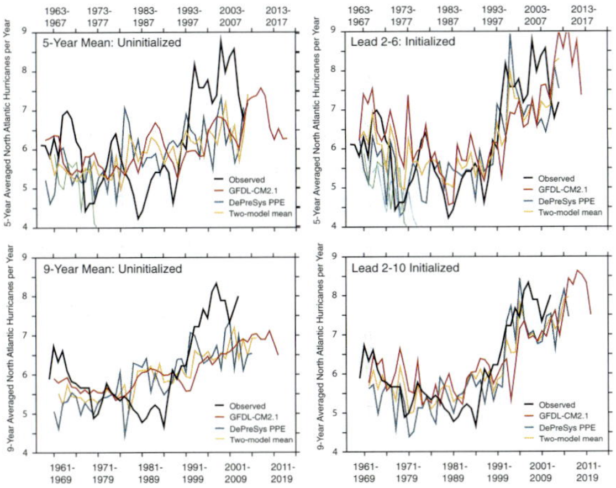

Fig. 9. From (Vecchi *et al.*, 2013) "© Copyright [August 2013] AMS": Retrospective and future forecasts of 5-yr mean and 9-yr mean hurricane frequency in uninitialized (left) and initialized (right) decadal prediction experiments. Black line shows the observed counts from the NOAA HURDAT database that includes the adjustement for observing inhomogeneity prior 1966 described in (Vecchi and Knutson, 2011). Red lines correspond to forecasts from the GFDL-CM2.1 system, blue lines show results from the UKMetOffice-DePreSys-PPE system, and yellow lines correspond to the two-system ensemble-mean.

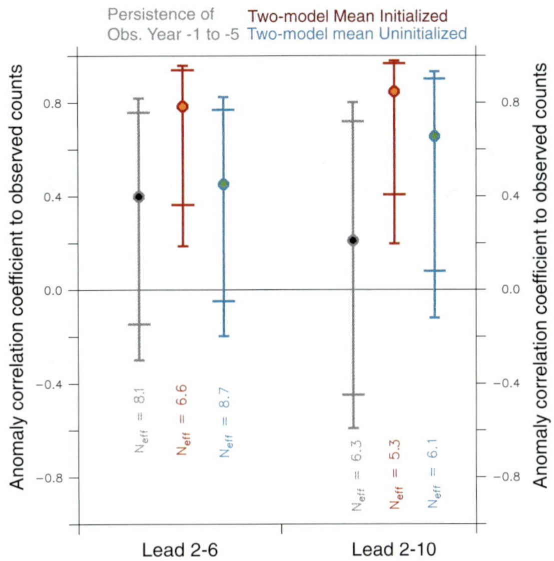

Fig. 10. (Adapted from (Vecchi *et al.*, 2013) "© Copyright [August 2013] AMS": Anomaly correlation of retrospective multi-year forecasts of North Atlantic hurricane frequency, for forecast leads of 2–6 years and 2–10 years. The values shown are averages of two prediction systems based on the GFDL and the UKMetOffice models. The sample correlation estimate is shown by a circle; the bars show the two-sided 90% and 95% uncertainty. Red bars correspond to the initialized predictions, blue bars to the uninitialized ones and gray bars indicate the correlation from a persistence forecast.

for the two-model mean (Figs. 9 and 10). The predicted time series follow quite closely the observed variations for several years. The correlations show a nominal improvement in the initialized experiments for both 5-yr and 9-yr mean predictions. However, the nominal improvement in the lead 2-to10-yr forecast arises only in the first part of the decade, suggesting potential multi-year forecast skill rather than decadal skill.

The improved accuracy in the initialized predictions was mainly attributed to a reduction of the conditional bias in SST that is large in the uninitialized predictions (Vecchi *et al.*, 2013). The impact of initialization on forecasts of the Atlantic MDR conditions appears key to the higher skill in the initialized forecasts. Though encouraging, these results need to be viewed with caution. Indeed the observational dataset is too short to confidently assess whether the additional correlation skill added by initializing with observations is statistically significant compared to the forced experiments as shown by the large confidence intervals in Fig. 10. Furthermore, the observed time series of North Atlantic hurricane frequency (Fig. 1) is characterized by an abrupt rise in 1995 leading to a positive trend over the 1961–2011 period. Vecchi *et al.* (2013) found that both the UKMetOffice and GFDL models show an increase in hurricane frequency in the mid-90s, leading to high correlations with observations. However they showed that these high correlations do not come from predicting the dynamical evolution of the climate system that leads to the mid-90s shift in hurricane frequency, but from "recognizing" that a climate shift has occurred around 1995 and persisting that shift.

Identifying the source of skill in retrospective predictions is key to assessing the likely success of future forecasts. The analysis of Vecchi *et al.* (2013) suggests a limited chance of predicting a similar positive or negative shift in hurricane frequency, were it to happen, using the current predictions systems. Identifying the origin of the observed upward shift in hurricane frequency in 1995 and assessing whether the modeled mechanisms are consistent with those in the real world are keys to interpreting any apparent skill in the retrospective forecasts. There is also a need to better understand the physical sources of predictability of Atlantic hurricane activity, whether remote as suggested by Dunstone *et al.* (2011) and Camargo *et al.* (2013) or local (Vecchi *et al.*, 2013). There are indications that changes in atmospheric aerosols could have influenced hurricane activity over the second half of the 20th century, with a decrease in Atlantic aerosol driving an increase in Atlantic hurricane activity (Mann and Emanuel, 2006; Evan *et al.*, 2009; Villarini and Vecchi, 2012). However, the discontinuities in the observing system for both hurricane counts and the observed ocean conditions used for the initialized decadal predictions make it challenging to attribute changes in hurricane activity to specific causes. There is a coincidence between the global implementation of Argo drifting floats in 2003 and the increase in hurricane activity predicted by the GFDL model (Fig. 9), suggesting an artifact of changes in the observing system. Vecchi *et al.* (2013) found a much reduced increase in hurricane activity over the next decade when the prediction experiments were repeated withholding Argo data, suggesting that enhanced observational sampling after 2003 may have led to a substantial change in the lead-dependent climatology. Additional methods to minimize the impact of observing system changes must be explored to improved future predictions of hurricane activity.

7. Conclusion

We reviewed in this chapter recent work on assessing past, present and future changes in North Atlantic hurricane activity. Although comprehensive observations of TCs have been collected in the Atlantic for more than a century, the original NOAA database almost certainly has important biases such as missing or miscategorized storms. Consequently, the dataset shows a (likely spurious) increase in the number of storms over the 20th century that could be misinterpreted as a response to anthropogenic warming. An estimated adjustment to the HURDAT database by (Vecchi and Knutson, 2008, 2011), has led to a revised estimate of past tropical storm and hurricane numbers and to the associated long-term trends. The updated record shows no significant trend, similar to U.S. landfalling hurricane counts, which also show no significant long-term trend, even without any adjustments applied. Our confidence in any climate change detection and attribution statements, or even in future projections, requires a sustained effort to observe the climate system, including tropical storms, along with a careful assessment of the available climate records. Recent changes have been documented in the observed record of Atlantic hurricane activity including an increase in potential intensity between about 1990 and 2005 (Kossin and Camargo, 2009). While such changes could be interpreted as the climatic impact of anthropogenic forcing on Atlantic hurricane activity, the attribution of these changes to climate forcing agents is complicated by the fact that decadal to multi-decadal changes in hurricane activity are closely linked to those of North Atlantic SST which presumably include contributions from anthropogenic forcing, natural forcing, and internal climate variability (Ting *et al.*, 2009; Terray, 2012). Model experiments have suggested that the late twentieth century increase in North Atlantic potential intensity includes a substantial contribution from internal variability and hence cannot be attributed purely to anthropogenic forcing Camargo *et al.* (2013). A number of studies also highlighted the importance of remote SST, showing that relative SST was a more robust indicator of hurricane changes than local MDR SST (Vecchi *et al.*, 2008; Ramsey and Sobel, 2011). From these studies, it appears that the recent increase in potential intensity is likely due to both internal and forced variability.

Statistical and dynamical models that reproduce quite well past hurricane activity have been used to project future hurricane activity. At present, we do not know with confidence even the sign of the expected changes in North Atlantic hurricane frequency over the coming century. However, the most likely response to increased greenhouse gases, according to current model projections, is a reduction in the overall frequency of North Atlantic tropical storms and hurricanes and an increase in the frequency of the most intense hurricanes, which means fewer but stronger storms in the tropical North Atlantic in a warmer climate (Bender *et al.*, 2010; Knutson *et al.*, 2013). Models project robustly that tropical storms in a warmer climate will be associated with higher rainfall rates than those in the present climate, in particular in the hurricane inner core where the projected increase exceeds what would be expected from a simple Clausius-Clapeyron atmospheric water vapor response.

Very intense hurricanes are relatively rare but are potentially very dangerous (Mendelsohn *et al.*, 2012; Peduzzi *et al.*, 2012). Pielke *et al.* (2008) stressed that although category 4–5 hurricanes account for only about 15% of US landfalling TCs, the damage they cause is considerable and disproportionate. Bender *et al.* (2010) estimated that by 2100 the overall destructive potential of hurricanes could increase by over 30% because of long-term climate warming leading to a greater frequency of category 4 and 5 hurricanes in the Atlantic. Higher cyclone wind speeds, along with increased sea level, would tend to increase the probability of storm surges. However, the net change in storm surge threat will depend on other factors as well, such as storm frequency changes and changes in preferred tracks, which are even less well-constrained.

Future hurricane activity will likely also depend on regional details of SST changes, which can differ substantially between the different climate model projections (Emanuel *et al.*, 2008; Villarini and Vecchi, 2012). The observed past TC changes and the model projections of hurricane activity are not spatially uniform (Vecchi and Soden, 2007; Vecchi and Knutson, 2008, 2011; Zhao *et al.*, 2009) and the spatial structure of these changes needs to be considered in relating future TC projections to impacts. Increasing the confidence in these projected regional details remains a challenge.

On the basis of current global climate model projections and our downscaling models, the forced response and internal variability uncertainties appear as dominant limitations in our ability to confidently project North Atlantic hurricane activity over the 21st century. Aerosol forcing and the climate response to aerosols also remain major sources of uncertainty. A better understanding of the climate forcings and the dynamical processes that will control changes in hurricane predictors like tropical Atlantic SST, over the 21st century, are key to reduce these uncertainties. A further source of uncertainty that cannot be eliminated, however, is the uncertainty that arises from the chaotic behavior of the atmosphere, with the limited predictability of weather patterns that can affect hurricane activity and vice versa.

Predicting hurricane frequency on multi-year timescales, and hence beyond the seasonal forecasts, is an emerging field of research. Recent studies suggest that coupled models that account for changes in both the initial state and external forcing can potentially lead to skillful multi-year predictions of Atlantic hurricane frequency (Smith *et al.*, 2010; Vecchi *et al.*, 2013). This improvement due to initialization should, however, be interpreted with caution given the relatively small effective number of degrees of freedom that results from the limited observed record and temporal persistence in the time series. Although models and observational techniques have improved and expanded greatly in recent years, there are still many challenges to face. It appears likely that additional improvements in dynamical models will improve TC representation and predictions. Landfalling hurricanes have not been discussed in detail in this chapter. While these are of even greater societal importance, they tend to be associated with larger uncertainties in terms of model projections and thus represent an additional long-term challenge.

Acknowledgments

We wish to thank Dr. Gabriele Villarini for valuable comments on a draft of this chapter. Dr. C.P. Chang and a number of professors, staff and students of the National Taiwan University are also greatly acknowledged for organizing the conference that led to very interesting exchanges of ideas around the science topics presented in this book.

References

Bender, M. A., T. R. Knutson, R. E. Tuleya, J. J. Sirutis, G. A. Vecchi, S. T. Garner, and I. M. Held, 2010: Modeled impact of anthropogenic warming of the frequency of intense Atlantic hurricanes, *Science* **327**, 454–458.

Blake, E. S., E. N. Rappaport, and C. W. Landsea, 2007: The deadliest, costliest, and most intense United States tropical cyclones from 1851 to 2006 (and other frequently requested hurricane facts). *NOAA Tech. Memo NWS TPC*-5, 454–458, 43 pp. Available online at http://www.nhc.noaa.gov/pdf/NWS-TPC-5.pdf.

Camargo, S. and A. G. Barnston, 2009: Experimental seasonal dynamical forecasts of tropical cyclone activity at IRI. *Wea. Forecasting.* **24**, 472–491, doi:10.1175/2008WAF2007099.1.

———, A. G. Barnston, P. Klotzbach, and C. W. Landsea, 2007: Seasonal tropical cyclone forecasts. *WMO Bull.* **56**, 297–309.

———, M. Ting, and Y. Kushnir, 2013: Influence of local and remote SST on North Atlantic tropical cyclone potential intensity, *Clim. Dyn.* **40**, 1515–1529, doi:10.1007/s00382-012-1536-4.

Chang, E. K. and Y. Guo, 2007: Is the number of North Atlantic tropical cyclones significantly underestimated prior to the availability of satellite observations, *Geophys. Res. Lett.* **34**, L14801, doi:10.1029/2007GL030169.

Chauvin, F., J.-F. Royer, and M. Dque, 2006: Response of hurricane-type vortices to global warming as simulated by ARPEGE-climat at high resolution, *Clim. Dyn.* **27**, 377–399.

DelSole, T., M. K. Tippett, and J. Shukla, 2011: A significant component of unforced multidecadal variability in the recent acceleration of global warming, *J. Clim.* **24**, 909–926.

Dunstone, N. J., D. M. Smith, and R. Eade, 2011: Multi-year predictability of the tropical Atlantic atmosphere driven by the high latitude North Atlantic, *Geophys. Res. Lett.* **38**, doi:10.1029/2011GL047949, l14701.

Emanuel, K. R., 2007: Environmental factors affecting tropical cyclone power dissipation, *J. Clim.* **20**, 5497–5509.

———, 2011. Global warming effects on US hurricane damage, *J. Clim.* **3**, 261–268.

———, S. Solomon, D. Folini, S. Davis, and C. Cagnazzo, 2013: Influence of tropical tropopause layer cooling on Atlantic hurricane activity, *J. Clim.* In press.

———, R. Sundararajan, and J. Williams, 2008: Hurricanes and global warming-results from downscaling IPCC AR4 simulations, *Bull. Amer. Met. Soc.* **89**, 347–367.

Evan, A. T., D. J. Vimont, R. Bennartz, J. P. Kossin, and A. K. Heidinger, 2009: The role of aerosols in the evolution of tropical North Atlantic ocean temperature, *Science* **324**, 778–781.

Goldenberg, S. B., C. W. Landea, A. M. Mestas-Nuez, and W. M. Gray, 2001: The recent increase in Atlantic hurricane activity: Causes and implications, *Science* **293**, 474–479.

———, and L. Shapiro, 1996: Physical mechanisms for the association of El Niño and West African rainfall with Atlantic major hurricane activity, *J. Clim.* **9**, 1169–1187.

Guilyardi, R., H. Bellenger, M. Collins, S. Ferrett, W. Cai, and Wittenberg, A. 2012: A first look at ENSO in CMIP5, *CLIVAR Exchanges* **17**, 29–32.

Guilyardi, R., *et al.*, 2009: Understanding El Niño in ocean-atmosphere general circulation models: Progress and challenges, *Bull. Amer. Met. Soc.* **90**, 325–340.

Hawkins, E. and R. Sutton, 2009: The potential to narrow uncertainty in regional climate predictions, *Bull. Amer. Met. Soc.* **90**, 1095–1107.

Holland, G. J., and P. J. Webster, 2007: Heightened tropical cyclone activity in the North Atlantic: natural variability or climate trend? *Philos. Trans. Roy. Soc. Math. Phys. Eng. Sci.* **365A**, 2695–2716.

Jarvinen, B. R., C. J. Neumann, and M. A. S. Davis, 1984: A tropical cyclone data tape for the North Atlantic basin, 1886–1983: Contents, limitations and uses. *Tech. Memo. NWS NHC, National Oceanic and Atmospheric Administration* **22**, 24pp.

Klotzbach, P. J., 2007: Revised prediction of seasonal Atlantic basin tropical cyclone activity from 1 August. *Wea. Forecasting* **22**, 937–949.

Knight, J. R., R. J. Allan, C. K. Folland, M. Vellinga, and M. E. Mann, 2005: A signature of persistent natural thermohaline circulation cycle in observed climate. *Geophys. Res. Lett.* **32**, L20708, doi:10.1029/2005GL024233.

———, C. K. Folland, and A. Scaife, 2006: A signature of persistent natural thermohaline circulation cycle in observed climate. *Geophys. Res. Lett.* **33**, L117706, doi:10.1029/2006GL026242, l17706.

Knutson, a. J. J. S., T. R., *et al.*, 2013: Dynamical downscaling projections of 21st century Atlantic hurricane activity: CMIP3 and CMIP5 model-based scenarios, *J. Clim.* Under review.

Knutson, T. R., *et al.*, 2008: Simulated reduction in Atlantic hurricane frequency under twenty-first century warming conditions, *Nature Geosci.* **1**, 359–364.

Knutson, T. R., *et al.*, 2010: Tropical cyclones and climate change, *Nature Geosci.* **3**, 157–163.

Kossin, J. P., and S. J. Camargo, 2009: Hurricane track variability and secular potential intensity trends, *Clim. Chan.* **97**, 329–337, doi: 10.1007/s10584-009-9748-2.

Landsea, C. W., 2007: Counting Atlantic tropical cyclones back in time, *EOS Trans. Am. Geophys. Union* **88**, 197–203.

Landsea, C. W., G. A. Vecchi, L. Bengtsson, and T. R. Knutson, 2010: Impact of duration thresholds on Atlantic tropical cyclone counts, *J. Clim.* **23**, 2508–2519.

LaRow, T. L., L. Stefanova, D.-W. Shin, and S. Cocke, 2010: Seasonal Atlantic tropical cyclone hindcasting/forecasting using two sea surface temperature datasets. *Geophys. Res. Lett.* **37**, L02804, doi:10.1029/2009GL041459, l02804.

Latif, M., N. Keenlyside, and J. Bader, 2007: Tropical sea surface temperature, vertical wind shear, and hurricane development, *Geophys. Res. Lett.* **34**, L01710, doi:10.1029/2006GL027969, l01710.

MacAdie, C., C. W. Landsea, C. J. Neumann, J. E. David, E. Blake, and G. R. Hammer, 2009: *Tropical cyclones of the North Atlantic Ocean*, 1851–2006, Tech. Memo., National Climatic Data Center, TCP/National Hurricane Center, 238 pp.

Mann, M. E., and K. A. Emanuel, 2006: Atlantic hurricane trends linked to climate change, *Eos Trans. Am. Geophys. Union* **87**, 233–244.

Mann, M. E., T. A. Sabbatelli, and U. Neu, 2007: Atlantic hurricane trends linked to climate change, *Geophys. Res. Lett.* **34**, doi:10.1029/2007GL031781.

Mendelsohn, R., K. Emanuel, S. Chonabayashi, and L. Bakkensen, 2012: The impact of climate change on global tropical cyclone damage, *Nature Climate Change* **2**, 205–209, doi: 10.1038/nclimate1357.

Murakami., H., *et al.*, 2012: Future changes in tropical cyclone activity projected by the new high-resolution MRI-AGCM, *J. Clim.* **25**, 3237–3260.

Oouchi, K., J. Yoshimura, H. Yoshimura, R. Mizuta, S. Kusunoki, and A. Noda, 2006: Tropical cyclone climatology in a global-warming climate as simulated in a 20 km-mesh global atmospheric model: Frequency and wind intensity analysis, *J. Meteorolog. Soc. Jpn.* **84**, 259–276, doi:10.2151/jmsj.84.259.

Peduzzi, P., B. Chatenous, H. Dao, A. D. Bono, C. Herold, J. Kossin, F. Mouton, and O. Nordbeck, 2012: Global trends in tropical cyclone risk, *Nature Climate Change* **2**, 289–294.

Pielke, R., J. Gratz, C. Landsea, D. Collins, M. Saunders, and R. Musulin, 2008: Normalized hurricane damages in the United States: 1900–2005, *Natural Hazards Review* **9**, 29–42.

Ramsey, H. A., and A. H. Sobel, 2011: The effects of relative and absolute sea surface temperature on tropical cyclone potential intensity using a single column model, *J. Clim.* **24**, 183–193.

Randall, D. A., *et al.*, 2007: Climate models and their evaluation in: *Climate change* 2007*: The physical science basis. contribution of working group I to the fourth assessment report of the intergovernmental panel on climate change,* eds. S. Solomon *et al.*, Cambridge University Press, pp. 591–648.

Saunders, M. A., and A. S. Lea, 2005: Seasonal prediction of hurricane activity reacing the coast of the United States, *Nature* **434**, 1005–1008.

Smith, D. M., R. Eade, N. Dunstone, D. Fereday, J. M. Murphy, H. Pohlmann, and A. Scaife, 2010: Skillful multi-year predictions of Atlantic hurricane frequency, *Nature Geosci.* **3**, 846–849.

Solow, A. R., and L. J. Moore, 2002: Testing for trend in North Atlantic hurricane activity, 1900–98, *J. Clim.* **15**, 3111–3114.

Sugi, M., H. Murakami, and J. Yoshimura, 2009: A reduction in global tropical cyclone frequency due to global warming, *SOLA* **5**, 164–167.

Sugi, M., A. Noda, and N. Sato, 2002: Influence of the global warming on tropical cyclone climatology: an experiment with the JMA global model, *J. Meteorolog. Soc. Jpn.* **80**, 249–272, doi:10.2151/jmsj.80.249.

Sugi, M. and J. Yoshimura, 2012: Decreasing trend of tropical cyclone frequency in 228-year high-resolution AGCM simulations, *Geophys. Res. Lett.* **39**, L19805, doi:10.1029/2012GL053360, l19805.

Swanson, K. L., 2008: Nonlocality of Atlantic tropical cyclone intensities, *Geochem. Geophys. Geosyst.* **9**, Q04V01, doi:10.1029/2007GC001844.

Taylor, K. E., R. Stouffer, and G. Meehl, 2012: An overview of CMIP5 and the experiment design, *Bull. Amer. Met. Soc* **93**, 485–498.

Terray, L., 2012: Evidence for multiple drivers of North Atlantic multi-decadal climate variability, *Geophys. Res. Lett.* **39**, L19712, doi:10.1029/2012GL053046.

Ting, M., Y. Kushnir, R. Seager, and C. Li, 2009: Forced and internal twentieth-century SST in the North Atlantic, *J. Clim.* **22**, 1469–1481, doi:10.1175/2008JCLI2561.1.

van Oldenborgh, G. J., F. J. Doblas-Reyes, B. Wouters, and W. Hazeleger, 2011: Decadal prediction skill in a multi-model ensemble, *Clim. Dyn.* **38**, 1263–1280.

Vecchi, a. R. M., G. A., *et al.*, 2013: Multi-year predictions of North Atlantic hurricane frequency: Promise and limitations, *J. Clim.* **26**, doi:10.1175/JCLI-D-12-00464.1.

Vecchi, G. A. and T. R. Knutson, 2008: On estimates of historical North Atlantic tropical cyclone activity, *J. Clim.* **21**, 3580–3600.

Vecchi, G. A. and T. R. Knutson, 2011: Estimating annual numbers of Atlantic hurricanes missing from the HURDAT database (1878–1965) using ship track density. *J. Clim.* **24**, 1736–1746.

Vecchi, G. A. and B. J. Soden, 2007: Effect of remote sea surface temperature change on tropical cyclone potential intensity, *Nature* **450**, 1066–1071.

Vecchi, G. A., K. L. Swanson, and B. J. Soden, 2008: Whither hurricane activity? *Science* **322**, 687–689, doi:10.1126/science.1164396.

Vecchi, G. A. and A. T. Wittenberg, 2010: El Niño and our future climate: Where do we stand? *Wiley Interdisciplinary Reviews: Climate Change* **1**, 260–270, doi:10.1002/wcc.33.

Vecchi, G. A., M. Zhao, H. Wang, G. Villarini, A. Rosati, A. Kumar, I. M. Held, and R. Gudgel, 2011: Statistical-dynamical predictions of seasonal North Atlantic hurricane activity, *Mon. Wea. Rev.* **139**(4), 1070–1082.

Villarini, G. and G. A. Vecchi, 2012: Twenty-first-century projections of North Atlantic tropical storms from CMIP5 models. *Nature Climate Change* **2**, 604–607.

Villarini, G. and G. A. Vecchi, 2013: Projected increases in North Atlantic tropical cyclone intensity from CMIP5 models. *J. Clim.* **26**.

Villarini, G., G. A. Vecchi, T. R. Knudson, and J. A. Smith, 2011a. Is the recorded increase in short duration North Atlantic tropical storms spurious? *J. Geophys. Res.* **116**, doi:10.1029/2010JD015493.

Villarini, G., G. A. Vecchi, T. R. Knudson, M. Zhao, and J. A. Smith, 2011b. North Atlantic tropical storm frequency response to anthropogenic forcing: projections and sources of uncertainty, *J. Clim.* **24**, 3224–3238.

Villarini, G., G. A. Vecchi, and J. A. Smith, 2010: Modeling of the dependence of tropical storm counts in the North Atlantic basin on climate indices, *Mon. Wea. Rev.* **138**(7), 2681–2705.

Vitart, F. and Coauthors, 2007: Dynamically-based seasonal forecasts of Atlantic tropical storm activity issued in June by EUROSIP. *Geophys. Res. Lett.* **34**, L16815, doi:10.1029/2007GL030740.

Wang, H., J.-K. E. Schemm, A. Kumar, W. Wang, L. Long, M. Chelliah, G. D. Bell, and P. Peng, 2009: A statistical forecast model for Atlantic seasonal hurricane activity based on the NCEP dynamical seasonal forecast, *J. Clim.* **22**, 4481–4500.

Zhang, R. and T. L. Delworth, 2006: Impact of Atlantic multidecadal oscillations on India/Sahel rainfall and Atlantic hurricanes, *Geophys. Res. Lett.* **33**, L17712, doi:10.1029/2006GL026267.

Zhang, R. and T. L. Delworth, 2009: A new method for attributing climate variations over the Atlantic hurricane basin's main development region, *Geophys. Res. Lett.* **36**, L06701, doi:10.1029/2009GL037260.

Zhang, R., *et al.*, 2012: Have aerosols forced the observed Atlantic multidecadal variability? *J. Atmos. Sci.* doi:10.1175/JAS-D-12-0331.1. 1/13, in press.

Zhao, M., I. M. Held, S.-J. Lin, and G. A. Vecchi, 2009: Simulations of global hurricane climatology, interannual variability, and response to global warming using a 50 km resolution GCM. *J. Clim.* **22**(24), doi:10.1175/2009JCLI3049.

Zhao, M., I. M. Held, and G. A. Vecchi, 2010: Retrospective forecasts of the hurricane season using a global atmospheric model assuming persistence of SST anomalies, *Mon. Wea. Rev.* **138**, 3858–3868.

Chapter 19

Observed Variations of Western North Pacific Tropical Cyclone Activity on Decadal Time Scales and Longer

Johnny C. L. Chan

Guy Carpenter Asia-Pacific Climate Impact Centre
School of Energy and Environment,
City University of Hong Kong, Hong Kong, China
johnny.chan@cityu.edu.hk
www.cityu.edu.hk/gcacic

This chapter reviews the observed variations of tropical cyclone (TC) activity in the western North Pacific on time scales of a decade or longer. It is shown that the number, intensity, tracks and landfall locations of TCs in this ocean basin have substantial variations on time scales from multidecadal to centennial, together with possible long-term trends, which, however, are not related to global warming. Many of these variations can be explained by similar variations in the atmospheric and/or oceanographic environments in which these TCs are embedded.

1. Introduction

Among all ocean basins in which tropical cyclones (TCs) occur, the western North Pacific (WNP) generally has the highest annual number of TCs, with an average of around 27 TCs reaching tropical storm intensity (hereafter, the term TC refers to one with tropical storm intensity or higher except for TC genesis). It is also the only basin in which TCs can occur throughout the year.

However, a large variability in this number is evident. During the last 50 years, the annual TC number ranges from 39 in 1964 to 14 in 2010 (Fig. 1). It is obvious from Fig. 1 that in addition to the large interannual variability, variations on longer time scales (decades) as well as a downward trend are present in the time series.

In this chapter, we will attempt to document the observed variations in TC number, intensity, tracks and landfall locations with a time scale of a decade or longer. Plausible explanations of these variations will also be discussed. Note that rainfall variations associated with TCs will not be described because of the large variabilities and the relatively few studies conducted. Hindcasts from global circulation models will also not be presented.

2. Variations in TC Number

The annual TC number during the last 50 years (Fig. 1) shows generally two "active" (1960–74 and 1989–97) and two "inactive" (1975–88 and 1998–2011) periods. The cut-off points were identified using a change-point analysis (Liu and Chan, 2013). A similar separation of the time series into active versus inactive periods was also found by Zhao and Chu (2010) using a Bayesian change-point analysis. Notice from Fig. 1 that during an active period, a La Niña year could have above-normal number of TCs while during an inactive period, even an El Niño year is likely to have below-normal number of TCs.

Climate Change: Multidecadal and Beyond
edited by Chih-Pei Chang et al.

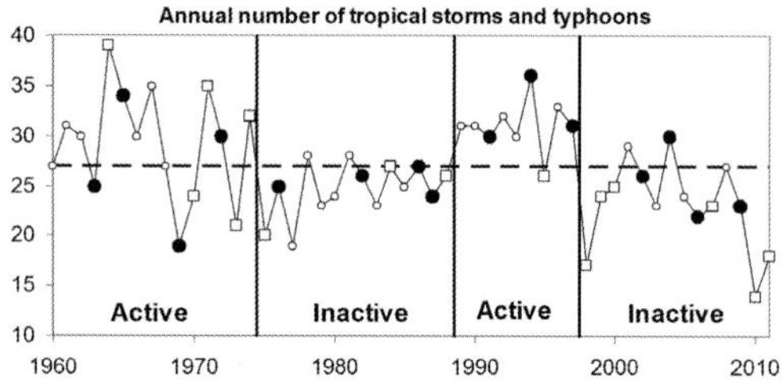

Fig. 1. Time series of the annual number of TCs in the WNP. Solid circle and open squares indicate the El Niño and La Niña years respectively. The horizontal dashed line indicates the normal number. The vertical lines divide the years 1960–2011 into the active and inactive periods. (From Liu and Chan, 2013).

These results are in contrast to the general conclusion that an El Niño year tends to have more TCs and the opposite is true for a La Niña year (Chan, 1985, 2000; Wang and Chan, 2002).

Yumoto and Matsuura (2001) examined TC activity in the WNP for the period 1951–1999 and also identified two high-activity (1961–72 and 1986–94) as well as two low-activity (1951–60 and 1973–85) periods through a smoothing of the time series. They, and later Matsuura *et al.* (2003), suggested that during a high-activity period, the sea-surface temperature (SST) east of 150°E is above normal, which enhances the convection there. As a result, the monsoon trough is extended eastward, and consequently more TC formation occurs.

Similarly, Liu and Chan (2013) found that during the active periods, the vertical wind shear in the tropics east of 150°E is weaker, and thus more TC formation is possible in this region (Fig. 2a). In addition, the subtropical high over the WNP is generally stronger in the inactive period (Fig. 2b), which leads to stronger trade winds and thus TC formation is not as favourable. During active periods, the number of TC formations in the southeastern part of the WNP is higher (Fig. 2c). Because TC formation in the western part generally does not vary substantially, a higher frequency of TC formation in the southeast would imply a higher frequency for the entire basin.

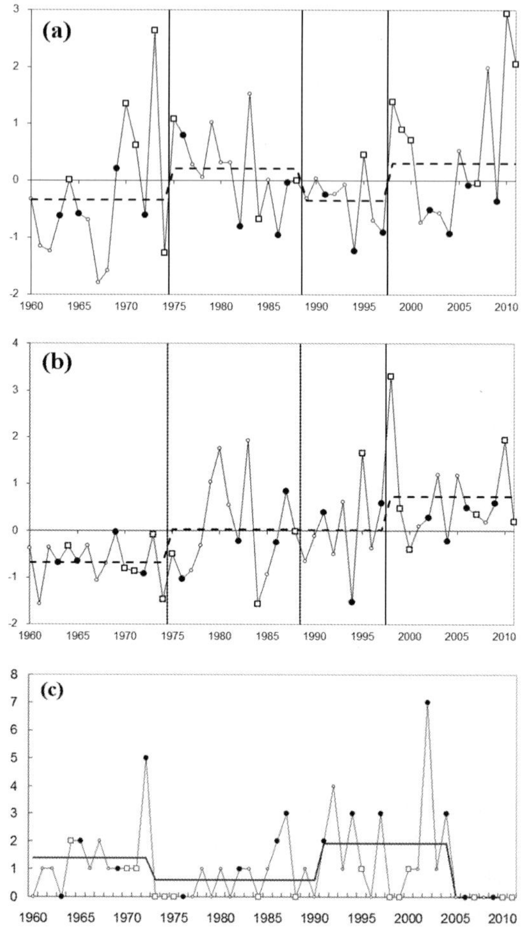

Fig. 2. Time series of (a) the vertical wind shear index, (b) the subtropical high area index and (c) the number of TC formations in the regions 10°–20°N, 160°–180°E. Solid circle and open squares indicate the El Niño and La Niña years respectively. The horizontal dashed lines in (a) and (b) indicate the mean values during the two active periods and the two inactive periods. The solid line in (c) indicates the mean in each period. See the definitions of the two indices in Liu and Chan (2013). (Adapted from Liu and Chan, 2013).

Thus, during the last 60 years (1951–2010), the number of TCs in the WNP has gone through three inactive and two active periods, each spanning 10 years or more. Such an oscillation of a period of around 20–30 years apparently results from changes in the atmosphere-ocean coupled system. Although the causes of such changes are not completely known, Chan (2008) suggested that perhaps the decadal signal of the El Niño/Southern Oscillation (ENSO) and the Pacific Decadal Oscillation (PDO) could be contributing factors. In addition, Fig. 1 appears to suggest a downward trend during the last 50 years. However, whether the downward trend is part of an even lower-frequency variation cannot be ascertained. To obtain such an answer would require at least another 50 years of data or more, which are not yet available.

3. Intensity

In operational forecasting, TC intensity is usually represented by its maximum sustained wind speed (MWS) near the earth's surface or the minimum sea-level pressure. To gauge how intense the TC had been throughout its life, one commonly used metric is the Saffir-Simpson hurricane scale that groups the TC intensity into different categories, with Category 5 being the strongest (see http://www.nhc.noaa.gov/aboutsshws.php). In climate studies, a TC is labeled by the highest category it reached during its lifetime.

To study the variability of TC intensity, the focus has been on the most intense TCs, i.e., those in Categories 4 and 5 (Cat45). The number of Cat45 TCs in the WNP during the period 1960–2010 again shows a large decadal variability (Fig. 3), in addition to the large interannual variability. Chan (2008) applied a wavelet analysis to the time series of this number for the period 1960–2005 and found a significant 16–32-year period. A similar oscillation period was also found by Lee *et al.* (2012) using the empirical mode decomposition method.

Chan (2008) reconstructed the time series based on this oscillation and obtained two periods (1960–70 and 1987–97) with above-normal and a period (1971–86) with below-normal number of Cat45 TCs. He also noted that the 16–32-year oscillation in this number correlates well with ENSO time series on the same time scale,

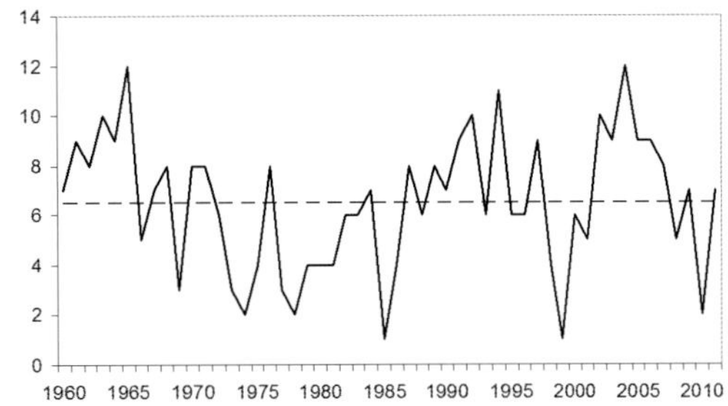

Fig. 3. Time series of the number of Cat45 TCs in the WNP during the period 1960–2011. The horizontal dashed line is the mean number of Cat45 TCs during this period. Data from the Joint Typhoon Warning Center website (http://www.usno.navy.mil/NOOC/nmfc-ph/RSS/jtwc/best_tracks/).

with a correlation coefficient of 0.60, which suggests that the decadal ENSO signal might be related to the decadal variation in the number of intense TCs.

The main difference between the above- and below-normal periods again lies in the vertical wind shear especially in the southeastern part of the WNP (Fig. 4) where it is smaller during the above-normal periods. In fact, throughout most of the basin where a majority of TCs traverse, the vertical wind shear is also smaller during these periods. The consequence is that more TCs can form in the southeastern part of the WNP. Because they form further to the southeast, they have a longer time over the ocean where the dynamic conditions remain favourable. As a result, they can become more intense.

Because TCs tend to move northwestward under the influence of the beta effect (Chan and Williams, 1987) and the steering effect of the subtropical ridge, the intense TCs that form in the southeastern part of basin during the above-normal periods have a higher chance of recurving and affect Korea and Japan. This is evident in taking the difference of the number of intense TCs between the above- and below-normal periods (Fig. 5). It is clear that more intense TCs tend to hit Korea and Japan during the above-normal period.

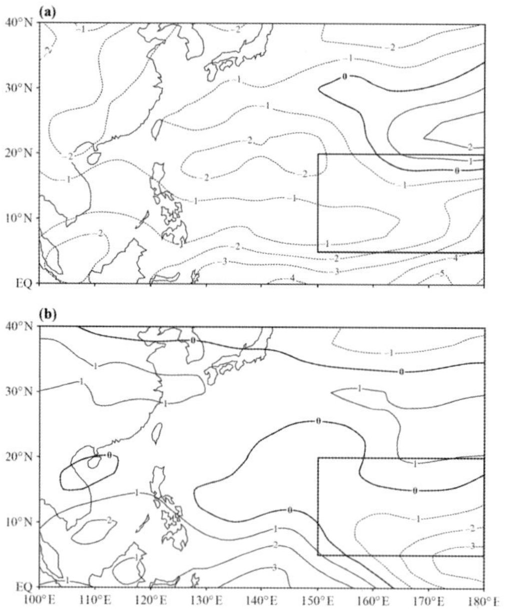

Fig. 4. Differences in May–November averaged vertical (200-hPa minus 850-hPa) zonal wind shear between the below-normal period of the number of Cat45 TCs (1971–86) and (a) first (1960–70), and (b) second (1987–1997) above-normal period. Unit: ms^{-1}. The thick rectangle indicates the area where the largest difference in the origin of the Cat45 TCs occurs. (From Chan, 2008).

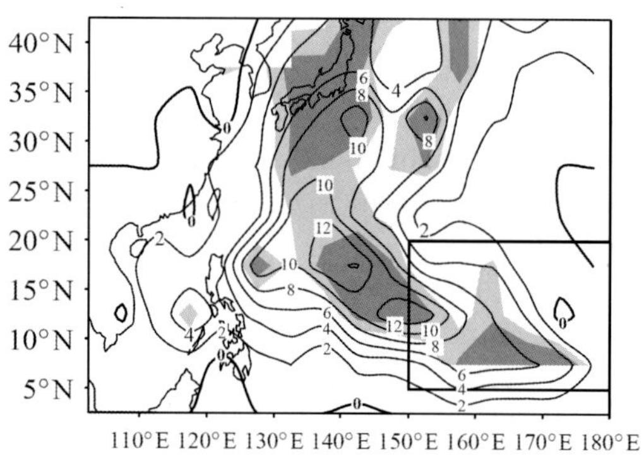

Fig. 5. Difference in the number of Cat45 typhoons per year (times 10) occurring in each 5°latitude × 5°longitude box between 1971–86 and 1987–1997. Positive numbers indicate less in the former period. Light and dark shadings indicate the difference is significant at the 90 and 95% significance level respectively. The thick rectangle indicates the area where the largest difference in the origin of the Cat45 TCs occurs. (From Chan, 2008).

Recently, Park *et al.* (2011) further quantified this result by examining the annual number of TCs making landfall in the Korean Peninsula and Japan, their intensity, duration and power dissipation index for the period 1977–2008 and found a significant increase in these TC characteristics during the latter ten years (1997–2008). In addition to weak vertical wind shear, they attributed this increase to warmer SST, higher atmospheric humidity and a stronger southerly steering flow. However, the enhanced thermodynamic factors could be related to global warming and not necessarily a cause.

Park *et al.* (2013) further investigated TC intensity variations over the entire WNP for the same period of 1977–2008 and found that while TC intensity increases in Korea and Japan, it decreases in the Philippine Sea area. An examination of the thermodynamic and dynamic parameters suggests that the latter appears to be dominant in explaining such a spatial distribution of intensity variations, which is consistent with the conclusion of Chan (2009).

To summarise, the number of intense TCs therefore again goes through large multi-decadal oscillations, with a period of about 20–30 years, which are driven probably by the same mechanisms as discussed in Section 2. It is also emphasized that the time series in Fig. 3 shows no trend at all. This reinforces the conclusion made by Chan (2006) that the increasing trends of intense TCs identified by Emanuel (2005) and Webster *et al.* (2005) are simply the rising branch of the multi-decadal oscillation. The multi-decadal variation is also evident in different parts of the WNP, with the oscillation in different parts of the basin being generally out of phase (Park *et al.*, 2013).

4. Tracks

Ho *et al.* (2004) examined the locations of TC occurrence during the period 1951–2001 and found that a significant difference occurs between the period 1951–1979 and 1980–2001

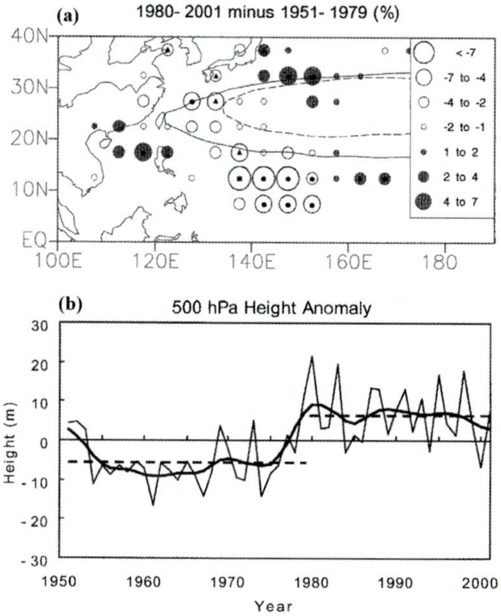

Fig. 6. (a) Geographical distribution of the difference in typhoon passage frequency between the periods 1980–2001 and 1951–79. Dotted and solid lines indicate the mean position of 5780 gpm for 1951–79 and 1980–2001, respectively. The symbols delta, circle and square represents the 90%, 95% and 99% confidence levels respectively. (b) Time series of the 500-hPa geopotential height anomaly in the region of 20–25°N and 125–140°E. Heavy solid line denotes 9-point Gaussian filtered values and heavy dotted lines denote the means for the periods 1951–79 and 1980–2001. (From Ho *et al.*, 2004).

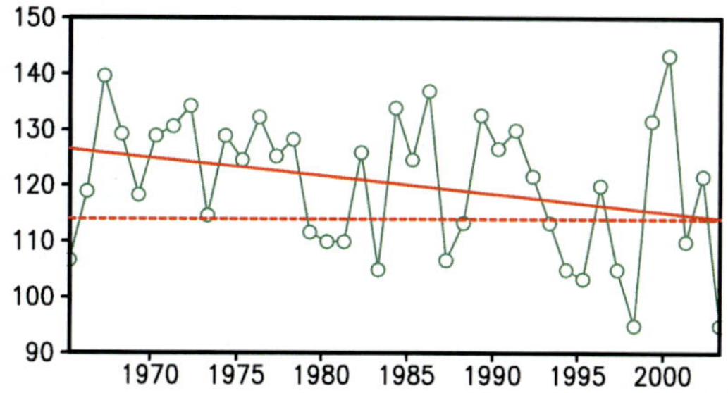

Fig. 7. Time series of the westernmost longitude of the July–September mean WNP subtropical high. The horizontal dashed lines indicate the minima of the linear fits. (From Wu *et al.*, 2005).

(Fig. 6a). During the first period, more TCs tend to move through the central part of the WNP so that fewer TCs occurred in the South China Sea or the eastern part of the WNP. They attributed this change in occurrence pattern to the strengthening of the subtropical high in the latter period (Fig. 6b) so that more TCs move westward to hit South China or to the east of Japan. In a later study, Wu *et al.* (2005) found a similar pattern shift in TC occurrence between the period 1965–83 and 1984–2003. They also documented a general westward extension of the subtropical high during the entire period of study (Fig. 7).

With such an extension of the subtropical high, one would expect the easterly flow to strengthen in the western part of the WNP. However, Chu *et al.* (2012) found a decrease in the easterly steering flow [defined as the 850–300 hPa pressure-weighted mean flow, following Chan and Gray (1982)] within a box southeast of Taiwan and the northern part of the South China Sea (Fig. 8a). They also showed that consistent with this reduction in the steering flow, the translation speed of the TCs has also decreased (Fig. 8b). In a study of the rainfall brought by TCs making landfall in Okinawa, Ikema *et al.* (2010) found an increase in heavy rain events and attributed a decrease in TC translation speed to be one of the contributing factors. Chu *et al.* (2012) further suggested that such the reduction in the steering flow could result in a longer lifetime of the TCs as well as a higher tendency to recurve, and argued for an increase in the number of TCs affecting the northern part of the South China Sea and Taiwan, which is similar to the pattern shown in Fig. 6a.

In a related study, Tu *et al.* (2009) found an abrupt jump in TC activity near Taiwan such that the average number of TCs in the vicinity of Taiwan increased from 3.3 per year during the period 1970–99 to 5.7 since 2000 (Fig. 9). They again attributed such a jump to a change in the intensity and pattern of the subtropical high.

In an earlier study to examine the spatial distribution of TC tracks, Liu and Chan (2008) performed an empirical orthogonal function (EOF)

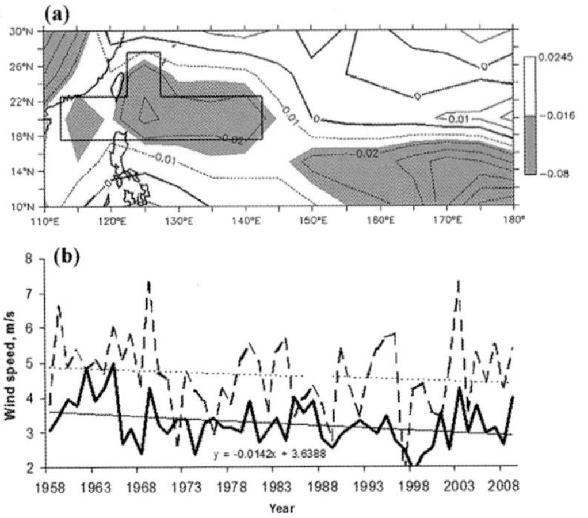

Fig. 8. (a) Trends in steering flows for July–September from 1958–2001 based on the NCEP/NCAR reanalysis data. Shading indicates statistical significance at 5% level. The box delineates the region used to produce the time series in (b). Broken lines denote negative trend. Unit of trend is $ms^{-1}yr^{-1}$. (b) Time series of area-averaged steering flows over the box indicated in (a) during 1958–2009 (solid line) and time series of area-averaged translation speed for the same period from the Regional Specialized Meteorological Centre in Tokyo (broken line). Linear trends from steering flows and translation speed are also shown, respectively, as thin solid line and thin dotted line. (Adapted from Chu *et al.*, 2012).

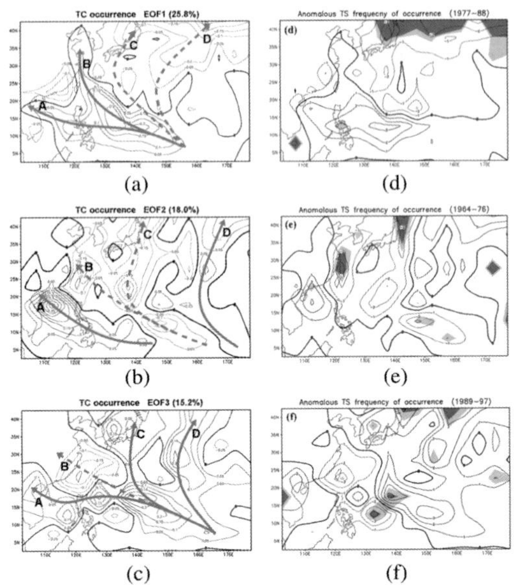

Fig. 10. Loading patterns of the annual TC occurrence pattern for the (a) first, (b) second and (c) third EOFs. The solid and dashed arrows (with alphabets) indicate the prevailing tracks with increased and decreased frequency respectively. Anomalous TC frequency of occurrence during the periods (d) 1977–88, (e) 1964–76 and (f) 1989–97. Dark and light shadings indicate the areas with anomalies significant at the 95% and 90% confidence level (Adapted from Liu and Chan, 2008).

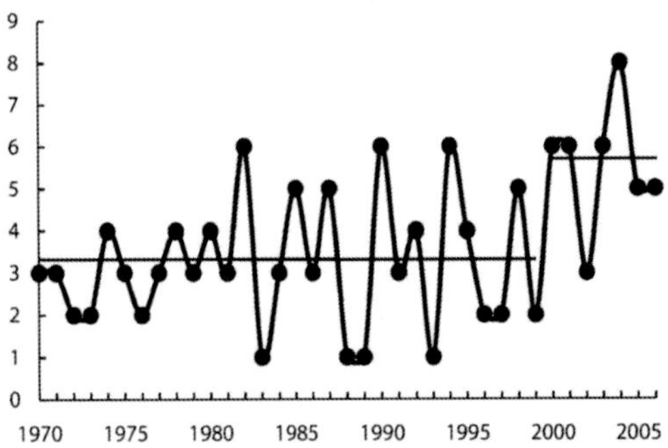

Fig. 9. Time series of seasonal (June–October) typhoon numbers passing the vicinity of Taiwan (21–26°N, 119–125°E from 1970 to 2006 as complied by the Central Weather Bureau of Taiwan. (From Tu *et al.*, 2009).

analysis on the annual number of TC occurrence in each grid box over the WNP during the period 1960–2005 and found three major principal components (Figs. 10a–c). The time series of the EOF coefficients associated with these three principal components have large variations, with the maximum amplitude in EOF1 occurring during the period 1977–88, EOF2 during 1964–76 and EOF3 during 1989–97 (not shown). A comparison of the plot of the anomalies of TC occurrence during each of these three periods against the EOF pattern shows that they replicate the EOF patterns very well (Figs. 10d–f). The authors attributed these variations in TC occurrence to changes in the strength of the subtropical high, which lead to changes in the zonal winds that steer the TCs. Some of these changes are also related to the PDO.

In a later study, Yeh *et al.* (2010) found the maximum TC track density over the WNP to have a similar shift, from the South China Sea during 1979–89 to the East China Sea and south of Japan during 1990–2000. The decadal variation in TC activity in the South China Sea was also noted by Goh and Chan (2010).

A recent study by Yokoi and Takayabu (2012) further confirmed the conclusion of Liu and Chan (2008) that on decadal time scales, changes in the track, rather than the genesis locations or the variations in basin-wide frequency of TC occurrence, are mainly responsible for the observed spatial distribution of TC track density.

All these studies suggest that the tracks of the TCs in the WNP also show significant variations on decadal time scales. Changes in the strength of the subtropical high apparently are largely responsible for such track changes.

5. Landfall

Because changes in TC track will lead to subsequent changes in the locations of TC landfall, it is useful to examine the variations in such locations. Here, the data are more abundant as each country would have a relatively good record of whether a TC has made landfall along its coast. Some of these records can go back for hundreds of years. We will therefore present results for three periods: post-1945, 1902–1939 and prior to 1900.

5.1. *Post-1945*

Chan and Xu (2009) examined the official records of TC landfall for the period 1945–2004 for the entire East Asia. They divided the landfall into three regions: South Asia that includes the Philippines, Vietnam and south coast of China (including Hainan Island), Middle Asia that includes the east coast of China and North Asia that includes the Korean Peninsula and Japan (Fig. 11). They then performed a wavelet analysis of the time series of the annual number of TCs that made landfall in each of the regions and found that for all the three regions, the two most prominent oscillations are 2–7 years and 16–32 years (Fig. 12). The former is attributed

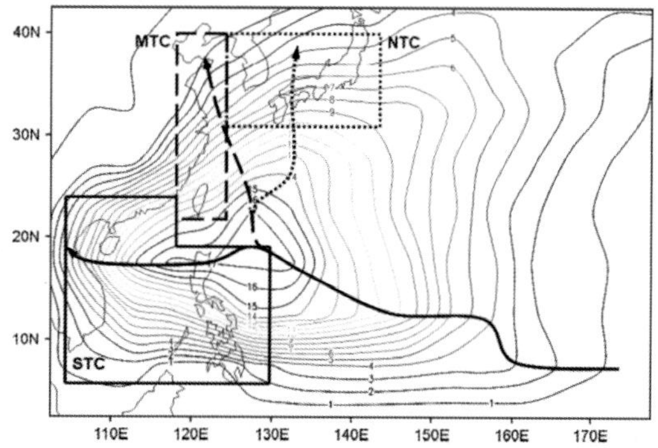

Fig. 11. June–October mean frequency of TC occurrence (unit: number per year) derived from the Joint Typhoon Warning Center best track data from 1965 to 2003 (adopted from Wu *et al.* 2005). The three boxes mark the different groups of landfalling TCs studied in the paper. STC — South Asia TCs, MTC — Middle Asia TCs, NTC — North Asia TCs. (from Chan and Xu, 2009).

mostly to ENSO while the latter might be related to the PDO. However, the phase of the multi-decadal oscillation is sometimes different among the regions, with some being in phase at times and out of phase in other times, as can be seen from Fig. 13. Note that because the majority of TCs make landfall in the South Asia region, the total number of landfalls as well as the total number of TCs have very similar phase as that for South Asia, a conclusion also drawn by Zhang *et al.* (2012) who examined the frequency of TCs making landfall in the Guangdong Province (in southern China) of China. In addition, they identified similar decadal variability in the frequency of landfall, with more in the 1960s and the 1990s and fewer in the 1980s (compare their Fig. 3 with Fig. 12a).

5.2. *1902–1939*

Prior to 1945, it was difficult to have data records from all countries. However, Kubota and Chan (2009) were able to retrieve data for landfall over the Philippines from the Monthly Bulletin of the Philippine Weather Bureau (MBP). An analysis of the number of landfall from

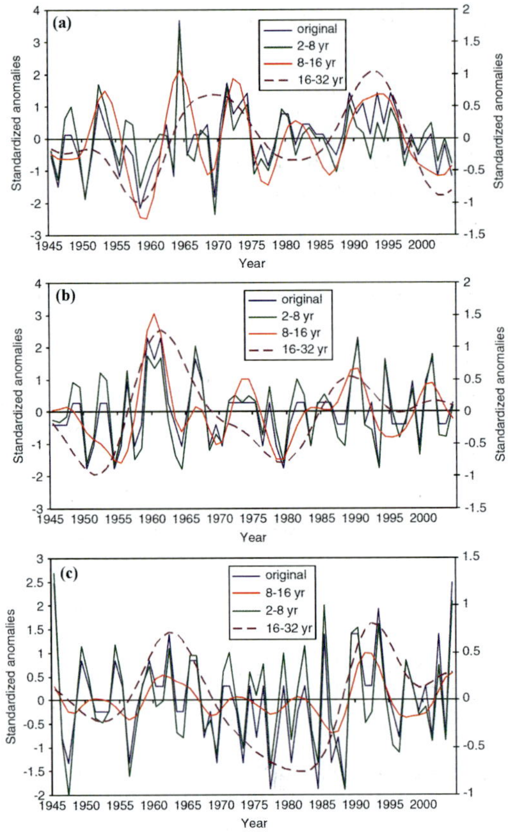

Fig. 12. Standardized number of landfalling TCs (blue) in (a) South Asia, (b) Middle Asia, and (c) North Asia (see Fig. 11 for the definition of the regions). Reconstruction of these time series based on a wavelet analysis for periods 2–8 years (green), 8–16 years (red, values multiplied by 5) and 16–32 years (blue dashed, values multiplied by 10) are also shown (scale on the right axis) (Adapted from Chan and Xu, 2009).

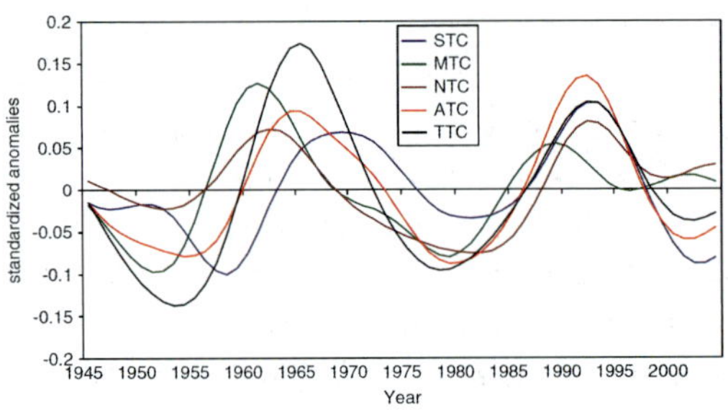

Fig. 13. 16–32-year reconstructed series of the standardized number of landfalling TCs in South Asia (STC), Middle Asia (MTC), North Asia (NTC), all of Asia (ATC), and of the total number of TCs in the WNP (TTC). (From Chan and Xu, 2009).

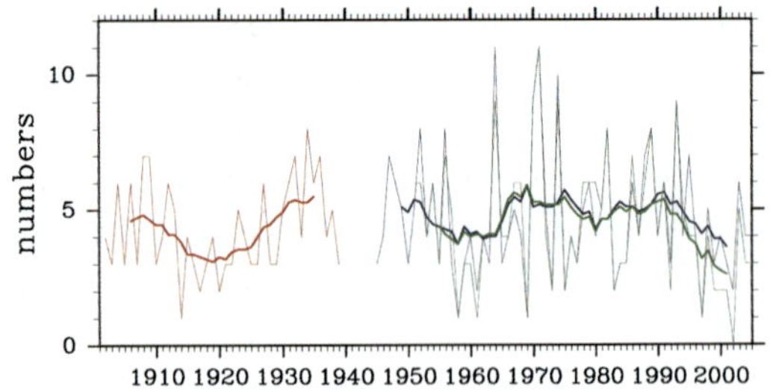

Fig. 14. Annual number of landfalling TCs in the Philippines: 1902–1939 (red, from MBP), 1951–2005 (blue from Joint Typhoon Warning Center, green from Japan Meteorological Agency). The solid lines of the same colour indicate the 10-year running mean (From Kubota and Chan, 2009).

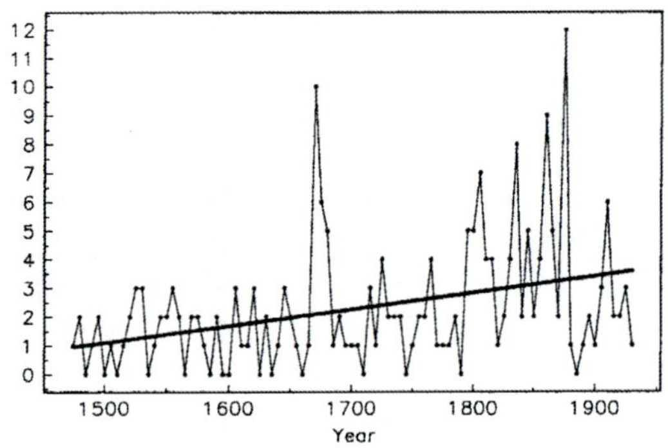

Fig. 15. Time series of the total number of TCs (assuming to be all typhoons) every five years making landfall over Guangdong Province between 1470–1931. (From Chan and Shi, 2000).

1902–1939 shows two maxima, one at around 1910 and another in the mid 1930s (Fig. 14). This periodicity of around 20–30 years is similar to the one shown in Fig. 12.

5.3. *Prior to 1900*

To go back even further, archives of ancient documents need to be retrieved. Chan and Shi (2000) were able to retrieve such documents for much of the Guangdong Province for the period 1470–1931 to identify instances of TC landfall. They found a positive linear trend as well as centennial and decadal oscillations, the most significant being around 140–210 years and 30–60 years, respectively (Fig. 15). The two periods of high activity during the second half of the 17th century (~1660–1680) and the first two-thirds of the 19th century, especially after 1850 (~1850–1880) are also found in a more detailed study by Liu *et al.* (2001).

Applying a wavelet analysis on the TC landfall frequency during the period 1450–1949 along the coast of East China (Fujian, Zhejiang and Jiangsu provinces), with data again obtained from historical government records, Chan *et al.* (2012) identified oscillations ranging from 4–8 years, 32–64 years and ~250 years, together with a positive linear trend (Fig. 16). Whether the trends in both studies are real or an artifact of the availability of data is uncertain.

Garcia-Herrera *et al.* (2007) studied the writings of a Spanish Jesuit who documented TC activity around the Philippines from the 16th century. Periods of high and low activity are found in the time series with a periodicity similar to the more modern-day record. The period of high activity in the 19th century also roughly coincides with that identified by Chan and Shi (2000) and Liu *et al.* (2001), which suggests the robustness of the results.

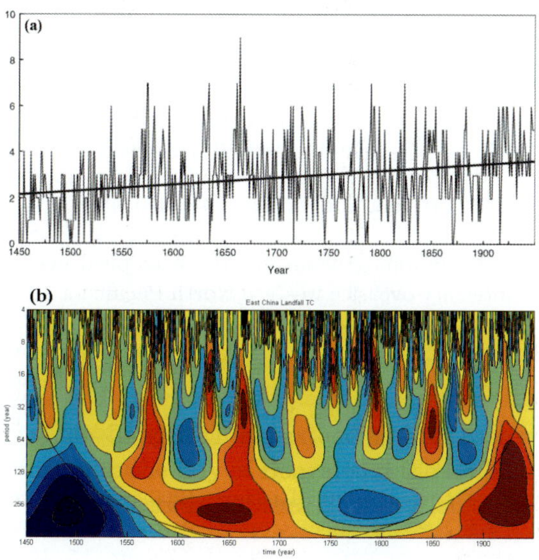

Fig. 16. (a) Annual number of TCs making landfall in East China during 1450–1949. The thick straight line represents the linear trend. (b) Wavelet analysis of the time series in (a) (From Chan *et al.*, 2011).

5.4. *Summary*

All the results on the frequency of landfalling TCs suggest that a multi-decadal signal is prevalent. Similar multi-decadal to centennial variations in TC activity are also found in paleoclimate studies (*e.g.*, Woodruff *et al.*, 2009; Nott, 2011; Nott and Forsyth, 2012). Results from these studies further suggested that variations in historical ENSO activity apparently correlate with those of TC activity, which is consistent with the conclusion of Chan *et al.* (2012). Such variations in ENSO activity modify the atmospheric flow patterns that lead to changes in the steering flow, hence the track types and landfall locations.

6. Summary

This paper summarizes some of the recent studies on the observed variations of TC activity in the WNP. Whether it is the TC number, intensity, track or landfall location, a strong decadal or multi-decadal signal is found. None of these TC characteristics during the last 100 years shows any significant trend, which strongly suggests that variations in WNP TC activity cannot be attributed to global warming, in contrast to conclusions of some earlier studies. The multi-decadal signals in the TC characteristics are found to be related to changes in the atmospheric and/or oceanographic conditions with variations on similar time scales.

References

Chan, J. C. L., 1985: Tropical cyclone activity in the northwest Pacific in relation to the El Niño/Southern Oscillation phenomenon. *Mon. Wea. Rev.*, **113**, 599–606.

———, 2000: Tropical cyclone activity over the western North Pacific associated with El Niño and La Niña events. *J. Climate*, **13**, 2960–2972.

———, 2006: Comment on "Changes in Tropical Cyclone Number, Duration, and Intensity in a Warming Environment". *Science*, **311**, 1713b.

———, 2008: Decadal variations of intense typhoon occurrence in the western North Pacific. *Proc. Royal Soc. A.*, **464**, 249–272.

———, 2009: Thermodynamic control on the climate of intense tropical cyclones. *Proc. Roy. Soc. A.*, **465**, 3011–3021.

———, and W. M. Gray, 1982: Tropical cyclone movement and surrounding flow relationships. *Mon. Wea. Rev.*, **110**, 1354–1374.

———, and R. T. Williams, 1987: Analytical and numerical studies of the beta-effect in tropical cyclone motion. Part I: Zero mean flow. *J. Atmos. Sci.*, **44**, 1257–1265.

———, and J.-E. Shi, 2000: Frequency of typhoon landfall over Guangdong Province of China during the period 1470–1931. *Int. J. Climatol.*, **20**, 183–190.

———, and M. Xu, 2009: Interannual and interdecadal variations of landfalling tropical cyclones in East Asia. Part I: Time series analysis. *Int. J. Climatol.*, **29**, 1285–1293.

———, K.-S. Liu, M. Xu and Q. Yang, 2012: Variations of frequency of landfalling typhoons in East China, 1450–1949. *Int. J. Climatol.*, **32**, 1946–1950.

Chu, P.-S., J.-H. Kim, and Y. R. Chen, 2012: Have steering flows in the western North Pacific and South China Sea changed over the last 50 years? *Geophys. Res. Lett.*, **39**, L10704, doi:10.1029/2012GL051709.

Emanuel, K. A., 2005: Increasing destructiveness of tropical cyclones over the past 30 years. *Nature*, **436**, 686–688.

Garcia-Herrera, R., P. Ribera, E. Hernandez, and L. Gimeno, 2007: Northwest Pacific typhoons documented by the Philippine Jesuits, 1566–1900. *J. Geophys. Res.*, **112**, D06108, doi: 10/1029/2006JD007370.

Goh, A. Z.-C., and J. C. L. Chan, 2010: Interannual and interdecadal variations of tropical cyclone activity in the South China Sea. *Int. J. Climatol.*, **30**, 827–843.

Ho, C.-H., J.-J. Baik, J.-H. Kim, D.-Y. Gong, and C.-H. Sui, 2004: Interdecadal changes in summertime typhoon tracks. *J. Climate*, **17**, 1767–1776.

Ikema, T., G. R. Bigg, and R. G. Bryant, 2010: Increasing rain intensity over Okinawa, 1982–2005, and the link to changes in characteristics of northwest Pacific typhoons. *J. Geophys. Res.*, **115**, D24121, doi: 10.1029/2010JD014428.

Kubota, H., and J. C. L. Chan, 2009: Interdecadal variability of tropical cyclone landfall in the Philippines from 1902 to 2005. *Geophys. Res. Lett.* **36**, L12802, doi:10.1029/2009GL038108.

Lee, H. S., T. Yamashita, and T. Mishima, 2012: Multi-decadal variations of ENSO, the Pacific Decadal Oscillation and tropical cyclones in the western North Pacific. *Prog. Oceanog.* **105**, 67–80.

Liu, K. B., C. M. Shen, and K. S. Louie, 2001: A 1,000 year history of typhoon landfalls in Guangdong, southern China, reconstructed from Chinese historical documentary records. *Ann. Assoc. Am. Geogr.*, **91**, 453–464.

Liu, K. S., and J. C. L. Chan, 2008: Interdecadal variability of western North Pacific tropical cyclone tracks. *J. Climate*, **21**, 4464–4476.

———, and J. C. L. Chan, 2013: Inactive period of western North Pacific tropical cyclone activity in 1998-2011. *J. Climate*, **26**, 2614–2630.

Matsuura, T., M. Yumoto, and S. Iizuka, 2003: A mechanism of interdecadal variability of tropical cyclone activity over the western North Pacific. *Climate Dyn.*, **21**, 105–117.

Nott, J., 2011: Tropical cyclones, global climate change and the role of Quaternary studies. *J. Quaternary Sci.*, **26**, 468–473.

———, and A. Forsyth, 2012: Punctuated global tropical cyclone activity over the past 5,000 years. *Geophys. Res. Lett.*, **39**, L14703, doi: 10.1029/2012GL052236.

Park, D.-S. R., C.-H. Ho, J.-H. Kim, and H.-S. Kim, 2011: Strong landfall typhoons in Korea and Japan in a recent decade. *J. Geophys. Res.*, **116**, D07105, doi:10.1029/2010JD014801.

———, ———, ———, and ———, 2013: Spatially inhomogeneous trends of tropical cyclone intensity over the western North Pacific for 1977–2010. *J. Climate*, in press.

Tu, J.-Y., C. Chou, and P.-S. Chu, 2009: The abrupt shift of typhoon activity in the vicinity of Taiwan and its association with western North Pacific-East Asian climate change. *J. Climate*, **22**, 3617–3628.

Wang, B., and J. C. L. Chan, 2002: How strong ENSO events affect tropical storm activity over the western North Pacific. *J. Climate*, **15**, 1643–1658.

Webster, P. J., G. J. Holland, J. A. Curry, and H.-R. Chang, 2005: Changes in tropical cyclone number, duration, and intensity in a warming environment. *Science*, **309**, 1844–1846.

Woodruff, J. D., J. P. Donnelly, and A. Okusu, 2009: Exploring typhoon variability over the mid-to-late Holocene: evidence of extreme coastal flooding from Kamikoshiki, Japan. *Quart. Sci. Rev.*, **28**, 1774–1785.

Wu, L., B. Wang, and S. Geng, 2005: Growing typhoon influence on east Asia. *Geophys. Res. Lett.*, **32**, L18703, doi:10.1029/2005GL022937.

Yeh, S.-W., S.-K. Kang, B. P. Kirtman, J.-H. Kim, M.-H. Kwon, and C-H. Kim, 2010: Decadal change in relationship between western North Pacific tropical cyclone frequency and the tropical Pacific SST. *Meteor. Atmos. Phys.*, **106**, 179–189.

Yokoi, S., and Y. N. Takayuba, 2012: Attribution of decadal variability in tropical cyclone passage frequency over the western North Pacific: A new approach emphasizing the genesis location of cyclones. *J. Climate*, **26**, 973–987.

Yumoto, M., and T. Matsuura, 2001: Interdecadal variability of tropical cyclone activity in the western North Pacific. *J. Meteor. Soc. Japan*, **79**, 23–35.

Zhang, Q., W. Zhang, X. Lu, and Y. D. Chen, 2012: Landfalling tropical cyclones in the south China: Intensifying or weakening? *Int. J. Climatol.*, **32**, 1815–1824.

Zhao, X., and P.-S. Chu, 2010: Bayesian change point analysis for extreme events (typhoons, heavy rainfall and heat waves): An RJMCMC approach. *J. Climate*, **23**, 1034–1046.

Chapter 20

Record-Breaking Increase of Tropical Cyclone Heavy Rainfall in Taiwan in the First Decade of 21st Century

Chih-Pei Chang, Hung-Chi Kuo, and Chung-Hsiung Sui

Department of Atmospheric Sciences, National Taiwan University, Taipei, 10617, Taiwan
cpchang@nps.edu
http://faculty.nps.edu/cpchang/

Record-breaking increase of tropical cyclone (TC)-induced heavy rainfall in the first decade of the 21st Century produced an apparent very large trend of extreme rainfall in Taiwan. This large trend has been suggested as a dramatic example of the effect of global warming on extreme rainfall. However, this trend is not observed in nearby regions of Southern China and Hainan Island where TC rainfall actually experienced a decreasing trend that masked the increasing trend of non-TC extreme rainfall. Detailed analyses show that most of the observed large increase in typhoon rainfall is the result of slowly moving TCs and their tracks relative to the meso-α scale terrain. Stronger interaction between typhoon circulation and southwest monsoon winds after the TC center moves into the Taiwan Strait create a trend of increasing TC rainfall intensity, which is not observed before the TC center exits Taiwan. These observations are consistent with the recently discovered multidecadal trend of intensifying subtropical monsoon and tropical circulations, which is believed to be related mainly to natural variations. There is also no evidence of a positive feedback between global warming-related water vapor supply and TC intensity, as the number of strong landfalling TCs has decreased significantly since 1960 and the recent heavy rainfall typhoons are all of weak to medium intensities.

1. Introduction

Taiwan, which is situated in one of the main paths of western North Pacific tropical cyclones (TCs), has experienced a series of TCs with extraordinary amount of rainfall since the late 1990s. In addition to an increase in the number of TCs, nine of the top 12 typhoons in total rainfall since hourly rainfall observations started in 1960 have occurred in the 21st century (Table 1). The most extreme case is the record-breaking Typhoon Morakot in 2009 (Ge *et al.*, 2010; Lee *et al.*, 2011; Chien and Kuo, 2011) that caused huge economic and life losses, and became the first natural disaster in Taiwan that actually triggered a change of government. Based on the rainfall trend created by these recent typhoons, stern warnings of a more than doubling of extreme rainfall per each 1 K global warming were issued by some Taiwan researchers, whose criticism of the rainfall forecasts of Typhoon Morakot for ignoring global warming effects was prominently highlighted by news media and some government reports (Hsiao, 2009).

Whether the heavy rainfall associated of these heavy rainfall typhoons is an impact of global warming, and in particular whether global warming-enhanced water vapor capacity caused a positive interaction between water vapor supply and TC that led to the observed

Climate Change: Multidecadal and Beyond
edited by Chih-Pei Chang et al.

Table 1. The twelve typhoons in 1960–2011 with total rainfall over Taiwan exceeding 3500 mm during the three phases. The eight since 2004 are highlighted in boldface.

Rank	Year	Typhoon Name	PR	OL	EX	Total (h)	Rainfall (mm)	Track type
1	**2009**	**Morakot**	**12**	**15**	**18**	**45**	**8996**	**CWB 3 (C)**
2	2001	Nari	10	51	14	75	8108	CWB Special
3	**2008**	**Sinlaku**	**16**	**10**	**22**	**48**	**8105**	**CWB 2 (N)**
4	**2005**	**Haitang**	**11**	**9**	**12**	**32**	**5589**	**CWB 3 (C)**
5	1996	Herb	5	7	4	16	4836	CWB 2 (N)
6	1989	Sarah	5	20	13	38	4655	CWB 3 (C)
7	1960	Shirley	3	11	10	24	4637	CWB 2 (N)
8	**2007**	**Krosa**	**12**	**1**	**10**	**23**	3936	**CWB 2 (N)**
9	**2004**	**Mindulle**	**16**	**18**	**7**	**41**	**3856**	**CWB 6**
10	**2008**	**Jangmi**	**4**	**13**	**8**	**25**	**3800**	**CWB 2 (N)**
11	**2008**	**Kalmaegi**	**8**	**10**	**5**	**23**	**3763**	**CWB 2 (N)**
12	**2005**	**Talim**	**4**	**9**	**4**	**17**	**3526**	**CWB 3 (C)**

large increasing trend of heavy rainfall, is the subject of intense debate (Hsu *et al.*, 2011). This chapter summarizes the results of recent research that are relevant to these questions. The main reference is Chang *et al.* (2013), which is a detailed analysis of the hourly rainfall of all the typhoons that made landfall in Taiwan since 1960. Also included are results from Chang *et al.* (2012) that examined the contribution of TC rainfall in the trend of extreme rainfall in the China summer monsoon region since the middle 20th century; Hsu *et al.* (2013) that studied the asymmetry of typhoon translation speed across the island, and Wang *et al.* (2013) that investigated the change of Northern Hemisphere monsoon and tropical circulation in the recent decades.

2. Effect of TCs on the Trend of Extreme Rainfall in the China Summer Monsoon Region

A thermodynamic theory of global warming suggests that rainfall intensity will increase with global temperature due to the increased atmospheric water vapor capacity (Trenberth *et al.*, 2007). However, efforts to find observational evidence of this relationship has been more difficult in the Asian monsoon region than in other Northern Hemisphere land areas (Min *et al.*, 2011). An important reason for this is that in the Asian monsoon region, significant amounts of extreme rainfall can be due to TCs (Ren *et al.*, 2006; Kim *et al.*, 2006, 2012), whose development and characteristics are subject to many dynamic and thermodynamic controls over tropical oceans before they approach the land areas (Emanuel, 2005; Knutson *et al.*, 2010). These controls include sea-surface temperature (SST), vertical wind shear, static stability, and tropopause temperature, so the intensity, frequency and amount of extreme rainfall over the Asian monsoon land areas can be affected by conditions over the tropical oceans.

In evaluating the contributions of western North Pacific TCs on the extreme rainfall over the China summer monsoon region, Chang *et al.* (2012) separated the daily rainfall data from 499 stations into TC and non-TC (monsoon) components for 53 summers (JJA) between 1958–2010. The separation was based on the Objective Synoptic Analysis Technique (Ren *et al.*, 2006) that uses the distance from TC center and the closeness and continuity between neighboring raining stations to trace TC-influenced rain belts that may extend from 500 km to 1100 km away from a TC center. Their results show that over most of the China summer monsoon region, the trends of extreme rainfall, defined as the highest 90th percentile of rainfall events, have been distorted by western North Pacific typhoons, which bring rainfall with decreasing frequency and increasing intensity. Typically the latter is not sufficient to overcome the former, so TC-related extreme rainfall trend is smaller than monsoon-related extreme rainfall. The net impact underestimates the increasing trend and overestimates the decreasing trend in monsoon extreme rainfall over most areas.

The impact of the typhoon varies significantly. The effect is minimal in the Meiyu rainbelt region, but reaches 30% in northeastern and southern China. The largest distortions occur on offshore islands in the main TC paths that underwent significant decadal variations. In Hainan, the −3%/decade trend becomes +7%/decade if typhoon rainfall is removed (Fig. 1). An exception to the negative impact of typhoons on the trend of extreme rainfall occurs in Taiwan (Fig. 2). Here the observed total extreme rainfall trend is near 20% over the 53 years (3.8%/decade). After removing typhoon rainfall, the monsoon component has a trend of only 11% (2%/decade), and thus there is an 80% overestimate of the extreme rainfall trend due to typhoons. For the 98th percentile extreme rainfall (Fig. 2b), which corresponds to the 90th percentile rainfall amount, the observed

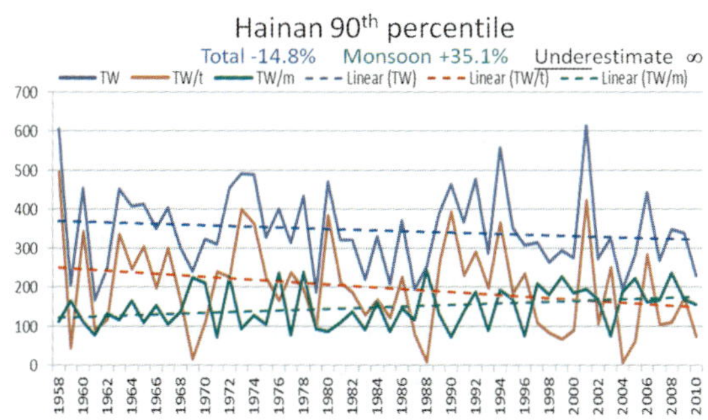

Fig. 1. The time series and linear trends of total (blue), TC (red) and monsoon (green) extreme rainfall (90th percentile by rain events) for in Hainan Island. Texts indicate the numerical values of the linear trends and the percentage reduction of the monsoon extreme rainfall due to typhoons.

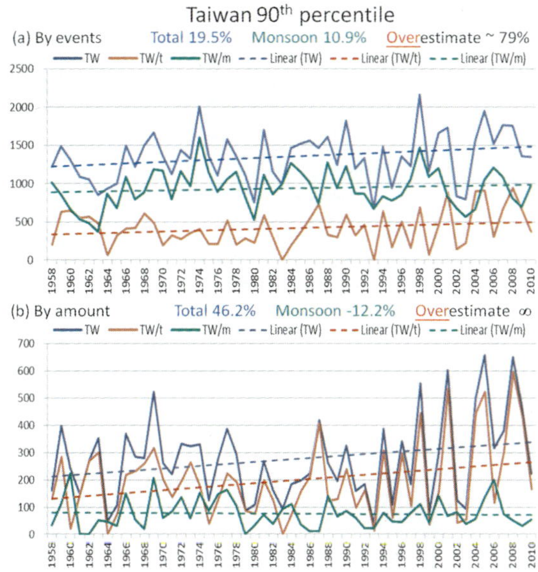

Fig. 2. Same as in Fig. 1 except for (a) Taiwan with the extreme rainfall threshold determined by the 90th percentile by rain events, and (b) Taiwan with the extreme rainfall threshold determined by the 90th percentile of rainfall amount.

increasing trend of 46% (8.7%/decade) totally masked the −12% (−2.3%/decade) trend for the non-typhoon rain. These opposite drastic effects of typhoons on the trends of extreme rainfall in Hainan and Taiwan clearly indicate the danger of attributing the large increase in Taiwan extreme rainfall to global drivers such as global warming.

3. The Role of Taiwan Terrain on Typhoon Rainfall Intensity

The terrain of Taiwan (Fig. 3a) is dominated by the Central Mountain Range. It has a significant effect on TC structure and rainfall (Chang *et al.*, 1993; Ge *et al.*, 2010). Since terrain is invariable in time, Chang *et al.* (2013) investigated the role terrain plays in the TC rainfall in order to isolate factors that may lead to insights on the mechanisms involved in the apparent large increase of Taiwan TC rainfall in the recent decades.

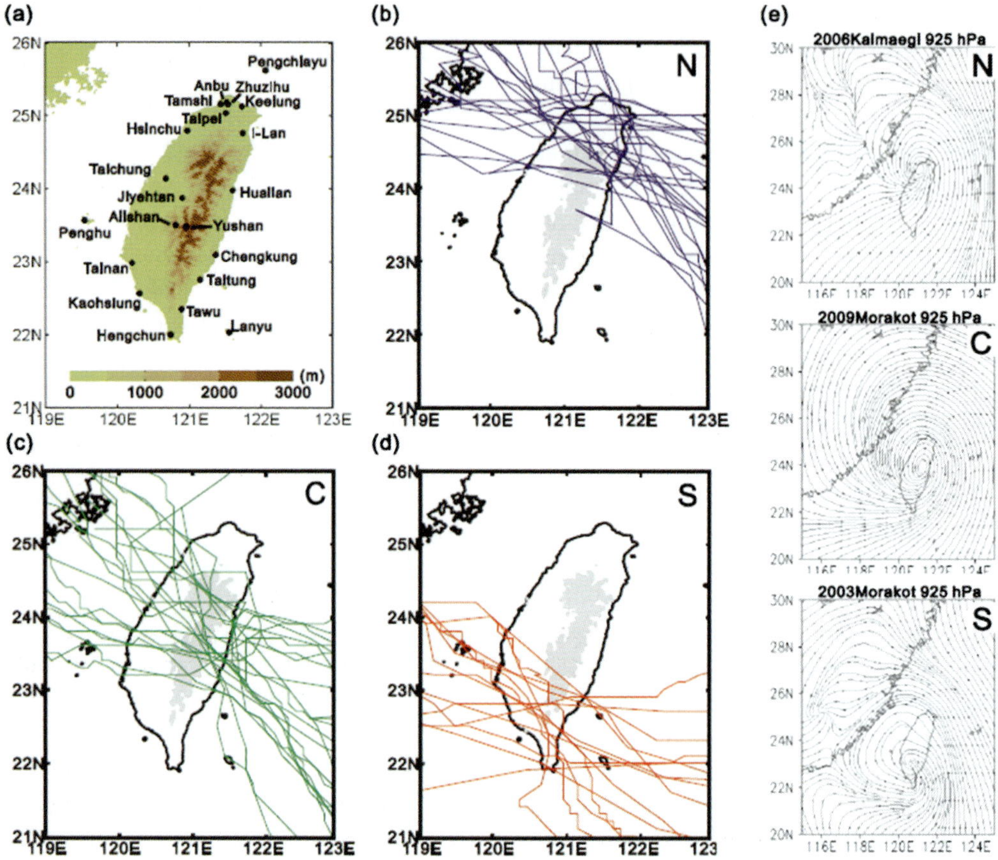

Fig. 3. (a) Taiwan topography and rainfall stations. (b) The northern (N) type tracks. (c) The central (C) type tracks. (d) The southern type (S) tracks. (e) Examples of 925 hPa streamlines for the three track types: N: Typhoon Kalmaegi (2006), C: Typhoon Morakot (2009), and S: Typhoon Morakot (2003).

The Central Weather Bureau of Taiwan classifies the 84 landing typhoons during 1960–2011 into nine track types. Chang *et al.* (2013) focused on the three leading types that directly cross the island whose maximum width is 144 km: Northern type (N, 26 cases, Fig. 3b); Central type (C, 23 cases, Fig. 3c); and Southern type (S, 14 cases, Fig. 3d). The average separation between tracks of the Northern and Central types and those of the Central and South types is about 100 km, which is about one half of the length of the Central Mountain Range. Examples of the 925 hPa streamlines for each of the three types based on the European Centre for Medium-range Weather Forecasts (ECMWF) — Tropical Ocean Global Atmosphere Global Advanced Analysis are shown in Fig. 3e.

To study the effect of the terrain, three phases of each track is selected. The first is the pre-landing phase (PR) that starts when the typhoon center moves to within 100 km from the nearest coastline point and ends when the center made landfall. The next one is the overland phase (OL), which starts at the end of the PR phase and ends when the typhoon center exits the coast. This is followed by the exit phase (EX), which starts at the end of the OL phase and ends when the typhoon center reaches 100 km away from the nearest coastline.

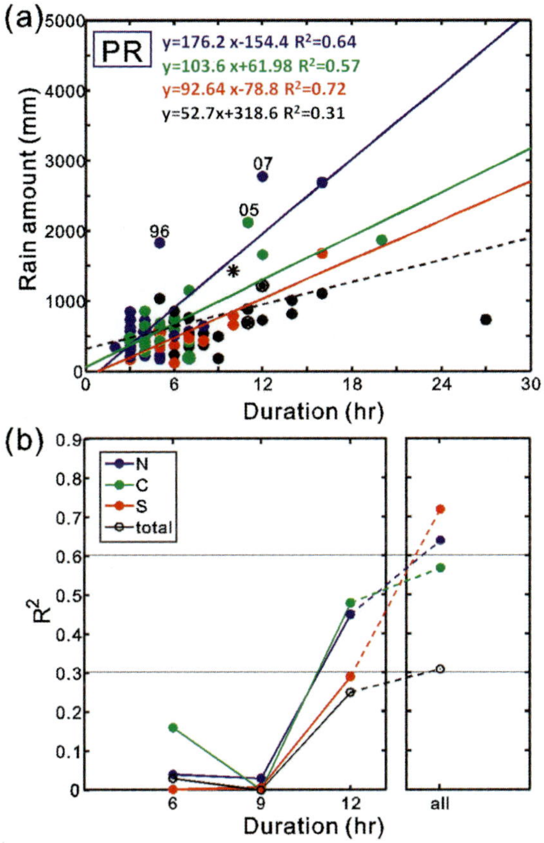

Fig. 4. (a) Rainfall amount (mm) versus duration (hr) during the pre-landing (PR) phase. The blue, green, and red dots mark typhoons of the three leading track types, N, C, and S, respectively. Black dots mark typhoons of the other six track types that made landfall in Taiwan during 1960–2011. Circled dots indicate two overlapping dots. The star symbol indicates Typhoon Nari (2001), an unusual case that broke the record of overland duration. The linear fitting lines and formulas for the N, C, S and All types are colored in blue, green, red, and black. The All type includes all 84 typhoons in the nine track types. (b) The coefficient of determination (R2) for typhoons with pre-landing durations up to six hours, nine hours, and 12 hours, and all cases.

Table 2. The 12-hour rainfall intensity (mm) for the three track types during the pre-landing (PR), over land (OL) and exit (EX) phases.

Phase Track Type	PR	OL	EX
N	1960	2132	1458
C	1305	1558	1467
S	1033	812	572

A scatter plot of the accumulated total rain amount of the 21 rain-gauge station network versus the duration of that typhoon is shown in Fig. 4a for the pre-landing phase. The linear fits are separately computed for the N, C, S types and the All type (all nine track types with a total of 84 cases). When all typhoons are considered, the variation of rainfall intensity is far more scattered with a coefficient of determination $R^2 = 0.31$ than those of the individual N ($R^2 = 0.64$), C ($R^2 = 0.57$), and S ($R^2 = 0.72$) types. All coefficients pass the 99% significance level with the Student's t-test. This large variation is mainly due to the large differences in the mean rainfall intensities among the three types (Table 2, column 1), which is highest for the N type (1960 mm/12 h), followed by the C type (1305 mm/12 h) and the S type (1033 mm/12 h). These results indicate that when the typhoons are stratified according to these track types, the rainfall intensity during the pre-landing phase changes over a far smaller range over the 52 years than might be inferred from the long-term rainfall time series of all typhoons.

These differences among the track types can be explained by the interaction of the TC circulation and the terrain (Fig. 3). During the most active period of the typhoon season (June-September), Taiwan is under the influence of the East Asian summer monsoon (EASM) so the quadrant of the TC with southwesterly winds will have an enhanced moisture flux. The moisture flux convergence resulting from interaction between these southwesterly winds and the mountain range produces more rainfall and mesoscale convective systems than do the winds in other quadrants. Prior to landfall, a typhoon situated to the north will have the most exposure of westerly and southwesterly winds against the terrain, but this exposure will be decreased if the typhoon is farther south.

Within each track type, linear fittings are also computed for all typhoons and for rainfall durations up to the first six, nine, and 12 hours in a phase. During the pre-landing phase, the coefficients of determination (Fig. 4b) demonstrate that tightening of the range of rainfall intensity variations over 12 h becomes evident earlier for the N and C types ($R^2 = 0.47$ and 0.49 at 12 h, respectively, significance level 99%) than for the S ($R^2 = 0.29$, significance level 95%) and All ($R^2 = 0.25$, significance level 99%) types. These results suggest that starting from 12 hours after a typhoon is within 100 km of the coast, the wind-terrain interaction has a modulating effect on the rainfall intensity that constrains its variations. However, this modulation effect is delayed for a typhoon of the S track type, which is attributed to the smaller wind-terrain interaction.

For the overland phase (Fig. 5), the rainfall intensities for the three track types are all clustered within tight ranges starting from the six-hour duration. This short duration is consistent with the expectation that the largest terrain effect should be in this phase since the typhoon centers are over land. This maximum terrain effect in regulating the TC rainfall intensity is also verified by the fact that the final coefficients of determination are the highest among the three phases for every track type, and particularly the All track type with a value of 0.74 that is much larger than the corresponding values of around 0.3 during the pre-landing and exit phases.

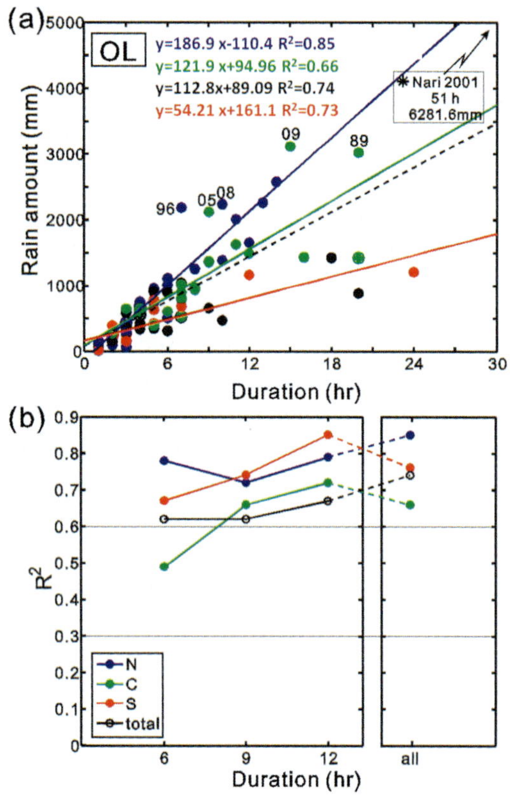

Fig. 5. Same as Fig. 2 except for the overland (OL) phase.

During the overland phase, the rainfall intensity (Table 2, column 2) again is highest for the N type (2132 mm/12 h), followed by the C type (1558 mm/12 h), and then for the S type (812 mm/12 h). The N and C types have higher rainfall intensities than the corresponding rainfall intensities during the pre-landing phase because the maximum wind-terrain interaction occurs while the center is over land, even though the TC intensity typically weakens over land. The larger rainfall intensities are also consistent with the stream flow in Fig. 3e, with westerly and southwesterly winds interacting with the terrain over the entire western coast of Taiwan for the N type and over the southern half of the coast for the C type. The wind-terrain interaction is the smallest for the S type, which actually has lower rainfall intensity over land than for the pre-landing phase. Rain amounts for the All type are substantially distorted by inclusion of the unusual Typhoon Nari (2001), which produced a record-breaking 6282 mm rainfall as it moved southward over nearly the entire length of the mountainous island.

For the exit phase (Fig. 6), the modulating effect of the terrain on the rainfall becomes smaller compared to the previous two phases. The largest decrease in the coefficient of determination is for the S type ($R^2 = 0.42$ versus 0.72 for PR and 0.73 for OL). The S type typhoons, which have the smallest terrain modulation effect on rainfall intensity, are moving

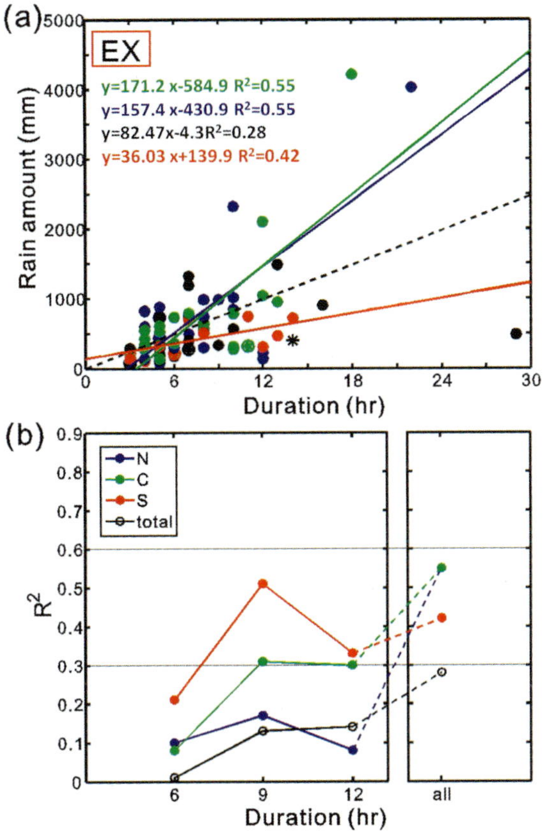

Fig. 6. Same as Fig. 2 except for the exit (EX) phase.

farther away into the South China Sea during the exit phase (Fig. 3d) and thus the terrain effect is even more diminished. As indicated in Table 2, the N and the S types also have significantly lower rainfall intensity in the exit phase (1458 mm/12 h and 572 mm/12 h, respectively) than during the other two phases. The rainfall intensity for the C type (1467 mm/12 h) is only slightly reduced from the peak value during the overland phase, and matches the rainfall intensity of the N type. This continued high rainfall intensity for the central track is due to the typhoon center being in the middle of the Taiwan Strait during the exit phase (Fig. 3c). As a result the C type typhoon circulation still has a robust wind — terrain interaction, and also has a stronger interaction with the southwest monsoon winds originating from the South China Sea than does the N type, which is moving northwestward away from the South China Sea (Fig. 3a).

The above results suggest that when stratified according to track types the rainfall intensity of typhoons crossing Taiwan has not changed greatly since 1960, especially during the pre-landing and over land phases when the scattering for each of the leading track type is quite small as indicated by coefficients of determination around or larger than 0.6. This suggests the possibility that the recent increase in typhoon rainfall may be mainly due to two factors: (1) a slowdown of the typhoon movement so more rainfall is produced due to the longer duration over Taiwan; and (2) a shift in the TC tracks such that the TC circulation-terrain effect leads to more rainfall. For example, the slow translation speed of Typhoon Morakot (2009) has been reported as an important factor for its record-breaking rainfall (Lee *et al.*, 2011). Indeed, eight of the nine heavy rainfall typhoons in the 21st century were slow-moving storms with a total duration ranging from slightly above to more than twice of the 22 h average for all 84 typhoons (Table 1). Seven of these nine typhoons are of either the N or C track types, and none was the S type. The two others are of less frequent track types with very long durations, including Typhoon Nari (2001) that produced the second highest total rainfall. Typhoon Nari is considered to be an exceptional case because of its extraordinarily slow and very rare southward movement that lasted 51 hours over land, which is a duration that is more than twice that of the second longest one in Table 1 — Typhoon Sarah of 1989, which was a C track type that lasted 20 hours over land.

4. Decadal Changes

Excluding Typhoon Nari (2001), the other eight heavy rainfall typhoons in the 21st century all occurred in or after 2004 (boldfaced in Table 1). The time series of typhoon occurrences and

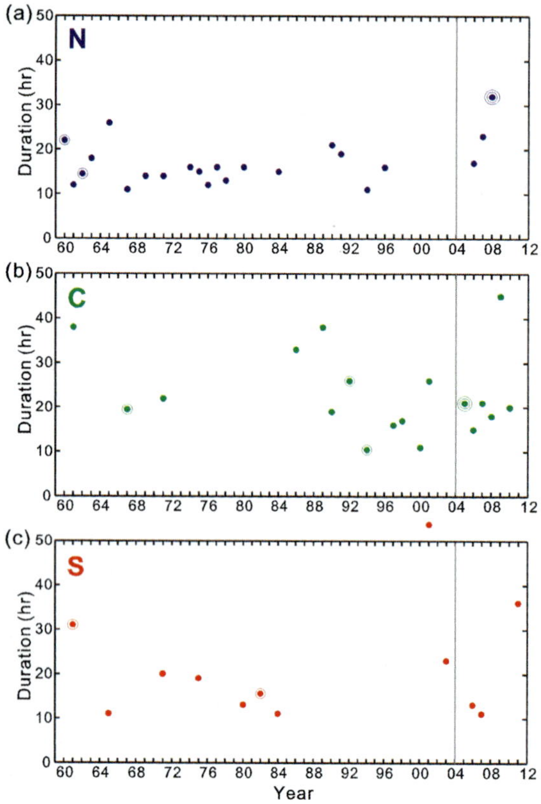

Fig. 7. Time series of average duration of typhoons from 1960 to 2011 for (a) north (N, blue), (b) central (C, green), and (c) south (S, red) track type typhoons. The thin vertical line inside each panel denotes year 2004. A dot with one circle indicates two typhoons and a dot with two circles indicate three typhoons.

Table 3. The frequency and duration of the N, C, and S typhoons before and after 2004.

	1960–2003	2004–2011
Frequency		
N	21	5
C	15	8
S	11	3
Duration (hour)		
N	15.9	24.0
C	23.9	23.3
S	21.7	20.0

durations for the N, C, and S track types are shown in Fig. 7, with the 52-year period divided at 2004. The frequencies and durations of typhoons of the three track types before and since 2004 are also compared in Table 3. When a year has more than one typhoon of the same type, the average duration is used. Overall, the averaged total number of TCs of all track types making landfall in Taiwan each year increased during recent decades. But this increasing trend in Taiwan does not reflect global or basin scale changes, as it is opposite to the decreasing trends of both the western North Pacific TCs (Lau and Zhou, 2012) and TCs affecting the EASM region (Chang *et al.*, 2012) since middle-20th century, and also the century-trend of decreasing number of landfalling typhoons on Taiwan (Lee and Chia, 2008).

The bulk of the increase of rainfall since 2004 was due to Types N and C. During the 44 years before 2004, N type typhoons occurred about once every two years, C type typhoons occurred about once every three years, and S type typhoons occurred about once every four years. For all three types, the number of typhoons per year increased after 2004, but the increase is minimum for the N and C types each of which has one extra typhoon over the eight-year period. On the other hand, the C type had a drastic increase from averaging one every three years to one every year, a 300% increase. While the frequency for the N type did not increase much, its average duration increased 51%. Since the N type has the highest rainfall intensity and the C type has the second highest rainfall intensity (and highest intensity during the exit phase), the longer-duration N type typhoons and the more frequent C type typhoons since 2004 are the reasons for the large increase in rainfall and rainfall intensity. The S type, on the other hand, has the lowest rainfall intensity, and its duration have decreased by 10%.

To examine possible decadal changes, the time series of typhoon rainfall intensity for the three phases for all 84 cases is plotted in Fig. 8. When a season has multiple TCs, the average rainfall intensity is used for that year. If only the weak to medium intensity TCs

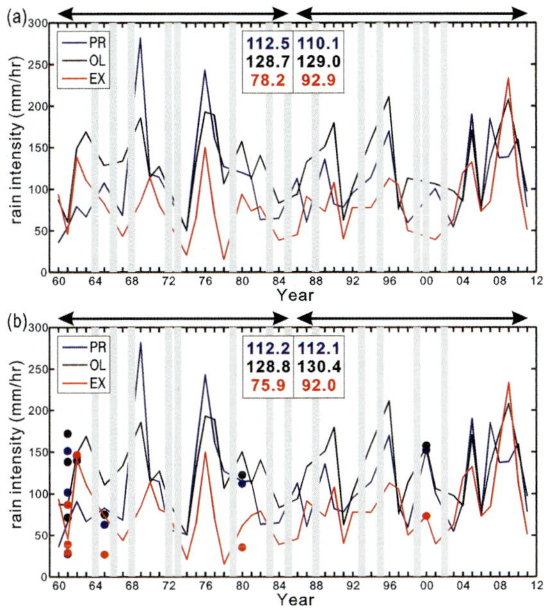

Fig. 8. (a) Time series of rainfall intensity from 1960 to 2011 for the pre-landing (PR, blue), overland (OL, black) and exit (EX, red) phases for weak and medium intensity (Category 1–3) typhoons. Grey vertical columns indicate years with no typhoons of the three leading track types. The table inside the panel lists the averaged rainfall intensities of the three phases during the first half (1960–1985, left column) and second half (1986–2011, right column) of the 52 year period. The two sub-periods are indicated by double end arrows above the panel. (b) Same as (a) except strong types (Category 4–5) are included. The rainfall intensity of each strong typhoon is indicated by the respective colored dots.

(Categories 1–3) are included (Fig. 8a), the pre-landing and over-land series have interannual and interdecadal variations, but a 52-year trend is not obvious. However, it is quite conspicuous that the exit time series has an increasing trend in addition to the interannual and decadal variations. This trend is verified by comparing the averages of each year's rainfall intensity between the first 26 years (1960–1985) and the last 26 years (1986–2011). While the changes in the average rainfall intensity for the pre-landing phase (from 112.5 mm/hr to 110.1 mm/hr) and the overland phase (from 128.7 mm/hr to 129.0 mm/hr) are both negligible, the exit phase has a relatively large increase from 78.2 mm/hr to 92.9 mm/hr (19%), although it is far below the 95% significance level due to the large interannual and decadal variations. It is interesting to note that this increasing trend is entirely related to typhoons that occurred in the last decade. The results are very similar if all TCs (Categories 1–5) are included (Fig. 8b).

These rainfall statistics are a strong indication that the effect of the wind-terrain interaction may constrain potential climate changes of typhoon rainfall intensity in Taiwan. That is, very small long-term trends are detected in the pre-landing and overland phases when the typhoon-terrain interaction is strong and more important in contributing to the heavy rainfall. The trend is positive in the exit phase after the typhoon leaves Taiwan. The Taiwan Strait and regions southwest of Taiwan are in the path of major southwesterly surges during the EASM. These moist monsoon surges are a primary mechanism to bring heavy rainfall to Taiwan (Johnson, 2006). The surges may be especially strengthened when interacting with a TC, and the resulting heavy rainfall often leads to major floods as in the case of Typhoon Morakot (2009) (Wu *et al.*, 2011; Chien and Kuo, 2011). In the exit phase, the west and southwest sides of the typhoon circulation come into closer contact with the southwest monsoon winds. Since the terrain effect may not be as strong, increased interaction with the monsoon and the resultant additional moisture supply may play a role in the positive trend of typhoon rainfall intensity during the exit phase.

This positive trend for the exit phase cannot be attributed to an increase in the TC intensity (Fig. 8). The exit-phase trend when strong typhoons (Categories 4–5) are included is slightly larger than when they are excluded (21% versus 19% without strong typhoons), but the number of strong typhoons actually decreased over the 52 years. This may seem counter-intuitive since it is generally accepted that strong TCs produce more intense rainfall. It is possible that this result may be influenced by the 21-station rainfall average that covers an area larger than the scale of the radius of maximum winds, while the heaviest rainfall that may be related to TC intensity occurs on this scale. It is also possible that the strong terrain effect weakens the relationship between rainfall intensity and TC intensity.

5. Discussions

The analyses reviewed here indicated that typhoon rainfall intensity in Taiwan is strongly regulated by the terrain effects, particularly as the typhoon approaches Taiwan from the east and during the period its center is over land. Because of the strong terrain constrains during these two phases, the long-term trend of rainfall intensity is negligible. Thus, the large increases in typhoon rainfall, and the apparent increase in rainfall intensity but not in TC intensity, during recent decades are attributed to the 50% increase in the duration of the northern track typhoons and the 300% increase in the frequency of the central track typhoons that have crossed the island.

Tu *et al.* (2009) and Wang *et al.* (2011) have suggested that changes in western Pacific sea-surface temperatures associated with global warming may have caused more typhoons to move northward toward midlatitude East Asia rather than westward to the South China Sea. However, changes in the relative frequencies among the three track types in this study are unlikely related to global warming, because the increase in tracks has been in the central type, rather than have been biased toward the north.

The increase in duration may be related to a change of the large-scale easterly steering flow. Chu *et al.* (2012) reported a slowing trend between 1959 and 2009 of the steering flow in the subtropical western North Pacific east of Taiwan and the South China Sea in both the National Center for Environmental Prediction/National Center for Atmospheric Research and the ECMWF 40-year (ERA-40) reanalyses. Although this would be consistent with the increase of average duration of typhoons, most of the slowing down in this study was for the northern type typhoons among the three leading tracks. The explanation may lie in the location of the maximum terrain interaction, which is in southern Taiwan behind the N type track in which the typhoons move in a prevailing northwestward direction. Hsu *et al.* (2013) used mesoscale model simulations to demonstrate that the topographically locked convection acts to slow down (speed up) the northern (southern) landfalling typhoons. A slower steering flow will slow down all three track types, but the effect will be largest on the N type due to this topographic locking effect. A slower-moving northern typhoon will have more time to experience a positive feedback mechanism, in the sense more prolonged precipitation and latent heat release in the windward side of the southern Taiwan mountains will result in further speed reductions. A similar effect involving the interaction of the Typhoon Morakot (2009) circulation with the southwest monsoon winds near southwestern Taiwan was shown by Wang *et al.* (2013) to contribute to a slow translation speed on a central track.

The only long-term trend in typhoon rainfall intensity detected between the first and second halves of the 1960–2011 study period was during the exit phase, when the TC circulation interacts strongly with the monsoon southwesterlies while experiencing less terrain effects. Wang *et al.* (2012, 2013) found from satellite rainfall and ERA-40 reanalysis wind data that the Northern Hemisphere subtropical monsoon and the Hadley and Walker circulations have a coherent increasing trend since the late 1970s. Over East Asia the subtropical westerlies penetrates more eastward into the western part of the western North Pacific, about 20 degrees longitude past Taiwan. An increasing subtropical monsoon trend offers a plausible explanation for the increasing trend of exit-phase typhoon rainfall, since the typhoon-monsoon interaction will be enhanced. Increased subtropical westerlies over East Asia would also be consistent with the slowing down of the easterly steering flow and typhoon movement during the recent decades.

Wang *et al.*'s (2013) results are opposite to theoretical and model studies of global warming effects which project a weakening of the monsoon and tropical circulations (Held and Soden, 2006; Vecchi *et al.*, 2006). Using coupled-model experiments, they attributed the main cause of the strengthening of the monsoon and tropical circulations to natural variations associated with multi-decadal variations of the Mega-El Nino Southern Oscillation (ENSO) and the Atlantic Multidecadal Oscillation (AMO). They concluded that the impacts of the Mega-ENSO and AMO are so strong that they would be larger than the anthropogenic effects during the recent decades that have been predicted in the global warming model studies.

6. Conclusion

While the possibility that increased water vapor capacity due to global warming may have some role cannot be ruled out, we are not aware of any research on global warming effects that can explain the changes in the typhoon tracks and durations observed in the most recent decade. The slowdown of the steering flow and the increased interaction with southwest monsoon winds may even be contrary to some modeling studies of the effects of global warming. It is also clear that the recent increase in typhoon-related rainfall in Taiwan cannot be attributed to a positive feedback between global warming-related water vapor supply and typhoon intensity. That is, the frequency of strong typhoons affecting Taiwan has actually decreased from five in the first 26 years of this study to just one during the last 26 years.

Acknowledgments

This work is supported in part by the National Science Council, R.O.C., and the National Taiwan University.

References

Chang, C.-P., T.-C. Yeh, and J. M. Chen, 1993: Effects of terrain on the surface structure of typhoons over Taiwan. *Mon. Wea. Rev.*, **121**, 734–752.

———, Y. Lei, C.-H. Sui, X. Lin, and F. Ren, 2012: Tropical cyclone and extreme rainfall trends in East Asian summer monsoon since mid-20th century. *Geophys. Res. Lett.*, **39**, L18702, doi: 10.1029/2012GL052945.

———, Y.-T. Yang, and H.-C. Kuo, 2013: Large increasing trend of tropical cyclone rainfall in Taiwan and the roles of terrain. *J. Climate*, **26**, 4138–4147, doi: 10.1175/JCLI-D-12-00463.1.

Chien, F.-C., and H.-C. Kuo, 2011: On the extreme rainfall of Typhoon Morakot (2009). *J. Geophys. Res.-Atmos.*, **116**, D05104.

Chu, P.-S., J.-H. Kim, and Y. R. Chen, 2012: Have steering flows in the western North Pacific and the South China Sea changed over the last 50 years? *Geophys. Res. Lett.*, **39**, L10704, doi: 10.1029/2012GL051709.

Emanuel, K., 2005: Increasing destructiveness of tropical cyclones over the past 30 years. *Nature*, **436**, 686–688, doi: 10.1038/nature03906.

Ge, X., T. Li, S. Zhang, and M. Peng, 2010: What causes the extremely heavy rainfall in Taiwan during Typhoon Morakot (2009)? *Atmos. Sci. Lett.*, **11**, 46–50.

Held, I. M., and B. J. Soden, 2006: Robust responses of the hydrological cycle to global warming. *J. Climate*, **19**, 5686–5699.

Hsiao, F., 2009: Fatal water. *CommonWealth Magazine*, **428**, Cover story, http://english.cw.com.tw/article.do?action=show&id=11260.

Hsu, H.-H. *et al.*, 2011: *Science Report of Climate Change in Taiwan 2011.* Available from National Science Council, 106, Sec. 2, Heping E. Rd., Taipei, Taiwan, R.O.C., 362 pp. (Chinese).

Hsu, L.-H., H.-C. Kuo, and R. G. Fovell, 2013: On the geographic asymmetry of typhoon translation speed across the mountainous island of Taiwan. *J. Atmos. Sci.*, **70**, 1006–1022.

Johnson, R. H., 2006: Mesoscale processes. *The Asian Monsoon*, B. Wang, Ed., Praxis-Publishing Ltd., 331–779.

Kim, J.-H., C.-H. Ho, M.-H. Lee, J.-H. Jeong, and D. Chen, 2006: Large increase in heavy rainfall associated with tropical cyclone landfalls in Korea after the late 1970s. *Geophy. Res. Lett.*, **33**, L18706, doi: 10.1029/2006GL027430.

———, C.-C. Wu, C.-H. Sui, and C.-H. Ho, 2012: Tropical cyclone contribution to interdecadal change in summer rainfall over South China in the early 1990s. *Terr. Atmos. Ocean. Sci.*, **23**, 49–58.

Knutson, T. R. *et al.*, 2010: Tropical cyclones and climate change. *Nat. Geosci.*, **3**, 157–163.

Lau, W. K. M., and Y. P. Zhou, 2012: Observed recent trends in tropical cyclone rainfall over the North Atlantic and the North Pacific. *J. Geophys. Res.-Atmos.*, **117**, D03104, doi: 10.1029/2011JD016510.

Lee, C.-S., C.-C. Wu, T.-C. Chen Wang, and R. L. Elsberry, 2011: Advances in understanding the "perfect monsoon-influenced typhoon": Summary from International Conference on Typhoon Morakot (2009). *Asia-Pac. J. Atmos. Sci.*, **47**(3), 213–222.

Lee, C.-T., and H. H. Chia, 2008: Variation of landfalling typhoons in Taiwan during the recent 100 years. *2008 Taiwan Climate Change Workshop, Central Weather Bureau, Taipei, Taiwan*, 25–26 August 2008. Figs. 2–5.

Min, S.-K., X. Zhang, F. W. Zwiers, and G. C. Hegerl, 2011: Human contribution to more-intense precipitation extremes. *Nature*, **470**, 378–381, doi: 10.1038/nature09763.

Ren, F. *et al.*, 2006: Changes in tropical cyclone precipitation over China. *Geophys. Res. Lett.*, **33**, L20702, doi: 10.1029/2006GL027951.

Trenberth, K. E. *et al.*, 2007: Observations: Surface and atmospheric climate change. *Climate Change 2007: The Physical Science Basis. Contribution of Working Group I to the Fourth Assessment Report of the IPCC*, S. Solomon *et al.*, Ed., Cambridge Univ. Press, Cambridge, U.K., 235–336.

Tu, J.-Y., C. Chou, and P.-S. Chu, 2009: The abrupt shift of typhoon activity in the vicinity of Taiwan and its association with western North Pacific–East Asian climate change. *J. Climate*, **22**, 3617–3628, doi: 10.1175/2009JCLI2411.1.

Vecchi, G. A., B. J. Soden, A. T. Wittenberg, I. M. Held, A. Leetmaa, and M. J. Harrison, 2006: Weakening of tropical Pacific atmospheric circulation due to anthropogenic forcing. *Nature*, **441**, 73–76, doi: 10.1038/nature04744.

Wang, B., J. Liu, H.-J. Kim, P. J. Webster, and S.-Y. Yim, 2012: Recent changes of the global monsoon precipitation (1979–2008). *Clim. Dyn.*, **39**, 1123–1135, doi: 10.1007/s00382-011-1266-z.

———, ———, ———, ———, ———, and B. Xiang, 2013: Northern Hemisphere summer monsoon intensified by mega-El Niño/southern oscillation and Atlantic multidecadal oscillation. *Proc. Natl. Acad. Sci.*, **110**(14), 5347–5352, doi: 10.1073/pnas.1219405110.

Wang, R., L. Wu, and C. Wang, 2011: Typhoon track changes associated with global warming. *J. Climate*, **24**, 3748–3752.

Wu, L., J. Liang, and C.-C. Wu, 2011: Monsoonal influence on Typhoon Morakot (2009). Part I: Observational analysis. *J. Atmos. Sci.*, **68**, 2208–2221.

Chapter 21

Multi-Decadal Variability in Indian Summer Monsoon Rainfall Using Proxy Data

Bhupendra N. Goswami, Ramesh H. Kripalani, Hemant P. Borgaonkar, and Bhaskar Preethi

Indian Institute of Tropical Meteorology,
Pune, 411008, India
goswami@tropmet.res.in
www.tropmet.res.in

Some enticing evidence of multi-decadal variability of the Indian Summer Monsoon Rainfall (ISMR) during 1871–2000 has been linked to the Atlantic Multi-decadal Oscillation (AMO, Goswami *et al.*, 2006). However, neither the periodicity and its stationarity nor its origin could be established unambiguously from the relatively short record. Using three different proxies of monsoon rainfall for a duration of over 500 years, we find that the Asian monsoon has a multi-decadal oscillation with period between 50 years and 80 years that change in time in episodic manner. Using Intrinsic Mode Function (IMF) analysis, we show that the 50–80 year oscillation could be considered as a 'mode' of variability of the Asian monsoon. We further find that, AMO, El Niño Southern Oscillation (ENSO) as well as the Pacific Decadal Oscillation (PDO) also have such a 50–80 year multi-decadal variability. Coherence between Asian monsoon multi-decadal mode and that of AMO, ENSO and PDO shows that they all are likely to be integral part of a global multi-decadal mode with periodicity of 50–80 years. We speculate that such a multi-decadal mode of variability may arise from ocean-atmosphere-land interactions.

1. Introduction

Nearly 80% of the annual rainfall over major parts of India occurs during the summer monsoon period (June through September: JJAS). The Indian summer monsoon rainfall (ISMR) has exhibited variability on a variety of time scales, from intra-seasonal, inter-annual to inter-decadal time scales. Various other components of the Indian monsoon also have shown variability on inter-decadal time scales. The space-time structure of the monsoon inter-decadal variability is poorly documented as compared to the inter-annual variability and the mechanisms responsible are also poorly understood (Goswami, 2006). This problem is partly related to the lack of good quality observational data for a sufficiently long period. The instrumental record for surface climate, including precipitation, could only be extended for about 150 years.

Epochal amplitude modulation of the Indian monsoon with alternate three decades of below normal seasonal rainfall followed by three decades of above normal seasonal mean rainfall has been known (see Goswami, 2006 and references therein). A hypothesis was put forth and it was shown that the inter-decadal variability of El Niño Southern Oscillation (ENSO) phenomenon was strongly negatively correlated with the inter-decadal variability of the ISMR (Krishnamurthy and Goswami, 2000). Thus the inter-decadal variability of the Indian monsoon

Climate Change: Multidecadal and Beyond
edited by Chih-Pei Chang et al.

and that of ENSO are thought to be a manifestation of a global-scale oscillation on a quasi-60-year time scale (see Goswami, 2006 for more details). These inferences and hypothesis were based on the data period 1871–2000. However, during the decade 2001–2010, India experienced three severe droughts during the summer monsoons of 2002, 2004 and 2009. Incidentally, the rains have been below normal (92% of long period average) during the recent summer monsoon of 2012. Furthermore, the ISMR has been indicating a declining trend during the recent decades (see Section 2). Under these circumstances, it becomes imperative to reexamine the lack of trend in ISMR, and the robustness of the quasi-60-year periodicity using updated data up to 2012.

2. Inter-Annual and Decadal ISMR Variability

Figure 1 (green bars) shows the standardized ISMR time series depicting the inter-annual variability. On inter-annual time scale, the most prominent variations are between the so-called good monsoon seasons with above average rainfall and the poor monsoons with deficient rainfall. These can be inferred from Fig. 1 as years in which the standardized anomaly is >1 or <−1 respectively.

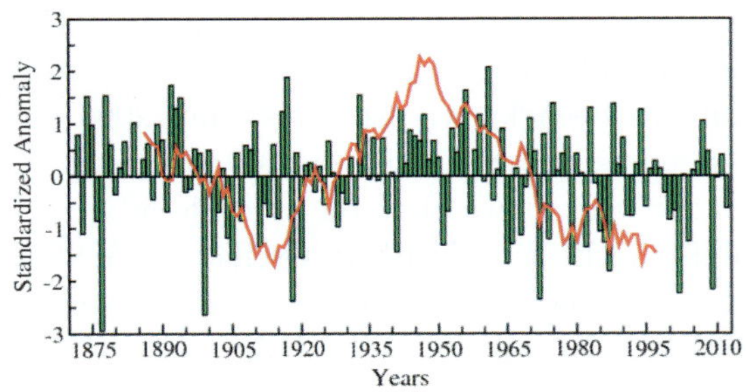

Fig. 1. Inter-annual (green bars) and epochal (red solid line) variability of Indian summer monsoon rainfall (1871–2012). Epochal variability is obtained by using 31 year running mean. Inter-annual as well as low pass filtered anomalies are normalized by their own standard deviations.

The short-term decadal fluctuations are examined by applying a simple low-pass filter super-imposed on the standardized values (Fig. 1: red curved line). The most striking features are the alternate epochs of approximately 3-decades of above and below normal rainfall, as mentioned earlier. Thus there were clear alternate regimes of good and poor monsoons. This inter-decadal variability in the ISMR has been linked, not only to the inter-decadal fluctuations of Pacific sea surface temperatures (Krishnan and Sugi, 2003; Goswami, 2006) but also with the Atlantic Multi-decadal Oscillation (AMO: Goswami *et al.*, 2006; Lu *et al.*, 2006; Zhang and Delworth, 2007). The leading mode of low-frequency North Atlantic (0°–70°N) sea surface temperature variability has been termed as AMO (Kerr, 2000). Over the instrumental period (1856-present), the AMO exhibited a 65–80 year cycle, similar to the periodicity exhibited by the observed ISMR. The warm phase of AMO occurred during 1925–1965 (Fig. 1: above normal ISMR regime) and cold phases during 1900–1925 and 1965–1995 (Fig. 1: below normal ISMR regimes). However, in recent times the AMO entered into a warm phase since 1995 (*e.g.*, Gray *et al.*, 2004; Zhang and Delworth, 2007), while the ISMR still continues to reel under a below normal rainfall regime (see Fig. 1).

The above mentioned alternate 3-decade regimes and the AMO entering a warm phase during mid-1990s suggested a turning point around 1990. However, the expected turning point around 1990 towards an above normal epoch did not occur for ISMR. In fact the ISMR appears to be in the below normal epoch for the last 6–7 decades and exhibiting a significant decreasing trend (also see Kripalani and Kulkarni, 2012). This decreasing trend appears to have commenced from the late 1940s (Fig. 1). Based on the above observations, following questions arise:

(i) Is the Indian monsoon entering into a phase having much longer period?

(ii) Whether the Indian monsoon was dominated by such longer periods during the pre-observational data period i.e., earlier than the 19th century?

As noted above, reliable observational data based on instrumental record are available only for about 150 years and hence not useful to settle the above questions. To gain some insight into these questions, we need to turn to available proxy records of ISMR. While the answer to question (ii) could be ascertained through the past proxy sources, high power computer simulations using coupled ocean-atmosphere climate models are needed to answer question (i). Hence our focus here is to seek answer to question (ii) only. Multi-century ISMR reconstructions derived from proxy records of precipitation are vital to assess pre-instrumental patterns of ISMR variability.

3. Long Proxy ISMR Records

In recent times, many attempts have been made to draw inferences of past monsoonal variations using proxy records such as lake sediments, archaeological material, oxygen isotopes, tree-rings and historical accounts from region to region.

3.1. *Tree-rings*

Research over the past few decades has demonstrated that quantitative reconstruction of the past Indian monsoonal climate over the period of a few centuries is possible through dendroclimatology i.e., through tree-ring based analysis. Recently, a 523-year (AD 1481–2003) tree-ring width chronology of teak from Southern India has been constructed (Borgaonkar *et al.*, 2010). It was found that this chronology and the observed ISMR for the instrumental period (1871–2003) on inter-annual time scale showed a significant positive correlation coefficient (CC = 0.32) and a much stronger relationship on a decadal time frame (CC = 0.52).

Buckley *et al.* (2007) developed a 448-year (AD 1558–2005) teak chronology from Northwest Thailand to assess past changes in the strength of the summer monsoon. Through correlation analysis they confirmed that this chronology is a proxy for the summer monsoon strength and duration over Northwest Thailand. Studies have shown that the summer monsoon rainfall variations over India and Northwest Thailand are related on inter-annual as well as on inter-decadal time scales (Kripalani *et al.*, 1995; Kripalani and Kulkarni, 1997).

3.2. *Oxygen Isotope in speleothems*

A near-annually resolved record of the ISMR variation has been presented for the core monsoon region of India that spans from AD 625 to 2007 (1383 years) from a composite of ^{230}Th-dated stalagmite oxygen isotope composition (δ^{18}O) of speleothem record from Dandak (19°00′N, 82°00′E, 400 m; AD 625 to 1561) and Jhumar (18°52′N, 81°52′E; 600 m; AD 1075 to 2007) caves located 20 km apart over Central India (Sinha *et al.*, 2007, 2011a; Berkelhammer *et al.*, 2010). The Jhumar stalagmite record show a high degree of replication of Dandak cave record for the overlapping period. Speleothems can be excellent archives of monsoon rainfall variability (*e.g.*, Burns *et al.*, 2002). The oxygen isotope composition (δ^{18}O) of speleothem calcite from tropical and monsoon location is primarily controlled by the δ^{18}O value of precipitation, which in turn varies inversely with rainfall amount (Dansgaard, 1964). This proxy has a significant negative correlation of −0.46, at 95% confidence level, with the summer monsoon precipitation over India based on the instrumental data period AD 1903–2005.

Statistical techniques are applied on the long period data of each of the three proxies to draw inferences on the change in the character (if any) in the multi-decadal variability of the ISMR.

4. Analysis of Long Indian Monsoon Proxies

The standardized time series for the 3 proxies are displayed in Fig. 2. These standardized proxy indices, in general vary between plus/minus 3 (Fig. 2), and are comparable to the variability of the observed ISMR (Fig. 1). The short-term decadal fluctuations have been examined by applying a simple low-pass filter super-imposed on the annual values. As for the ISMR, alternate regimes of above and below normal values prevail lasting for several decades. It is interesting to note that the amplitude of inter-annual variability in the RWI from India (Fig. 2a) and that over Thailand (Fig. 2b) is much larger than that in the δ^{18}O proxy (Fig. 2c). This may be related to the way the proxies have been constructed. In the case of δ^{18}O from speleothems, samples represented periods ranging from 3–4 years to less than a year. Yearly values are obtained by using a spline fitting of the data. Therefore, a certain amount of implicit smoothing has gone into generating the annual time series in this case. Hence the weaker inter-annual variance.

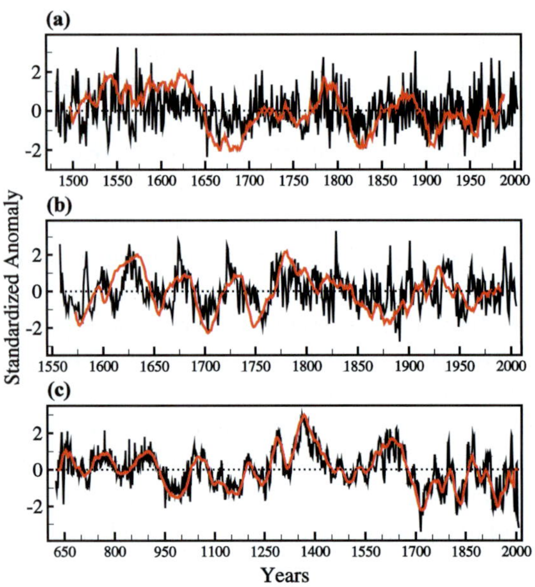

Fig. 2. Interannual (black) and epochal (red) variability of (a) Tree Ring Width Index chronology (RWI) from India (1481–2003); (b) RWI-Thailand (1558–2005) and (c) Oxygen isotopic composition (δ^{18}O) of speleothem from Central India (625–2007). Epochal variability is obtained by using 31 year running mean. Inter-annual as well as low pass filtered anomalies are normalized by their own standard deviations.

A growing body of evidence indicates that during the past 1500 years centennial-scale natural climate oscillations affected broad areas of the earth, including the Little Ice Age (LIA) during AD 1400–1850 and the Medieval Warm Period (MWP) during AD 900–1300 (*e.g.*, Sinha *et al.*, 2007). Sediment records in the Arabian Sea show that centennial-scale ISMR variability coincided with the changes in the North Atlantic region during the Holocene including a weaker summer monsoon during the LIA and an enhanced summer monsoon during the MWP (Altabet *et al.*, 2002; Gupta *et al.*, 2003). These changes might also been associated with a reduction in the Atlantic Meridional Overturning Circulation (AMOC) during the LIA (Lund *et al.*, 2006).

It is interesting to note that the tree-ring proxy over India depicts a decreasing trend during part of the LIA (Fig. 2a: AD 1600–1700). Even the tree-ring information over Thailand indicates similar features (Fig. 2b). Periods of severe droughts lasting decades occurred during the 14th and the 15th century and coincided with several of India's most devastating famines during LIA. Devastating Durga Devi famine occurred during AD 1396–1409 (Sinha *et al.*, 2007). The oxygen isotope series displays negative anomalies during MWP (Fig. 2c: AD 950–1250) and in general positive values during the LIA (Fig. 2c: AD 1400–1700). The oxygen isotopes and rainfall amounts in general vary inversely. Thus the oxygen isotope series indicates that an enhanced monsoon activity could have prevailed during the MWP and a suppressed monsoon during LIA. Furthermore, Sinha *et al.* (2011a) using the same proxies, showed oscillatory behavior of precipitation variability between periods with persistently "active-dominated" regime (AD ~1700 to 2007; negative values in Fig. 2c) and a "break-dominated" (AD 1400 to 1700; positive values in Fig. 2c) regime during the LIA. The periodic locking of the monsoon into this break monsoon mode — a mode that favors reduced precipitation over the Indian subcontinent and Southeast Asia and enhanced precipitation over the equatorial Indian Ocean — may have been provided by the cooler temperatures during the LIA at the extra tropical latitudes in the northern hemisphere, generating an inter-hemispheric thermal gradient, which could have forced the mean position of the ITCZ (Inter-tropical Convergence Zone) further southward in the Indian Ocean (Sinha *et al.*, 2011b). The switch between these regimes occurs abruptly (within decades) at a time (AD 1650–1700) when a proxy record of upwelling from the Arabian Sea suggested an abrupt increase in the monsoon winds. Similar inferences have been drawn from proxy reconstructions from Southern Vietnam and North-Central China (Sinha *et al.*, 2011b).

The weakening of the Indian and also the East Asian summer monsoon during the LIA is also indicated in the millennial-scale abrupt climate changes recorded in the Greenland ice cores (Zhang and Delworth, 2005) and paleo proxies from the Indian Ocean (Altabet *et al.*, 2002) and the Hulu Cave in Eastern China (Wang *et al.*, 2001). Mann *et al.* (2009) used a global proxy network to reconstruct surface temperature pattern for the past 1500 years. They found that the MWP was marked by a tendency for a La Niña-like conditions (in general favorable for the monsoon) in the tropical Pacific, and El Niño-like conditions (unfavorable) during the LIA.

Based on the above information and discussion, it is apparent that the inter-annual and multi-decadal variability in the proxy sources, tree-rings and oxygen isotopes over the Indian domain (Fig. 2), do indicate that a weaker monsoon during the LIA and a relatively stronger monsoon during the MWP, confirming that the past ISMR variability is reasonably well represented by the variability in the proxy sources examined here.

4.1. *Periodicities of the Asian monsoon*

The power spectra for the ISMR (1871–2012) depicts a significant (at 95% confidence level) periodicity around 2–3 years, possibly related with the tropical-biennial oscillation (TBO), and an insignificant periodicity with periodicity around 60 years and beyond (Goswami and Kripalani, 2012; Fig. 3a). While the power spectra for the tree-ring index over India shows significant periods in the range of 2–3 years, 10–20 years, 70–80 years and 150–200 years (Fig. 3b); the tree-ring index over Thailand shows significant periods in the 2–3 years, 4–5 years, 20–30 years 50–70 years. An insignificant peak around 150 years is also indicated (Fig. 3c). The oxygen isotope index (Fig. 3d) displays significant periods in the range 10–20 years and 50–70 years and insignificant period around 100 years. It is interesting to note that the quasi-biennial period is robust in both the long records of RWI consistent with the instrumental data of the ISMR. However, it is absent in the $\delta^{18}O$ proxy (Fig. 3d) due to the smoothing on inter-annual time scale. Consistent with the RWI-India time series, the $\delta^{18}O$ record also shows significant peaks with 10–20 year and 50–60 year periods. The power spectra for the oxygen isotope for the entire available period (AD 625–2007) also show significant peaks with longer periods around ~200 years and around ~600 years (Fig. 3e), besides the periods noted above. These ultra-slow modes (200 years, 600 years), may be due to a long term decreasing trend observed in the oxygen isotope series (Fig. 2c: decreasing during AD 1350–1950). As noted earlier, the ISMR reveals a decreasing trend since the mid-20th century (Fig. 1). Interestingly, the oxygen isotope series reveals an increasing trend during the same period (Fig. 2c). Thus the proxies do indicate

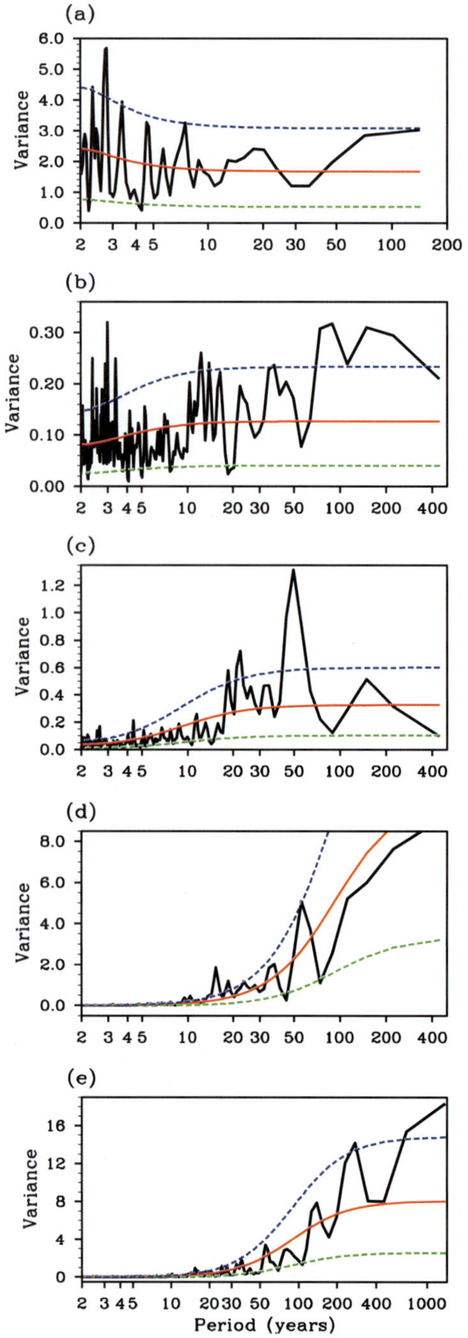

Fig. 3. Spectral analysis of detrended (a) ISMR (1871–2012) (b) RWI-India; (c) RWI-Thailand; (d) δ^{18}O-Central India, for the common period (1558–2003) and (e) δ^{18}O-Central India for the period (625–2007). Red noise continuum (red line) and associated 10 (green dash line) and 90 (blue dash line) percent confidence limit are added.

possible presence of periods with longer period, in particular in the 50–80 and the 100–200 years range.

Xu *et al.* (2012) reconstructed the ISM intensity during the past 270 years from tree ring δ^{18}O at Hongyuan, located at the eastern edge of the Tibetan Plateau. Significant periods of 110.9, 59.0 and 22.6 years in this reconstructed series were detected. This 110.9 year component shows similar trend with ISM record from Oman (Burns *et al.*, 2002). This typical period of ISM variation is also detected in speleothem records of Dongge Cave (Wang *et al.*, 2005) and sediment records of Arabian Sea (*e.g.*, Anderson *et al.*, 2002), suggesting that such multi-decadal oscillations are common in the Asian monsoon region.

It is interesting to note that a 50–80 year multi-decadal oscillation is common to all the three proxies. We propose that this is an intrinsic multi-decadal mode of variability of the Asian monsoon. However, it is also noted that the preferred periodicities range from 50 to 80 years indicating that the mode may not be stationary. To test the epochal variation of the mode, we carry out wavelet analysis.

4.2. *Changes in periodicities*

Wavelet transforms is able to determine both the dominant modes of variability and how these modes vary in time (Torrence and Compo, 1998). The Morlet wavelet spectra for the three proxies for the common period AD 1558–2003 are shown in Fig. 4a, b, c. For the tree-ring index over India, the wavelet spectrum clearly indicates that the maximum variance is contained around the ~150 year period during the earlier two centuries AD 1550–1750 (Fig. 4a: red shades). Thereafter the variance is around ~80 year period during the period AD 1700–1900. A periodicity around ~50 years prevails during the late 19th century and the early 20th century. Finally during the second half of the 20th century a periodicity around ~30 years is noticed.

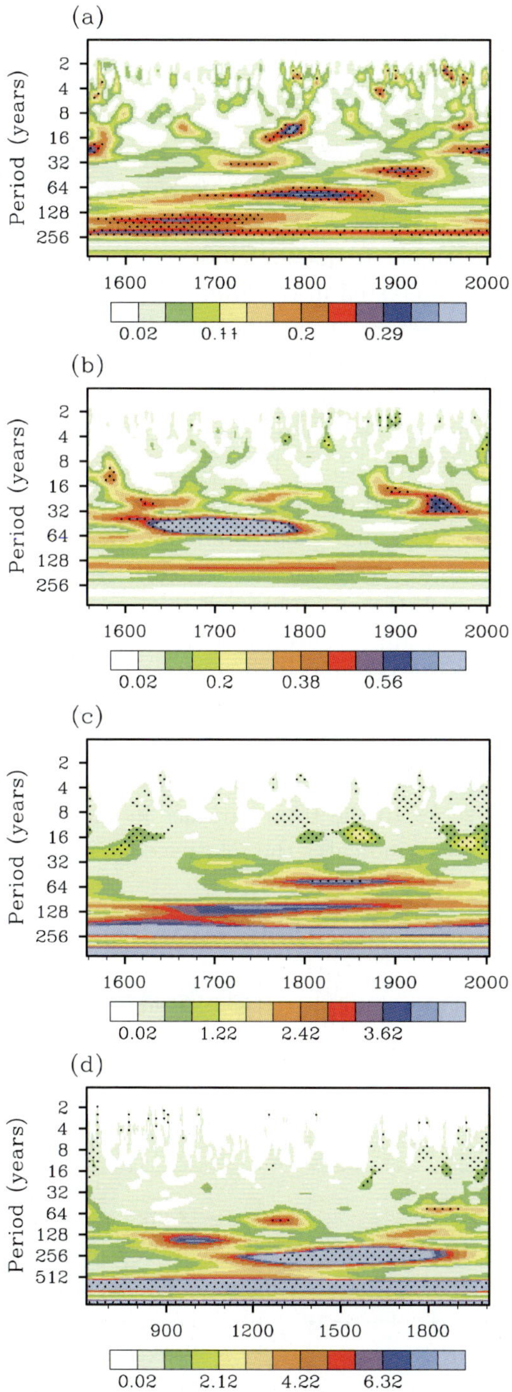

Fig. 4. Wavelet power spectrum of detrended (a) RWI-India; (b) RWI-Thailand; (c) δ^{18}O-Central India, for the common period (1558–2003) and (d) δ^{18}O-Central India for the period (625–2007). Dots represents significance at 95% confidence level.

There appears to be a clear systematic decrease in the periodicity length from ~150 years to ~30 years as one proceeds from the 15th century to the 20th century. Thus this dataset indicates that the 70–80 year multi-decadal peak seen in Fig. 3b is non stationary.

On the other hand the spectrum of the tree-ring index over Thailand depicts a significant periodicity ~150 years throughout the entire period (Fig. 4b). The ~50 year multi-decadal mode has been quite strong from AD 1600 and 1800. That changed to a slightly longer period of ~70 years in the recent years with a rather quiescent period between AD 1800 and 1900. This is consistent with getting a strong spectral peak around 50 years in Fig. 3c. During the second half of the 20th century, a periodicity in the 20–30 years range is also noticed. For the same data period, the wavelet spectrum for the oxygen isotopes (Fig. 4c) reveals a periodicity ~150 years up to the mid 18th century and a periodicity in the 40–60 years range during AD 1760 to 1920. Finally the spectrum for oxygen isotopes for the entire period (AD 625–2007) again reveals a monotonic decrease in periodicity from ~150 years to ~60 years from the 10th century to the 19th century. Thus, all the three proxies show multi-decadal variation in within 50 to 80 years, indicating a common mode of variability. However, the periodicity of the multi-decadal mode is non stationary and varies from time to time. In order to have confidence in attributing this as a 'mode' of variability, we employ the Empirical Mode Decomposition (EMD) method to these proxies.

5. Empirical Mode Decomposition

Climate data are full of noise from all kind of sources. A new method, EMD, has been introduced to separate noise from signal and also to identify the cycles of change from the trend (Huang *et al.*, 1998; Huang, 2012). The EMD proposed by Huang *et al.* (1998) is an adaptive time-frequency data analysis method

(Wu and Huang, 2009). The EMD works in temporal space directly rather than in the corresponding frequency space and is intuitive, direct and non-parametric (Huang and Wu, 2008). The EMD is therefore a powerful method of decomposing a data time series into Intrinsic Mode Functions (IMFs) each associated with a unique frequency. The amplitude and frequency of IMFs vary slowly with time and their instantaneous frequencies lie in a narrow band and are nearly orthogonal to each other. Thus from a data time series, each mode can be separated and physical significance attached to each of these temporal modes.

We now apply the EMD technique, according to Huang *et al.* (1998), for decomposing the time series of the three proxies into finite number of IMFs and determine the dominant modes associated with these IMFs, in particular the multi-decadal mode.

5.1. *IMFs of ISMR proxies*

The IMFs of the observed ISMR during the instrumental record period (1871–1990) have been presented by Iyengar and Raghu Kanth (2005), hence not shown here. The tree-ring data series over India and Thailand are decomposed into 8 EMFs each; for the oxygen isotopes the series for the common period is decomposed into 7 IMFs and for the entire period into 8 IMFs. As the first few IMFs in each time series represent higher frequency modes and we are primarily interested in the lower frequency modes, the time series for each of the IMFs from IMF3 through IMF6 are shown in Fig. 5. The temporal amplification/weakening of each of these modes over the analyzed time period (AD 1558 to 2003) can be inferred from Fig. 5a–c. The amplitude and the period of an IMF slowly vary.

To determine the dominant periodicities associated with each of these IMFs, the time series of IMFs are subjected to spectral analysis. The results of spectral analysis on each of the IMF series (IMF3–IMF6) are displayed in Fig. 6. Each IMF so obtained is a narrowband spectrum within which the oscillations take place. The ratio of the variance contained in each of the IMF series to the total variance in the original data series indicates the relative contribution of

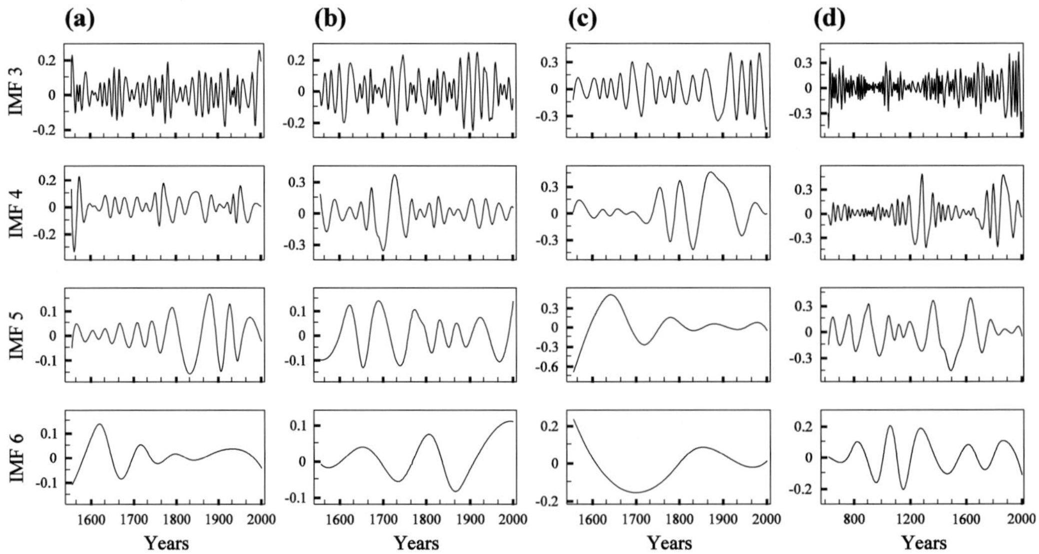

Fig. 5. Intrinsic mode funtions (IMFs) of (a) RWI-India; (b) RWI-Thailand; (c) δ^{18}O-Central India, for the common period (1558–2003) and (d) δ^{18}O-Central India for the period (625–2007).

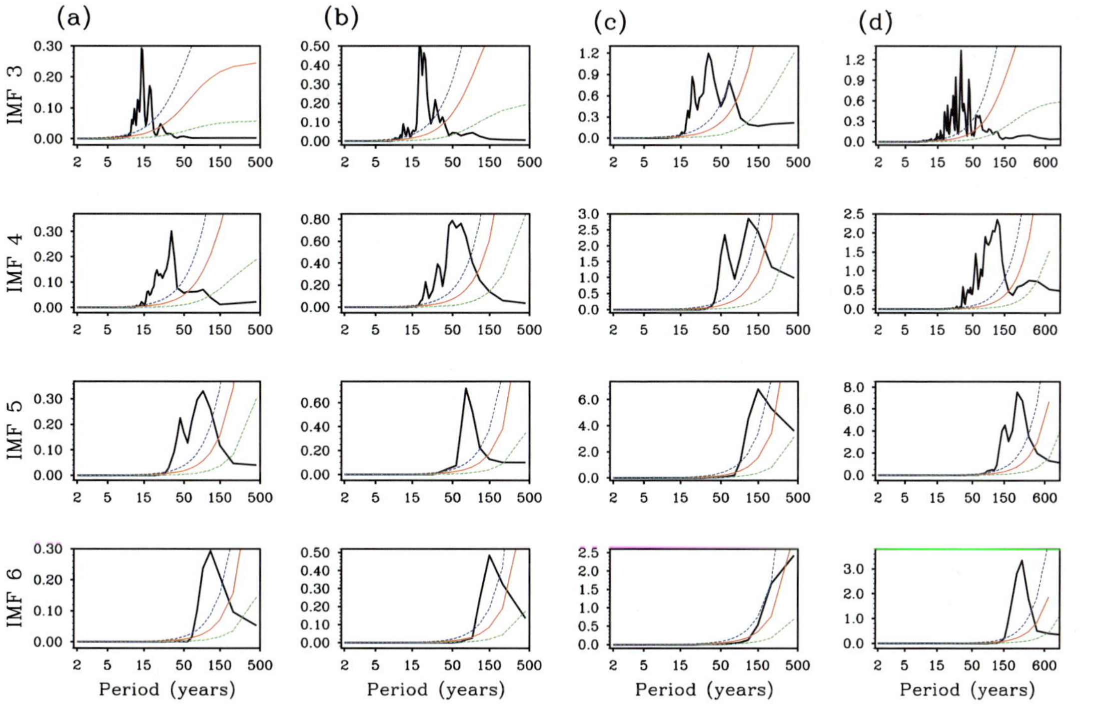

Fig. 6. Spectral analysis of the IMFs of (a) RWI-India; (b) RWI-Thailand; (c) δ^{18}O-Central India, for the common period (1558–2003) and (d) δ^{18}O-Central India for the period (625–2007). Red noise continuum (red line) and associated 5 (green dash line) and 95 (blue dash line) percent confidence limit are added.

an IMF to the total variability. It is interesting to note that IMF5 of RWI-India and IMF5 of RWI-Thailand, IMF4 of δ^{18}O-Central India for the period AD 625–2007 as well as for period AD 1558–2003 all have 50–80 year as the dominant period. As this IMF is orthogonal to all others, the 50–80 year oscillation may be considered as a distinct mode of variability of all the proxies. Thus, we conclude that 50–80 year mode is a natural multi-decadal mode of variability of the Asian monsoon.

The maximum variance for the tree-ring series is contained in the periods with shorter wavelengths (2–5 years). Tree-ring series over India explains 57% of the variance in this band, while the Thailand series 28% of variance is explained by this high frequency mode. As stated above, this mode may be related with the TBO and the ENSO mode. The second IMF mode (5–15 years) also explains substantial variance, 23% by the tree-ring data over India and 18% over Thailand. Thus the first two IMF modes of the tree-ring time series explain nearly 80% of the original variance over India and nearly 45% over Thailand. It is interesting to note that the multi-decadal mode, namely IMF4, of tree-ring data over Thailand explains nearly 30% of the variance. Furthermore even the IMF4 of oxygen isotope series explains nearly 20% of the total variance. However, the multi-decadal mode of RWI-India and the RWI-Thailand, namely IMF5 explains about 10% of total variance respectively. This does not mean that this multi-decadal mode is not important in RWI time series. As noted in Section 4, these time series has large inter-annual variance resulting in IMF1 explaining 57% and 28% respectively. As a result, IMF5 explain relatively smaller variance.

5.2. *Significance of IMFs of ISMR proxies*

The significance of the IMFs in particular the 50–80 year multi-decadal mode is now examined. We follow the method adopted in Wu and Huang (2005). In brief, the method is as follows: using EMD, the characteristics of Gaussian white noise are used for determining the relationship between the mean energy density of IMFs and its corresponding mean period. These are verified using Monte Carlo method. Based on this, the significance test is designed using the bounds for the energy density spread function of the IMFs of Gaussian white noise. The IMFs with the energy located above the upper bound (95% confidence interval) or below bound (5%) should be considered as containing signal information at the selected confidence level. Details of the mathematical procedure in assigning statistical significance to the IMFs can be found in Wu and Huang (2005). We have used the software codes for extraction of IMFs and the related significance tests available at the website http://rcada.ncu.edu.tw/research1.htm (courtesy Norden E. Huang).

It is interesting to note that the first 3 IMFs of RWI–India representing higher frequency modes do not exceed the 95th percentile (Fig. 7). Only the 4th, 5th and the 6th IMF mode exceed the 95th percentile, in particular the 5th mode representing the multi-decadal mode in the 50–80 year period, suggesting that this mode is statistically significant (red circle in Fig. 7a). While all the modes for RWI-Thailand exceed the 95th percentile, the maximum significance is shown by the 4th and the 5th mode, representing the multi-decadal mode of our interest (red circle in Fig. 7b). Similar inferences can be drawn from the significance test on the IMFs of Oxygen-isotopes (Fig. 7c, d). Furthermore, it is interesting to note that the IMFs with lower frequency are more significant compared to their higher frequency counterpart. Thus, the mode

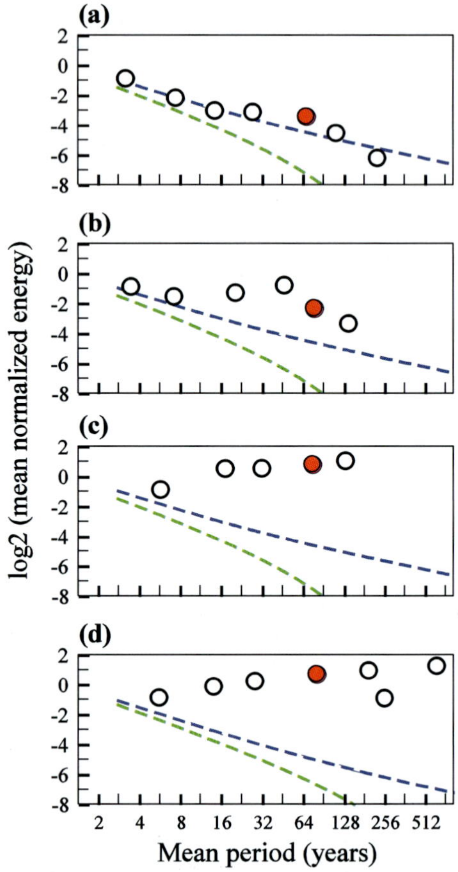

Fig. 7. Statistical significance test of the IMFs of the (a) RWI-India; (b) RWI-Thailand; (c) δ^{18}O-Central India, for the common period (1558–2003) and (d) δ^{18}O-Central India for the period (625–2007). The dashed lines are the 5th (green) and 95th (blue) percentiles. Circles correspond to the pairs of mean normalized energy and the averaged mean period of the IMFs. The IMF representing the 50–80 year multi decadal mode is highlighted by red circle.

representing the 50–80 year period is unique and significant.

In order to see whether the multi-decadal mode of RWI-India is coherent with that from RWI-Thailand and δ^{18}O-Central India, lead-lag correlations between IMF5 of RWI-India and IMF5 of RWI-Thailand and IMF4 of δ^{18}O-Central India during the same period are calculated (Fig. 8). The periodic nature of the lead-lag correlations within the period indicates

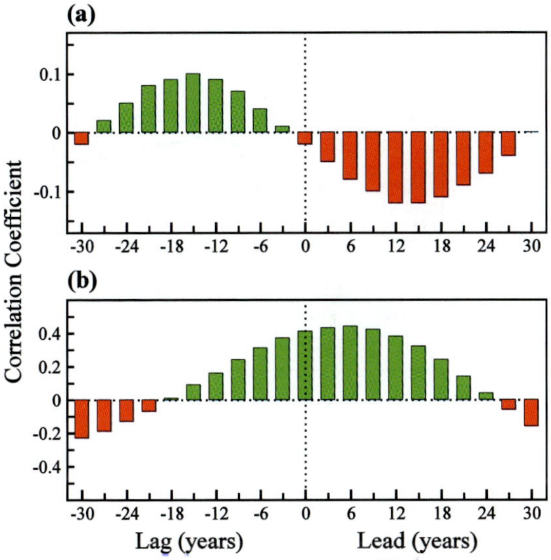

Fig. 8. Lag/Lead Correlation Coefficient of RWI India (IMF5) with proxies of (a) Thailand (IMF5) and (b) δ^{18}O-Central India (IMF4) precedes/follows, for the common period (1558–2003).

that the time series are coherent but phase shifted as a result of spatial pattern of the multi-decadal mode.

Is the multi-decadal mode of Asian monsoon a regional mode or an integral part of a global multi-decadal mode of variability? In an attempt to arrive at some answer to this question, we examine some long term proxy time series of AMO, ENSO and Pacific Decadal Oscillation (PDO).

6. Analysis of Long Global Proxies: AMO, ENSO, PDO

Coupled air-sea interaction and ocean dynamics in the tropical Pacific are important for connecting Atlantic changes with the Asian summer monsoon variation (Zhang and Delworth, 2005, 2006). Multi-decadal variability in the North Atlantic Ocean can cause the observed multi-decadal variation of the ISMR (Goswami *et al.*, 2006). The positive AMO phase, characterized by anomalous warm North Atlantic and cold South Atlantic, leads to a strong Southeast and East Asian summer monsoon and late withdrawal of the Indian summer monsoon. The changes of monsoon are manifested through the West Pacific and the Indian Ocean and the temperature changes over Eurasia (Goswami *et al.*, 2006; Lu *et al.*, 2006). Besides AMO, ENSO also causes climate extremes over the Indian subcontinent and has been well documented since the time of Walker. PDO also has impact on the ISMR variability (Krishnan and Sugi, 2003).

6.1. *Data: SST indices over North Atlantic, Equatorial Pacific and North Pacific*

Since long time series of traditional indices of AMO, ENSO and PDO are not available, in this study, we use SST averaged over the north Atlantic, Eastern Equatorial Pacific and North Pacific as proxies for AMO, ENSO and PDO respectively. The large-scale surface temperature reconstructions using global proxy data spanning the ocean and land regions of both hemispheres are available for the past 1500 years (Mann *et al.*, 2009). The proxy data contains decadal-resolution data in-addition to annual-resolution data, hence only inter-decadal and longer-term variations are meaningfully resolved (Mann *et al.*, 2009). The present study uses indices of SST averaged over AMO as defined by Kerr (2000), Niño3 (2.5°S–2.5°N; 92.5°W–147.5°W) as an index for ENSO and North Pacific region (22.5°N–57.5°N; 152.5°E–132.5°W) as an index of PDO (http://www.ncdc.noaa.gov/paleo/pubs/mann2009b/mann2009b.html).

6.2. *Inter-annual and multi-decadal variability*

While the inter-annual variations show random fluctuations, all the 3 proxies clearly indicate an increasing trend commencing from the 19th century (Fig. 9). Vance *et al.* (2013) report a statistically significant link between ENSO and

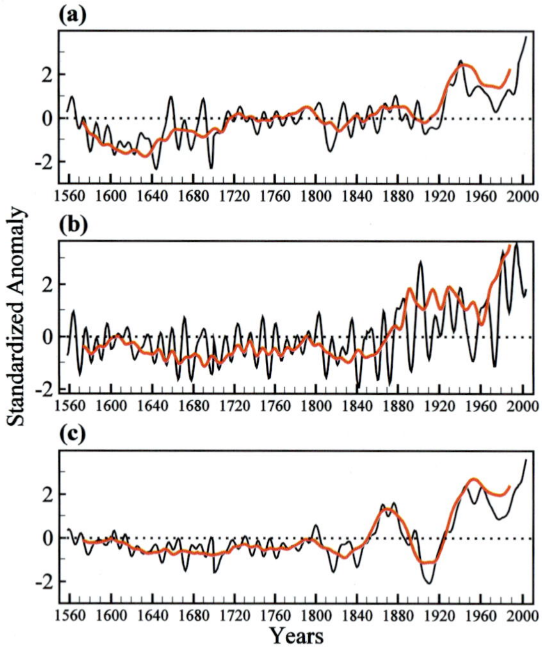

Fig. 9. Same as Fig. 2, but for (a) AMO, (b) ENSO and (d) PDO for the common period (1558–2003).

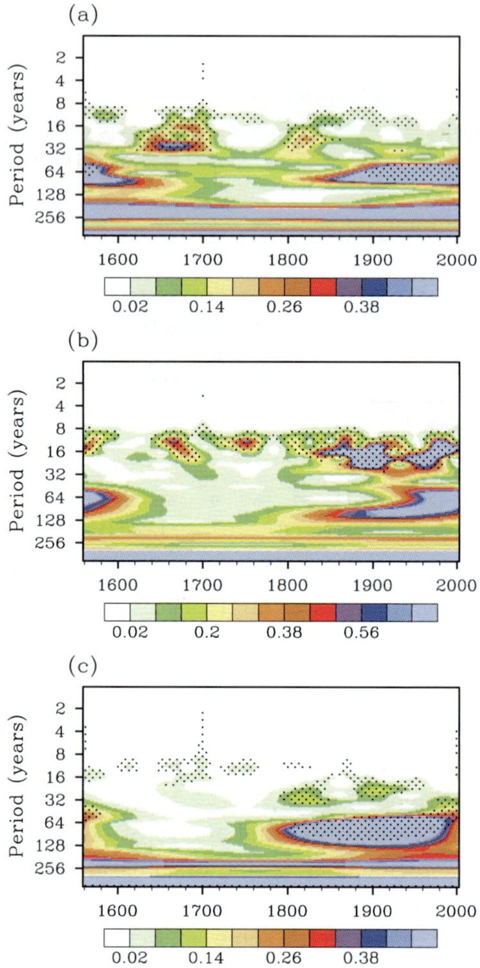

Fig. 10. Wavelet power spectrum of the detrended reconstructed indices of (a) AMO; (b) ENSO and (c) PDO for the period 1558–2003. Dots represents significance at 95% confidence level.

sea salt deposition during summer from the Law Dome (LD) ice core in East Antarctica. The reduced (enhanced) summer sea salt deposition at LD indicates El Niño (La Niña) conditions. They further report that over the last 1010 years, the LD summer sea salt record has exhibited two below-average (El Niño-like) epochs, AD 1000–1260 and 1920–2009 and a longer above average (La Niña-Like) epoch for AD 1260 to 1860. The La Niña (El Niño) — like conditions during AD 1620–1860 (1900 to present) can be inferred from Fig. 9b. Both the AMO (Fig. 9a) and PDO (Fig. 9c) are in the positive phase since the last two decades. The positive phase of AMO (ENSO, PDO) is known to be favorable (unfavorable) for the increasing ISMR variability. Thus the recent behavior of the AMO, ENSO, PDO and the ISMR suggests that the multi-decadal variability in the ISMR could probably be more linked with the events over the Pacific Ocean rather than over the Atlantic Ocean.

6.3. *Dominant periodicities and temporal modes*

The wavelet spectra for these 3 proxies are shown in Fig. 10. Most noteworthy feature is the dominance of the periodicity in the 60–80 years range during the 16th century and mid 19th century to present, as displayed by all these 3 proxies. Another noteworthy feature inferred from this figure is the near absence or existence of weak 60–80 year periodicity during the LIA (17–18th century) in all the 3 proxies. The

AMO, which fluctuates in a period of 60–70 years, has been found, connected with many observed climate fluctuations (Hsu and Wu, 2012). In addition, the ENSO index (Fig. 10b) also displays waxing and waning of the periodicity in the 8–16 years range throughout the analyzed period.

Some indications of a periodicity around 150 years prevalent in all the 3 proxies is also seen. More details can be inferred from Fig. 10. The high frequency modes (less than a decade) are not discernible since the variations on inter-decadal and longer-time scales only are resolved in this data (Mann *et al.*, 2009).

It is clear from Figs. 9 and 10 that all three SST time series has linear increasing trends. Therefore, EMD technique is used to extract the IMFs after removing the linear trend from each time series. It is interesting to note that a total of 6 IMFs for AMO, 7 IMFs for ENSO and PDO account for most of the variability in the original data in the respective time series. As the first IMF in each case represents a high frequency mode, the time series of IMF2 through IMF5 are shown in Fig. 11. Amplitude variation of individual IMFs over the time period considered here can be inferred from Fig. 11. The spectral analysis on the time series of IMFs (Fig. 12) shows that the IMF2 of AMO, PDO and IMF3 of ENSO represent the 50–80 year mode. As IMFs are nearly orthogonal, the 50–80 year oscillation in these fields also represents a mode of variability. The multi-decadal mode of AMO (IMF2) explains 46.0% of variance while that of ENSO (IMF3) explains 6.3% of total variance and that of PDO (IMF2) explains 32.1% of total variance.

6.4. *Significance of multi-decadal modes of AMO, ENSO and PDO*

To examine the statistical significance of these IMF modes, similar procedure and methodology is used as done for the rainfall proxies (Section 5.2). These are displayed in Fig. 13 which reveals that most of these IMF modes exhibit statistical significance exceeding the 95th percentile. Again the significance for the slower modes, in particular the mode representing the 50–80 year periodicity is higher in general (red circle).

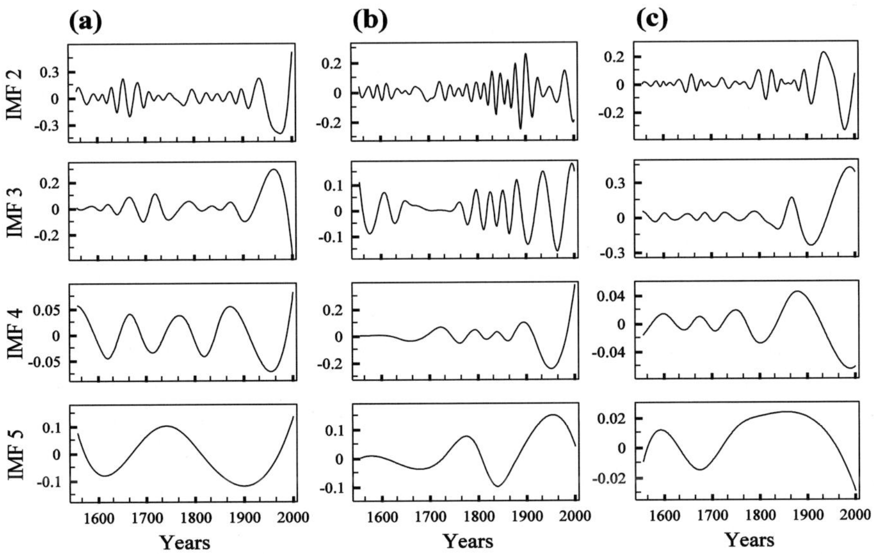

Fig. 11. IMFs of the detrended reconstructed (a) AMO, (b) ENSO and (c) PDO indices for the period 1558 to 2003.

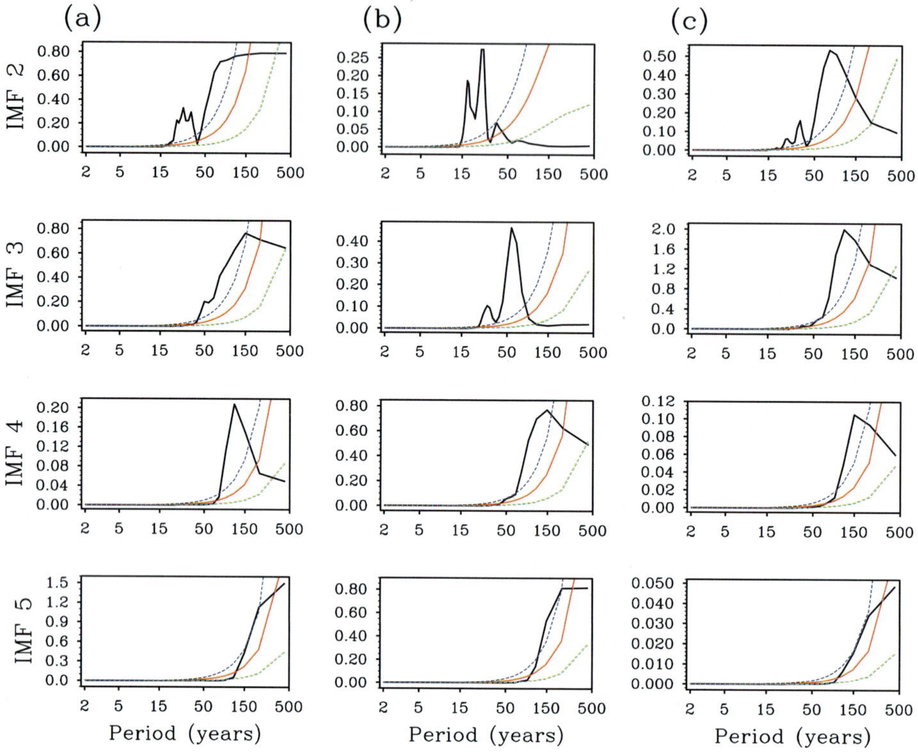

Fig. 12. Spectral analysis of IMFs of the detrended reconstructed (a) AMO, (b) ENSO and (c) PDO indices for the period 1558 to 2003. Red noise continuum (red line) and associated 5 (green dash line) and 95 (blue dash line) percent confidence limit are added.

Thus the rigorous statistical significance test (Wu and Huang, 2005) reveals that the relevant EMD components in ISMR proxies and the large-scale modes over the Pacific as well as the Atlantic are significant at the 95% significance level, in particular the 50–80 year period mode, the mode of our interest.

The collective results of several studies (*e.g.*, Gray *et al.*, 2004; Wu *et al.*, 2011; Walsh and Chapman, 2012; Wyatt *et al.*, 2012) point to a multi-decadal variability as a key component of the North Atlantic. While there is a large spectrum of variability (20–120 years), an outstanding feature is the prevalence of periodicities in the time scales of 50–80 years, also noticed in the ISMR proxies.

Thus the existence of a multi-decadal 50–80 year AMO has been drawn while we present evidence here that a similar multi-decadal mode is also present in ENSO and PDO. It has been argued (Delworth and Mann, 2000) that AMO is a mode of variability arising from ocean-atmosphere interaction on a global-scale. The finding here that a multi-decadal variability with same time scale exists for ENSO and PDO as well lends support to a hypothesis that all these are manifestation of a global mode of variability possibly arising from ocean-atmosphere-land interaction. Preliminary analysis of multi-decadal mode of ISMR (Goswami *et al.*, 2006) using relatively short record (1871–2000) indicates that 50–80 year multi-decadal variability of ISMR is coherent with that of AMO. Is the 50–80 year ISMR multi-decadal variability coherent with that of AMO, ENSO and PDO? To examine this aspect, we calculated

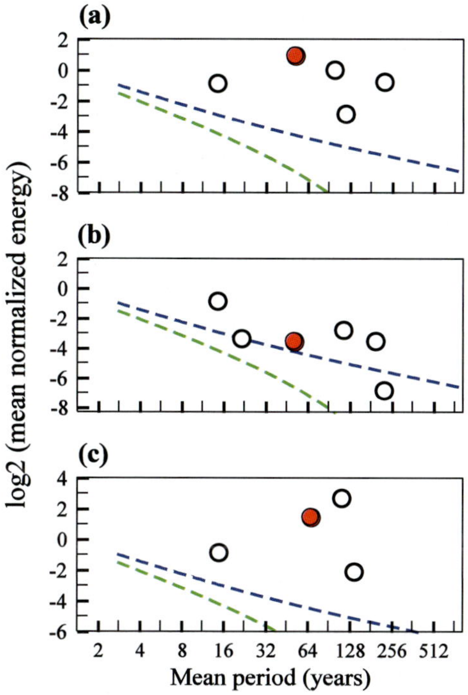

Fig. 13. Same as Fig. 7, but for (a) AMO, (b) ENSO and (c) PDO for common period (1558–2003).

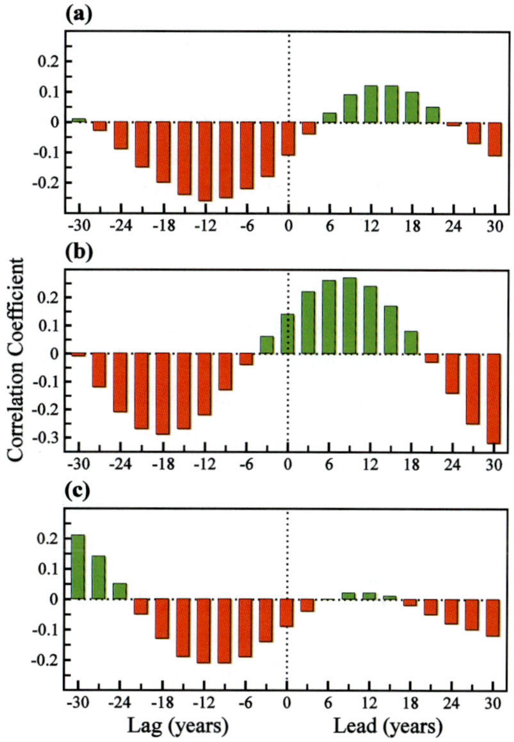

Fig. 14. Lag/Lead Correlation Coefficient of RWI India (IMF5) with proxies of (a) AMO (IMF3), (b) ENSO (IMF3) and (c) PDO (IMF2) precedes/follows, for the common period (1558–2003).

lead-lag correlations between IMF5 of RWI-India and IMF2 of AMO and PDO and IMF3 of ENSO (Fig. 14).

It is interesting to note that the lead-lag correlations between IMF5 of RWI and IMF2 of AMO and PDO and IMF3 of ENSO show a periodic association with a period around 65 years (half period between 30 and 35 years) which is also the central period of each of these modes. Although the peak correlation coefficients are not highly statistically significant due to decreased degrees of freedom as a result of implicit filtering, these correlations indicate that the 50–80 year monsoon mode is likely to be an integral part of the global 50–80 year mode. In each case, there are regional modes with higher frequency that interact with the lower frequency mode (~65 year) and make the spectrum of the lower frequency mode broad (50–80 year). The 'noise' component of the low frequency modes is completely uncorrelated with each other resulting in the low correlations between the IMFs.

7. Discussion and Summary

On inter-annual time scales, the existence of a tropical biennial oscillation with ~2.5 year period and the ENSO signal with 3.5 year period are well known in the Indian monsoon rainfall during the period of instrumental record (Shukla, 1987; Webster *et al.*, 1998). In addition, existence of an 11 year cycle possibly related to the sunspot number variation is also known (Iyengar and Raghu Kanth, 2005). While the ISMR records between 1871 and 2012, does indicate the existence of a multi-decadal variability (Fig. 1), the robustness of the periodicity could not be established from the short record.

The question whether there exists a multi-decadal 'mode' of variability certainly could not be addressed by such a short record.

The possible existence of a multi-decadal mode of variability of the Asian monsoon is investigated using three long term proxies of Asian monsoon rainfall extending for a period of 500 years or more. One of these proxies is tree-ring-width index (RWI) from Southern India while another is RWI from Thailand. The third proxy is δ^{18}O from speleothem in Dandak and Jhumar caves in Central India. While the annual variations are well resolved in the RWI indices resulting in large inter-annual variance in these time series, the δ^{18}O from speleothem could not resolve annual signal throughout the entire length of the time series (Sinha *et al.*, 2007). As a result, the inter-annual variability is weak in δ^{18}O time series. However the decadal and inter-decadal variabilities are well resolved in the data.

Spectral analysis of the three proxy time series of rainfall indicates that an oscillation with time period between 50 and 80 years is present in all time series. However wavelet analysis reveals that the oscillation is not stationary but varies in intensity and period in different epochs. Whether the multi-decadal oscillation of the Asian monsoon could be considered as a multi-decadal mode of variability was examined using IMF decomposition of all the three proxy time series and found that one IMF of each of the time series has 50–80 year periodicity. Lead-lag correlation between these IMFs also indicate that they are coherent, providing strong support that the 50–80 year multi-decadal oscillation is indeed a mode of variability of the Asian monsoon.

In order to explore whether the Asian monsoon multi-decadal mode has counterpart in other parts of the globe, IMF decomposition was carried out on long proxies of AMO, ENSO and PDO. It has been revealing to find that all these three time series also have one IMF representing 50–80 year periodicity. These IMFs are also found to be coherent amongst them as well as with IMF of Asian monsoon representing the 50–80 year mode. These results lead us to conclude that the 50–80 year multi-decadal mode of Asian monsoon variability is likely to be an integral part of a global mode of multi-decadal variability with same time line.

As this mode is intimately related to the AMO, we may speculate that the origin of this global mode of multi-decadal variability is due to coupled ocean-atmosphere-land interactions as proposed for AMO (Delworth and Mann, 2000). The broad band nature of the mode with periodicities ranging from 50 years to 80 years is a manifestation of this non stationary behaviour of the oscillation and has implication in limiting the predictability of the same. What is the potential limit on predictability of this multi-decadal mode? As on inter-annual time scale such as the ENSO, the broad band nature of the mode limits the potential predictability (Goswami and Shukla, 1991), the broad band nature of the multi-decadal mode is also expected to limit its potential predictability. With a coupled model that simulates this multi-decadal variability reasonably well, it may be worthwhile to carry out predictability studies to estimate potential limit on predictability of the mode.

The decreasing trend of ISMR over the past six decades has been a cause of concern in the Asian monsoon community as it provides fuel to a speculation whether the South Asian monsoon may have tipped to a completely different climate. However our study shows that such a long excursion of the South Asian monsoon on the negative side has been common in the past (Fig. 2a) and thus could still be part of the natural 50-80 year mode of multi-decadal variability.

Acknowledgments

The authors and Indian Institute of Tropical Meteorology are fully funded by Ministry of Earth Sciences, Govt. of India. We thank

B. M. Buckley, Tree-Ring Laboratory, Lamont-Doherty Earth Observatory, NY, USA, for providing tree-ring-width index over Northwest Thailand and A. Sinha, Department of Earth Science, California State University, California, USA, for providing composite of oxygen isotope composition (δ^{18}O) of speleothem record from Dandak and Jhumar caves over Central India. We also thank N.E. Huang, National Central University, Taiwan and Z. Wu, Florida State University, USA, for providing the software codes for computing IMFs and their significance.

References

Altabet, M. A., M. J. Higginson, and D. W. Murray, 2002: The effect of millennial-scale changes in Arabian Sea denitrification on atmospheric CO_2. *Nature*, **414**, 159–162.

Anderson, D. M., J. T. Overpeck, and A. K. Gupta, 2002: Increase in the Asian SW monsoon during the past 4 centuries. *Science*, **297**, 596–599.

Berkelhammer, M., A. Sinha, M. Mudelsee, and K. G. Cannariato, 2010: Persistent multidecadal power in the Indian Summer Monsoon. *Earth Planet. Sc. Lett.*, **290**, 166–172.

Borgaonkar, H. P., A. B. Sikder, Somaru Ram, and G. B. Pant, 2010: El Niño and related monsoon Droughts signals in 523-year-long ring width records of teak (Tectona grandis L.F.) trees from South India. *Palaeogeogr. Palaeocli.*, **285**, 74–84.

Buckley, B. M., K. Palakit, K. Duangsathaporn, P. Sanguanthan, and P. Prasomsin, 2007: Decadal scale droughts over Northwest Thailand over the past 448 years: Links to the tropical Pacific and the Indian Ocean sector. *Clim. Dyn.*, **29**, 63–71.

Burns, S. T., D. Fleitmann, M. Mudelsec, U. Neff, A. Matter, and A. Mangini, 2002: A 780-year annually resolved record of Indian Ocean monsoon precipitation from speleothem from South Oman. *J. Geophys. Res. Atmos.*, doi: 10.1029/2001JD001281.

Dansgaard, W., 1964: Stable isotopes in precipitation. *Tellus*, **16**, 436–468.

Delworth, T. L., and M. E. Mann, 2000: Observed and simulated multi-decadal variability in the Northern Hemisphere. *Clim. Dyn.*, **16**, 661–676.

Goswami, B. N., 2006: The Asian monsoon: Inter-decadal variability. *The Asian Monsoon*, B. Wang, Ed., Praxis Springer Berlin Heidelberg, 295–328.

———, and R. H. Kripalani, 2012: Multi-decadal variability in Indian Summer Monsoon Rainfall: Some new perspectives. *NTU International Science Conference on Climate Change: Multi-decadal and beyond*, National Taiwan University, Sep 17–21, 159–167.

———, M. S. Madhusoodanan, C. P. Neema, and D. Sengupta, 2006: A physical mechanism for North Atlantic SST influence on the Indian summer monsoon. *Geophys. Res. Lett.*, **33**, L02706, doi: 10.1029/2005GL024803.

———, and J. Shukla, 1991: Predictability of a coupled ocean–atmosphere model. *J. Climate*, **4**, 3–22.

Gray, S. T., L. J. Graumlich, J. L. Betancourt, and G. T. Pederson, 2004: A tree-ring based reconstruction of the Atlantic Multidecadal Oscillation since 1567 A.D. *Geophys. Res. Lett.*, **31**, L12205, doi: 10.1029/2004GL019932.

Gupta, A. K., D. M. Anderson, and J. T. Overpeck, 2003: Abrupt changes in the Asian southwest monsoon during Holocene and their links to the North Atlantic Ocean. *Nature*, **421**, 354–357.

Hsu, H. H., and R. J. Wu, 2012: AMO-like interdecadal variability in the CMIP5 — Are models oversensitive to prescribed forcing? *NTU International Science Conference on Climate Change: Multi-decadal and beyond*, National Taiwan University, Sep 17–21, 123–125.

Huang, N. E., 2012: On noise, cycle and trend in climate data. *NTU International Science Conference on Climate Change: Multi-decadal and beyond*, National Taiwan University, Sep 17–21, 116–120.

———, Z. Shen, S. R. Long, M. C. Wu, H. H. Shih, Q. Zheng, N.-C. Yen, C. C. Tung, and H. H. Liu, 1998: The empirical mode decomposition method and the Hilbert spectrum for non-stationary time series analysis. *Proc. R. Soc. Lond.* **454A**, 903–995.

———, and Z. Wu, 2008: A review on Hilbert–Huang transform: Method and its applications to geophysical studies. *Rev. Geophys.*, **46**, RG2006, doi: 10.1029/2007RG000228.

Iyengar, R. N., and S. T. G. Raghu Kanth, 2005: Intrinsic mode functions and a strategy for forecasting Indian monsoon rainfall. *Meteorol. Atmos. Phys.*, **90**, 17–36.

Kerr, R. A., 2000: A North Atlantic climate pacemaker for the centuries. *Science*, **288**, 1984–1986.

Kripalani, R. H., and A. Kulkarni, 1997: Rainfall variability over Southeast Asia — connections with Indian monsoon and ENSO extremes: New perspectives. *Int. J. Climatol.*, **17**, 1155–1168.

———, and ———, 2012: Monsoon aspects related to climate change. *Met Monographs*, Vol. II, India Meteorological Department, 505–559.

———, S. V. Singh, N. V. Panchawagh, M. Brikshavana, 1995: Variability of the summer monsoon rainfall over Thailand-comparison with features over India. *Int. J. Clim.*, **15**, 657–672.

Krishnamurthy, V., and B. N. Goswami, 2000: Indian monsoon-ENSO relationship on inter decadal time scales. *J. Climate*, **13**, 579–595.

Krishnan, R., and M. Sugi, 2003: Pacific decadal oscillation and variability of the Indian summer monsoon rainfall. *Clim. Dyn.*, **21**, 233–242.

Lu, R., B. Dong, and H. Ding, 2006: Impact of the Atlantic Multidecadal Oscillation on the Asian summer monsoon. *Geophys. Res. Lett.*, **33**, L24701, doi: 10.1029/2006GL027655.

Lund, D. C., J. Lynch-Stieglitz, and W. B. Curry, 2006: Gulf Stram density structure and transport during the past millennium. *Nature*, **444**, 601–604.

Mann, M. E., Z. Zhang, S. Rutherford, R. S. Bradley, M. K. Hughes, D. Shindell, C. Ammann, G. Faluvega, and F. Ni, 2009: Global signature and dynamical origin of the Little Ice Age and Medieval Climate Anomaly. *Science*, **326**, 1256–1260, doi 10.1126/science.1177303.

Shukla, J., 1987: Interannual variability of monsoons. *Monsoons*, J. S. Fein, and P. L. Stephens, Eds, Wiley and Sons, 399–463.

Sinha, A., M. Berkelhammer, L. Slott, M. Mudelse, H. Cheng, and J. Biswas 2011a: The leading mode of Indian summer monsoon precipitation during the last millennium. *Geophys. Res. Lett.*, **38**, L15703, doi: 10.1029/2011GL047713.

———, K. G. Cannariato, L. D. Stott, H. Cheng, R. L. Edwards, M. G. Yadava, R. Ramesh, and I. B. Singh, 2007: A 900-year (600 to 1500 A.D.) record of Indian summer monsoon precipitation from the core monsoon zone of India. *Geophys. Res. Lett.*, **34**, L16707, doi: 10.1029/2007GL030431.

———, L. Slott, M. Berkelhammer, H. Cheng, R. L. Edwards, B. Buckley, M. Aldenderfer, and M. Mudelsee, 2011b: A global context for megadroughts in monsoon Asia during the past millennium. *Quat. Sci. Rev.*, **30**, 47–62.

Torrence, C., and G. P. Compo, 1998: A practical guide to wavelet analysis. *Bull. Amer. Meteorol. Soc.*, **79**, 61–78.

Vance, T. R., T. D. van Ommen, M. A. J. Curran, C. T. Plummer, and A. D. Moy, 2013: A millennial proxy record of ENSO and eastern Australian rainfall from the Law Dome ice core, East Antarctica. *J. Climate*, **26**, 710–725.

Walsh, J. E., and W. L. Chapman, 2012: Variability of sea ice cover over decadal and longer timescales. *NTU International Science Conference on Climate Change: Multi-decadal and beyond*, National Taiwan University, Sep 17–21, 193–200.

Wang, Y., H. Cheng, R. L. Edwards, Y. He, X. Kong, Z. An, J. Wu, M. J. Kelly, C. A. Dykoski, and X. Li, 2005: The Holocene Asian monsoon: Links to solar changes and north Atlantic climate. *Science*, **308**, 854–857.

Wang, Y. J., H. Cheng, R. L. Edwards, Z. S. An, J. Y. Wu, C.-C. Shen, and J. A. Dorale, 2001: A high resolution absolute-dated late Pleistocene monsoon record from Hulu Cave, China. *Science*, **294**, 2345–2348.

Webster, P. J., V. O. Magaña, T. N. Palmer, J. Shukla, R. A. Tomas, M. Yanai, and T. Yasunari, 1998: Monsoons: Processes, predictability, and the prospects for prediction. *J. Geophys. Res. Oceans*, **103**, 14451–14510.

Wu Z., and N. E. Huang, 2005: Statistical significance test of intrinsic mode functions. *Hilbert-Huang Transform and Its Applications*, N. E. Huang, and S. S. P. Shen, Eds., World Scientific Publishing Co. Pvt. Ltd., Singapore 596224, 107–125.

———, and ———, 2009: Ensemble empirical mode decomposition: A noise assisted data analysis method. *Advances in Adaptive Data Analysis*, **1**(1), 1–41.

———, ———, J. M. Wallace, B. V. Smoliak, and X. Chen, 2011: On the time-varying trend in global-mean surface temperature. *Clim. Dyn.*, **37**, 759–773.

Wyatt, M. G., S. Kravtsov, and A. A. Tsonis, 2012: Atlantic Multidecadal Oscillation and Northern Hemisphere's climate variability. *Clim. Dyn.*, **38**, 929–949.

Xu, H., Y. Hong, and B. Hong, 2012: Decreasing Indian summer monsoon intensity after 1860 AD in the global warming epoch. *Clim. Dyn.*, **39**, 2079–2088.

Zhang, R, and T. L. Delworth, 2005: Simulated tropical response to a substantial weakening of the Atlantic thermohaline circulation. *J. Climate*, **18**, 1853–1860.

———, and ———, 2006: Impact of Atlantic Multi-decadal Oscillations on India/Sahel rainfall and Atlantic hurricanes. *Geophys. Res. Lett.*, **33**, L17712, doi: 10.1029/2006 GL026267.

———, and ———, 2007: Impact of the Atlantic Multidecadal Oscillation on North Pacific climate variability. *Geophys. Res. Lett.*, **34**, L23708, doi: 10.1029/2007GL031601.

Chapter 22

The South-Flood North-Drought Pattern Over Eastern China and the Drying of the Gangetic Plain

Sumant Nigam, Yongjing Zhao and Alfredo Ruiz-Barradas
Department of Atmospheric Science,
University of Maryland, College Park, MD 20742, USA

Tianjun Zhou
LASG, Institute of Atmospheric Physics,
Chinese Academy of Sciences, Beijing, 100029, China
nigam@umd.edu

The Indo-Gangetic Plain and the lowlands/plains eastward of the Tibetan Plateau exhibit substantial — at times, precarious — trends in summer precipitation since the mid-20th century. These include the declining rainfall over the Gangetic Plain, and a dipole trend pattern over eastern China that is referred as the South-Flood North-Drought (SFND) pattern. The trends have been attributed to increased aerosol and dust loadings, significant land-use land-cover change, and increased greenhouse gas emissions, among others. Interestingly, multidecadal natural variability is not a commonly cited cause. The presence of oppositely-signed trends in the first-half of the 20th century in some of the same regions prompted this reassessment of the role of multidecadal SST variability in generation of rainfall trends over monsoon Asia.

Analysis of the century-long precipitation and SST records suggests that the post-1950 decline in rainfall over the northwestern-central Gangetic Plain and northeastern China is linked with North Pacific decadal SST variability. The rainfall decline over the Himalayan foothills and northeastern Gangetic Plain, and increasing rainfall over the lower Yangtze River basin (the southern SFND center), on the other hand, are found linked with SST Secular Trend (or secular variability) in this analysis.

The SST anomalies associated with North Pacific decadal variability extend well beyond the midlatitude Pacific — into the tropical Indian and Pacific basins where oppositely signed SST anomalies are present. It is hypothesized that the tropical Indian Ocean SST anomalies modulate the meridional ocean-continent thermal contrast, and thus the large-scale distribution of monsoon rainfall over the Asian continent. The findings need corroboration from additional analyses of the observed circulation and thermodynamics fields, and the hypothesis support from controlled climate model experiments.

1. Introduction

The Indo-Gangetic Plain extending from eastern Pakistan to the Bay of Bengal and the lowlands and plains eastward of the Tibetan Plateau are home to two billion people, and the agricultural heartland of Asia. Rapid population and economic growth have led to significant land-use land-cover change, heightened demand for fresh water and energy, degraded environment from dust and aerosol loadings, and increased greenhouse gas emissions.

Climate Change: Multidecadal and Beyond
edited by Chih-Pei Chang et al.

Regional hydroclimate (precipitation, surface air temperature, soil moisture, subsurface water) exhibits substantial — at times, precarious — trends across the second half of the 20th Century. The post-1950 summer rainfall trend (Fig. 1) is impressive over the Ganges River basin as it represents a decline of 2 mm/day over a 60-year period against climatology of 8–10 mm/day (not shown). A coherent meridional dipole over eastern China, with significant drying in the middle and lower reaches of the Yellow River (where climatological rainfall is less than 4 mm/day) and increasing rainfall across and to the south of the Yangtze River is also evident. The anomalous rainfall pattern is often referred as the South-Flood North-Drought (SFND) pattern (Yu and Zhou, 2007; Zhou *et al.*, 2009a).

The SFND pattern and the drying of the Gangetic Plain are robust trends given their unambiguous presence in independent analyses of observed precipitation. Figure 1 shows the trends in CRU TS3.1 (0.5°, Mitchell and Jones, 2005), WCRP GPCC (0.5°, Schneider *et al.*, 2011), University of Delaware (0.5°, Willmott and Matsuura, 1995), Chinese Meteorological Administration (station data), and the Aphrodite (0.5°, Yatagai *et al.*, 2009) analyses. The Gangetic Plain trend is more variable of the two, and likely overestimated in the CRU analysis. The CRU data is nonetheless analyzed in view of our interest in the surface air temperature trends as well.

The SFND pattern is manifest also in the Yellow and Yangtze River discharge and other hydroclimate variables despite anthropogenic impacts (*e.g.*, water withdrawal). The annual river discharge at downstream stations in eastern China is shown in Fig. 2, while recent trends in the surface water equivalent (SWE) from the GRACE satellite gravity solutions are shown in Fig. 3; both supporting the SFND rainfall trend pattern. The GRACE solutions also capture the declining water stores in the Gangetic Plain.

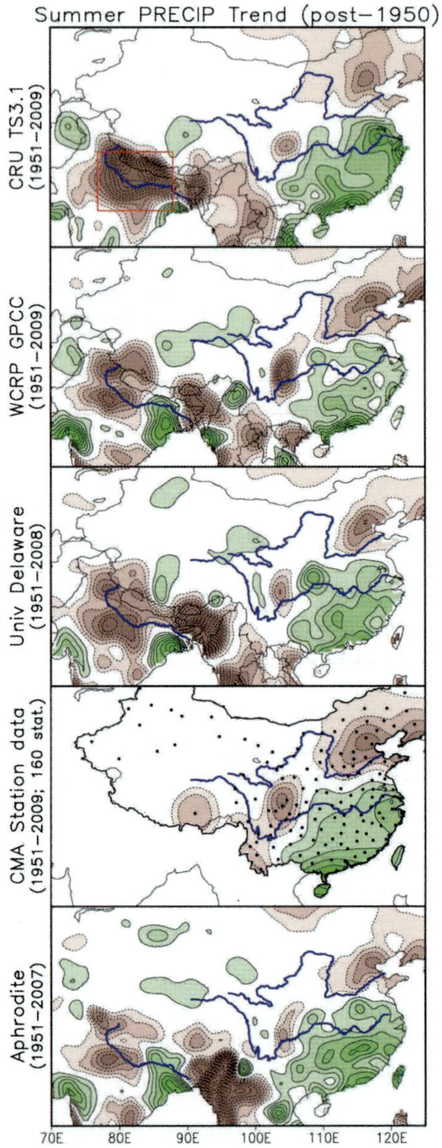

Fig. 1. The linear trend in summer (JJA) rainfall in the post-1950 period in the CRU TS 3.1, WCRP GPCC, University of Delaware, Chinese Meteorological Administration (CMA), and the Aphrodite analysis of observed precipitation. The trend over the recent 60-year period is shown; the exact period is stated in the left labels. The trend is plotted at 0.5 mm/day/century interval, with deficit/surplus shown in brown/green, and the zero-contour omitted. The Ganges River in the Indian subcontinent and the Yangtze and Yellow rivers in China are marked in blue for regional reference. The South-Flood North-Drought (SFND) pattern over eastern China, and the drying of the Gangetic Plain is manifest in various analyses.

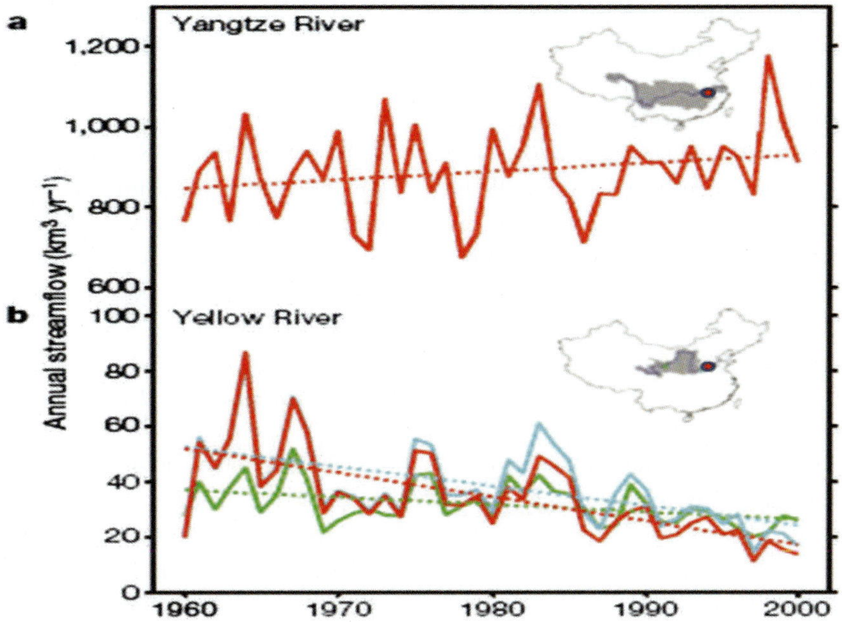

Fig. 2. Interannual variations in annual runoff during 1960–2000 at the Datong station in the lower Yangtze River (top), and Gaocun station in the lower Yellow River (bottom), both in red; from Piao *et al.* (2010); the stations are marked by solid dots. The green and blue curves in the lower panel are for the upstream stations on the Yellow river.

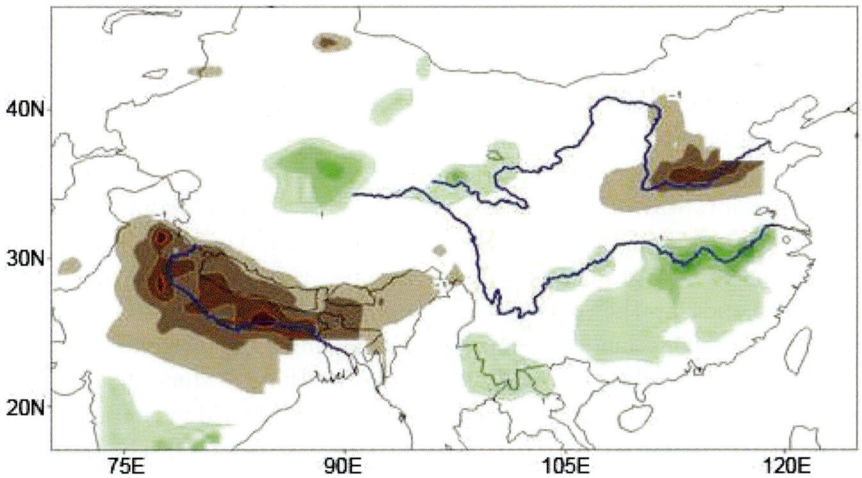

Fig. 3. Linear trend in the summer terrestrial water equivalent thickness during 2003–2010 from the GRACE satellite gravity solutions. Brown (greens) denote depletion (accumulation) of ground water; contour interval is 1 cm/year and the zero-contour omitted. Note the declining trend over the Gangetic Plain and the South-Flood North-Drought pattern over eastern China.

The summer circulation over the Asian continent and the surrounding oceans also exhibits significant trends during the same period, *e.g.*, westward expansion of the Western Pacific Subtropical High (WPSH) since the 1970s. The expansion was attributed to the warming of the Indian Ocean and the Western Pacific, from modeling of the atmospheric response of specified SST anomalies (Zhou *et al.*, 2009b). This modeling analysis (with idealized SST anomalies) was however unable to generate the SFND rainfall pattern or the drying of the Gangetic Plain in a convincing manner.

Is the steep decline in summer monsoon rainfall over the Gangetic Plain and northeastern China indicative of accelerated anthropogenic effects (from human imprints on greenhouse gas emissions, aerosol/dust loadings, land-use land-cover degradation, and groundwater irrigation), multidecadal ocean variability, or both?

2. Possible Mechanisms

2.1. *Aerosols*

Increasing aerosol loadings over South Asia since the 1950s have been advanced as one cause of the declining summer precipitation trend over the Gangetic Plain. Ramanathan *et al.* (2005) have emphasized the short wave (SW) absorption of black carbon aerosols, and related surface 'dimming', which impacts SST gradients in the northern Indian Ocean and thus the South Asian monsoon; Meehl *et al.* (2008) modeling analysis corroborates this assessment. The aerosol SW absorption also leads to heating of the lower-to-mid troposphere, altering static stability, tropospheric temperature gradients, and thus the monsoon (Menon *et al.*, 2002; Lau *et al.*, 2006). Bollasina *et al.* (2011), on the other hand, argue for the importance of aerosol SW reflection and the resulting cooling of the Asian continent to be the cause of declining summer rainfall over South Asia in the post-1950 period.

The aerosols generate declining summer rainfall also over East Asia in the Bollasina *et al.* analysis, in contrast with the observed increasing trend in regional rainfall (the South-Flood part of the SFND rainfall trend pattern). Atmospheric model simulations produced with historical forcing (prescribed SSTs, greenhouse gases, and direct aerosol effects) have also been unsuccessful in generating the SFND pattern (Li *et al.*, 2010).

2.2. *Land-use land-cover change*

Rainfall decline over the Gangetic Plain has also been attributed to increasing desertification over northwestern India. Bollasina and Nigam (2011) found an expanded Thar Desert linked with reduced summer monsoon rainfall over eastern India in a modeling analysis of regional hydroclimate change over the Indian subcontinent. The low-level anticyclone forced by increased desert subsidence generates northwesterly flow over the Gangetic Plains, weakening the climatological monsoon circulation (and moisture transports) over eastern India.

2.3. *Eurasia and Tibetan Plateau*

Springtime snow cover and snow depth over Eurasia and the Tibetan Plateau are also known to influence the South Asian (Bamzai and Shukla, 1999) and East Asian (Liu and Yanai, 2002b; Zhang *et al.*, 2004) summer monsoons. For example, increase in the spring snow depth over the Plateau since the late 1970s was linked with the westward expansion of the WPSH, and in turn to increased summer rainfall over the Yangtze River basin. As westward expansion of the WPSH was attributed to the Indian Ocean and Western Pacific warming from modeling studies (Zhou *et al.*, 2009b), the SST-warming in these basins has been suggested as a cause of the spring snow depth increase over Tibet (Zhang *et al.*, 2004). The weakened heating of the Tibetan Plateau contributes to weakening of the East Asian summer monsoon circulation by reducing ocean-continent thermal contrast (Ding *et al.*, 2007; Duan and Wu, 2008).

2.4. *Low-frequency climate variability*

The climate system, including the Asian monsoon, exhibits multidecadal variability. Wang *et al.* (1981) and Zhu and Wang (2002) analyses of observed and reconstructed summer rainfall reveal the existence of an approximately 80-year variability timescale. Ding *et al.* (2007) have argued that the rainfall decline over northeastern China in the second half of the 20th century is a reflection of this intrinsic multidecadal variability. This idea has some support from the diagnostic analysis of Lei *et al.* (2011) who find natural decadal variability to be generally dominant. However, the increasing intensity of rainfall throughout China and the decrease in light rainfall days, particularly in the north, is, at least, partially of anthropogenic origin according to their analysis. The Atlantic Multidecadal Oscillation may also contribute to the multidecadal variability of the East Asian monsoon (Lu *et al.*, 2006).

3. Simulation of Monsoon Rainfall Trends by Climate System Models

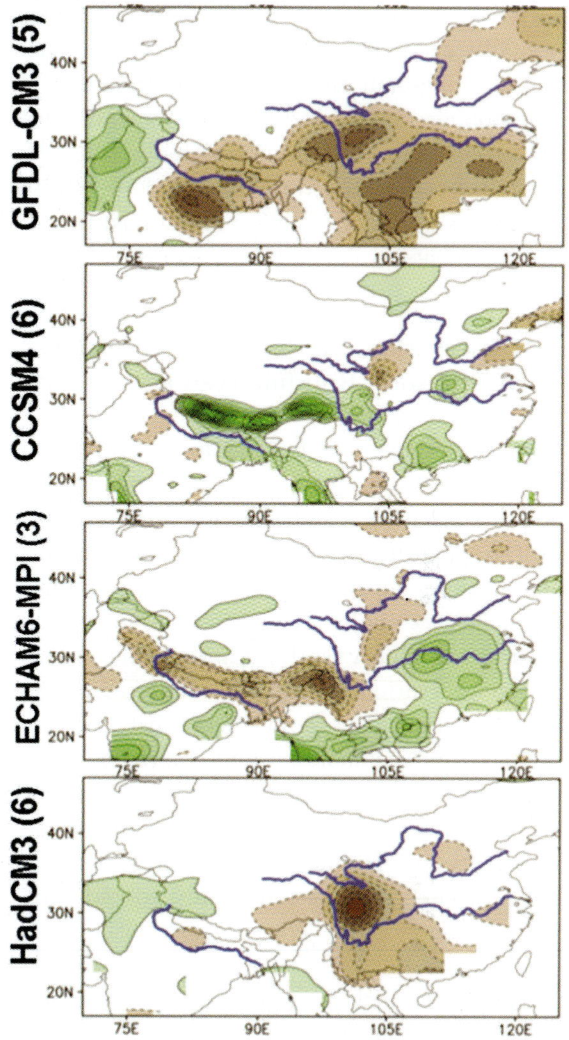

Fig. 4. The 1951–1999 linear trend in summer precipitation in the ensemble-mean CMIP5 simulations of 20th century climate by four leading IPCC-AR5 models. The number of ensemble members averaged in producing the ensemble-mean is indicated in brackets next to the model name. The trend is plotted at 0.5 mm/day/century interval, with deficit/surplus shown in brown/green, and the zero-contour omitted.

The declining rainfall over the Gangetic Plain and the SFND trend over eastern China results from multiple influences, including anthropogenic ones. The relative contribution of these influences can be best assessed from experiments with the climate system models provided the modeled 20th century climate (forced by time-dependent greenhouse gas concentrations, aerosols and dust loadings, volcano injections, land-use land-cover, etc.) realistically captures the above-described subcontinental-scale, multidecadal monsoon precipitation trends.

The IPCC-AR5 models, unfortunately, do not rise to this challenge even via their ensemble-means. As shown in Fig. 4, only the GFDL-CM3 model captures the drying of the Gangetic Plain but it is unable to generate the SFND trend over eastern China. In fact, this model produces severe drying over the South-Flood region! There is a weak hint of the SFND trend in the ECHAM6-MPI simulation, which also contains elements of the Gangetic Plain rainfall trends. The climate system models limited skill in simulating the long-term trends in monsoon circulation and rainfall stems, in part, from their inability to generate realistic modes of multidecadal SST variability and/or related circulation and hydroclimate teleconnections to the continents.

4. Observationally Rooted Attribution of Monsoon Rainfall Trends

Climate system models are rapidly improving but evidently not ready for providing insights into the origin of regional hydroclimate trends. In the interim, analysis of the 20th century observations can perhaps provide a preliminary assessment of the contribution of multidecadal natural variability and anthropogenic change in the generation of rainfall trends over Asia. Such an assessment would necessarily have to be based on a climate system component with large thermal inertia (*e.g.*, ocean heat-content, or, perhaps, even just SST) so that multidecadal variability and secular change components can be unraveled. An observationally rooted analysis strategy is, of course, not without its own caveats, including the assumption of non-interaction (and thus separability) of the non-stationary secular trend and natural variability components.

4.1. *Spatiotemporal SST analysis*

Pacific and Atlantic SSTs were analyzed to obtain refined evolution descriptions from contextual separation of natural variability and the secular trend without advance filtering (and potential aliasing) of the SST record. The extended-EOF analysis technique (EEOF, Weare and Nasstrom, 1982) yields robust characterization of all non-seasonal modes of SST variability, including secular trend, from a single analysis of unfiltered data (Guan and Nigam, 2008). Mode physicality was assessed via correlations with the Pacific fish recruitment records, following Hare and Mantua (2000). The correlations were found comparable, if not larger, than those obtained by these authors; analog counts were also used in assessing mode physicality (Guan and Nigam, 2008). Atlantic SSTs were similarly analyzed but after removing the influence of Pacific SST variability and the SST Secular Trend from the Atlantic SST anomalies, leading, among others, to a clarified structure of Atlantic Multidecadal Oscillation (AMO; Guan and Nigam, 2009).

The Pacific SST principal components are shown in Fig. 5. By focusing on spatial and temporal recurrence but without imposition of periodicity constraints, EEOF analysis discriminates between biennial, El Nino Southern Oscillation (ENSO), and multidecadal variabilities in the Pacific: Canonical ENSO variability is captured as two modes: growth (ENSO) and decay (ENSO+). Deviation from canonical development, especially in the 1976/77-onward period is identified as a distinct mode, the ENSO non-canonical mode (ENSONC). Pacific decadal variability is resolved into two modes, Pan Pacific (PDVPP) and North Pacific (PDVNP). The former exhibits connections to the tropical-subtropical Atlantic resembling some AMO features. The latter captures the 1976/77 climate-shift, and is very close to Pacific Decadal Oscillation (PDO, Mantua *et al.*, 1997) in structure but with interesting links to the North Atlantic as well as the western tropical Pacific and Indian Ocean (Guan and Nigam, 2008). The nonstationary secular trend (SST-Trend) consists of widespread but non-uniform warming of all basins along with a sliver of cooling in the central equatorial Pacific.

4.2. *Rainfall reconstruction*

Reconstruction is based on the linear, summer regressions of the 11 SST PCs (7 Pacific and 4 Atlantic) on contemporaneous rainfall; the

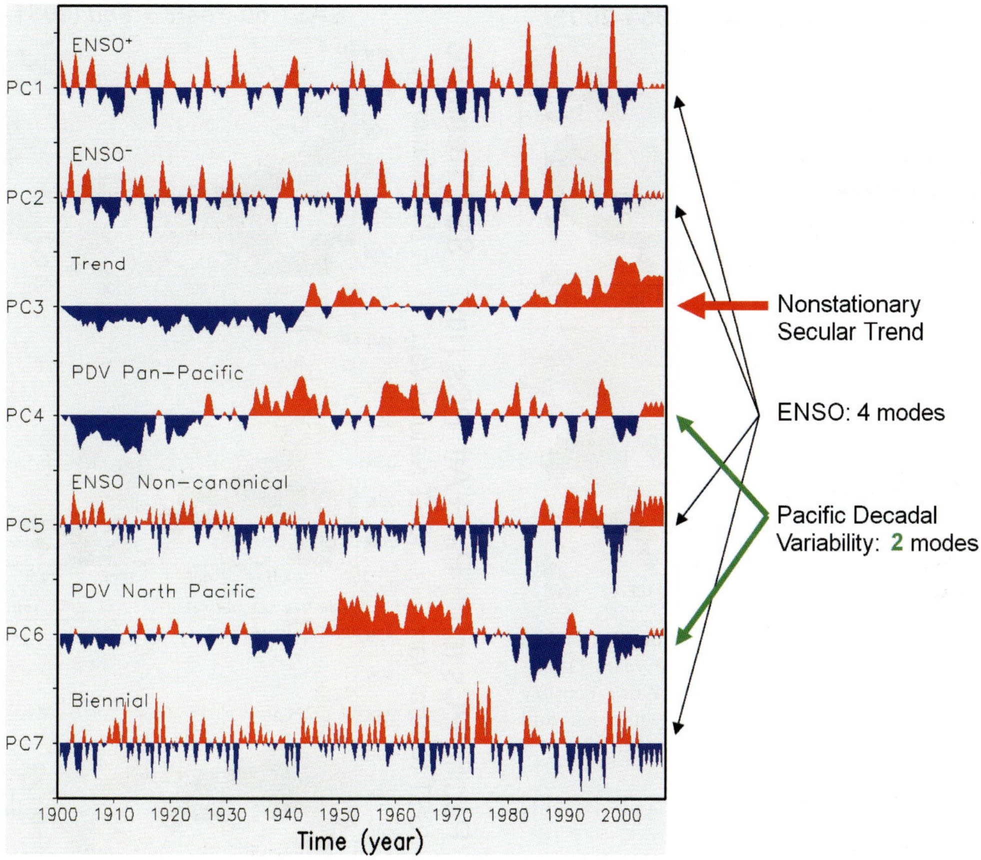

Fig. 5. SST principal components (PCs) from the extended-EOF (EEOF) analysis of HadISST 1.1 data. Seasonal anomalies during January 1900 — August 2009 were analyzed in the Pacific basin (20S-60N, 120E-60W) using 5-season-long anomaly sequences. The 7-leading PCs were rotated using the VARIMAX method. The EEOF technique is equivalent to multichannel singular spectrum analysis. See Guan and Nigam (2008, 2009) for more details.

regression period is summer 1901 to summer 2009, consistent with the CRU TS3.1 data span. The intra-basin PCs are temporally orthogonal (assured by the analysis method) while the inter-basin ones are nearly so (ensured by the filtering of Pacific's influence from Atlantic SSTs prior to their analysis; largest inter-basin PC correlation is 0.08), facilitating reconstruction of the rainfall anomalies. Multiplication of each SST PC (time varying) with its summer rainfall regression pattern (time invariant), and summing the 11 contributions yields the summer rainfall anomalies in this reconstruction. A similar strategy was recently used to reconstruct tropical cyclone counts in the Atlantic (Nigam and Guan, 2011) and precipitation variations (droughts) over the Great Plains (Nigam *et al.*, 2011).

The linear trend in the observed and reconstructed summer rainfall in the post-1950 period is compared in Fig. 6. The key trend features, including declining rainfall over the Gangetic Plain and the SFND pattern over eastern China, are robustly manifest in the reconstruction. Rainfall trends in the region south of the Yangtze River and further south over Indo-China are however somewhat weaker in the reconstruction.

Is the notable reconstruction of rainfall trends a result of over-fitting data with 11 SST

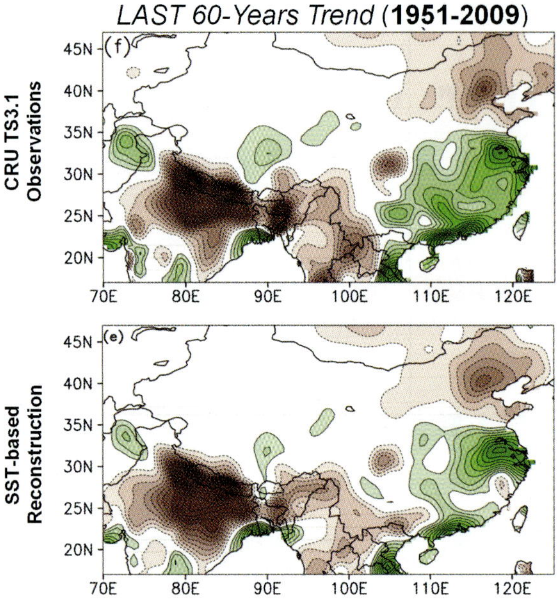

Fig. 6. The observed and reconstructed linear trend in summer monsoon rainfall over the last 60 years (1950–2009). The trend is plotted at 0.375 mm/day/century interval, with deficit/surplus shown in brown/green, and the zero-contour omitted.

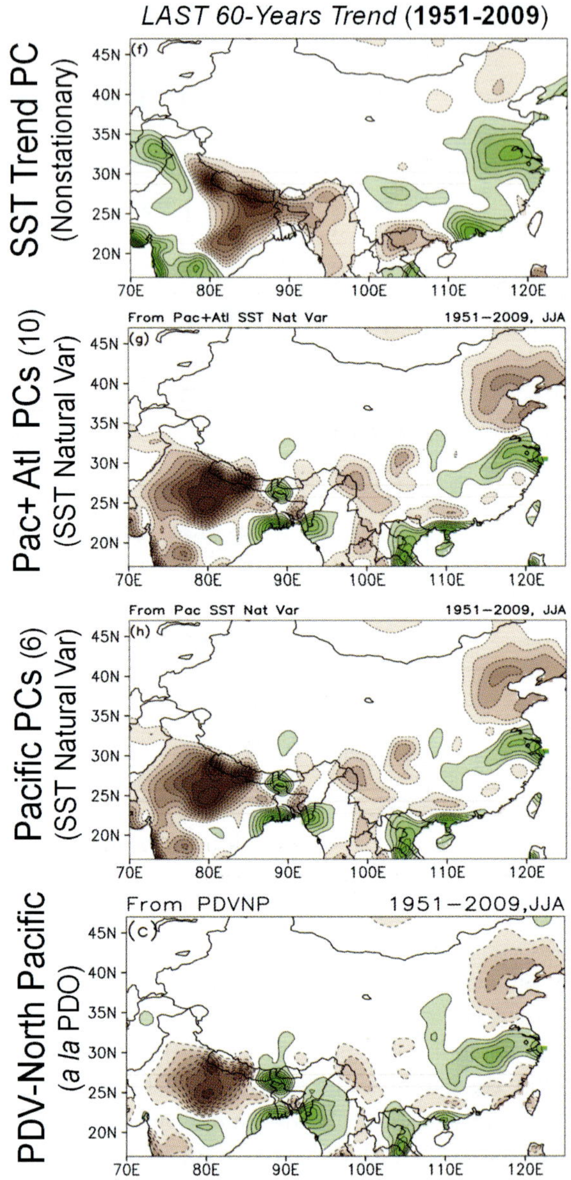

Fig. 7. Analysis of the reconstructed summer rainfall trend shown in Fig. 6 (bottom panel). The contribution of the SST Secular Trend is shown in the upper panel while that of SST natural variability in the Pacific and Atlantic basins in the second-from-top panel. Contribution of just Pacific SST natural variability is shown in the second-from-bottom panel. The trend contribution of just North Pacific decadal SST variability is shown in the bottom panel. The trend is plotted at 0.375 mm/day/century interval, with deficit/surplus shown in brown/green, and the zero-contour omitted.

PCs, or are just a few SST PCs key to this reconstruction?

4.3. *Analysis of reconstructed rainfall trend*

The reconstructed rainfall trend in the post-1950 period is parsed into 4 components in Fig. 7: The SST Secular Trend contribution is shown in the top panel while that of the remaining 10 SST PCs (referred together as Pacific and Atlantic SST natural variability) is in the second-from-top panel. The characterization of AMO as natural variability (*e.g.*, as above) has been challenged (Booth *et al.*, 2012) but this issue is moot in context of the present analysis in view of AMOs negligible contribution in the reconstruction of multidecadal precipitation trends (as shown next).

Comparison of the two indicates the considerable influence of both components. The

SST Secular Trend PC is linked with the drying of the Himalayan foothills and northeastern Gangetic Plain, and with increasing rainfall in the western subcontinent. Over China, this PC is linked with increasing rainfall over the lower Yangtze River basin, i.e., over the southern center of the SFND dipole. Its influence over northeastern China is however modest.

The Pacific and Atlantic SST natural variability is associated with strongly declining rainfall over the northwestern-central Gangetic Plain. Over eastern China, this SST variability component is linked with the robust drying of the northeast, i.e., with the northern center of the SFND dipole. It is also linked with increasing rainfall but only over the immediate Yangtze River basin in its lower reaches.

The second-from-the-bottom panel of Fig. 7 shows the monsoon rainfall trends associated with SST natural variability in just the Pacific basin (or 6 SST PCs). Comparison of the two middle panels shows Atlantic SST variability to be minimally influential on the South Asian and East Asian summer-monsoon rainfall trends.

The notable reconstruction of summer rainfall trends (Fig. 6) thus results from contemporaneous regressions of just 7 SST PCs (SST Secular Trend and 6 Pacific PCs). Further analysis of the Pacific contribution (Fig. 7, bottom panel) shows that it results almost entirely from the influence of North Pacific decadal variability, referred as PDV-North Pacific or just PDV-NP in Guan and Nigam (2008) and in the figures here; its PC closely tracks the PDO index, as noted earlier. The post-1950 rainfall trends over monsoon Asia can thus be reconstructed from just 2 SST PCs, obviating any concerns regarding over-fitting of data.

5. North Pacific Decadal SST Variability and Asian Summer Monsoon Rainfall Trends

In view of the singular importance of North Pacific decadal SST variability (PDV-NP) for the summer monsoon rainfall trends over South Asia and East Asia (cf. Fig. 7), its SST and rainfall signatures are shown in Fig. 8, along with those of the PDO index. It is noteworthy that these were obtained from regressions on summer (JJA-averaged) anomalies over the full century (1901–2009 period), and not just the post-1950 period over which trends were analyzed. The PDV-NP links are shown for its negative phase to facilitate comparison with the PDO ones and the earlier displayed rainfall trends. A cooler midlatitude Pacific, with a band of cold SSTs extending eastward from Japan (until 140°W), characterizes the SST signal. The contemporaneous precipitation links depict the unmistakable rainfall deficit over the northwestern-central Gangetic Plain and northeastern China. An excessive rainfall signal over the immediate lower Yangtze River basin is also manifest in the PDV-NP regressions but not as robustly in the PDO ones.

The SST and precipitation anomalies shown in Fig. 8 are contemporaneous, leaving the cause and effect question open. Even if SSTs were assumed the cause, elaboration on the mechanism by which midlatitude Pacific SST anomalies impact rainfall over monsoon Asia would be challenging, as SST anomalies in only the Tropics are generally considered causative. Interestingly, the PDV-North Pacific is not without tropical links, notwithstanding its name. Guan and Nigam (2008) showed PDV-NP linked with the tropical Indian and Pacific basins; see also Deser *et al.* (2004). Figure 12 of Guan and Nigam shows SST regressions and correlations of the PDV-NP SST principal component. The correlation map readily reveals the tropical links, especially in the Indian Ocean and Western Pacific where smaller amplitudes of SST variability create a detection challenge, but not for the correlation statistic.

The tropical SST links of PDV-NP are, moreover, shown to be of opposite sign to the midlatitude ones (Guan and Nigam, 2008; Fig. 12). Declining summer rainfall over the

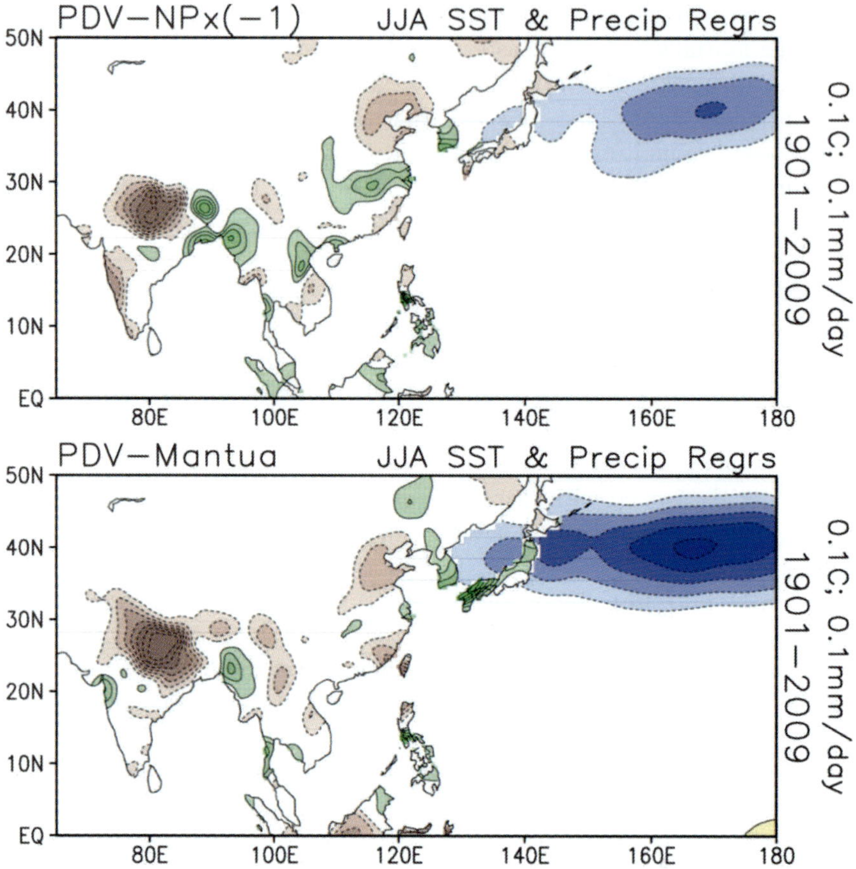

Fig. 8. The summer (JJA) precipitation and SST regressions of Pacific Decadal Variability (PDV) for the 1901–2009 period. The top panel shows regressions of the PDV-North Pacific principal component (for the opposite phase shown in Fig. 5), while the lower panels shows the same for the Pacific Decadal Oscillation (PDO) index, a commonly used marker for decadal SST variability in the North Pacific. The SST regressions are contoured with 0.1K interval, with cold/warm anomalies shown in blue/red. Precipitation regressions are contoured with a 0.1 mm/day interval, with deficits/surplus shown in brown/green. The zero-contour is omitted, as before.

northwestern-central Gangetic Plain and northeastern China is thus also linked with a warmer Indo-Pacific basin to the immediate south of the Asian Continent. Although the mechanisms by which the multi-basin SST footprint of PDV-NP influences regional summer rainfall over the Asian Continent remains to be elucidated, at least the sign of the SST anomalies in the tropical Indian and Western Pacific basins is consistent with the basic understanding of the monsoon mechanism — that large-scale land-ocean thermal contrast is the driver. Anomalously warm SSTs to the south of the Asian landmass would weaken the summertime contrast, leading to large-scale reduction in monsoon rainfall. The observed summer rainfall trends however exhibit complex spatial structure, reflecting the significance of regional interactions rooted in heterogeneous terrain and coastlines, and massive orographic complexes, including Tibetan Plateau. Clearly, an investigation of the origin of the post-1950 summer rainfall trends over the Asian Continent will require controlled modeling experiments with a climate system model that can generate realistic patterns of seasonal, interannual,

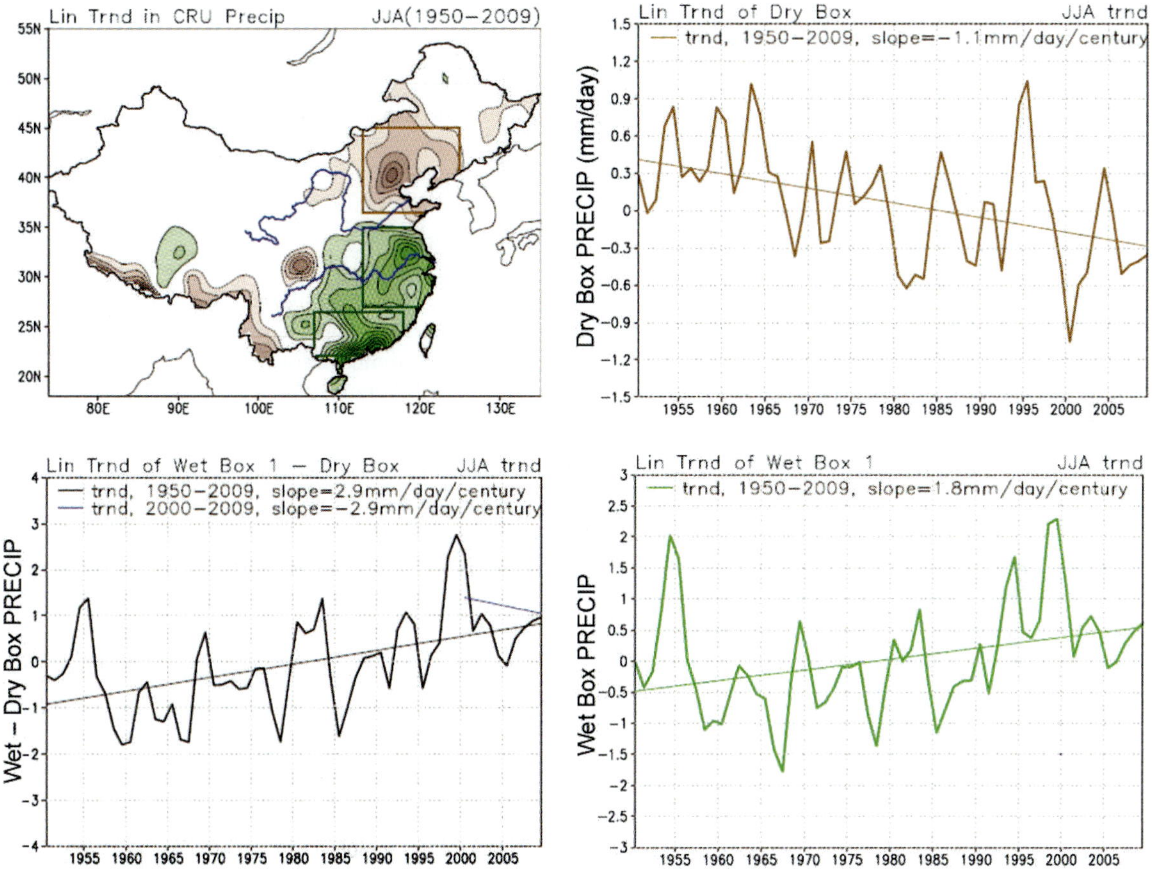

Fig. 9. Area-averaged summer (JJA) precipitation anomaly with respect to the 1901–2009 mean in the North-Drought (Dry Box, top-right panel) and South-Flood (Wet Box 1, bottom-right) core region of the SFND trend pattern. The boxed regions are marked with brown and green lines in the top-left panel which shows the 1950–2009 period linear trend in summer rainfall at intervals of 0.5 mm/day/century; Wet Box 1 refers to the box encompassing the lower Yangtze River. The Wet Box 1 minus Dry Box precipitation anomaly a marker of the strength of the SFND pattern is plotted in the bottom-left panel. All precipitation anomalies are plotted in mm/day units, and the post-1950 period linear trend is plotted in each panel with the slope noted. The recent 10-year (2000–2009) slope is also shown in the bottom-left panel as its negative sign has prompted some to suggest that the SFND pattern is weakening; however such short-period weakening is not uncommon in the considered period (*e.g.*, 1955–1965).

and decadal/multidecadal variability in regional rainfall — a challenging target for the current model class (IPCC-AR5), as shown in Fig. 4.

6. Concluding Remarks

Analysis of the century-long precipitation and SST records suggests that the recent (post-1950) decline in rainfall over the northwestern-central Gangetic Plain and northeastern China is linked with North Pacific decadal SST variability (closely related to the PDO). The rainfall decline over the Himalayan foothills and northeastern Gangetic Plain, and increasing rainfall over the lower Yangtze River basin (the southern SFND center), on the other hand, are found linked with SST Secular Trend (or secular variability).

An elucidation of the mechanisms generating the indicated links would require elaborate observational analyses keyed to lead-lag regressions on 3D thermodynamic and circulation fields, many of which are unfortunately unavailable for a significant portion of the 20th century. The SST links of the declining summer rainfall over the Gangetic Plain are however broadly consistent with the rudimentary monsoon mechanism, imparting some dynamical support for the observational findings. Controlled experiments with climate system models that replicate Asian precipitation and circulation climatology, interannual variability, and multidecadal trends can be potentially insightful but such models are presently unavailable, as noted earlier.

This investigation of the role of decadal SST variability in generation of multidecadal rainfall trends was instigated by the findings of both increasing and decreasing trends over the Gangetic Plain; the former in the first-half of the 20th century. The analysis of increasing trends in the first half-century is not reported for reasons of space, but reconstruction of rainfall trends in this period suggests a significant role of the Pan-Pacific mode of decadal SST variability (PDV-PP; see Fig. 11 and related discussion in Guan and Nigam, 2008). The PDV-PP has robust footprints also in the western Atlantic resembling AMOs tropical structure; the AMO and PDV-PP are correlated at $\sim$0.45.

Finally, inspection of the rainfall record has led some to conclude that the SFND trend pattern has weakened recently. The area-averaged rainfall anomaly over the SFND core regions is plotted in Fig. 9, separately for the North-Drought and South-Flood sectors to investigate this issue. The right panels in Fig. 9 do not show any notable modulation of the decreasing and increasing rainfall trends; nor is trend modulation apparent in the wet-minus-dry index (bottom-left panel). To be sure, this index exhibits decadal fluctuations, including periods of abatement but not just in the recent decade; for example, note the index decline during 1955–1965.

The present analysis is focused on the half-century-long trends; not decadal ones which could easily result from the interference of interannual and decadal variabilities.

References

Bamzai, A., and J. Shukla, 1999: Relation between Eurasian Snow Cover, Snow Depth, and the Indian Summer Monsoon: An Observational Study. *J. Climate*, **12**, 3117–3132.

Bollasina, M. A., Y. Ming, and V. Ramaswamy, 2011: Anthropogenic aerosols and the weakening of the South Asian summer monsoon. *Science*, **334**, 502–505 (28 October 2011).

———, and S. Nigam, 2011: Modeling of Regional Hydroclimate Change over the Indian Subcontinent: Impact of the Expanding Thar Desert. *J. Climate*, **24**, 3089–3106.

Booth, B. B. B., N. J. Dunstone, P. R. Halloran, T. Andrews, and N. Bellouin, 2012: Aerosols implicated as a prime driver of twentieth-century North Atlantic climate variability. *Nature*, **484**, 228–232 (12 April 2012).

Deser, C., A. S. Phillips, and J. W. Hurrell, 2004: Pacific Interdecadal Climate Variability: Linkages between the Tropics and the North Pacific during Boreal Winter since 1900. *J. Climate*, **19**, 3109–3124.

Ding, Y. H., Z. Wang, and Y. Sun, 2007: Interdecadal variation of the summer precipitation in East China and its association with decreasing Asian summer monsoon. Part I: Observed evidences. *Int. J. Climatol.*, **28**, 1139–1161, doi:10.1002/joc.1615.

Duan, A. M., and G. X. Wu, 2008: Weakening Trend in the Atmospheric Heat Source over the Tibetan Plateau during Recent Decades. Part I: Observations. *J. Climate*, **21**, 3149–3164.

Guan, B., and S. Nigam, 2008: Pacific Sea Surface Temperatures in the Twentieth Century: An evolution-centric analysis of variability and trend. *J. Climate*, **21**, 2790–2809.

———, 2009: Analysis of Atlantic SST variability factoring inter-basin links and the secular trend: Clarified structure of the Atlantic Multidecadal Oscillation. *J. Climate*, **22**, 4228–4240.

Hare, S. R., and N. J. Mantua, 2000: Empirical evidence for North Pacific regime shifts in 1977 and 1989. *Prog. Oceanogr.*, **47**, 103–145.

Lau, K.-M., and K.-M. Kim, 2006: Observational relationships between aerosol and Asian monsoon rainfall, and circulation, *Geophys. Res. Lett.*, **33**, L21810, doi:10.1029/2006GL027546.

Lei, Y., B. Hoskins, and J. Slingo, 2011: Exploring the interplay between natural decadal variability and anthropogenic climate change in summer rainfall over China. Part I: Observational evidence. *J. Climate*, **24**, 4584–4599.

Li, H., A. Dai, T. Zhou, and J. Lu, 2010: Responses of East Asian summer monsoon to historical SST and atmospheric forcing during 1950–2000. *Climate Dyn.*, **34**, 501–514.

Liu, X. D., and M. Yanai, 2002: Influence of Eurasian spring snow cover on Asian summer monsoon rainfall. *Int. J. Climatol.*, **25**, 1075–1089.

Lu, R., B. Dong, and H. Ding, 2006: Impact of the Atlantic Multidecadal Oscillation on the Asian summer monsoon. *Geophys. Res. Lett.*, **33**, L24701, doi:10.1029/2006GL027655.

Mantua, N. J., S. R. Hare, Y. Zhang, J. M. Wallace, and R. C. Francis, 1997: A Pacific interdecadal climate oscillation with impacts on salmon production. *Bull. Amer. Meteor. Soc.*, **78**, 1069–1079.

Meehl, G. A., J. M. Arblaster, and W. D. Collins, 2008: Effects of black carbon aerosols on the Indian monsoon, *J. Climate*, **21**, 2869–2882.

Menon, S., J. Hansen, L. Nazarenko, and Y. Luo, 2002: Climate effects of black carbon aerosols in China and India, *Science*, **297**, 2250–2253.

Mitchell T. D., and P. D. Jones, 2005: An improved method of constructing a database of monthly climate observations and associated high-resolution grids. *Int. J. Climatol.*, **25**, 693–712.

Nigam, S., B. Guan, and A. Ruiz-Barradas, 2011: Key role of the Atlantic Multidecadal Oscillation in 20th century drought and wet periods over the Great Plains. *Geophys. Res. Lett.*, **38**, L16713, doi:10.1029/2011GL048650.

______, and B. Guan, 2011: Atlantic Tropical Cyclones in the 20th Century: Natural Variability and Secular Change in Cyclone Count. *Climate Dynamics*, **36**, 2279–2293, doi: 10.1007/s00382-010-0908-x.

Piao, S., *et al.*, 2010: The impacts of climate change on water resources and agriculture in China. *Nature*, **467**, 43–51.

Ramanathan, V., and Coauthors, 2005: Atmospheric Brown Clouds: Impacts on South Asian Climate and Hydrological Cycle. *PNAS*, **102**, 5326–5333.

Schneider, U., T. Fuchs, A. Meyer-Christoffer, and B. Rudolf, 2008: Global precipitation analysis products of the GPCC. Global Precipitation Climatology Centre (GPCC), DWD, *Internet Publication*, 1–12.

Wang, S. W., Z. Zhao, and Z. Chen, 1981: Reconstruction of the summer rainfall regime for the last 500 years in China. *Geophys. J.*, **52**, 117–122.

Weare, B. C., and J. S. Nasstrom, 1982: Examples of extended empirical orthogonal function analyses. *Mon. Wea. Rev.*, **110**, 481–485.

Willmott, C. J., and K. Matsuura, 1995: Smart Interpolation of Annually Averaged Air Temperature in the United States. *J. Appl. Meteorol.*, **34**, 2577–2586.

Yatagai, O. Arakawa, K. Kamiguchi, H. Kawamoto, M. I. Nodzu, and A. Hamada, 2009: A 44-year daily precipitation dataset for Asia based on dense network of rain gauges. *SOLA*, **5**, 137–140, doi:10.2151/sola.2009-035.

Yu, R. C., and T. J. Zhou, 2007: Seasonality and three-dimensional structure of the interdecadal change in East Asian monsoon, *J. Climate*, **20**, 5344–5355.

Zhang, Y. S., T. Li, and B. Wang, 2004: Decadal change of the spring snow depth over the Tibetan Plateau: The associated circulation and influence on the East Asian summer monsoon. *J. Climate*, **17**, 2780–2793.

Zhou, T., D. Gong, J. Li, and B. Li, 2009a: Detecting and understanding the multidecadal variability of the East Asian summer monsoon — Recent progress and state of affairs. *Meteor.Zeitschrift*, **18**, 4, 455–467.

______, R. Yu, J. Zhang, H. Drange, C. Cassou, C. Deser, D. L. R. Hodson, E. Sanchez-Gomez, J. Li, N. Keenlyside, X. Xin, and Y. Okumura, 2009b: Why the western pacific subtropical high has extended westward since the late 1970s. *J. Climate*, **22**, 2199–2215.

Zhu, J., and S. Wang, 2002: 80 yr oscillation of summer rainfall over North China and East Asian Summer Monsoon. *Geophys. Res. Lett.*, **14**, 1672, doi:10.1029/2001GL013997.

Chapter 23

Impacts of Aerosols on the Asian Monsoon — An Interim Assessment

William K. M. Lau* and Kyu-Myong Kim

Earth Science Division, NASA/Goddard Space Flight Center
Greenbelt, MD 20771, USA
Earth System Science Interdisciplinary Center (ESSIC), University of Maryland,
College Park, MD 20742, USA
wkmlau@umd.edu

In this chapter, we assess the current state of knowledge of impacts of aerosols on the Asian monsoon. Evidences from observations and modeling studies are presented showing that aerosols can alter monsoon water cycle through direct (radiative) effects and indirect (microphysical) effects. For the South Asian monsoon, the solar dimming (SDM) by aerosols may reduce meridional surface thermal gradient, increase atmospheric stability and suppress rainfall, thereby weaken the monsoon. However atmospheric heating and induced dynamical feedback by absorbing aerosols (dust and black carbon) via the so-called Elevated Heat Pump (EHP) can strengthen the meridional tropospheric temperature gradient and enhance monsoon rainfall in northern India in May–June, subsequently altering the evolution of the monsoon. For the East Asian monsoon, studies have suggested that aerosol direct radiative forcing may have contributed to north/dry-south/wet pattern in recent decades, and that SDM effects may be intertwined with downstream teleconnections emanating from the EHP effects over northern India. Both monsoons are likely to be strongly affected by sea surface temperature changes induced by aerosol SDM, as well as from long-term natural and anthropogenic climate forcing. For the Asian monsoon as a whole, aerosol impacts on regional and local monsoon water cycle and climate are highly significant, if not more so compared to warming by greenhouse gases.

1. Introduction

Floods and droughts cause major loss in lives and properties, crop yield, and gross national productivity. Aerosols associated with air and water pollution lead to harmful effects on human and plant health, and degradation of the environment. In recent decades, the rapid industrialization and modernization in Asian countries, have greatly increased the loading of aerosols in the atmosphere, and aggravated the adverse effects of aerosols on humans and the environment. The IPCC report (IPCC 2007) and many studies have shown that increased floods and droughts occurrence in different parts of the world can be attributed to anthropogenic activities. Recently, a growing number of studies (Ramanathan *et al.* 2005, Lau *et al.* 2008, Ramanathan and Carmichael 2008, Meehl *et al.* 2008) have found that aerosols can affect the regional and global water cycle, especially in Asian monsoon regions. The field of aerosol-monsoon climate studies has been growing

*Corresponding author.

Climate Change: Multidecadal and Beyond
edited by Chih-Pei Chang et al.

rapidly in recent years. As the knowledge base expands, so are associated uncertainties. An assessment of the subject matter is opportune. However a comprehensive review including exploration of new frontiers of research for aerosol-monsoon interactions will entail far too much effort and space than is appropriate as a book Chapter here. Moreover, such a review at this time will likely become obsolete very soon. Hence in this Chapter, we will provide an interim assessment based mostly on the authors' viewpoint, focusing on aspects of interactions of aerosol with rainfall and circulation for the boreal summer monsoons over South Asia and East Asia, where the most active recent research have been conducted.

2. Fundamentals of Monsoon–Aerosol Interaction

It is well known that the monsoon is governed by the seasonal variation of incoming solar radiation reaching the earth surface (Wang, 2006; Webster *et al.*, 1998; Wu *et al.*, 2007; Krishnamurti, 1985; Lau and Li, 1984). The differential surface heating between continental land and the oceans, through the interaction of surface turbulent heat flux, and latent heating associated with rainfall and deep convection produces strong meridional tropospheric temperature gradient which drives the monsoon circulation. In this regard, the Tibetan Plateau plays an important role as a source of elevated thermal and mechanical forcing in driving the monsoon circulation and moisture convergence patterns, exerting strong control on the timing and duration of the Asian monsoon (Yanai and Wu, 2006; Li and Yanai, 1996). Recently there is a renewed debate regarding the relative importance of thermal vs. mechanical forcing by the Tibetan Plateau and the Himalayas in giving rise to the monsoon climate system of today (Boos and Kuang, 2010; Wu *et al.*, 2012) by comparing the states of Asian monsoon during the pre- and post-Tibetan Plateau geological eras. However, this topic is outside the scope of this review.

Aerosols are suspended solid particles or liquid droplets in the atmosphere, derived from both natural and anthropogenic sources. Examples of natural aerosols are desert dust, sea salt, and smoke from forest fire. Major species of anthropogenic aerosols include sulfates, soot, also known as black carbon (BC). In particular, BC aerosols which result from incomplete combustion of hydrocarbons, *e.g.*, from internal combustion engines, coal firing power plants, slash and burn agricultural practice, smoke from cooking, are in increasing abundance in Asia as more and more energies are consumed and demanded by a rapidly growing population. The mixture of these different aerosol species scatter and attenuate incoming solar radiation creating the appearance of a semi-opaque yellowish smoke sometimes referred to as Atmospheric Brown Clouds (ABC) prevailing over urban, and mega-metropolises of the South and East Asian regions (Ramanathan *et al.*, 2007a,b).

Aerosols are transported by monsoon winds from the source locations to nearby and remote regions, depending on the size of the aerosols species. Dry deposition by gravitational settling and wet deposition by rainfall (wash-out) are major sinks of aerosols. In heavily populated monsoon region, it is common experience that even with wash-out by heavy monsoon rain which occurs intermittently, aerosol can build up to high concentration within hours or days, because of the continuous high rate of local emissions, as well as remote transport and trapping of aerosols by local topography. Natural aerosols such as desert dusts are mobilized, and emitted from the desert surface by changing surface winds, and transported over long distance from deserts to monsoon regions (Chin *et al.*, 2007). The typical aerosol residence times in the atmosphere are days to weeks, depending on particle sizes, and atmospheric conditions. Aerosols lofted into the upper troposphere and stratosphere by volcanic eruptions, or jetstream dynamics can remain for months to years before settling to earth, and have been known to contribute to cooling of global climate (IPCC 2077).

Aerosols can affect the net radiation balance of the atmosphere and underlying surface by direct (radiation) effects and indirect (microphysical) effects. For direct effect, a layer of aerosols between the sun and the earth surface blocks off incoming sunlight by scattering and/or absorption, causing cooling of the surface — the so-called solar dimming (SDM) effect. Sulfate aerosols are highly scattering, and therefore produce net cooling of the surface but negligible heating of the atmosphere. On the other hand, absorbing aerosols (desert dust and BC) are effective absorbers of solar radiation, and heats the atmosphere in addition to cooling the surface. On microphysical scales, through indirect effects, aerosols act as cloud condensation nuclei essential for the formation of clouds and precipitation. Increased loading of fine aerosol increases the scattering cross-section, inhibits the rates of coalescence and collision of cloud drops, resulting in brighter and more abundance of water clouds and suppressing the formation of warm rain (Twomey, 1977; Albrecht, 1989; Lohman and Feichter, 2005). In an environment of abundant moisture and strong convective potential such as the monsoon region, the delayed initiation of rain, may lead to smaller super-cooled particles reaching, higher altitudes where ice nucleation occurs, invigorating deep convection and ice-phase precipitation (Rosenfeld *et al.*, 2008; Li *et al.*, 2011). The altered rain and cloud populations associated with changing convection types may further modulate the radiation balance of the atmosphere-earth surface system. While there is now a large body of work on microphysical effects on generic deep tropical convection, mostly from idealized numerical simulations using cloud resolving models (Feingold *et al.*, 1999; Khain *et al.*, 2001; Fan *et al.*, 2012), the possible influence of aerosol microphysics on the system-wide Asian monsoon are still mostly unknown. To limit the scope of this assessment, we will focus mainly on aerosol radiative effects, which is likely to have first order impacts on monsoon climate (Lau *et al.*, 2008).

3. Aerosol Radiative Forcing

The advent of NASA earth observing satellites (EOS) since the 2000, together with improved physics and assimilation techniques in global climate models have enable reliable estimate of the direct aerosol radiation forcing (DARF) of the atmosphere and the earth surface on regional and global scales. It is well known that globally DARF is negative (cooling) and thus masks or diminishes the potential warming by greenhouse gas (GHG) (IPCC 2007). However, regionally aerosols can do much more than simply masking GHG warming. Large positive atmospheric DARF (Fig. 1c) signifying strong atmospheric heating during the boreal summer season (JJA) are found over regions of high aerosol optical depth (AOD) coinciding with aerosol emission sources over East Asia and South Asia (Fig. 1a). The large positive atmospheric DARF is primarily due to absorbing aerosols such as black carbon and dust (Ramanathan and Carmichael, 2008). Equally pronounced is the negative surface DARF (Fig. 1d) indicating surface cooling co-located with the atmospheric heating. This almost equal and opposite DARF between the atmosphere and the surface give rise to only relative small DARF at the top of the atmosphere (Fig. 1b). Additionally, large heating differential ($\sim$30–40 $\mathrm{Wm^{-2}}$) between the high and low DARF regions could change pressure gradient, and induce changes in the large scale circulation, influencing moist convergence and rainfall, as well as aerosol emission (in the case of desert dust and sea salt) and transport (Lau *et al.*, 2008).

During the INDOEX field campaign, DARF of the atmospheric column over the Indian Ocean in the vicinity of the urban aerosol outflow from the Indo-Gangetic Plain (IGP) has been estimated to be of the order of +14 $\mathrm{Wm^{-2}}$ (Ramanathan *et al.*, 2001). This is almost

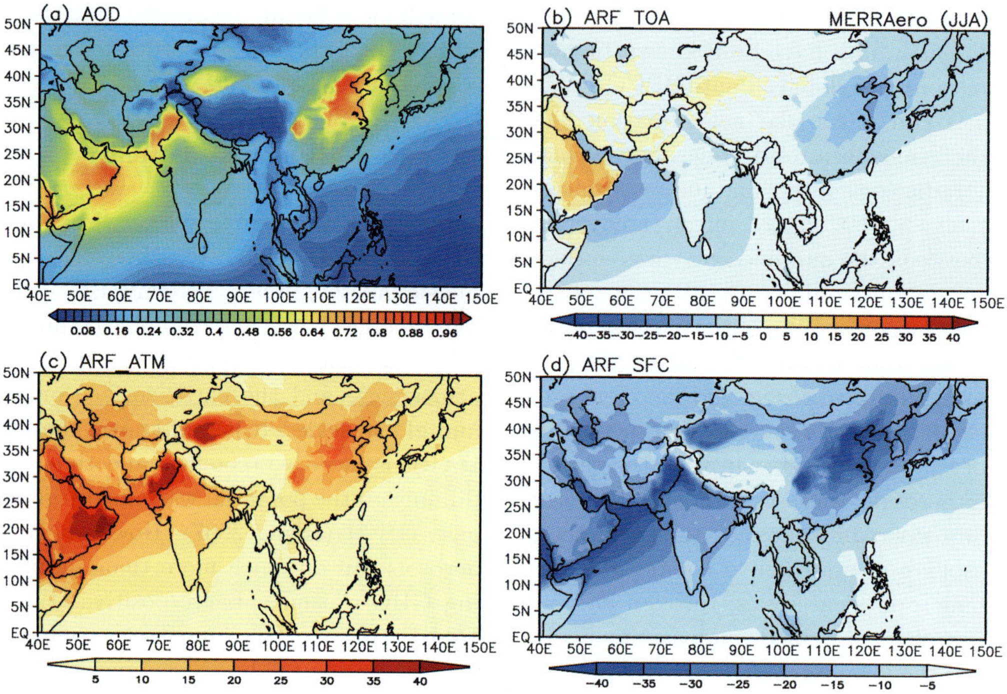

Fig. 1. JJA distribution of (a) AOD, and ADRF for (b) TOA, (c) atmosphere and (d) surface, from GEOS-5 assimilated aerosol reanalysis. Units for ADRF in $\mathrm{Wm^{-2}}$.

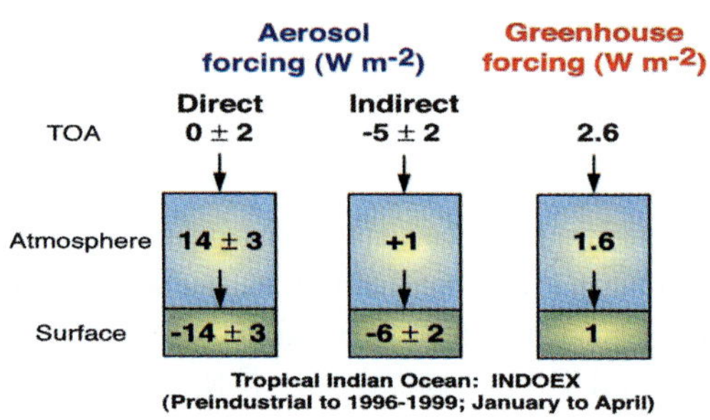

Fig. 2. Estimates of aerosol direct, indirect, and GHG radiative forcing at TOA, atmosphere and surface respectively relative to pre-industrial era, during INDOEX. Adopted from Ramanathan *et al.* (2001).

completely compensated by the cooling (negative DARF) at the surface (Fig. 2). By comparison, indirect aerosol radiative forcing arising from aerosol-cloud interaction is relatively small ($+1\,\mathrm{Wm^{-2}}$) in the atmosphere and at the surface ($-6\,\mathrm{Wm^{-2}}$). Most important, the magnitude of the atmospheric heating and cooling at the surface due to GHG radiative forcing over the region is negligibly small compared to DARF. Clearly DARF is an important first order climate forcing on regional monsoon scales (Pielke *et al.*, 2009).

From basic columnar one-dimensional considerations, all aerosols will cool the earth surface via SDM, and atmospheric heating by absorbing aerosols will warm the atmosphere above the surface. The warm-above cool-below condition will increase stability and reduce convective potential and hence may lead to reduction in rainfall locally. Additionally, heating of the atmosphere by absorbing aerosols may "burn off" low clouds i.e., the semi-direct effect, increase solar radiation reaching the surface, and thus warm the surface, working against the SDM effect (Hansen *et al.*, 1997; Koch and Del Genio, 2010). However, as shown below, in the monsoon environment where there is an abundance of moisture, and hence high potential convective instability, and propensity for large scale dynamical feedback, the columnar consideration may not be applicable.

While there has been an increasing solar SDM trend (global dimming) measured in terms of shortwave radiance flux reach the earth surface up to the 1970s, a decreasing SDM trend (global brightening) in recent decades has been reported over North America and Europe due to the Clean Air Act (Wild *et al.*, 2005). However, over the densely populated mega-cities and vicinity of Asia, the SDM has been trending upward in recent decades, as the aerosol loading increases due to the rapid pace of modernization (Che *et al.*, 2007; Kaskaoutis *et al.*, 2012; Sahu *et al.*, 2008; Srivastava *et al.*, 2012).

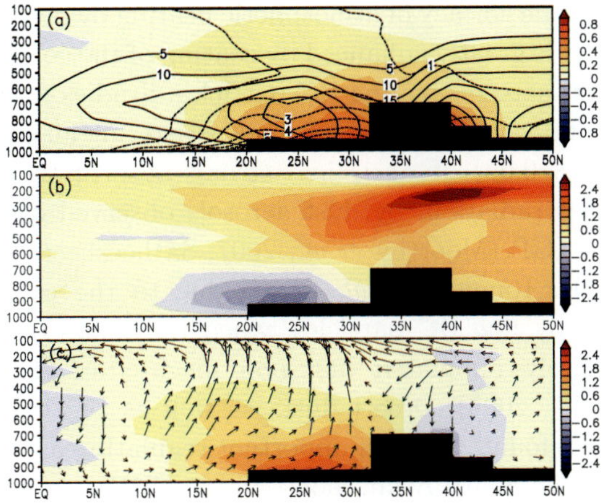

Fig. 3. Latitude-height profile averaged over the Indian subcontinent in May for (a) desert dust (solid contour) BC (dashed contour) and induced shortwave atmospheric heating (shading, Wm^{-2}), (b) aerosol induced temperature change (°C) and (c) aerosol induced changes in moisture ($g\,kg^{-1}$) and meridional circulation in wind vectors.

4. The South Asian Monsoon

Studies have shown that strong SDM effect over the aerosol source regions over the Asian continent relative to the adjacent oceans may lead to reduction in meridional surface temperature gradient, and surface evaporation leading to a weakening of the South Asian monsoon in recent decades (Ramanathan *et al.*, 2005; Chung and Ramanathan, 2006). Bollasina *et al.* (2012) argued from global warming simulations of the GFDL coupled model that the remote forcing by global aerosols could weaken the South Asian monsoon. However, the role of regional distribution of aerosols, and possible impacts were not addressed in their study.

Atmospheric heating by deep layers of dust and black carbon accumulated over the Indo-Gangetic Plain (IGP) and the Himalaya foothills can induce atmospheric moisture convergence feedback — the so-called Elevated Heat Pump (EHP) effect (Lau *et al.*, 2006), leading to increased precipitation in northern India and the foothill regions during the early monsoon (May–June) seasons. Specifically, the EHP postulates that shortwave absorption by accumulation of BC and dust over the northern and southern slopes of the Himalayas over the longitudes of the Indian subcontinent during pre- and early monsoon period (May–June) heats the lower and middle troposphere around the Tibetan Plateau (see Fig. 3a, b). The heating induces rising motion and increases rainfall over northern India and the foothill regions. Latent heating from the increased rain enhances the developing southwesterly monsoon flow, bringing more moisture and rainfall to the IGP and the Himalayan foothills. As a result, the monsoon rainfall is shifted northward, accompanied by an anomalous meridional circulation with enhanced moisture convergence and ascending air over northern India, and compensating subsidence over the Northern Indian Ocean (0–10 N) and the Tibetan Plateau (Fig. 3c). Upper troposphere warming becomes widespread due to adiabatic warming from the descending air, and warm advection (Fig. 3b). For the EHP mechanism to work, the atmospheric heating over the IGP and Himalaya regions has to be sufficiently strong and over a deep layer so as to overcome the stabilizing effect of SDM which tends to suppress convection.

The efficacy of atmospheric heating by desert dust is greatly enhanced by coating of the coarse dust aerosols by fine BC particles. Studies have found very low values of single scattering albedo (0.88–0.96) of dust, i.e., strong absorption of solar radiation, by dust aerosols observed over the IGP (Moorthy *et al.*, 2007; Satheesh *et al.*, 2007; Ramana *et al.*, 2007) due to the coating by fine BC particles, when they pass over densely populated monsoon mega-cities (Eck *et al.*, 2005). Desert dusts as well as BC have been found to be an important regulating agent for the South Asian monsoon via EHP or other atmospheric heating mechanisms on intraseasonal to decadal time scales (Lau and Kim, 2006; Lau *et al.*, 2009; Manoj *et al.*, 2011). The coating of dust by BC mixes natural and anthropogenic effects and adds further complexity to monsoon climate change in response to aerosol forcing.

Large-scale circulation features associated with EHP effect include: (a) an anomalous upper tropospheric anticyclone over the southern Tibetan Plateau, (b) increased low-level westerly flow, and moisture transport to northern India and the Bay of Bengal in May–June, and (c) subsequent reduction of the monsoon rainfall associated and weakening of the peak Indian monsoon in July–August (Lau and Kim, 2006). Additionally, studies have also suggested atmospheric heating by BC and dust, and snow-darkening by deposition of BC and dust on snow surface, could accelerate the melting of seasonal snowpack in Himalayas and Tibetan Plateau, and subsequently influence monsoon hydrology and climate (Ramanathan *et al.*, 2007b, Flanner *et al.*, 2007; Lau *et al.*, 2010; Qian *et al.*, 2009; Yasunari *et al.*, 2009).

Different flavors of the atmospheric heating effects have been found by several GCM experiments, generally indicating that heating due to BC aerosol may increase monsoon rainfall in spring, but increased cloudiness from the increase in early monsoon rain may induce a negative feedback that reduce rainfall during the peak summer monsoon season (Meehl *et al.*, 2008; Collier and Zhang, 2009; Randles and Ramaswamy, 2008; Mahmood and Li, 2012). The northward shift of the summer monsoon rainfall has also been related to the increase in low level moist static energy over land due to black carbon heating of over the IGP (Wang *et al.*, 2009). Others have suggested that semi-direct effects and induced land surface feedback may also play a role (Bollasina *et al.*, 2008).

The relative roles of the SDM vs. EHP in affecting the South Asian monsoon has been investigated in experiments using the NASA GCM coupled to a mixed-layer ocean (Lau and Kim, 2007). Comparing results of a 100-year coupled mixed layer model control run, to an anomaly experiment in which the aerosol forcing are withheld, the EHP effect is evident in the increased rainfall over northern India and Himalaya foothills in May–June, (20–30°N) (Fig. 4 upper panel). Likewise, the SDM effect in reducing all-India monsoon rainfall in July–August is also evident (Fig. 4, lower panel). Essentially, under an idealized setting when the EHP and SDM are allowed to compete for dominance in the absence of other climate forcing, the atmospheric heating via EHP induces an increased north-south tropospheric temperature gradient, which enhances the early monsoon, while the SDM and related atmospheric feedback reduce the land-sea thermal contrast because the land surface cools faster than the surrounding oceans.

Although far from definitive, observational evidence consistent with model simulations of aerosol SDM and EHP impacts on monsoon are emerging. A surface temperature cooling of about 0.3°C over the Indian Ocean and most of South Asia since the 1970s has been attributed to absorbing aerosols (Krishnan and Ramanathan, 2002). From measurements with unmanned aerial vehicle, it has been found that warming trend in the atmosphere is significantly enhanced by ABC in Asian monsoon regions (Ramanathan *et al.*, 2007b). From microwave satellite data, an accelerated

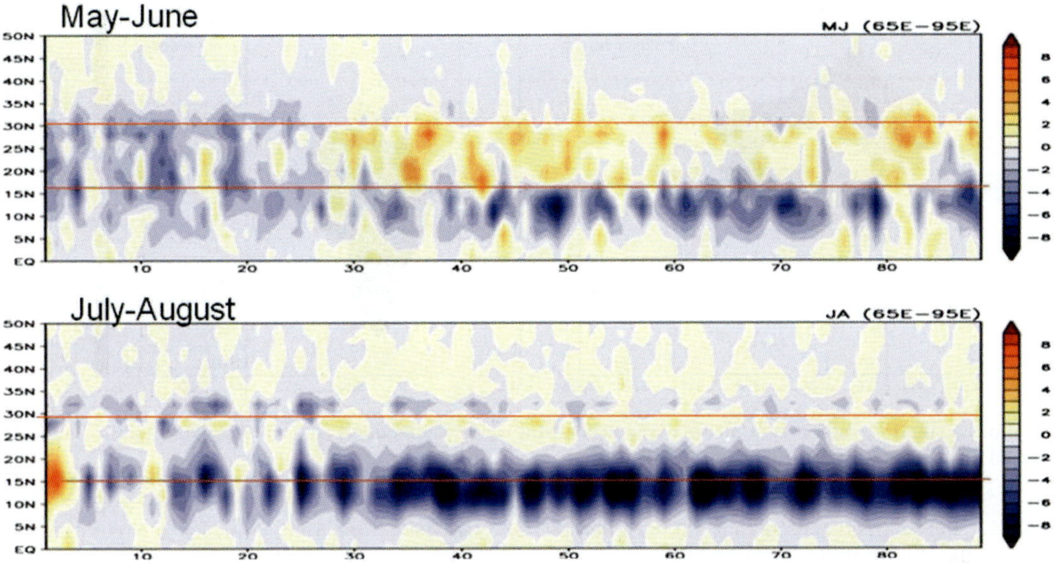

Fig. 4. Time-latitude section of precipitation anomalies (mm day^{-1}) induced by absorbing aerosols during (a) May–June and (b) July–August, from a 90-year coupled atmosphere-ocean (mixed-layer) model integration, with seasonally prescribed aerosol forcing. Region of northern India (15 N–30 N) are highlighted within red lines.

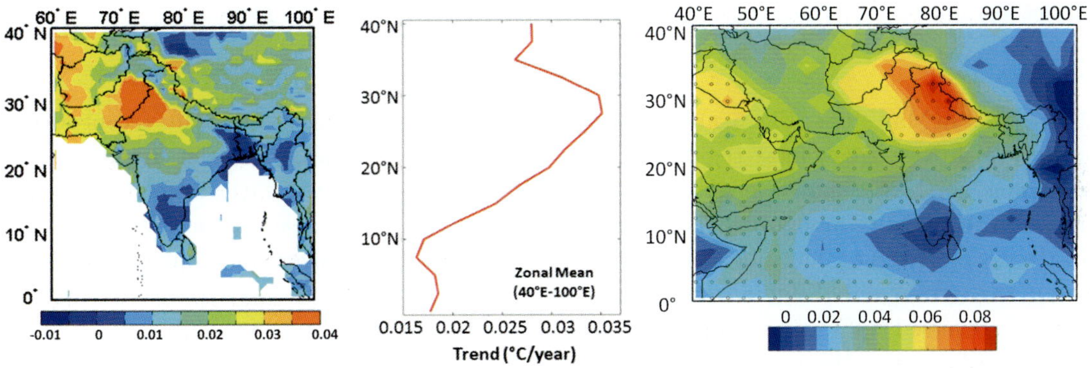

Fig. 5. Linear trend patterns of TOMS Aerosol Index (left panel), MSU tropospheric (4–7 km) temperature, and zonal averaged tropospheric temperature profile for the month of May during the period 1979–2007 in units of °C year^{-1} (Adopted from Gautam *et al.*, 2009b).

pre-monsoon warming trend (0.21°C/decade ±0.08°C/decade) has been found in the lower and middle troposphere in northwestern India, and the Himalayan foothills far exceeding the GHG warming rates (Fig. 5). The warming pattern represents an increased in tropospheric temperature gradient of approximately 0.2–0.3°C per decade between northern Indian and regions to the south. This warming is accompanied by an increasing rainfall trend over northern India in June from 1979–2007 (Gautam *et al.*, 2009a, b; Gautam *et al.*, 2010).

Using division rainfall data (1961–2004) over India, Lau and Kim (2010) found that for all regions over India, there are positive trends in rainfall, most pronounced (99% confidence level) in the northwest and central northeastern India, yielding an increase in All-India rainfall of

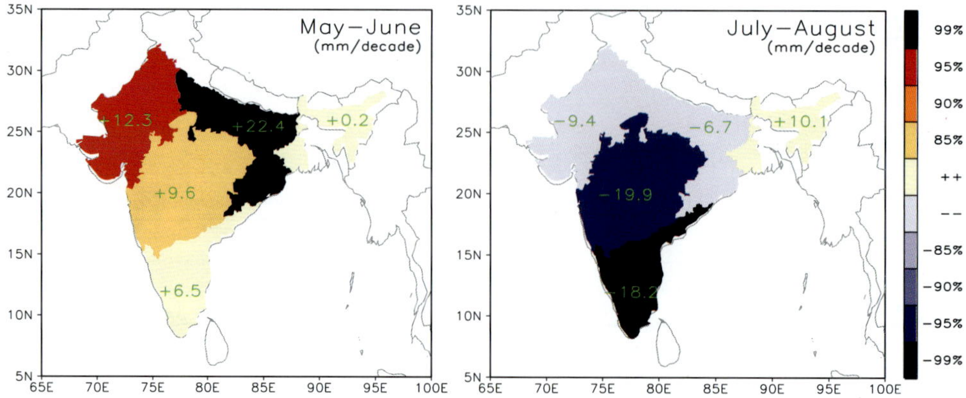

Fig. 6. Linear trend of Indian rainfall by division for May–June (left panel) and July–August (right panel) for the period 1961–2004. Units in mm/decade.

Table 1. Linear trend (mm decade^{-1}) of Indian rainfall by division from 1961–2004, for different averaging periods. Trend magnitude exceeding 95% significance are bolded.

	AIMR	Northwest	West Central	Central Northeast	Northeast (Assam)	Peninsular
May–June	**11.4**	**12.3**	9.6	**22.4**	0.2	6.5
Jul–Aug	−11.9	−9.4	**−19.9**	−6.7	10.1	**−18.2**
MJJAS	−6.6	−7.9	−23.4	13.6	27.4	−15.1
JJAS	−10.2	−10.7	−23.5	6.1	13.0	−15.7

11.4 mm/year in May–June (Fig. 6, and Table 1). However, by July–August, most regions, except the extreme northeast, show a negative rainfall trend, with the most pronounced signal over the central western, and the peninsular region. Interestingly, the sign of the all-Indian rain trend is completely reversed (−11.9 mm/year) compared to the May–June, but the magnitude does not rise above the 95% confidence level.

Conventional season averages *e.g.*, June through September, and May through September, yield negative rainfall trends in all-India rainfall, with the largest contribution from western central and peninsular regions. However, the full-season rainfall trends are not statistical significant at the 95% level and mask the positive rainfall trend over northern Indian during the early season (Table 1) The aforementioned results suggest that coupled ocean-atmosphere-land feedback processes induced by EHP and SDM may play important roles in the regional rainfall trends over India in recent decades, especially with regard to the change in sign of the trends between early and peak monsoon. However the full reasons for the decline in the seasonal rainfall over India are still not known.

5. The East Asian Monsoon

There has been increasing number of recent studies of possible responses of the East Asian monsoon to aerosol forcing. However, because of different data periods used for observations, different model experimental designs, and uncertainties in model physics, especially with regard to aerosol indirect effects, results have been as confusing as they are informative. Most studies have focused on the causes for the observed major decadal scale precipitation anomaly pattern known as the north-dry/south-wet (NDSW) pattern (Lau and Weng, 2001; Wang and Zhou, 2005) (see Fig. 7b). Menon

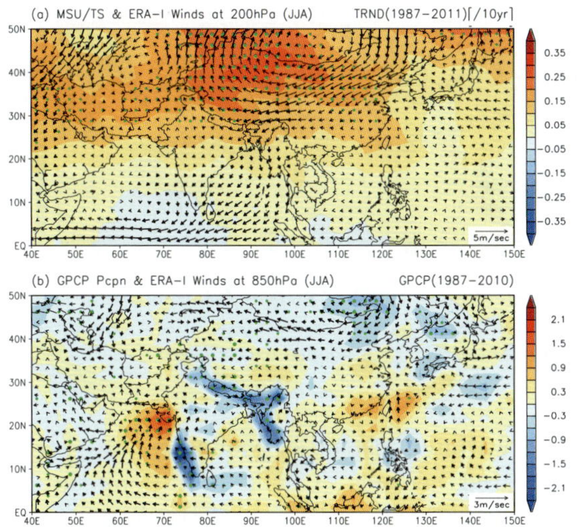

Fig. 7. Linear trend for the period 1987–2011 of (a) tropospheric (5–7 km) temperature from the NOAA Microwave Sounding Unit (MSU) in °C decade-1, and 200 hPa wind, (b) rainfall from the Global Precipitation Climatology Project (GPCP) in mm day^{-1} decade^{-1}, and 850 hPa wind, for June–July–August mean. Wind data are from the ERA-Interim, in ms^{-1} decade^{-1}. Wind vectors exceeding 95% confidence level are bolded.

et al. (2002) showed from a coarse resolution GCM simulation that direct effects from prescribed increasing absorbing aerosol (mainly BC) over India and China can induce a rainfall anomaly that roughly resemble the NDSW pattern over East Asia. However, their result has not been reproduced by other model experiments. Other GCM experiments have shown reduced rainfall in northern China due to surface cooling by increased loading of sulfate aerosol, and increased rainfall in precipitation in the southern part of China in July due to induced strengthening of the local Hadley circulation (Gu *et al.*, 2006). They found that heating of the air column by BC and dust particles in the middle to high latitudes tend to move the simulated precipitation inland. However, the inclusion of black carbon in their simulations did not produce the observed NDSW precipitation pattern. Furthermore the SDM effect of sulfate aerosol can reduce atmospheric baroclinicity, causing a deceleration of the East Asia jet stream and consequent rainfall reduction over East Asia in late spring and early summer (Kim *et al.*, 2007). In one study, the combined effect of BC and sulfate aerosols was found to be largely dominated by sulfates in producing a weakened East Asian monsoon for both summer and winter season (Liu *et al.*, 2009). Through semi-direct and indirect effects, BC aerosols have been shown to cause a reduction in cloud and rainfall in Southern China, while increased rainfall over northern China through increase moisture transport by the West Pacific Subtropical High (Zhang *et al.*, 2009). However, this result is contrary to the NDSW pattern.

The studies cited above represent only small samples of a large number of studies attempting to attribute aerosol impact on aspects of the East Asian long-term rainfall change. However, because of the multiple factors controlling the long-term variations of the East Asian monsoon, including flow variability downstream of the Tibetan Plateau, the southwest monsoon flow of South Asia, the displacement of West Pacific Subtropical High, and strong remote forcing from sea surface temperature (SST) anomalies associated with climate variability and change (Lau *et al.*, 2000), studies of attribution of aerosol impact on variation of the East Asian monsoon have so far yielded diverse and largely inconclusive results.

Given the complex regional and global controlling factors, it is important that possible aerosol impacts on the East Asian monsoon rainfall have to been examined in the context of the entire Asian atmospheric circulation and climate system from observations and modeling studies over common periods. To utilize observed upper tropospheric temperature from MSU, we focused on the decadal trend between 1987 and 2011. As an example, the June–July–August mean wind and temperature decadal (1987–2011) trend patterns in the upper troposphere over the broad-scale Asian monsoon region [Fig. 7, upper panel] shows a pronounced widespread warming of upper troposphere over northern central Asia, and deceleration of the East Asian jetstream (30–40 N, 90E–120E), in association with the formation of an anomalous blocking high at 500 hPa (not shown) over central Siberia and northern China. Rainfall is reduced over northern and northeastern China, but increased over southern China and the East China Sea (20°N–30°N) [Fig. 7 lower panel], reflecting the well-known NDSW pattern previously discussed. The rainfall suppression in the north may have to do with the anomalous anti-cyclonic flow advecting colder drier air from the land regions, thus preventing the *Mei-yu* front from advancing northward during the mature monsoon season (July–August). Additionally, rainfall is reduced over the deep tropics (10 N–20 N), appearing as a continuation of the NDSW pattern over East Asia land. This low-high-low rainfall pattern is characteristics of the multi-cellular meridional circulation structures characteristic of the East Asian monsoon (Lau *et al.*, 2000). Also, noteworthy is the overall reduction of rainfall over the India subcontinent, consistent with the discussion in the previous section on the possible aerosol impacts on seasonal mean rainfall over India. The trend patterns over monsoon land regions, with dense gauge networks, i.e., India subcontinent, and East Asia are quite robust. However, over remote high mountain and oceanic regions, the trend patterns have large uncertainties, with wide variations among different rainfall products. (Annamalai *et al.*, 2013).

Clearly, the impacts of aerosols on the East Asian monsoon cannot be fully comprehended without better understanding of the role of SST forcing. The SST trend for the same period shows pronounced warming over the northwestern Pacific, and much of the tropical oceans except for the coastal regions of the Bay of Bengal, the South China Sea, and the western Pacific (Fig. 8). It is possible that the SST cooling may be related to in part due to the SDM effect from aerosol outflow from the IGP, Southeast Asia and South China. However,

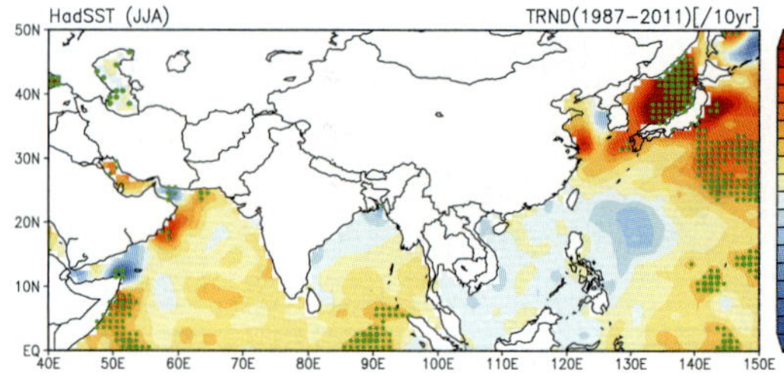

Fig. 8. Linear trend of SST in Asian monsoon regions during 1987–2011. Units in °C decade^{-1}. Green dots denote statistical confidence exceeding 95%.

SST warming instead of cooling expected from increased aerosols (Chung and Ramanathan, 2006 and others) is observed in the Arabian Sea. During the summer monsoon, aerosols over the Arabian Sea is dominated by desert dusts transported by the low level monsoon southwesterlies from the Arabian Peninsula. If aerosol has an impact on the Arabian Sea SST warming, it will likely be due to a *reduction* in dust aerosol, from reduced transport by the diminished southwesterlies of a weakened Indian summer monsoon. The diminished southwesterlies could also cause SST warming by reducing surface evaporation, and suppressing coastal and/or Ekman oceanic upwelling. At this time, it is not clear how the aforementioned rainfall and climate change patterns are forced by SST changes over Arabian Sea and the adjacent monsoon oceans, and how much of the SST changes can be attributed to aerosol SDM effects, and to coupled ocean-atmosphere dynamical processes.

The linear trend pattern for the pre-monsoon season April–May [Fig. 9 upper panel] reveals that the JJA upper tropospheric warming trend may have originated from strong warming over the Tibetan Plateau, which could be linked to EHP-like atmospheric heating induced by aerosols, and/or GHG warming (Lau *et al.*, 2010; Duan and Wu, 2006). During April–May, tropospheric heating over the Tibetan Plateau and land is consistent with the development of a large scale anticyclone anomaly, signaling the strengthening the climatological

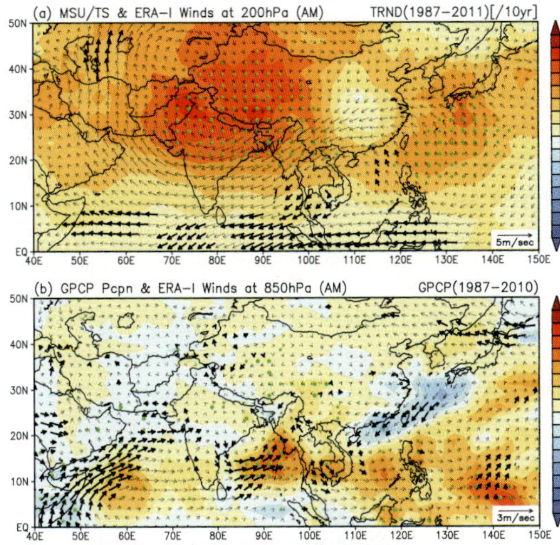

Fig. 9. Same as in Fig. 7, except for trend patterns of (a) MSU temperature and 200 hPa wind, and (b) GPCP rainfall and 850 hPa for April–May.

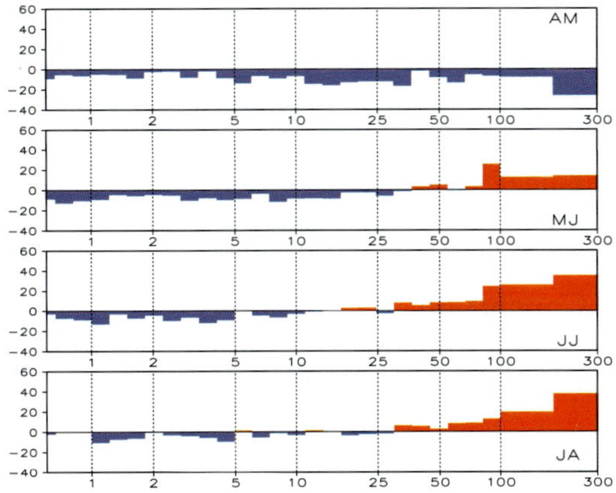

Fig. 10. Normalized changes in rainfall probability distribution by rain rate over Southern China for (a) April–May, (b) May–June, (c) June–July and (d) July–August between the period 1961–1984 and 1986–2009.

South Asian High. The thermal and circulation anomalies increase the tropospheric meridional temperature gradient between the Tibetan Plateau and regions to the south. This increase in tropospheric temperature gradient may have a contribution from EHP-related pre-monsoon atmospheric heating induced by aerosols over the IGP and Himalayan foothills (see discussion in Section 2). The circulation patterns in Fig. 9 shows increased easterly vertical wind shear over India subcontinent and the northern Indian Ocean as required by thermal wind balance for increased tropospheric temperature gradient, signaling an increased broad-scale Asian monsoon in its early formative stage (Lau *et al.*, 2000; Webster and Yang, 1992). During April–May, rainfall is enhanced over a long span from the eastern Indian Ocean to the western Pacific, but reduced over southern China and the East China Sea [Fig. 9 lower panel]. The latter represents a reversal of the JJA trend over the same region.

Observations and numerical simulations have shown that there has been a decreased frequency of light rain over China, and suppressed rain over the East China Sea in the last 50 years, attributable in part to the warm rain suppression by aerosols (Qian *et al.*, 2009; Bennartz *et al.*, 2011). Quantile regression analysis of daily station rainfall probability distribution function over southern China shows that for April–May, the long-term trend is suppression of rainfall for all rain types (Fig. 10). However, as the season progress the trend pattern evolves into one with increased frequency of heavy and very heavy rain, but reduced in moderate to light rain in July–August. Thus the reversal in sign in the total rain trend is mainly due to the changes in heavy to very heavy rain. Here, an argument can be made that the rainfall suppression in the late spring season may be related to increased biomass burning from Indo-China during the pre-monsoon period April–May, when warm rain and low stratus clouds dominate, and control by dynamics is steady but weak. As the monsoon season advances in April–June, the strong southwest monsoon flow brings copious moisture to southern China, and the East China Sea, the suppressed and delayed warm rain, may cause sporadic invigoration of deep convection and ice-phase precipitation by aerosols (Rosenfeld *et al.* 2008). In the mature monsoon season (JA), it is likely that strong monsoon circulation dynamics, possibly in response to large scale forcing such as GHG warming and global aerosol forcing will be important in producing the NDSW pattern. Aerosol may also play a role in the late monsoon season, but identifying impacts will be challenging, because of the prevalence of strong dynamical feedback.

6. Conclusion

Modeling and observational analyses have provided valuable information regarding possible aerosol impacts on the Asian monsoon. However, large uncertainties and unknowns remain. For the South Asian monsoon, heating of the atmosphere from build-up of dust and BC during the pre-monsoon season may induce atmospheric feedback through the EHP effect, advancing the onset of the rainy season in northern India and the Himalayan foothills in May–June. Subsequently in July–August, rainfall over the Indian subcontinent is reduced. The reduction could be due to the SDM effect which tends to cool the land and ocean surface near the aerosol source region, reduces the meridional surface thermal gradient, and weakens the monsoon through a spin down of the local Hadley circulation (Ramanathan *et al.*, 2005; Bollasina *et al.*, 2012). However, the peak monsoon may also be weakened by the increased cloudiness and associated dynamical feedback from EHP, which may hasten the termination of the monsoon season (Lau *et al.*, 2008; Meehl *et al.*, 2008). Conventional averaging for the entire monsoon season will mask much of the signal of aerosol impacts in early season. Aerosol indirect effects can also modulate monsoon clouds and convections, but their impacts on the overall South Asian monsoon are not clear.

For the East Asian monsoon, studies so far have yielded mixed results, suggesting influences from both aerosol direct and indirect effects as well as atmospheric feedback but with large uncertainties. No dominant discriminating signals of aerosol impacts have been found. There are evidences of SDM effects on surface temperature, but relationships with monsoon rainfall anomalies remain elusive. In addition to impacts by local emissions of aerosols, the East Asian monsoon appears also to be affected by downstream teleconnections emanating from the EHP effects over South Asia. At this time, possible contributions of aerosols to the NDSW decadal anomalies by single model studies are likely to be model dependent and speculative at best. This is due to the highly chaotic nature of the dynamics of the East Asian monsoon and the plethora of other possible first-order climate forcings ranging from downstream influences from the Tibetan Plateau, variability of moisture transport from the South Asian monsoon, and the West Pacific Subtropical High, as well as land use and change, regional SST anomalies (which may or may not be related to aerosol SDM), remote forcing from ENSO, decadal scale oscillations and GHG warming. Attributions of aerosols impact from observations and model outputs for the East Asian monsoon require extreme caution in discriminating response to aerosol forcing from other climate forcings. For better understanding of possible climatic impacts of aerosol on East Asian monsoon, multi-dimensional empirical and model data analyses involving temperature, wind, rainfall types, and aerosols properties, and pre-monsoon aerosol forcing over the broad scale Asian monsoon regions are appropriate.

Aerosol impacts on monsoon are strongly dependent on aerosol types, and their mixing states. As a rule of thumb, scattering aerosol such as sulfate is likely to reduce precipitation and weaken the Asian monsoon through the SDM effects. On the other hand, absorbing (dust and BC) aerosol induces atmospheric feedback via the EHP can redistribute rainfall, and affect the monsoon evolution including timing and duration of the monsoon rainy season. Aerosol indirect effect is the ultimate "wild card", which may further modulate monsoon rainfall.

Detecting aerosol indirect effects in the Asian monsoon regions is greatly hampered by the lack of long-term and detailed records of aerosol properties during different phases of the monsoon. This calls for continued effort in remote sensing, enhanced international coordination of field observations over targeted regions, such as the IGP, northern and southern East Asia and Southeast Asia. Supporting evidence of aerosol indirect effects on monsoon may also be obtained through analyses of changes in rainfall and cloud characteristics (types, intensity, frequency and duration) from historical rainfall and cloud data, and from climate model outputs (Li *et al.*, 2011; Lau and Wu, 2006, 2011; Lau *et al.*, 2013).

Undoubtedly, modeling has played a critical role in advancing theory, and understanding of aerosol impacts on monsoon dynamics. Particularly, CMIP (Coupled Model Intercomparison Project) model projection using prescribed emission scenarios have provided useful information regarding possible global impacts of aerosols, and guidance for action by policy makers and stakeholders. As such, CMIP/IPCC projection experiments are not optimal from the perspective of unraveling complex earth system interactions. Given the complex responses of the Asian monsoon water cycle to the multitude of climate forcings, detecting and understanding aerosol impacts are very challenging, requiring well designed climate model experiments for hypothesis testing, with sufficient high resolution and realistic aerosol physics, in conjunction with a more nuanced and detailed approach to tease out signals from model outputs and observations.

References

Albrecht, B., 1989: Aerosols, cloud microphysics, and fractional cloudiness. *Science*, **245**, 1227–1230.

Annamalai, H., J. Hafner, K. P. Sooraj, and P. Pillai, 2013: Global warming shifts the monsoon circulation, drying South Asia. *J. Climate*, **26**, 2701–2718.

Bennartz, R., J. Fan, J. Rausch, L. Y. R. Leung, and A. K. Heidinger, 2011: Pollution from China increases cloud droplet number, suppresses rain over the East China Sea. *Geophys. Res. Lett.*, **38**, L09704, doi: 10.1029/2011GL047235.

Bollasina, M., S. Nigam, and K. M. Lau, 2008: Absorbing aerosols and summer monsoon evolution over South Asia: An observational portrayal. *J. Climate*, **21**, 3221–3239.

Bollasina, M. A., Y. Ming, and V. Ramaswamy, 2012: Anthropogenic aerosols and the weakening of the South Asian summer monsoon. *Science*, **334**, 502–505.

Boos, W. R., and Z. M. Kuang, 2010: Dominant control of the South Asian monsoon by orographic insulation versus plateau heating. *Nature*, **463**, 218–223.

Che, H., X. Zhang, Y. Li, Z. Zhou, and J. Qu, 2007: Horizontal visibility trend in China 1981–2005. *Geophys. Res. Lett.*, **34**, L24706, doi: 10.1029/2007GL031450.

Chin, M., T. Diehl, P. Ginoux, and W. Malm, 2007: Intercontinental transport of pollution and dust aerosols: Implications for regional air quality. *Atmos. Chem. Phys.*, **7**, 5501–5517.

Chung, C., and V. Ramanthan, 2006: Weakening of the North Indian SST gradients and the monsoon rainfall in India and the Sahel. *J. Climate*, **19**, 2036–2045.

Collier, J. C., and G. J. Zhang, 2009: Aerosol direct forcing of Indian summer monsoon as simulated by NCAR CAM3. *Clim. Dyn.*, **32**, 313–332, doi: 10.1007/s00382-008-0464-9.

Duan, A., and G. Wu, 2006: Change of cloud amount and the climate warming on the Tibetan Plateau. *Geophys. Res. Lett.*, **33**, L22704, doi: 10.1029/2006GL027946.

Eck, T., B. N. Holben, O. Dubovik, A. Smirnov, P. Goloub, H. B. Chen, B. Chatenet, L. Gomes, X. Y. Zhang, S. C. Tsay, Q. Ji, D. Giles, and I. Slutsker, 2005: Columnar aerosol optical properties at AERONET sites in central eastern Asia and aerosol transport to the tropical mid-Pacific. *J. Geophys. Res.*, **110**, D06202, doi: 10.1029/2004JD005274.

Fan, J., D. Rosenfeld, Y. Ding, L. R. Leung, and Z. Li, 2012: Potential aerosol indirect effects on atmospheric circulation and radiative forcing through deep convection. *Geophys. Res. Lett.*, **39**, L09806, doi: 10.1029/2012GL051851.

Feingold, G., W. R. Cotton, S. M. Kreidenweis, and J. T. Davis, 1999: The impact of giant cloud condensation nuclei on drizzle formation in stratocumulus: Implications for cloud radiative properties. *J. Atmos. Sci.*, **56**, 4100–4117.

Flanner, M. G., C. S. Zender, J. T. Randerson, and P. J. Rasch, 2007: Present-day climate forcing and response from black carbon in snow. *J. Geophys. Res.*, **112**(D11), D11202, doi: 10.1029/2006JD008003.

Gautam, R., N. C. Hsu, and K. M. Lau, 2010: Premonsoon aerosol characterization and radiative effects over the Indo-Gangetic Plain: Implications for regional climate warming. *J. Geophys. Res.*, **115**, D17208, doi: 10.1029/2010JD013819.

———, ———, ———, and M. Kafatos, 2009a: Aerosol and rainfall variability over the Indian monsoon region: Distributions, trends and coupling. *Ann. Geophys.*, **27**, 3691–3703.

———, ———, ———, S. C. Tsay, and M. Kafatos, 2009b: Enhanced pre-monsoon warming over the Himalayan-Gangetic region from 1979 to 2007. *Geophys. Res. Lett.*, **36**, L07704, doi: 10.1029/2009GL037641.

Gu, Y., K. N. Liou, Y. Xue, C. R. Mechoso, W. Li, and Y. Luo, 2006: Climatic effects of different aerosol types in China simulated by the UCLA general circulation model. *J. Geophys. Res.*, **111**, D15201, doi: 10.1029/2005JD006312.

Hansen, J., M. Sato, and R. Ruedy, 1997: Radiative forcing and climate response. *J. Geophys. Res.*, **102**, 6831–6864.

Intergovernmental Panel on Climate Change. *Climate Change, 2007: The Physical Science Basis. Contribution of Working Group I to the Fourth Assessment Report of the IPCC*, Cambridge University Press, Cambridge, U.K., 996 pp.

Kaskaoutis, D. G., R. P Singh, R. Gautam, M. Sharma, P. G. Kosmopoulos, and S. N. Tripathi, 2012: Variability and trends of aerosol properties over Kanpur, northern India using AERONET data (2001–10). *Environ. Res. Lett.*, **7**, 024003, doi: 10.1088/1748-9326/7/2/024003.

Khain, A. P., D. Rosenfeld, and Aa. Pokrovsky, 2001: Simulating convective clouds with sustained supercooled liquid water down to $-37.5°$C using a spectral microphysics model. *Geophys. Res. Lett.*, **28**, 3887–3890.

Kim, M. K., K. M. Lau, K. M. Kim, and W. S. Lee, 2007: A GCM study of effects of radiative forcing of sulfate aerosol on large scale circulation and rainfall in East Asia during boreal spring. *Geophys. Res. Lett.*, **34**, L24701, doi: 10.1029/2007GL031683.

Koch, D., and A. D. Del Genio, 2010: Black carbon semi-direct effects on cloud cover: review and synthesis. *Atmos. Chem. Phys.*, **10**, 7685–7696.

Krishnamurti, T. N., 1985: The summer monsoon experiment — A review. *Mon. Wea. Rev.*, **113**, 1590–1626.

Krishnan, R., and V. Ramanathan, 2002: Evidence of surface cooling from absorbing aerosols. *Geophys. Res. Lett.*, **29**, 1340, doi: 10.1029/2002GL014687.

Lau, K. M., and K. M. Kim, 2006: Observational relationships between aerosol and Asian monsoon rainfall, and circulation. *Geophys. Res. Lett.*, **33**, L21810, doi: 10.1029/2006GL027546.

———, and ———, 2007: Does aerosol strengthen or weaken the Asian monsoon? *Mountains: Witnesses of Global Change*, R. Baudo, G. Tartari, and E. Vuillermoz, Eds., Elsevier B. V., 13–22.

———, and ———, 2010: Fingerprinting the impacts of aerosols on long-term trends of the Indian summer monsoon regional rainfall. *Geophys. Res. Lett.*, **37**, L16705, doi: 10.1029/2010GL043255.

———, ———, C. Hsu, and B. Holben, 2009: Possible influences of air pollution, dust and sandstorms on the Indian monsoon. *WMO Bulletin*, **58**(1), 22–30.

———, ———, and S. Yang, 2000: Dynamical and boundary forcing characteristics of regional components of the Asian summer monsoon. *J. Climate*, **13**, 2461–2482.

———, M. K. Kim, and K. M. Kim, 2006: Aerosol induced anomalies in the Asian summer monsoon: The role of the Tibetan Plateau. *Clim. Dyn.*, **26**(7–8), 855–864, doi: 10.1007/s00382-006-0114-z.

———, ———, ———, and W. S. Lee, 2010: Enhanced surface warming and accelerated snowmelt in the Himalayas and Tibetan Plateau induced by absorbing aerosols. *Environ. Res. Lett.*, **5**, 025204, doi: 10.1088/1748-9326/5/2/025204.

———, and M. T. Li, 1984: The monsoon of East-Asia — A survey. *Bull. Amer. Meteor. Soc.*, **65**, 114–125.

———, V. Ramanathan, G. X. Wu, Z. Li, S. C. Tsay, C. Hsu, R. Sikka, B. Holben, D. Lu, G. Tartari, M. Chin, P. Koudelova, H. Chen, Y. Ma, J. Huang, K. Taniguchi, and R. Zhang, 2008: The Joint Aerosol–Monsoon Experiment: A new challenge in monsoon climate research. *Bull.*

Amer. Meteor. Soc., **89**, 369–383, doi: 10.1175/BAMS-89-3-369.

———, and H. Weng, 2001: Coherent modes of global SST and summer rainfall over China: An assessment of the regional impacts of the 1997–98 El Niño. *J. Climate*, **14**, 1294–1308.

———, and H. T. Wu, 2006: Trends in tropical rainfall characteristic, 1979–2003. *Int. J. Climatol.*, **27**, 979–988, doi: 10.1002/joc.1454.

———, and ———, 2011: Climatology and changes in tropical oceanic rainfall characteristics inferred from Tropical Rainfall Measuring Mission (TRMM) data (1998–2009). *J. Geophys. Res.*, **116**, D17111, doi: 10.1029/2011JD015827.

———, ———, and K. M. Kim, 2013: A canonical response in rainfall characteristics to global warming from CMIP5 model projections. *Geophys. Res. Lett.*, **40**, doi: 10.1002/grl.50420.

Li, C. F., and M. Yanai, 1996: The onset and interannual variability of the Asian summer monsoon in relation to land-sea thermal contrast. *J. Climate*, **9**, 358–375.

Li, Z., F. Niu, J. Fan, Y. Liu, D. Rosenfeld, and Y. Ding, 2011: Long-term impacts of aerosols on development of clouds and precipitation. *Nature Geoscience*, **4**, 888–894, doi: 10.1038/NGEO1313.

Liu, Y., J. Sun, B. Yang, 2009: The effects of black carbon and sulfate aerosols in China regions on East Asian monsoon. *Tellus*, **61B**, 642–656, doi: 10.1111/j.1600-0889.2009.00427.x.

Lohman, U., and J. Feichter, 2005: Global indirect aerosol effects: A review. *Atmos. Chem. Phys.*, **5**, 715–737.

Mahmood, R., and S. Li, 2012: Delay in the onset of South Asia summer monsoon induced by local black carbon in an AGCM. *Theor. Appl. Climatol.*, doi: 10.1007/s00704-012-0681-3.

Manoj, M. G., P. C. S. Devara, P. D. Safai, B. N. Goswami, 2011: Absorbing aerosols facilitate transition of Indian monsoon breaks to active spell. *Clim. Dyn.*, **37**, 2181–2198, doi: 10.1007/s00382-010-00971-3.

Meehl, G. A., J. M. Arblaster, and W. D. Collins, 2008: Effects of black carbon aerosols on the Indian monsoon, *J. Climate*, **21**, 2869–2882, doi: 10.1175/2007JCLI1777.1.

Menon, S., A. Del Genio, D. Koch, G. Tselioudis, 2002: GCM simulations of aerosol indirect effects: sensitivity of cloud parameterization and aerosol burden. *J. Atmos. Sci.*, **59**, 692–713.

Moorthy, K. K., S. S. Babu, S. K. Satheesh, J. Srinivasan, and C. B. S. Dutt, 2007: Dust absorption over the "Great Indian Desert" inferred using ground-based and satellite remote sensing. *J. Geophys. Res.*, **112**, D09206, doi: 10.1029/2006JD007690.

Pielke, R., K. Beven, G. Brasseur, J. Calvert, M. Chahine, R. Dickerson, D. Entekhabi, E. Foufoula-Georgiou, H. Gupta, V. Gupta, W. Krajewski, E. P. Krider, W. K. M. Lau, J. Mcdonnell, W. Rossow, J. Schaake, J. Smith, S. Sorooshian, E. Wood, 2009: Climate change: The need to consider human forcing besides greenhouse gases. *EOS Newsletter*, **90**, 10 November issue.

Qian, Y., D. Gong, J. Fan, L. R. Leung, R. Bennartz, D. Chen, and W. Wang., 2009: Heavy pollution suppresses light rain in China: Observations and modeling. *J. Geophys. Res.*, **114**, D00K02, doi: 10.1029/2008JD011575.

Ramana, V., V. Ramanathan, D. Kim, G. C. Roberts, and C. E. Corrigan, 2007: Albedo, atmospheric solar absorption and heating rate measurements with stacked UAVs. *Q. Roy. Meteo. Soc.*, **133**, 1913–1931.

Ramanathan, V., and G. Carmichael, 2008: Global and regional climate change due to black carbon. *Nature Geoscience.*, **1**, 221–227.

———, C. Chung, D. Kim, D. T. Betge, L. Buja, J. T. Kiehl, W. M. Washington, Q. Fu, D. R. Sikka, and M. Wild, 2005: Atmospheric brown clouds: Impacts on South Asian climate and hydrological cycle. *P. Natl. Acad. Sci., USA*, **102**(15), 5326–5333, doi: 10.1073/pnas.0500656102.

———, P. J. Crutzen, and J. T. Kiehhl, and D. Rosenfeld, 2001: Aerosols, climate, and the hydrological cycle. *Science*, **294**, 2119–2124, doi: 10.1126/science.1064034.

———, F. Li, M. V. Ramana, P. S. Praveen, D. Kim, C. E. Corrigan, H. Nguyen, E. A. Stone, J. J. Schauer, G. R. Camichael, B. Adhikary, and S. C. Yoon., 2007a: Atmospheric brown clouds: Hemispherical and regional variations in long-range transport, absorption, and radiative forcing. *J. Geophys. Res.*, **112**, D22S21, doi: 10.1029/2006JD008124.

———, M. V. Ramana, G. Roberts, D. Kim, C. Corrigan, C. Chung, and D. Winker, 2007b: Warming trends in Asia amplified by brown cloud solar absorption. *Nature*, **448**, 575–578.

Randles, C. A., and V. Ramaswamy, 2008: Absorbing aerosols over Asia: A GFDL general circulation model sensitivity study of model response to aerosol optical depth and aerosol absorption. *J. Geophys. Res.*, **113**, D21203, doi: 10.1029/2008JD010140.

Rosenfeld, D., U. Lohmann, G. B. Raga, C. D. O'Dowd, M. Kulmala, S. Fuzzi, A. Reissell, M. O. Andreae, 2008: Flood and drought: How do aerosols affect precipitation. *Science*, **321**, 1309, doi: 10.1126/science.1160606.

Sahu, S. K., G. Beig, and C. Sharma, 2008: Decadal growth of black carbon emissions in India. *Geophys. Res. Lett.*, **35**, L02807, doi: 10.1029/2007GL032333.

Satheesh, S. K., C. B. S. Dutt, J. Srinivasan, and U. R. Rao, 2007: Atmospheric warming due to dust absorption over Afro-Asian regions. *Geophys. Res. Lett.*, **34**, L04805, doi: 10.1029/2006GL028623.

Srivastava, A. K., K., Ram, P. Pant, P. Hegde, and H. Joshi, 2012: Black carbon aerosols over Manora Peak in the India Himalaya foothills: Implication for climate forcing. *Environ. Res. Lett.*, **7**, 014002, doi: 10.1088/1748-9326/7/1/014002.

Twomey, S. A., 1977: The influence of pollution on the shortwave albedo of clouds. *J. Atmos. Sci.*, **34**, 1149–1152.

Wang, B., 2006: *The Asian Monsoon.* Springer Praxis, New York, 787 pp.

Wang, C., D. Kim, A. M. L. Ekman, M. C. Barth, and P. J. Rasch, 2009: Impact of anthropogenic aerosols on Indian summer monsoon. *Geophys. Res. Lett.*, **36**, L21704, doi: 10.1029/2009GL040114.

Wang, Y., and L. Zhou, 2005: Observed trends in extreme precipitation events in China during 1961–2001 and the associated changes in large-scale circulation. *Geophys. Res. Lett.*, **32**, L09707, doi: 10.1029/2005GL022574.

Webster, P. J., V. P. Magana, T. N. Palmer, J. Shukla, R. A. Tomas, 1998: Monsoons: Processes, predictability, and the prospect for prediction. *J. Geophys. Res.*, **103**, 14451–14510.

———, and S. Yang, 1992: Monsoon and ENSO: Selectively interactive systems. *Quart. J. Roy. Meteor. Soc.*, **118**, 877–926.

Wild, M., H. Gilgen, A. Roesch, A. Ohmura, C. N. Long, E. G. Dutton, B. Forgan, A. Kallis, V. Russak, and A. Tsvetko, 2005: From dimming to brightening: Decadal change in solar radiation at the Earth's surface. *Science*, **308**, 847–850, doi: 10.1126/science.1103215.

Wu, G. X., Y. M. Liu, B. He, Q. Bao, A. Duan, and F. F. Jin, 2012: Thermal controls on the Asian summer monsoon. *Scientific Reports*, **2**, doi: 10.1038/srep00404.

———, ——— T. M. Wang, R. J. Wan, and X. Liu, 2007: The influence of mechanical and thermal forcing by the Tibetan Plateau on Asian climate. *J. Hydrometeorol.*, **8**, 770–789.

Yanai, M., and G. X. Wu, 2006: Effects of the Tibetan Plateau. *The Asian Monsoon*, B. Wang, Ed., Springer Praxis, 513–549.

Yasunari, T. J., P. Bonasoni, P. Laj, K. Fujita, E. Vuillermoz, A. Marinoni, P. Cristofanelli, R. Duchi, G. Tartari, and K. M. Lau, 2009: Estimated impact of black carbon deposition during pre-monsoon season from Nepal Climate Observatory–Pyramid data and snow albedo changes over Himalayan glaciers. *Atmos. Chem. Phys.*, **10**, 6603–6615, doi: 10.5194/acp-10-6603-201.

Zhang, H., Z. L. Wang, P. W. Guo, and Z. Z. Wang, 2009: A modeling study of the effects of direct radiative forcing due to carbonaceous aerosols on the climate in East Asia. *Adv. Atmos. Sci.*, **16**(1), 57–66, doi: 10.1007/s00376-009-0057-5.